INJECTION MOLDING

Theory and Practice

SPE MONOGRAPHS

Injection Molding

Irvin I. Rubin

Injection Molding

Theory and Practice

IRVIN I. RUBIN

Robinson Plastics Corporation

A Wiley-Interscience Publication

JOHN WILEY & SONS,　New York　·　London　·　Sydney　·　Toronto

Library of Congress Cataloging in Publication Data:

Rubin, Irvin I   1919 –
   Injection molding theory & practice.

   (SPE monographs, v. 1)
   Includes bibliographical references.
      1. Plastics – Molding. I. Title. II. Series:
Society of Plastics Engineers. SPE monographs, v. 1.

TP1150.R78   668.4'12   73-5
ISBN 0-471-74445-X

Printed in the United States of America

10   9   8   7   6   5   4   3

TO MY FAMILY _____

SERIES PREFACE _____

The Society of Plastics Engineers is dedicated to the promotion of scientific and engineering knowledge of plastics and to the initiation and continuation of educational activities for the plastics industry.

An example of this dedication is the sponsorship of this and other technical books about plastics. These books are commissioned, directed, and reviewed by the Society's Technical Volumes Committee. Members of this committee are selected for their outstanding technical competence; among them are prominent authors, educators, and scientists in the plastics field.

In addition, the Society publishes the *SPE Journal, Polymer Engineering and Science (PE&S)*, proceedings of its Annual, National and Regional Technical Conferences (*ANTEC, NATEC, RETEC*) and other selected publications. Additional information can be obtained by writing to the Society of Plastics Engineers, Inc., 656 West Putnam Avenue, Greenwich, Connecticut 06830.

William Frizelle,

Chairman, Technical Volumes Committee
Socity of Plastics Engineers
St. Louis, Missouri

PREFACE

This book is written for the following groups:

1. Injection molding engineers.
2. Operating personnel in injection molding plants.
3. Engineers in the plastics industry serving either in supplying machinery, materials, molds, equipment, and such, or in allied production fields such as blow molding, forming, and extrusion.
4. Nonplastic engineers who deal with plastics.
5. Purchasing agents who buy plastics.
6. End users who are not engineers but need knowledge of the process, molds, and materials.
7. Education, in the following:

   a. Technical training programs.
   b. In-house technical training programs.
   c. In college courses as a textbook.
   d. Background material for those attending lectures and courses on injection molding.

To accommodate readers with such varied interests, the material is organized so that the thrust of the information is stated clearly in nontechnical language, so that the scientific and technical parts can be utilized as required.

The information for this book was gathered from my practical experience operating an injection molding plant since 1940, constant communications with knowledgeable people in plastics, and theoretical information generated for courses and lectures given over the years.

Strong emphasis is placed not only on theory, but also on practical material, such as the 73 points checklist for mold designing and the chapter on correcting molding faults.

It might seem unusual that there is no book published in the United States about injection molding, since this process is one of the major methods of plastics fabrication and its parts are used in almost every type of manufactured product. The major problem in this type of book is the selection of material, as a book could easily be written on the subject of any of the chapters. For this reason, extensive bibliographies are provided at the end of each chapter which cover English language periodicals until approximately the beginning of 1972. These periodicals were selected because of the information they offer, and to assist those wishing to expand their knowledge of a subject.

I am most grateful to the Society of Plastics Engineers, Inc., its officers, committee members, and staff (past and present) for establishing the intellectual climate which led to the writing of this book. The Society, too, is responsible for a large part of the literature and educational activities found in our industry. I am indebted to Louis I. Naturman of the Business Communications Co., Stamford, Conn., for reviewing critically the manuscript. The cooperation of the publishers of the various plastic periodicals, the material suppliers and machinery manufacturers, as well as the customers of Robinson Plastics Corp. is acknowledged.

Finally, I am more than appreciative of my wife, Laura, and children, Jesse and Julie, who "did without" many times while I was writing the book.

Irvin I. Rubin

Brooklyn, New York
January 1973

CONTENTS _____

ABBREVIATIONS USED IN REFERENCES_____

BP	British Plastics
Chem. Eng. News	Chemical and Engineering News
IPE	International Plastics Engineering
Jap Pl.	Japan Plastics
MP	Modern Plastics
MP Enc.	Modern Plastics Encyclopedia
PDP	Plastics Design & Processing
PES	Polymer Engineering & Science (formerly SPE Transactions)
Plastics	no abbreviation
PT	Plastics Technology
Plastics World	no abbreviation
SPE-J	Society of Plastics Engineers-Journal
SPE-Tech Pap.	Society of Plastic Engineers Technical Papers
SPE Trans.	SPE Transactions (now PES)

CHAPTER 1

The Injection Molding Machine

Injection molding is a major processing technique for converting thermoplastics, and now thermosetting materials, into all types of products. Approximately 25% of the 13 billion pounds of thermoplastics sold in the United States in 1971 were injection molded, and about 36% (4320) of the 12,000 processing plants in the United States injection molded (1). Furthermore, in 1970 about 5000 injection molding machines were purchased in this country which brought the total of injection machines in-place to about 58,000. Since there were 130,000 processing machines, injection machines represent close to 45% of all processing units.

Sixty percent of the machines use a reciprocating screw, 35% a plunger (concentrated in the smaller machine sizes), and 5% a screw pot. This reinforces the fact that only 20% of those responding to the survey indicated that obsolescence was important.

The average plant had 12.5 machines. Custom molding plants tended to cluster in 12 machine units (12, 24, and 36 machine plants), as this is an economical management unit. The average age for smaller machines was 6.3 years, for medium size machines 5.0 years, and for larger machines 4.8 years. Also 52% of injection molders processed plastics by another method, and 20% used two or more different processes. In view of the high capital cost of injection machinery, it is surprising that for all plastic processing plants, 24% worked one shift, 18% two shifts, and 58% three shifts.

The process is not new. John and Isiah Hyatt received a patent in 1872 for an injection molding machine, which they used to mold camphor-plasticized cellulose nitrate (celluloid). In 1878 John Hyatt introduced the first multicavity mold. In 1909 Leo H. Baekeland introduced phenol-formaldehyde resins which are now injection moldable with the screw molding machine.

The experimental and theoretical works of Wallace H. Carothers led to a

1

general theory of condensation-polymerization that provided the impetus for the production of many polymers, including nylon. At the end of the 1930's modern technology began to develop and great improvements in materials permitted injection molding to become economically viable. A similar advance in machine technology is developing now.

There are both advantages and disadvantages to injection molding.

Advantages of Injection Molding

1. Parts can be produced at high production rates.
2. Large volume production is possible.
3. Relatively low labor cost per unit is obtainable.
4. Process is highly susceptible to automation.
5. Parts require little or no finishing.
6. Many different surfaces, colors, and finishes are available.
7. Good decoration is possible.
8. For many shapes this process is the most economical way to fabricate.
9. Process permits the manufacture of very small parts which are almost impossible to fabricate in quantities by other methods.
10. Minimal scrap loss result as runners, gates, and rejects can be reground and reused.
11. Same item can be molded in different materials, without changing the machine or mold in some cases.
12. Close dimensional tolerances can be maintained.
13. Parts can be molded with metallic and nonmetallic inserts.
14. Parts can be molded in a combination of plastic and such fillers as glass, asbestos, talc, and carbon.
15. The inherent properties of the material give many advantages such as high strength-weight rates, corrosion resistance, strength, and clarity.

Disadvantages and Problems of Injection Molding

1. Intense industry competition often results in low profit margins.
2. Three shift operations are often necessary to compete.
3. Mold costs are high.
4. Molding machinery and auxiliary equipment costs are high.
5. Process control may be poor.
6. Machinery that is not consistent in operation, and whose controls are not directly related to the end product.
7. Susceptibility to poor workmanship.
8. Quality is often difficult to determine immediately.
9. Lack of knowledge about the fundamentals of the process causes problems.

10. Lack of knowledge about the long term properties of the materials may result in long-term failures.

Many of the above problems can be ameliorated if the processor understands the operation of the machines and the plastics processing principles discussed in this text.

MACHINE OPERATION

Figure 1-1 shows a reciprocating screw injection molding machine with a clamping capacity of 250 tons, using a 2-in. reciprocating screw which delivers a maximum of 13 oz of polystyrene per shot.

The machine basically is a tool for the following.

1. Raising the temperature of the plastic to a point where it will flow under pressure.

2. Allowing the plastic to solidify in the mold, which the machine keeps closed.

3. Opening the mold to eject the plastic.

Figure 1-2 shows a schematic representation of the clamping end of an hydraulic machine, and Figure 1-3 shows the injection end of an inline reciprocating screw plasticizing unit. The injection side of the mold is clamped to the stationary platen, and the ejection side of the mold is clamped to the moving platen. The mold has an empty space in the configuration of the part to be molded. This empty space is filled with melted plastic under high pressure.

The moving platen rides on four steel bars called tie rods (or tie bars). The clamping force is generated by the hydraulic mechanism pushing against the moving platen and stretching the tie rods.

Basic Principles Reciprocating Screw Machine

The steps of the molding process for a reciprocating screw machine with an hydraulic clamp follow (Figures 1-2 and 1-3):

1. Plastic material is put into the hopper. (The virgin powder is normally granulated to 1/8 to 3/16 spheres or cubes.)

2. Oil behind the clamp ram moves the moving platen, closing the mold. The pressure behind the clamp ram builds up, developing enough force to keep the mold closed during the injection cycle. If the force of the injecting plastic material is greater than the clamp force, the mold will open. Plastic will flow past the parting line on the surface of the mold, producing "flash" which either has to be removed or the piece has to be rejected and reground.

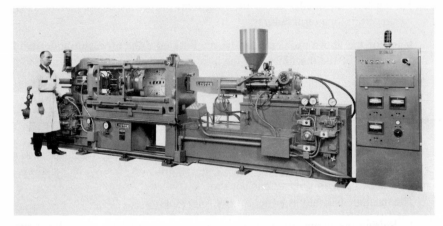

Figure 1-1 A 2-in. 250-ton reciprocating screw injection molding machine (Lester Engineering Company).

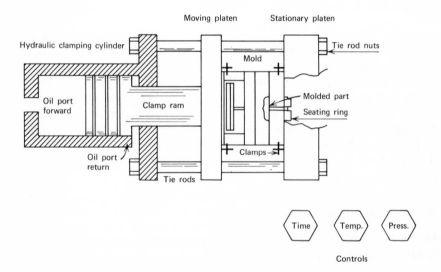

Figure 1-2 Schematic drawing of hydraulic clamping system.

3. The material is melted primarily by the turning of the screw which converts mechanical energy into heat. It also picks up some heat from the heating bands on the plasticizing cylinder (extruder barrel). As the material melts, it moves forward along the screw flights to the front end of the screw. The pressure generated by the screw on the material forces the screw, screw drive system, and the hydraulic motor back, leaving a reservoir of plasticized

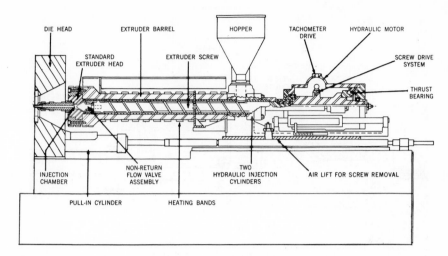

Figure 1-3 Schematic drawing of an injection end of reciprocating screw machine (HPM Division of Koehring Co.).

material in front of the screw. The screw will continue to turn until the rearward motion of the injection assembly hits a limit switch, which stops the rotation. This limit switch is adjustable, and its location determines the amount of material that will remain in front of the screw (the size of the "shot").

The pumping action of the screw also forces the hydraulic injection cylinders (one on each side of the screw) back. This return flow of oil from the hydraulic cylinders can be adjusted by the appropriate valve. This is called "back pressure", which is adjustable from zero to about 400 psi.

4. Most machines will retract the screw slightly at this point to decompress the material so that it does not "drool" out of the nozzle. This is called the "suck back" and is usually controlled by a timer.

5. The two hydraulic injection cylinders now bring the screw forward, injecting the material into the mold cavity. The injection pressure is maintained for a predetermined length of time. Most of the time there is a valve at the tip of the screw that prevents material from leaking into the flights of the screw during injection. It opens when the screw is turning, permitting the plastic to flow in front of it.

6. The oil velocity and pressure in the two injection cylinders develop enough speed to fill the mold as quickly as needed and maintain sufficient pressure to mold a part free from sink marks, flow marks, welds, and other defects.

7. As the material cools, it becomes more viscous and solidifies to the point where maintaining injection pressure is no longer of value.

8. Heat is continually removed from the mold by circulating cooling media

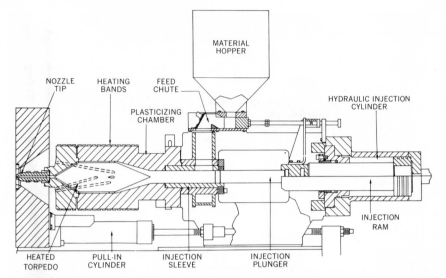

Figure 1-4 Schematic drawing of injection end of a single stage plunger machine (HPM Division of Koehring Co.).

(usually water) through drilled holes in the mold. The amount of time needed for the part to solidify so that it might be ejected from the mold is set on the clamp timer. When it times out, the moveable platen returns to its original position, opening the mold.

9. An ejection mechanism separates the molded plastic part from the mold and the machine is ready for its next cycle. During the regular molding cycle steps 3 and 4 occur after step 7.

Basic Principles-Plunger Machine

A Plunger machine (Figure 1-4), which is discussed on p. 13, heats the material by conduction and convection only and has very different characteristics.

In a plunger machine, the cold granules in the rear of the cylinder are compressed as the plunger comes forward; thus the machine starts to fill the mold more slowly than a reciprocating screw machine in which the plunger acts directly on the plasticized material. As soon as the mold fills, the pressure builds up inside the mold.

In a plunger machine, there is a large pressure loss in the cylinder because of the nature of its design. Therefore, the pressure at the end of the plunger is considerably higher than the pressure at the nozzle. In a screw machine, the ram pressure and nozzle pressure are almost the same.

Figure 1-5 shows the injection end of a 16-oz. 425-ton clamp plunger

Figure 1-5   Photograph of 16-oz. 425-ton plunger machine (Robinson Plastics Corp.).

machine. The operator has his left hand on the safety gate, which has been extended upward for safety. He opens the gate and removes the parts with his right hand, lays them on the table for cooling and inspection, and separates the molded parts from the runner. In the rear is a multiple head drill press for post finishing operations. Next to the press are cabinets containing all the spare parts pertaining to that particular machine. A 1-ton hoist rides on an *I* beam that is central over the length of the machine. It is used for mold changes and repairs. Note the large nuts which hold the four tie rods to the stationary platen. The light signals an obstruction on the mold and is connected in parallel with a bell which is used when the machine runs automatically. A weighing device weighs the material for each shot.

To facilitate maintenance and the use of auxiliary equipment 110-V AC outlets were installed on the machine; thus wires were not put on the floor where they might get wet or become exposed. To the left of the outlets is a manually controlled valve which moves the whole injection carriage back for purging or nozzle change. The wheel next to it controls the injection pressure. The 10-hp motor and pump maintain the clamping pressure. The main motor and pump are behind the machine. The gauges for the injection pressure, clamp pressure, pilot pressure, and oil temperature are found on a panel beneath the controls.

The wheel directly beneath the gate is used for "inching" or bypassing most of the oil to allow very slow movement of the machine for setup and checking. Immediately above the wheel is a gate safety limit switch attached in series with another one activated by the back of the gate. The gate also activates an hydraulic safety. This machine was subsequently converted to a two-stage screw-plunger machine.

Figure 1-6 shows a cutaway of a 2½-in. reciprocating screw 300-ton machine with an injection capacity of 24-oz. of polystyrene. As with all machines, the injection end is to the right of the operator. This permits him to open the safety gate with his left hand and remove the molded piece with his right.

The Injection Cycle

Figure 1-7 shows a schematic representation of a single cycle for a plunger-type machine and a screw-type machine. A plunger machine heats the material by conduction from the heated cylinder wall. A screw machine plasticizes the material by the shearing action of the rotating screw.

Pressure is resistance to flow. Therefore, as the clamp ram closes, there is very little resistance to the flow of the hydraulic oil so that the pressure remains very low. As soon as the mold clamps resistance builds up quickly, the clamp pressure goes up rapidly and remains at a steady level until released; it then drops to zero, and the mold is opened.

Once the mold is initially filled, additional material is added to the mold by the injection pressure to compensate for shrinkage as the material cools. Adding too much material is called "over packing," which results in highly stressed parts and may cause ejection problems. Insufficient material causes short shots, poor surface, sink marks, welds, and other defects. Material will continue to flow into the mold as long as there is injection pressure, provided the gate has not been sealed by the material solidifying. Once the gate seals or the injection pressure ceases, no more material enters the mold, and the contraction of the cooling material causes a decay of mold pressure.

In any given molding operation there is a maximum residual cavity pressure above which the part cannot be satisfactorily ejected. The operator must adjust his time, temperature, and pressure controls to attain a residual pressure below this maximum.

Estimating Cycle Time

A molder will look at a plastic part and estimate its production, for example, as a 30-sec cycle or 120 shots per hour. He will be making a judgement based on

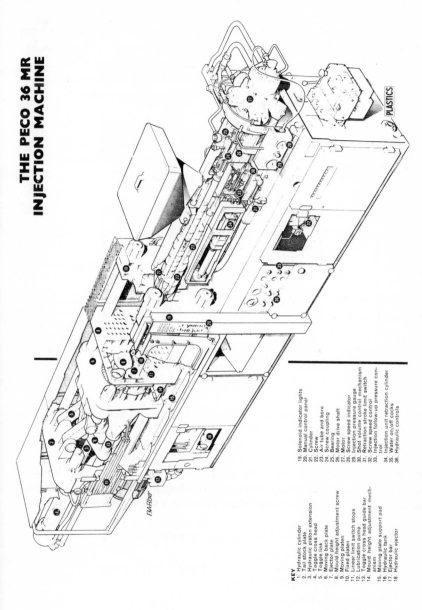

THE PECO 36 MR INJECTION MACHINE

KEY

1. Hydraulic cylinder
2. Tail stock plate
3. Hydraulic piston extension
4. Toggle cross head
5. Toggle link
6. Moving back plate
7. Ejector plate
8. Mould height adjustment screw
9. Moving platen
10. Fixed platen
11. Linear limit switch stops
12. Lubrication pump
13. Toggle cross head guide bar
14. Mould height adjustment mechanism
15. Moving plate support pad
16. Hydraulic tank
17. Ejector bar
18. Hydraulic ejector

19. Solenoid indicator lights
20. Manual control panel
21. Cylinder
22. Screw
23. Air tube and bore
24. Screw coupling
25. Bearing
26. Motor drive shaft
27. Motor
28. Screw speed indicator
29. Injection pressure gauge
30. Shot volume control mechanism
31. Retraction stroke limit switch
32. Screw speed control
33. Injection follow-up pressure control
34. Injection unit retraction cylinder
35. Water on/off cocks
36. Hydraulic controls

Figure 1-6 Cutaway view of 2½-in. 300-ton reciprocating screw machine (British Plastics).

9

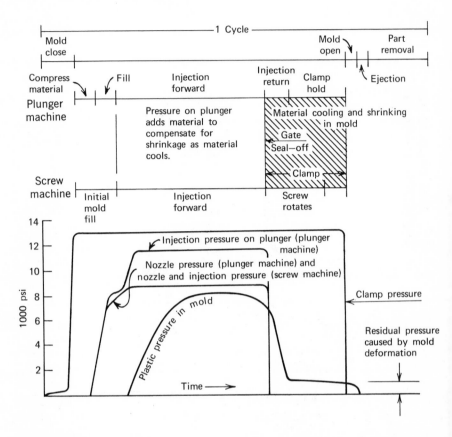

Figure 1-7 Schematic representation of cycle time of plunger and reciprocating screw injection molding machines.

the design of the part, the material in which it is molded, the tolerance requirements, the machine on which it is molded, the mold, and, on occasion, even the operator. An hypothetical illustration for a screw machine follows:

Part weight = 210 g, GP-PS (general purpose polystyrene)
*Injection rate = 12 in.^{3} sec
    (for polystyrene 12 in.^{3}/sec × 17.4 g/in.^{3} = 210 g/sec)
*Screw recovery rate (PS) = 40 g/sec
*Mold closing = 20 in./sec
*Mold opening = 26 in./sec
*Machine specification.

Calculation

Mold closing (10 in.) =	0.5 sec
Mold slow down =	0.5 sec
Mold filling time = weight part/injection rate = $\dfrac{210 \text{ g/sec}}{210 \text{ g/sec}}$ =	1 sec
Injection hold timer =	5 sec
Clamp timer =	16 sec
(Screw recovery rate = 40 g/sec = 5.2 sec is less than clamp hold time so it will not restrict cycle.)	
Mold opening plus ejection slow down =	2 sec
Part removal =	5 sec
Total = 120 shots/hr $\cong$	30 sec

The machine specifications assume that the hydraulic system is in good condition. Loss of pumped oil due to internal leakage can significantly slow the machine. The estimated injection hold timer and clamp timer (which controls the cooling time) settings are almost always based on experience with similar parts and materials. Attempts to compute them mathematically have not yet been successful (2).

Material Feed Control

In a straight plunger machine (Figure 1-4) the amount of feed can be controlled volumetrically. An adjustable chamber is filled with cold material every time the injection plunger reciprocates. This does not compensate for different particle sizes which change the bulk density, filling the chamber with different weights of material. An improved method uses a scale to weigh a predetermined charge of powder and dumps it, via a bucket, into the feed opening. A simple control system automatically compensates for incorrect feed (Figure 1-5).

Drying Material

The injection end has a hopper for holding the molding material. Some materials, such as nylons, polycarbonates, acrylics, acrylonitriles, and acetates, are hygroscopic and require drying before molding. Material is usually dried in an oven specifically designed for plastics. The oven is filled with trays so that the plastic can be spread out in thin layers, about 3/4 to 1 in. deep. Hot air is forced through the material by blowers. A small amount of warm air is bled out of the oven to prevent the accumulation of moisture. Raising the temperature of the air in the oven increases the diffusion of the water from the material. It also increases the capacity of the air to hold water. For example, at 70°F air holds 0.0158 lb of water per pound of dry air. At 175°F air holds 0.53 lb of water per pound of dry air. Air entering at 100°F and 90% relative humidity contains only 0.038 lb of water per pound of dry air. For most materials the water equilibrium level is sufficiently high so that dehumidifying the intake air is unnecessary.

For nylons and polycarbonates the statement above may not be true. In general the moisture content of nylon should not be greater than 0.28%. Above that level it will tend to react with the nylon during plasticizing and reduce the molecular weight, thus reducing the physical properties. Additionally, as with other materials, excess water results in splay marks and the possibility of internal bubbles caused by steam. Drying nylon at 175°F using intake air at 90°F and 70% relative humidity will give an equilibrium of 0.42% water. Trying to "dry" nylon as supplied by the manufacturer under these conditions would result in increasing the water content of the nylon to a point where it could not be molded properly. The air would have to be dehydrated by mechanical refrigeration or chemical desiccants that are regenerated by heating. Usually two sets are used, one in the oven and the other in the regenerating process. The length of time for drying depends on the type of material, its wetness, the circulation of the system, and the humidity and temperature of the makeup air. Normally 3 to 4 hr in an oven will dry the material (3).

Filtered hot air systems can be attached to the hopper to blow heated and dehumidified air through the plastic pellets. These are called hopper dryers. The intake air is filtered and dehumidified if required.

Hygroscopic material from the manufacturer in airtight containers can be used without drying if too much material is not placed in the hopper. Some machines are equipped with infrared heating devices to keep the material warm. A hopper dryer is always valuable when molding these materials.

In the extrusion process the removal of volatiles, including moisture, can be accomplished by using two screws in tandem on one barrel. A venting valve, connected to the atmosphere, is between them. The first section picks up material from the hopper and feeds it to the compression section where it is melted. The material is decompressed to atmospheric pressure at the point under the vent valve, thus permitting volatiles to escape. It is then recompressed and the balance of the screw acts like a single screw (p. 27)(3a, 3b). This removes the need for drying.

Keep Foreign Matter Out of the Hopper

Regardless of the care exercised by management there is a possibility of screw drivers, nuts and bolts, scales, and other metallic items entering the plasticizing chamber. These objects can cause major damage to the screw and barrel. A magnet is placed in the throat of the hopper to catch ferrous material, since this is the most common metal. Brass and copper are usually soft enough to be "molded" without damaging the machine.

TYPES OF INJECTION MACHINES CLASSIFIED BY INJECTION COMPONENT OR "END"

An injection machine basically consists of a clamping portion that contains the mold and the injection end which feeds, melts, and meters the plastic.

Four major types of injection ends in use today follow:

1. The *single stage plunger* (Figure 1-4, 1-10) uses a plunger to force material over a spreader or torpedo. The heat is supplied by resistance heaters. The material is heated by conduction and convection.

2. *The two stage plunger-plunger* (two stage plunger or plunger into a "pot") type machine uses a single stage plunger to plasticize the material and force it into a second cylinder (Figure 1-8). The second cylinder shoots the material into the mold.

3. *The two stage-screw plunger* (screw-pot) is essentially similar to the plunger-plunger machine except that a fixed screw is used for plasticizing instead of a plunger (Figure 1-9).

4. *A reciprocating screw* (In-line screw) uses a rotating screw to plasticize the material (Figure 1-3). As the screw turns, the plasticized material is forced in front of it, pushing the screw back. The material is injected by bringing the screw forward, which then acts as a plunger. More reciprocating screws are sold than any other machine today.

Single Stage Plunger

Approximately one third of the machines in operation in 1972 are of the plunger type, even though very few are now being manufactured. Therefore, a limited discussion of its operation is presented.

The single stage plunger (Figure 1-4) uses a plunger to force the material over a spreader or torpedo. The heat is supplied by resistance heaters, and the material is melted primarily by conduction and also by convection. The flow in the cylinder is basically laminar, giving the least homogeneous melt of all plasticizing systems.

The essentially laminar flow pattern can be shown by changing colors while the cylinder is in operation and cross-sectioning the extrudate at that time. The fin sections are clearly outlined as the material flows more slowly around them.

The plunger cylinder also transmits the pressure from the injection plunger to the molded material at the nozzle end. As mentioned previously pressure losses are considerable. Under flow conditions similar to molding pressure drops as high as 80% were recorded.

The pressure loss depends on the injection pressure, the barrel temperature, and to a lesser extent the design at the fin end of the torpedo. The major pressure loss occurs in the cold granular area.

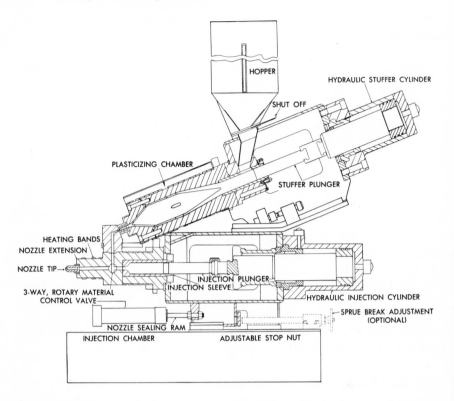

Figure 1-8 Schematic drawing of injection end of two stage plunger machine (HPM Division of Koehring Co.).

Figure 1-10 shows a cross section of a typical straight bore cylinder. The heat is supplied by resistance bands separated into three zones, each controlled by its own thermocouple. The adaptor and nozzle have their own heating band under separate control. Cold material in granular form is compacted by the plunger and heated by the cylinder wall. It is spread around a torpedo which is rigidly supported at the nozzle end by an integral slab through which are drilled holes connecting the plasticizing chamber to the nozzle area. The rear of the torpedo is supported by several fins touching the bore of the heating cylinder. The torpedo is heated by conduction through the flange and fin area, both of which are in contact with the heated cylinder walls. This cylinder is easily cleaned by unscrewing the adaptor and hitting the fin end of the torpedo. A disadvantage of this design is the relative coolness of the fin end of the torpedo.

The second popular design has the torpedo attached near the fin section by a heavy flange, through which holes are bored for the plastic flow. This keeps the fin end warmer. Its disadvantage is its difficulty to disassemble and clean.

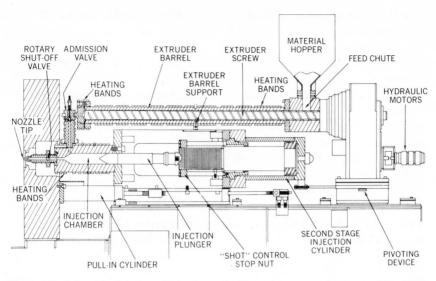

Figure 1-9 Schematic drawing of injection end of two stage screw-plunger machine (HPM Division of Koehring Co.).

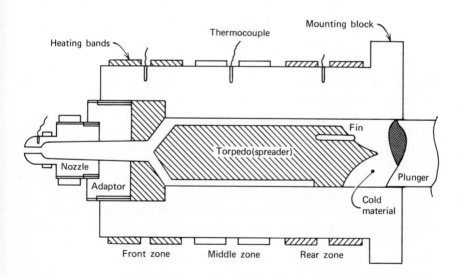

Figure 1-10 Straight bore plunger cylinder.

Efficiency calculations for the plunger machine.   There is no acceptable way to rate the output of a plunger machine. Specifications are usually the manufacturers optimistic evaluation under the best possible conditions. If this specification is important, it is best to consult with someone who has the identical equipment.

A rigorous analysis of the plunger cylinder including temperature and pressure measurements, temperature variations, and the effect of design factors on heater performance was given at the annual National Technical Conference of the Society of Plastic Engineers, Inc. in January 1955 (4). In analyzing the process the plastic was considered a tube with the cylinder diameter being the outside diameter and the torpedo diameter being the inside diameter. The heat transfer equations for an infinite slab will give sufficiently accurate approximations.

The material enters the chamber at a temperature T_0. The highest temperature it could reach at the nozzle end is the wall temperature T_w. In practice the melt temperature T_m will be considerably lower. The efficiency of the cylinder can be expressed as the ratio of these two temperature differences.

$$E = \frac{T_w - T_0}{T_m - T_0} \tag{1-1)]}$$

E = heating efficiency
T_0 = temperature of entering material
T_w = temperature of cylinder wall
T_m = temperature of exiting material

The heating efficiency should not be confused with the machine efficiency. The latter is the ratio of the amount of heat acquired by the plastic over the amount of heat supplied by the electrical heaters.

While attempts have been made to use induction heating as a source of energy for injection molding thermoplastic (5,6), all cylinders for thermoplastics use electrical heating. Cylinders for injection molding thermosetting materials are usually heated with water on the barrel and electrically on the nozzle. The wattage of the heating elements of the cylinder are given for the purpose of connecting the machine electrically. It is assumed they are sufficient for the cylinder. Obviously a significant amount of heat is lost by radiation. To maintain a plastic temperature of $400°$, which corresponded to an output of 6500 Btu/hr, a nonshielded cylinder required an input of 20,800 Btu/hr for a loss of 14,300 Btu/hr. Shielding with an asbestos lined aluminum cover reduced the heat loss to 6900 Btu/hr. This is a very significant reduction in operating costs. On the other hand, complete insulation of the heating cylinder is not desirable because of the difficulty of good temperature control.

Some of the factors that affect the efficiency and output of a plunger cylinder follows:

1. Thermal diffusivity. This is the rate at which a temperature change at one

point in a body travels to another point. It is defined as the thermal conductivity divided by the product of the density and specific heat.

$$\alpha = \frac{K}{\rho C}$$

K = thermal conductivity   (1-2)

ρ = density

C = specific heat

Since plastic is an excellent insulator, it will take approximately 4½ hr to bring the average temperature of a 4-in.-diameter cylinder of plastic to 80% of the wall temperature.

2. The length of time that the plastic is in the cylinder. This is the product of the number of shots stored in the cylinder multiplied by the time of a single shot.

3. The ratio of the surface area of the heated section of the cylinder to the volume of the plastic. Obviously the larger the surface area, the more material that can be plasticized. This can be increased by fluting the torpedo. However, the gain would be counterbalanced by the extra thickness of the plastic.

4. The thickness of the plastic. The thinner the plastic, the more rapidly the heat will be transmitted. If the torpedo is internally heated, the thickness of the infinite slab is effectively cut in half. If it is just heated from the outer wall by conduction through the metal, the effective thickness of the slab is between that of one heated on both sides and heated on only one side. A cylinder with an internally heated torpedo at an efficiency (E) of 0.8 plasticized 73 lb/hr. When the torpedo heat was turned off, the rate dropped to 39 lb/hr. The production rate of a nonheated torpedo with a large contact area near the fin is superior because of the better heat transfer into the torpedo.

5. The flow path. The flow path through the cylinder should be clear with minimum spots for material hangup. If all material does not move through at a reasonably uniform velocity, some material will remain in longer than others and contribute to a very wide temperature variation. This gives different densities in the material, causes large stresses in the molded part and can also degrade the material. Measurements show that the material that goes by the fin area (velocity is slowed down) is higher in temperature than the other material in the cylinder.

It was found that with no lubricant and a 20,000-psi ram pressure, the pressure drop was 16,600 psi. Adding 200 ppm of lubricants reduced the pressure loss to 7800 psi. Anything that will reduce the granular volume, such as torpedo heat and prepacking the cylinder, will reduce the pressure loss.

Alternatives to the Conventional Plunger Machine

A number of other methods were tried but did not achieve commercial acceptance. They included a rotating torpedo with fins on it (7), the purpose of which was to introduce mixing and break up the laminar flow. A melt extractor (8) wherein the material was plasticized through a strainer-like device and sent into a shooting chamber was also tried. A third method was a vented reverse flow cylinder. The material made three passes through the cylinder in grooved channels. At the end of the second pass, gases were vented back to the cold material (8a).

Preplasticizing Machines

The single stage plunger machine has some severe limitations which are overcome by using a preplasticizing system (9). Here the material is melted in one chamber and transferred into a second chamber. In a typical two stage machine — plunger-plunger or screw-plunger — the plunger or screw plasticizes the material in one cylinder and transfers it into another cylinder where the material is forced into the mold by the direct action of another plunger on the melted material. In a reciprocating or in-line screw, the material is plasticized and forced in front of the screw, which is in effect another chamber. The material is injected by bringing the screw forward which acts like a plunger. A check valve on the front end of the screw is used with nonheat sensitive thermoplastics.

Preplasticizing gives rise to a number of significant advantages.

1. The two basic requirements of a plunger cylinder are in conflict. Effective heat transfer requires small clearances between the plunger and torpedo and a long cylinder to permit gradual and full heat transfer. Good pressure transmission requires large channels or clearance between the wall and the torpedo and a short cylinder. This conflict is overcome in a preplasticizing system by separating the two functions. The material is heated in one chamber and injected in another.

2. The melt is more homogeneous because it is mixed as it passes through the small opening connecting the two chambers. This significantly reduces the high temperature differential (150°F) and the resultant stresses often found in a plunger machine.

3. Better shot weight control is possible. The plasticized material forces the injection plunger back until it reaches a predetermined location. The amount of material to be injected, therefore, has been volumetrically determined on the material to be injected at the temperature of injection. This contrasts with weighing cold pellets to determine the charge in a plunger machine. Accurate shot weight control is important in that it prevents underfilling — with its shorts and sink marks — and overfilling with its packing and difficult ejection. The bulk factor, specific gravity, and temperature difference from shot to shot does not affect the accuracy of the preplasticizing feed.

4. The injection plunger exerts pressure directly on the material which is advantageous. The pressure on the material can be determined accurately by a pressure gauge in the hydraulic injection system. In contrast, the true injection pressure on a plunger machine can only be read with a pressure sensing device on the nozzle. The injection pressure in a preplasticizing machine can be controlled accurately and generally remains constant. In a plunger machine, the pressure will change depending on the compressibility and temperature of the granules at the feed end.

5. Faster injection is possible. This is very important in controlling the quality of the molded parts and the length of the cycle.

6. Injection pressures of 20,000 psi on the material are required. Because of the large pressure losses in a plunger cylinder, a condition not encountered in a two stage machine, the cylinder and hydraulic system of the plunger machine must be significantly larger and more costly to produce 20,000 psi at the nozzle.

7. It is relatively easy to increase the shot capacity, particularly in two stage machines. Increasing the volume of the shooting cylinder is very inexpensive compared to the massive cylinders that would be required for large shots in a plunger machine. For this reason most plunger machines now in use are in the 12 oz or lower capacities.

8. The lack of pressure loss in the injection system reduces the horsepower requirements and the cost of operating the machine.

9. A preplasticizing unit permits molding at lower pressure because of the rapid injection rate and the better homogeneity of the melt. This lowers the mold clamping force per unit area and permits a larger projected area of molding on the machine.

10. With proper design of two stage machines, plasticizing can take place throughout most of the cycle, yielding more pounds per hour per dollar of machine cost.

There are a number of disadvantages to preplasticizing systems, primarily concerned with the additional maintenance of the valves, shooting cylinders, and reciprocating portions and mechanism of an inline machine. Another disadvantage is the leakage around the injection ram. Leakage adversly affects the accuracy of the injection. In an inline screw the leakage is internal as it is past the check valve on the tip of the screw. In a two stage machine the leakage is less critical, but improper wiping of the plunger tip may cause material to hang up and degrade.

Plunger-plunger Machine. Figure 1-8 illustrates a plunger-plunger machine. The preplasticizing chamber is that of Figure 1-4. It is connected by a three way valve shown in the plasticizing position. As the stuffer plunger reciprocates, plasticized material is forced in front of the injection plunger. The injection plunger moves back a predetermined distance until it hits a limit switch (not shown) which stops the preplasticizer. At the appropriate time the rotary valve is turned, and the injection plunger comes forward, filling the mold.

A two stage plunger machine will almost always give better results than a single stage one. A superior way to plasticize is to use a screw. There is no advantage in using a plunger as a preplasticizer. For this reason, the plunger-plunger type of machine is no longer being manufactured. Many plants have converted plunger preplasticizers either to screw plunger machines or to reciproacting screw machines. This is not particularly difficult to do (10).

HEATING CYLINDER AND SCREW CONSTRUCTION AND MAINTENANCE

Heating cylinders of injection molding machines have their internal bore finished either by nitriding or by adding another metallic liner. Cylinders for nitriding are made of a special steel with a high aluminum content and chromium and molybdenum. The steel is first hardened, and the surface must be completely cleared of all traces of decarburization. If this is not done, the nitriding layer will be very brittle and peel off in operation. The cylinder is heated from $930°F$ to $1200°F$ in an atmosphere of ammonia. This results in a thin case, approximately 0.020 in deep. This surface has to be finished further, and the final depth of hardness may be considerably less. The hardness of the lining decreases with its depth, thus wear further reduces the surface hardness and accelerates itself.

Corrosion resistance depends primarily on the chromium content of the steel which ranges from 9 to 32% in bimetallic cylinders compared to a maximum of 2% in nitrided cylinders. Field surveys indicate most bimetallic cylinders outlast nitrided cylinders by factor of approximately 3 to 1. Even though they are more expensive, the bimetallic cylinders are recommended for molding equipment.

Bimetallic cylinders are made in two ways. The sleeve can be made separately and shrunk into a prebored barrel. This method allows for higher operating pressures. The second way involves centrifugally casting the molten lining material at speeds to produce up to 75 G. While welding techniques can deposit lining material, they are not used for injection cylinders.

The major lining material used is manufactured by the Xaloy, Inc. Their standard material is used for abrasion resistance and their Xaloy 306® for corrosion resistance. The first is an iron based alloy and the second a nickel cobalt based alloy. They are cast either into a low carbon steel shell, 1020, or an alloy steel shell, 4140. The latter is strongly recommended, because it almost doubles the tangential bore stress. This figure is 51,000 psi for standard Xaloy in a 4140 shell and 38,000 psi for Xaloy 306. The standard lining has a hardness of 60-63 R_c and the 306 a hardness of 48-52 R_c. The cylinder bore can be chrome plated (^{C}11). Most screws are made from 41^{C}40 steel which are flame hardened and chrome plated. They can also be carburized and nitrided, or have a hard surface material, by spraying of metal. Worn screws can be rebuilt several times (12).

Why Cylinders Breakdown

Major causes for breakdowns of plunger cylinders follow:

1. Fatigue failure, which is primarily caused by thermal shock conditions aggravated by improper starting and stopping procedures.
2. Corrosion from the thermoplastics used.
3. Excessive molding of metal parts, screw drivers, and miscellaneous trash. Steel can be caught by the use of magnets. Other material damages are reduced by good management procedures.
4. Prolonged use of high injection pressure. Two stage injection pressure capability will substantially reduce this cause of failure.

Corrosion

With corrosive materials such as PVC, polycarbonates, and butyrates, the cylinder should be purged if there is to be a prolonged shutdown. Polystyrene is an excellent purging material, as is reground cast acrylic (see Ref. 13 for a list of manufacturers' recommendations for purging).

Corrosion is a serious problem with certain materials (14, 14a, 14b). Chrome plating and chemically deposited electroless nickel coating minimizes the problem. Molding glass-filled thermoplastics in screw machines has caused considerable wear. It was thought the wear came from erosion at the feed end of the screw. Evidence has shown that wear occurs more in the return valve section at the front of the screw. This wear was probably caused by corrosion from the decomposition of the wetting agent between glass and plastic. The use of large runners and gates reduced the need for higher temperatures which in turn reduced this screw wear.

Cylinder Repair

Cylinders are repaired by disassembling and smoothing out surface imperfections to prevent material hangup. Small cracks can be welded and finished. Leakage from cylinders requires refitting. Occasionally a soft copper gasket will effect a seal.

Screw barrels seem to have less mechanical problems but are subject to considerable corrosion damage. This is due to the molding of corrosive materials which are difficult, if not impossible, to mold in plunger equipment. Wear caused by the sliding of the nonreturn valve has not been a major problem. Screws are relatively easy to disassemble and clean. The barrel is best cleaned with a wire brush attached to a long broom handle. Worn barrels can be rebored and the screw size increased to the necessary dimensions (12).

Purging and Startup

By purging we mean the cleaning of one color or type of material from the barrel by forcing this material out with a new color or material. When a machine starts up from cold, enough time should be allowed for the heating cylinder to completely reach molding temperature. Many molders allow a "soak-in" period of 20 min after the pyrometers indicate molding heat. This substantially lowers the stress in the barrel when starting. Purging at start-up should be done under low pressure. In screw machines the pressure is only in front of the tip. When a machine is shut down, it should be purged until no more material comes out. The plunger should be left in the forward position.

CYLINDER NOZZLES

The nozzle is a tube which provides a mechanical and thermal connection from the hot cylinder to the much colder injection mold with a minimum pressure and thermal loss. There are three types of nozzles:

1. An open channel with no mechanical valve between the cylinder and mold.
2. An internal check valve held closed either by an internal or external spring and opened by the injection pressure of the plastic.
3. A cutoff valve operated by an external source such as a cylinder.

The land length in any nozzle is kept to a minimum consistent with the strength requirements of the nozzle. In nozzle construction (Figure 1-11) a straight hole is bored down the length of the nozzle. This bore is usually about 1/2 in. diameter as this permits quick melting and rapid flow. It is then tapered out to meet the diameter of the hole in the cylinder. Since this measurement is difficult to make, it is better to have the nozzle hole larger than the cylinder hole (if they cannot be perfectly matched) to minimize degradation and hangup. There should also be a small straight section so that if the nozzle is refaced the matching dimension will not change. There is a taper of about $1/2°$ to permit better sealing on the nozzle seat.

The nozzle need not be in one piece. Very often the tip is made replaceable by screwing it into the nozzle body. This makes replacement and repair considerably less expensive.

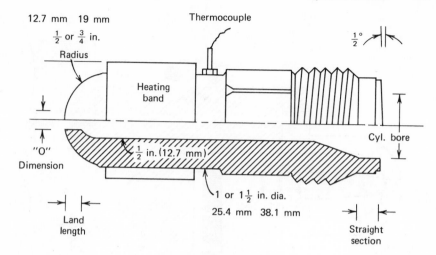

Figure 1-11 Standard nozzle.

Nozzles that extend directly into the surface of a cavity are more properly considered under gating. As a general rule it is desirable to minimize the length of the sprue. The nozzle must be extended as long as possible. Therefore, heating bands have to be used and there must be enough clearance in the central platen area to permit them to extend into the mold.

There are many specialty nozzles such as one with a screen pack to trap small foreign matter that plug hot runner and insulated runner type molds. Reference 15 is an excellent source of information about nozzles.

Nozzles that shut off mechanically are necessary if the screw in a reciprocating machine is rotating while the mold is open. There are two types of self-operating valves both using the same principle. One type has a spring loaded ball-type check valve which is placed internally at the tip of the nozzle. Injection pressure opens the valve and the spring closes it. The disadvantages of this valve are the restriction of flow and the possibility of material hangup.

An antisuck back nozzle is designed to prevent material flowing backwards from the mold into the cylinder. This action, the reverse of the shut-off nozzle, is accomplished by using a ball without a spring which is forced forward while the mold is under pressure. As soon as the pressure is released any suckback or material ejection from the cavity will force the ball back against the seat.

Figure 1-12 shows a needle shut-off nozzle. An external spring forces a bar toward the mold. This bar is attached to a needle-shaped piston which seals off

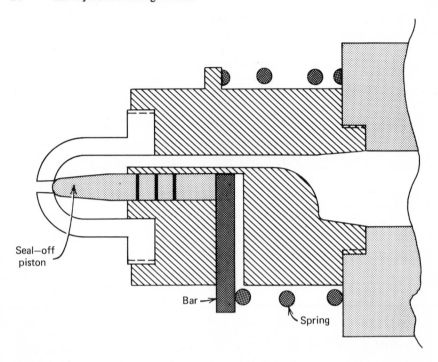

Figure 1-12   Needle type shut-off nozzle.

the front of the nozzle. When injection pressure is applied, it hydraulically forces the needle back (compressing the spring) and permits the material to flow into the mold. When the pressure drops to a value determined by the spring, the bar forces the needle forward sealing off the opening to the mold. The pistons can valve has the disadvantage of possible leakage around the needle.

The third type of seal-off is obtained as the result of an external motion. The simplest and most expensive way is to use a flat faced nozzle and sprue bushing, and move the injection cylinder off center, sealing off the opening. A second method uses the sprue break motion of the machine to move part of the nozzle. This can take the form of a sliding seal in the nozzle or a linkage-operated sliding pin whose axis is perpendicular to the nozzle bore.

The most common method is to move an internal needle or slide valve by an hydraulic cylinder through linkages. It is preferable to operate the cylinders pneumatically, as a broken hydraulic hose in this area constitutes a fire hazard. As with all devices sealing off through molten plastic, wear and subsequent contamination can be a problem.

Nylon molding presents special problems in drooling. DuPont developed a reverse taper nozzle (Figure 1-13) which, when coupled with enough heat and

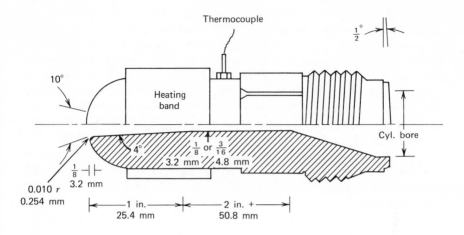

Figure 1-13 Nylon nozzle.

good temperature control, gives clean sprue removal. If the temperature is too low, the material will seal in the nozzle, preventing the next shot. If the nozzle is too hot drooling will occur and the part may be damaged. The opening has a 10° taper for 1/8 in. with a 0.010 radius for strength. A 4° taper continues for 3/8 to 1 in. This is the point at which the sprue separates from the nozzle. A 1/8 to 3/16-in. hole extends for 1 to 2 in., ending in a taper which adapts itself to the cylinder bore.

Nozzle Alignment

If the nozzle does not seat correctly into the cylinder, streaking, burning, and black specks will appear. Additionally, material will hang up and possibly corrode the cylinder and nozzle. If the sprue tip is burned or discolored, the probable cause is misalignment or heat control. If a long streak of dirt occurs in the same place, it can be caused by a poorly seated nozzle, a foreign substance in the nozzle, a foreign substance in the barrel, or, least likely, a cracked barrel. Nozzles are kept streamlined to prevent pressure loss and hangup of material and subsequent degradation. Since the nozzle is a conveyor, restriction should be as limited as possible. A simple way to check restriction is to time the ram forward motion when the mold is on cycle. This time should be compared with the ram forward time for the same feed with the machine open, and it should be less than the ram forward time during molding. There is usually a 25 to 50% decrease, but if the time is the same, the nozzle is too restricted or foreign material is obstructing part of the passage. Pressure sensors are available which can be fitted into the nozzle. These are expensive at the moment but are valuable in automatically controlling the molding cycle.

The opening leading to the mold, "0" dimension (Figure 1-11), comes in increments of 1/32 in. It should be slightly smaller than the opening of the sprue bushing so that the sprue can be pulled. If the hole is misaligned or the seat poor, plastic material can solidify around the 1/2 or 3/4-in. radius and cause the sprue to stick. Misalignment, which is the most frequent cause of nozzle damage can be checked by closing the carriage assembly with a piece of paper between the nozzle and sprue. If the paper is cut on one side misalignment is evident. Another way to check for misalignment is to rub an oil suspension of a blue pigment on the nozzle and close the machine. When the machine is backed away, wherever the nozzle touched the sprue, the pigment will be transferred. Ball swivel-type nozzles are available to compensate. A better solution to misalignment is to realign the press.

Control of Nozzle Temperatures

It is impossible to mold correctly without control of the nozzle temperature. If the nozzle is very short, many times the heat conduction from the cylinder will be enough to maintain it at the proper temperature. Usually nozzles are long enough to require external heat. The heating elements should be independently controlled, and never attached to the front heating bands of the cylinder. The cylinder requirements are completely different than those of the nozzle. Overheating the nozzle may burn or degrade the plastic. Underheating may result in a cold nozzle plugging up the cylinder.

Nozzle heating elements can be controlled with a variable transformer or proportional timer. The best way is to insert a thermocouple and use a controlling pyrometer.

The usual way of heating nozzles is with mica resistance heating bands. The life of these units is relatively short since they come in contact with hot plastic or they are mechanically abused. An improvement over the heating band is a tubular heater.

Cartridges inserted into the nozzle are now available and give longer life and up to four times the amount of heat. Hot circulating liquids are also used. This method gives better heat distribution but has the disadvantages and dangers of a circulating hot fluid. Inexpensive recorders are available to record the nozzle temperature. This can be very important in trouble shooting and precision molding. Thermocouples are also available to enter into the melt stream, but are basically used for research.

Lubricant Requirements

A lubricant should be used when screwing in nozzles. Molybdenum-type grease, graphite, silicone, and copper flake lubricants are used; the latter is preferable because of its high heat and minimum contamination properties.

SCREW PLASTIFICATION

The use of a screw for preplasticizing and injecting material for injection molding is not new. A patent was granted in 1927 for a nonintermeshing twin screw extruder which fed directly into the mold. Maximum injection pressure for this type of system is in the order of 3 to 4000 psi. In 1932 a patent was granted for a stationary single screw extruding into a pot perpendicular to the screw. A plunger came down sealing off the screw and forcing the material into the mold. This is essentially the two stage screw pot system of today. In the early 1950s the inline system for preplasticizing was patented and in the middle of that decade the inline screw, as we know it today, became available.

The task of the injection molding screw (Figure 1-14) is to take cold pellets at the hopper end, compact the material in the feed section, degas and plasticize the material in the transition section, and pump it in the metering section.

Characteristics of the Injection Molding Screw

In an inline screw the barrel must be strong enough to maintain the full injection pressure. In a two stage screw it need only be as strong as required to maintain the pressure generated by the screw itself, rarely more than 8000 psi. In the

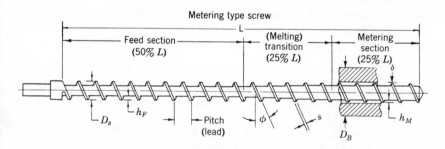

Figure 1-14   Typical screw used in injection molding.

8	D_B = diameter-barrel
2½	D_S = diameter screw (nominal)
17.8°	ϕ = helix angle (one turn per screw diameter)
0.250	s = land width
0.350	h_F = flight depth (feed)
0.105	h_M = flight depth (metering)
50	L = overall length
0.005	δ = flight clearance (radial)
20:1	L/D = ratio of length to diameter
3.3	h_F/h_M = compression ratio
	N = revolutions per minute

event of a failure of the connecting valve between the screw and the shooting cylinder during injection, full injection pressure would be transmitted to the extruder. To ensure safety and to prevent damage to the barrel a safety plug is installed which will blow open before dangerous conditions can occur. A pressure transducer with an adjustable electrical contact can be used. The contact is connected with the control circuit so that the machine is automatically stopped when the pressure approaches the danger point. If it is set just below the safety plug setting, replacement, which can take several hours, is eliminated. The diameter of the barrel (D_B), in turn, is determined by the nominal diameter of the screw (D_S).

Screws used in molding machines have a constant pitch. *The helix angle* (θ) affects the conveying and the amount of mixing in the channel. Experience has shown that a helix which advances one turn per nominal screw diameter gives excellent results. This corresponds to an angle of 17.8°, which has been universally adopted.

The land width (S) is 10% of the diameter. The radial flight clearance (δ) is the clearance between the screw flight and the barrel. It is specified considering the following effects:

1. The amount of leakage flow over the flights. This affects the output.
2. The temperature rise in the clearance. The heat is generated in shearing the plastic. The amount of heat generated is related to the screw speed and the nature of the material.
3. The scraping ability of the flights in cleaning the barrel.
4. The eccentricity of the screw and barrel.
5. Manufacturing costs.

For most materials, the radial clearance is the screw diameter times 2×10^{-3}. Using a 2 1/2 in. extruder at 72 rpm the output with a 0.0025 radial clearance was 130 lb/hr; with 0.0055 radial clearance was 126 lb/hr; and with 0.0105 radial was 118 lb/hr (16).

The length of the screw (L) is the axial length of the flighted section. An important criterion of a screw design is the ratio of the length over the diameter of the barrel (L/D).

Long screws with a 20:1 L/D are used. Some of the advantages of a long screw are as follows.

1. The larger the L/D ratio, the more the shear heat can be uniformly generated in the plastic without degradation. For best performance in the metering section, a uniform melt is needed from the transition section. Because the shear rate is relatively constant over the channel depth, the conversion of mechanical energy to heat by shearing the plastic is distributed evenly through the material.

2. Only one velocity component is directed toward the screw forward direction. Another component causes lateral velocity in the channel which in turn causes circulation. The material near the wall flows to the screw root and back up again. The longer screw introduces a longer flow path resulting in about the same output as a shorter screw but with fewer pressure variations in the melt. The larger the L/D the more opportunity for mixing and consequently the better homogeneity of the melt.
3. As seen in (equation 1-4) the forward or drag flow is not affected by the length. The pressure flow or resistance to flow (equation 1-6) decreases with the length. The higher the ratio the more closely the pumping section approaches the theoretical output.
4. Larger L/D ratios have more frictional area. Beyond approximately 24 to 1 the increase does not justify the additional cost and maintenance.

The channel depth is an important specification relating to the output rate and melt quality. Since the feed section of the screw is basically a conveyor, the deeper the channel, the larger the volume between the flights, and the larger the output. However, there are other considerations one should be aware of in selecting the channel depth. In the metering section one such consideration is shear rate.

The shear rate is defined as the surface velocity at the barrel wall divided by the channel depth.

$$\gamma = \frac{DN}{h} \tag{1-3}$$

where γ = shear rate
D = diameter screw
N = rate of screw rotation
h = channel depth

All materials have a maximum limiting shear rate, beyond which they degrade. The more heat sensitive they are, the lower the permissible shear rate. It is more desirable to decrease this rate by increasing the channel depth rather than decrease the screw speed. Increasing channel depth increases the undesirable negative output component, called the pressure flow [see (1-6)]. Pressure flow varies with the cube of the channel depth. Deep channels mean relatively poor circulation within the flight, poor mixing, and low thermal diffusivity which in turn result in increased temperature variation and lower homogeneity of the melt. The overall effect of flight depth on output is that the deeper the flight, the more rapid the decrease in output with pressure. A shallow screw, while it has a much lower output, is relatively insensitive to the back pressure setting of the molding machine. The flight depth in the metering section of five different 2 1/2-in. screws (L/D 20:1) designed for different materials are 0.075, 0.085, 0.100, 0.105, and 0.150. For a 3 1/2-in. screw (L/D 20:1) they are 0.095,

0.105, 0.120, 0.130, and 0.200.

The central section of the screw, which is approximately one fourth of the screw length, has the function of compacting the material. When the plastic enters the extruder, it is granular and full of air. When it leaves the extruder, it is a viscous liquid. This means that the volume of the screw channel must decrease somewhere to compensate for the increased density. This is done in the transition or compression section. In a 20:1 L/D screw it compromises approximately one fourth of the length or five turns. The compression ratio is defined as the volume of a unit length in the feed section divided by the volume of an equivalent length in the metering section. Since injection molding metering screws have a constant pitch the ratio of the channel depths is used, even though it is not precisely correct. For the 2 1/2-in. screw mentioned above the flight depths in the feed sections were 0.350, 0.320, 0.390, 0.350, and 0.420. The compression ratios for those screws were 4.6, 3.1, 3.9, 3.3, and 2.8, respectively. Compression ratios vary from 2 to 5 in screws used for molding. For the 3 1/2-in. screw they were 0.400, 0.350, 0.470, 0.400, and 0.520.

Screw Conveying – Basic Principle

If one puts some string in a jar of honey and pulls it out slowly, he sees the honey sticking to the string. Imagine an endless belt of string going through a vertical tube of honey. This tube has seals on the top and bottom, a reservoir on the bottom, and an outlet perpendicular to the tube at the very top. As the string is rotated upward, the friction of the honey on the string would cause an upward flow. The honey would be forced out of the top opening. The reason for the movement is that the honey is sticking to the string and sliding on the walls of the tube. In effect this is a friction pump identical in principle to the extruder, which is also known as an axial flow pump.

The string is now in the shape of a helix. The motion is relative to the barrel wall. Most of the motion using the 17.8° helix angle is in the direction of the barrel axis. The other component gives rise to circulation in the channel and promotes mixing. If the coefficient of friction between the plastic and the screw and the plastic and the barrel were identical, there would be no flow of material and it would just rotate as a plug within the flights of the screw.

To move forward the material must stick more to the barrel than the screw. It is the same principle as tightening a nut and bolt. If the nut is turned without holding the bolt, there is no relative motion. Such motion occurs only when one of the components is held. Obviously the larger the frictional difference between the plastic and the screw and the plastic and the barrel, the higher the output. (This is one reason too, why longer barrels are desirable.) A very useful determination of the effects of friction at different temperatures between steel and low density polyethylene, high density polyethylene, pypropylene, ABS, polystyrene, and ionomer resins is found in Ref. 17.

Figure 1-15 shows a graph of the coefficient of friction of polystyrene at different entering stock temperatures. Note the rapid lowering of friction at about 375°F. If the rear zone is overheated the coefficient of friction will be so low that a plug will form and feed will be blocked. On the contrary, the curve of high density polyethylene showed that a high barrel temperature at the feed will promote feeding. If ABS is preheated in drying to 200°F the coefficient of friction will probably cause feeding problems. If it is at 150°F, a broad latitude exists in the barrel temperature.

As more glass fiber and other reinforcements are used in injection molding, the molder will be forced to pay more attention to the feed system. The relationship of screw design, hopper throat and shape, feed hopper design, speed, effective back pressure, and the effect of forced feeding on the single screw extruder are important (see Refs. 18 and 19). Reference 20 evaluates the effect of particle size on molding. It was found that using beads of an average size of 0.004 in. gave better polymer melt, faster flow, and better pressure transmission than the conventional 1/8-in. beads. Some feed systems are now equipped with a variable opening of the hopper throat.

The feed section is located under the hopper and in the rear section of the screw. The flight depth is at its maximum, and the material from the hopper fills the flight of the screw. The feed section has a constant channel depth throughout its entire length. Since the conveying action is caused by the difference in friction between the plastic and the barrel wall and the plastic and the screw, the screw is always more highly polished than the barrel. Normally, the barrel temperature is higher than the screw temperature. Consequently, the material adheres to the barrel as it softens and slip upon the cooler screw. The material is then compacted in the feed section and begins to melt. The majority of the melting occurs in the compression or transition section. In most molding metering screws, the feed section is approximately half the screw length.

Transition Section

The transition or compression section where the channel depth is continually decreased completes the compacting and heating of the plastic granules. Here all the remaining air is released as a result of heat supplied by the cylinder heaters and mechanical energy supplied by the rotation of the screw.

Normally a transition zone encompasses approximately 25% of the screw length. When the material leaves the area under the hopper, it only partially fills the screw. After about four turns the material is fully compacted. The material touching the barrel melts by conduction (Figure 1-16). This melt (area 1) is scraped off by the forward or advancing edge of the screw where it starts a circulating pattern (area 2). Area 3 contains pellets which are sufficiently

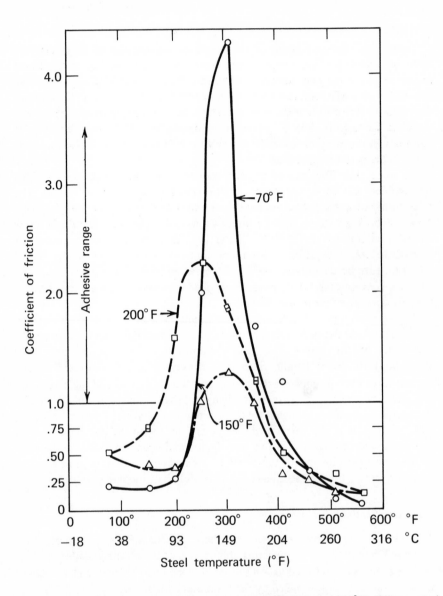

Figure 1-15 Coefficients of friction for polystyrene (at 70, 150, and 200°F) sliding on steel surfaces (at temperatures ranging from 70 to 550°F). Polystyrene was Dow Chemical Co.'s Styrene 666U. Rapid reduction of friction between polystyrene at room temperature (70°F) and hot steel (over 375°F) can cause slippage on screw (Ref. 17).

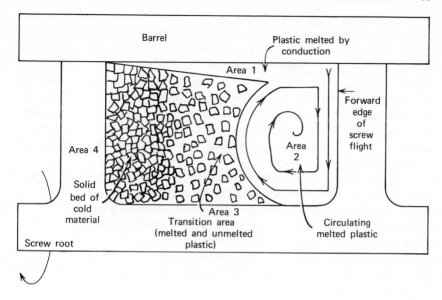

Figure 1-16   Melting of plastic in a screw.

warmed so that they stick together and adhere to the circulating melted mass of area 2. At the transitional sections between area 2 and 3 cold pellets are being melted and absorbed into area 2. Area 4 contains cold pellets which are conveyed as solids down the length of the screw.

Two types of heating are occurring. One is the convection of heat from the heater barrel. The second is the conversion of mechanical energy from turning the screw, into heat energy, by shearing the plastic.

It would seem logical that heating the screw would add more energy to the system and increase the output. This has been tried (21a) with significant output increases, lower temperature deviations, and of course, lower screw drive horsepower requirements. Boring the screw for the heater, maintenance, extra control parameter, and connection of the power leads may pose some problems. No field experience has been reported yet.

As more material is melted, area 2 increases until finally at the metering section area 2 completely fills the flights. To keep the plasticizing process going there must be a constant source of cold material in area 4. The longer the plasticizing zone, the more material that can be plasticized in the screw until such time as there is no more cold material. Once the maximum length is reached, where all the unmelted particles have been plasticized when traveling at their maximum speed, a further increase in the screw length will not increase output. The other limiting factor would be reaching the maximum flow rate capacity of the metering section.

Since the same weight of material per unit cross section must flow through the whole length of the extruder, and since the bulk density of the unmelted portion is less than that of the melted portion, the unmelted portion must be moving at a faster rate. Normally this flow rate is larger than can be used by the melting mechanism. However, if irregular pellets are conveyed, agglomerates are formed that temporarily partially block the flow channel of the unmelted particles; screw output will be reduced, and pressure surges will occur. [A series of colored pellets were dropped at small intervals into a stream of clear extruding plastic (21). A series of colored photographs graphically show the flow of unmelted particles in area 4.]

It would seem logical that long (up to 5 diameters) transition sections increase output, reduce air entrapment, and reduce pressure temperature and output variations. This has been confirmed many times as, for example, in the evaluation of screws for nylon (22). The transition section serves as the feed for the metering section. Unless there is a relatively long feed section which is able to fill the metering section at discharge pressures over 1000 to 2000 psi, the efficiency of the pumping action in the metering section will be diminished giving lower output rates.

Surging will arise primarily as a result of excessive pressure development in the feed and transition sections which overdrive the metering section (23).

Metering Zone

The metering section of the screw acts as a pump, removing the material plasticized in the transition zone. There are two general theories for extrusion in this area, with the physical reality probably a combination of both. The adiabatic theory assumes that there is no heat conducted into the material or out of the material as it moves through the extruder. The temperature increase of the material is developed by the mechanical working of the material. The heat applied to the barrel only compensates for radiation. The isothermal theory assumes the material temperature to be constant the entire length of the metering section. No heat is applied to it and the barrel heat compensates for radiation. The heat generated by the mechanical working of the material is conducted out through the cylinder walls. The isothermal theory, with certain basic assumptions, have led to useful equations describing the output of the metering section (24). As mentioned later these have been refined so that they can accurately predict extruder performance by a computer program (25). An accuracy of about 85% is obtained using the following simplifying assumptions (26):

1. The material entering the metering zone is fully plasticized.
2. The material is isothermal implying no change in temperature, viscosity, or density.

3. The flow is steady state.
4. The flow is laminar.
5. The flow is Newtonian.
6. The material is incompressible.
7. The screw is of uniform pitch and untapered.
8. The curvature of the channel around the screw flights is ignored. The cross section is considered to be rectangular. The helix is unrolled and laid out flat.
9. The land volume is neglected.
10. The channel width is large, relative to the channel depth, so that the velocity distribution is uniform across the whole width of the channel. The flow can then be treated unidimensionally.
11. The transverse velocity which contributes to mixing rather than pumping is disregarded.
12. The radial gap is negligible.
13. The frictional effects of the side walls of the screw flights are negligible compared to the drag effects on the screw and barrel.

Screw Output. The output of the extruder (Q) is the result of three different types of flow. The transverse or circulating flow which contributes to mixing is not considered. The drag flow (Q_d) is the major component of the flow. It is the forward conveying action caused by the rotation of the screw relative to the cylinder. The layer of melt touching the stationary cylinder does not move very much. The layer adjacent to it is pulled by the screw and moves slightly faster. As we travel from the barrel surface to the screw surface, the speed increases until the layer at the screw root turns at the same speed as the screw. It is called drag flow because of the drag of the barrel surface on the plastic. The path of a particle within the stream is complex and can best be described as a spiral within a spiral.

Intuitively it should make no difference whether the screw rotates or the barrel rotates. For analytical purposes it is sometimes easier to visualize the performance of an extruder if the barrel was assumed to be rotating. This convention has been adopted by most authors.

The pressure flow (Q_p) is the component resisting flow in the system. It is sometimes called back flow which is a total misnomer. The material only flows in a forward direction. This has been shown most graphically in Ref. 27.

The output is also reduced by the material flowing between the clearance of the flight and the barrel (Q_s). Its influence is relatively small in well maintained extruders and is generally not calculated for injection molding applications. The output of the screw metering section (which is the output of the screw) therefore is

$$Q = Q_{drag} - Q_{pressure} - Q_{slippage}$$

Drag Flow

$$Q_{d_1} = \frac{\pi^2}{2} D^2 \, N \, h \, \sin \theta \, \cos \theta \qquad (1\text{-}4)$$

where Q_{d_1} = drag flow (in^3/min)
  D = barrel diameter (in.)
  N = screw speed (rpm)
  h = channel depth (in.)
  θ = helix angle

Good results were obtained with a helix that advances one turn per screw diameter. This angle, θ, is 17.8°. Converting (1-4) into consistent dimensions

$$Q_d = 3.1 \, D^2 \, h \, N \, \rho \qquad (1\text{-}5)$$

Q_d = output (lb/hr)

ρ = specific gravity

Inspecting (1-4) shows that the output is fundamentally a result of the geometry of the screw. It is in effect describing a positive displacement pump. Obviously the density of the material conveyed will be a nondesign factor in the output.

As an example, using a 2 1/2-in. screw with a flight depth h = 0.100, (1-5) becomes

$$Q_d = 1.94 \, N \, \rho$$

For polystyrene with ρ = 0.92 (at 400°F) and N = 100 rpm

$$Q_d = (1.94)(0.92)(100) = 178 \text{ lb/hr}$$

For polyethylene ρ = 0.72 (at 400°F) and N = 100 rpm

$$Q_d = (1.94)(0.72)(100) = 140 \text{ lb/hr}$$

Pressure Flow

$$Q_{p_1} = \frac{\pi D h^3}{12} \frac{\Delta P \sin^2 \theta}{\eta L} \qquad (1\text{-}6)$$

where Q_{p_1} = output (in^3/sec)
  D   = diameter of screw (in.)
  h   = channel depth (in.)
  ΔP = increase in pressure (psi)
  θ = helix angle (17.8°)
  η = viscosity (lb-sec/in.^{2})
  L   = length metering section (in.)

With the specific gravity, ρ

$$Q_p = 3.14 \frac{Dh^3 \; \Delta P \rho}{\eta L} \qquad \qquad (1\text{-}7)$$

viscosity (lb-sec/in.^{2})

$$Q_p = 2.2 \times 10^5 \frac{Dh^3 \; \Delta P \; \rho}{\eta L} \qquad \qquad (1\text{-}8)$$

viscosity (poises)

Q_p = output (lb/hr)

It must be emphasized again that (1-6) is resistance to flow and does not signify a physical motion toward the rear. In a given extruder, the pressure is highly sensitive to the viscosity. The viscosity range of plastics usable in an extruder is from 10^{-3} to 1.0 lb-sec/in.^{2}. This is about 7×10^3 to 7×10^6 times the viscosity of water. The viscosity of water is so low that (1-6) would become very sensitive to pressure. This would make it unsuitable as a water pump. While temperature does not show in any of the output equations the density and particularly the viscosity are highly temperature dependent. The pressure loss ranges from 5 to 10%. It has been assumed that the feed and transition section have a flow rate equal to the metering section. If the flow rate there is less, the metering section will starve. If the flow rate is too high, high pressure will build up at the end of the transition zone and carry forward into the metering section. This may overcome the pressure flow component and even turn it positive. However, the quality of the melt is unacceptable.

For the same screw and materials, using L = 10 in., ΔP = 500 psi:
For polystyrene η = 0.02 lb-sec/in.^{2} at 1000 sec^{-1}

$$Q_p = \frac{(3.14)\,(2.5)\,(0.1)^3\,(500)\,(0.92)}{(0.02)\,(10)} = 18 \text{ lb/hr}$$

Q_p = 10% loss

For polyethylene η = 0.06 lb-sec/in.^{2} at 1000 sec^{-1} :

$$Q_p = 4.7 \text{ lb/hr} = 3.4\% \text{ loss}$$

Leakage Flow

$$Q_s = \frac{\pi}{10} \frac{D^2 \; \delta^3 \; tan \; \theta \; \Delta P}{S \; L \; \eta} \qquad \qquad (1\text{-}9)$$

where Q_s = output (in.^{3}/sec)
$\qquad$ D$\quad$ = screw diameter (in.)
$\qquad$ $\delta$$\quad$ = flight clearance (in.)
$\qquad$ $\Delta P$$\quad$ = pressure drop (psi)
$\qquad$ $\theta$$\quad$ = helix angle (17.8°)
$\qquad$ S$\quad$ = flight width (in.)

L = metering section length (in.)

η = viscosity (lb-sec/in.^{2})

The slippage over the flights is normally disregarded in calculations. Its significance is that it demonstrates the importance of wear in a screw and permits the engineer to calculate the point at which maintenance is economic.

For a given screw all the constants can be lumped together and the output for the drag and pressure components can be written

$$Q = \alpha N - \beta \frac{P}{\eta} \qquad (1\text{-}10)$$

where alpha and beta represent the respective constants. This shows that the only two variables under the control of the operator are the screw speed and the back pressure. The barrel temperature is also a variable but is controlled by the material temperature, directly related to the screw speed.

Almost all of the literature relates to extrusion which operates continually. Injection molding utilizes the screw for a fixed period of time. Therefore, to reach a given stock temperature, the screw can be run quickly at high speeds or more slowly at lower speeds, with the corresponding change of the barrel temperature. Slower screw speeds will normally give a better melt.

The other independent variable is the back pressure. Higher back pressure increases the homogeneity of the melt and as might be expected lowers the output, by increasing the pressure flow and the slippage. Higher back pressures usually improve the physical properties of the part (28). Problems of dimensional control, shrinkage, warpage, and color dispersion are often helped or eliminated by higher back pressure.

Flow Patterns in the Screw Flights. The flow pattern in the screw flights changes with the back pressure. Figure 1-17a shows a simplified schematic drawing of the flow of a particle in the flights with open discharge. The particle is moving in a circulatory motion from flight to flight. In addition it is moving forward along the axial direction of the barrel toward the nozzle. In the blocked flow (Figure 1-17b) there is a similar circulatory motion between the flights, but no forward motion because the open end is closed. Obviously there is the greatest mixing when the flow is blocked. The only time there is a plane of stagnation (0 velocity) is in blocked flow, when both the axial and transverse velocities pass simultaneously through zero at a plane two thirds of the distance up from the channel root. This should not occur in injection molding. The importance of this flow concept is that it shows that the more blocked the flow, the better the mixing in the screw. Changing the outlet orifice is in effect changing the ΔP (1-6). The higher the pressure the greater the pressure flow and the lower the output. In injection molding this pressure corresponds to the back pressure setting of the machine. This is the reason that color dispersion is improved and homogeneity increased by raising the back pressure. Often

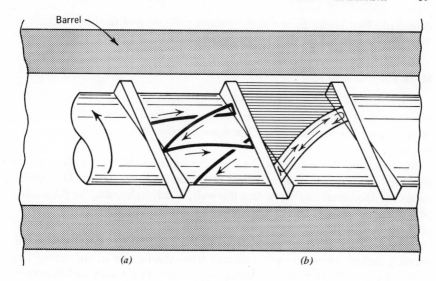

Figure 1-17   Flow in a screw: (*a*) open discharge and (*b*) blocked discharge.

warpage and shrinkage problems can be overcome in this manner.

Reference 27 is recommended for an excellent analysis of the flow patterns in a single screw extruder. There are photographs which clearly show that under no circumstances does there exist a local velocity directed backwards along the screw axis. Reference 29 has a more detailed mathematical analysis of channel flow.

Screw Conveying Considerations

Almost all the early theoretical and experimental work was concerned with the metering section of the screw. Recent work, primarily by Tadmor, Klein, and Marshall has resulted in an understanding and mathematical formulation of the melting or plasticating section of the screw. Combined with further studies of the metering section, a computer program for a mathematical model of screw conveying was developed. The program which corresponds remarkably well with experimental data is being continually updated and can now simulate screw design and plasticating performance without any need for empirical information. In addition the program can produce such information as the effect of screw cooling on the pressure profile, absence of barrel cooling can be simulated, determination of the optimum location for mixing devices, and barrel temperature for maximum rate of melting and the mechanisms of the reciprocating screw. Since understanding screw plasticizing is essential to knowing injection molding, it is strongly urged that Ref. 25, and 30 to 37 be consulted. Two books by the authors above (38, 39) give further information. The following section mentions some of the points made in Refs. 30-39.

Melting Mechanisms

The mechanism for melting is basically described in Refs. 21 and 35. The explanation of the mechanism was developed by putting in different colored particles into the screw, cooling the screw, disassembling, and unraveling the polymer.

Summarizing, the solid pellets are conducted from the hopper. They touch the barrel to form a thin film of melted plastic on the barrel surface. The relative motion of the barrel and screw drag this melt, which is picked up by the leading edge of the advancing flight of the screw. This flushes the polymer down in front of it, forming a pool which circulates. Heat is first conducted from the barrel through the film of plastic attached to it. Heat then enters the plastic by the shearing action, whose energy is derived from the turning of the screw. The width of the melted polymer increases as the width of the solid bed decreases. Melting is complete at the point where the width of the solid bed is zero.

It was found that the mechanism of melting does not immediately start in the first heating zone. One reason is that the barrel temperature there is kept below the melting point of the polymer to prevent plugging of the screw. Also even when the melting point is reached, the initial melt fills up some of the voids between the pellets before circulating. Until the thickness of the melt film on the barrel exceeds the radial clearance of the screw, the chances of circulation are minimal. This delay may continue until the material moves down 10% of the screw length.

The delay can be minimized by keeping the temperature as high as possible in the rear of the extruder, reducing the flight clearance, and preheating the material. Obviously the latter step will decrease the heat required from the screw for melting and give faster melting in the extruder. Care should be taken to prevent feeding problems caused by the temperature effect on the friction of the material to the barrel.

It was observed that the solid bed breaks up at certain points creating a gap which fills up with melted plastic. The external forces operating on the solid bed, such as the operating conditions, geometry, and physical properties of the polymer, create tensions which are larger than its tensile strength. This results in pieces breaking away and flowing into the channel of melted material. The bulk of the solid bed will continue to move ahead until another piece breaks away. This inherent unsteady state or condition is smoothed out in the relatively long transition sections of the screw. It should be noted that melting starts in the so-called feed section and often extends itself well into the metering section. If the screw is run at overcapacity, unmelted particles will appear in the extrudate.

Computer simulations, backed by experimental results, show that the length of the melting zone is very much affected by the flow rate. Keeping all other variables constant and increasing the flow rate lengthens the melting zone, giving a less homogeneous product or even an incompletely melted polymer. This might logically be expected as the faster the solid bed moves in a given geometry the less time it will have to melt and disappear. Increasing the flow rate also increases the delay in the start of melting because of the time dependence of heat transfer from the barrel to the plastic.

Output Rate and its Effect on Melt Properties.   Figure 1-18 (32) shows the effect of the output rate on the width of the solid bed. Low density polyethylene was extruded in a 2 1/2-in., 26 L/D extruder at a screw speed of 60 rpm and a barrel temperature of 400°F. Above the graph is a cross section of the extruder channel showing the depth and location of change. The material is fully plasticized when the solid bed width is zero. Curves 1 and 2 have a rate high enough so that unmelted particles appear at the nozzle end. It would appear that the maximum output for molding would be defined by curve 3, while a more homogeneous melt would come from operating the extruder under the conditions of curve 4. The table underneath shows the fraction of total heat required for melting which resulted from the shearing action of the screw. The fraction varies with the location on the barrel. The balance was provided by the barrel. The last column shows the shaft horsepower consumption for melting.

The effect of flow rate on melting is roughly linear. The effect of screw speed is the result of several factors. Keeping all other variables constant, an increase in the screw speed increases the shearing heat proportional to the square of the speed. It increases the cross channel velocity, which increases the rate at which the melt is being moved linearly. Both increase the rate of melting. Increased screw speed also reduces the delay in the start of melting. The first effect can be large enough so that the screw in Figure 1-18 can produce more heat than required by doubling its rpm. Practically, in injection molding, the improved melting rate caused by increased screw speed is counteracted by the reduced melt quality due to increased flow rate.

Barrel Temperature Effects.   The effect of elevating the barrel temperature is not completely clear. While an increase in barrel temperature increases heat transfer, the increase changes the velocity profile of the interface between the solid bed and the barrel; this tends to reduce the rate of melting. It also tends to reduce the viscosity of the melted film on the barrel wall reducing the heat generated by shearing. Experience has shown that there is an optimum barrel temperature which is determined empirically.

Effects of Physical Constants.   The effect of the physical constants of the screw such as channel depth, taper, helix angle, and flight clearance are found in Ref. 32. Aside from flight clearance their effect cannot be changed by the molder. As the flight clearance increases because of wear, the length of the screw

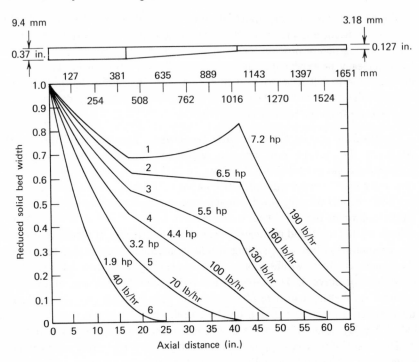

Figure 1-18 Effect of throughput on solid bed profile at 60 rpm screw speed and 400°F barrel temperature for low density polyethylene extruded in a 2½-in. single screw plasticating extruder. The attached table shows the effect of throughput on (*a*) the consumption of shaft-power for melting and (*b*) on the fraction of total heat required for melting which originates in viscous heat dissipation (Ref. 32).

	Output		(Heat by viscous dissipation)	
Curve	(lb/hr)	(kg/hr)	(Total heat for melting)	Power (hp)
1	190	86.2	0.37–0.58	7.19
2	160	72.6	0.34–0.63	6.49
3	130	59	0.33–0.66	5.51
4	100	45.4	0.36–0.70	4.40
5	70	31.8	0.38–0.73	3.20
6	40	18.1	0.40–0.77	1.92

required for melting increases. For example, in extruding low density polyethylene in a 2 1/2-in. screw, a flight clearance of 0.5% of the barrel diameter requires a melting length 1.75 times the length required if there were no flight clearance. If the wear doubles this flight clearance to 1%, an effective length of 2.25 times the no-flight clearance length would be required.

Peak Pressure. The maximum pressure developed in an extruder is realized toward the end of the transition section before the metering section. This has been confirmed many times (23, 25, and 40) and particularly in Ref. 33. Curves are given comparing the actual measured pressures and the computer calculated ones. Figure 1-19 shows part of their results using a 2 1/2-in., 26 *L/D*, 3:1 compression ratio screw, extruding low density polyethylene. The location of the peak pressure does not vary significantly with the screw speed or output. The material starts to melt between 4 1/2 and 6 turns from the feed section. As would be expected, the slower screw speeds plasticize more slowly, with the point at which half the channel is filled with molten plastic being just before the transition zone.

Temperature. Temperature fluctuations and temperature profiles in the extruder have been accurately measured (25). Figure 1-20 shows the temperature profile for a screw similar to the one of Figure 1-18. Polyethylene was extruded at 375°F against a head pressure of 1230 psi. The top curve shows the barrel temperature, and the second the screw temperature. The shaded portion represents the maximum and minimum temperatures, with the line in between them the time-average temperature. Temperature fluctuations were

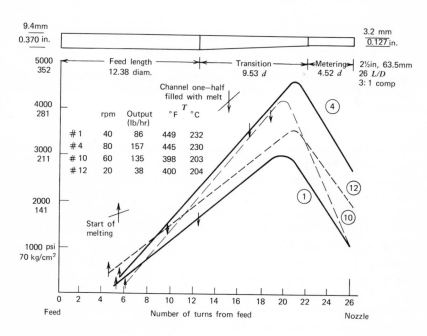

Figure 1-19 Barrel pressures for extrusion of LD - PE, with start of melting and point where channel is half full of melted polymer (Ref. 33).

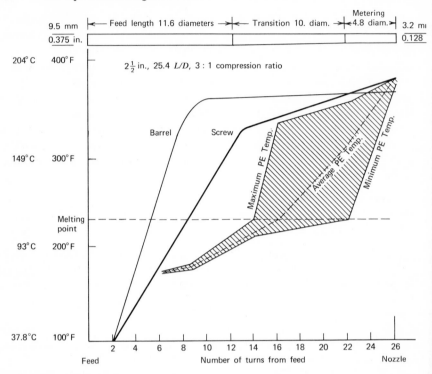

Figure 1-20 Barrel temperature, screw temperature, maximum, minimum, and average polyethylene temperatures, extruded at 375°F (190°C), 1230 psi (86 kg/cm²) head pressure (Ref. 25).

wide in the compression and metering zones, but rapidly diminished at the output end. When the screw was changed by tripling the metering length and correspondingly reducing the channel depth to 0.107, the profile changed radically. The longer, shallower, metering zone gave a very high back pressure, 6905 psi. The material was fully melted at the metering zone and gave a relatively uniform temperature. The shear heat developed caused the screw to rise 30° above the barrel temperature.

There is a relatively large amount of literature analyzing the single screw, which relates to theoretical considerations, experimental measurements, and the effect of the screw geometry and operating conditions on molding, and covers temperature profiles, pressure profiles, torque, and horsepower requirements. This literature is in addition to the previously mentioned references. A selected group are found in Refs. 41 to 49.

Twin Screws.   Because of the heat sensitivity of some materials as well as more precise production control requirements, there is renewed interest in low-shear screws for extruders and possibly for injection molding. Twin screws were used in reciprocating machines in the early 1960s, but they eventually gave way to the easier-to-make, more understandable, and less costly single screw.

The twin screw extruder with intermeshing screws rotating in the same direction is fundamentally a positive displacement pump. The material is transferred from screw to screw with relatively little mechanical energy converted into heat by shearing. Heat is supplied by the barrel. The output is controlled by controlling the amount of feed into the screw, which is always less than the capacity of the system. Because propulsion is almost entirely independent of friction and viscosity, the output, pressure, and material temperature are independent of the screw's speed. Since screw speed determine shear rate, shear sensitive material such as PVC, can be extruded most successfully, without reducing the output. By contrast, a single screw might have to be operated at reduced speed because of excess shear heat and in so doing output is significantly decreased. The desired output and pressure in a twin screw would be determined by the amount of feed. Some of the disadvantages of the twin screw follow:

1. Difficulty of controlling the feed.

2. The necessity of completely purging the machine before shutdown so that the extruder can be started up without a long wait for the heat to soak through the steel into the plastic.

3. The complicated design of the transitional area from the end of the twin screws to the circular exit at the nozzle.

4. The biggest historical disadvantages of the twin screw assembly is the difficulty in designing adequate thrust-bearing assemblies because of the limitations imposed by the screws being close together. This problem which is more evident in the reciprocating screws than in two stage machines is becoming less critical with experience in twin screw design.

5. The cost of a twin screw machine is almost double that of a single screw.

Notwithstanding, the use of twin screws intermeshing and rotating in the same direction for injection molding machines will probably increase somewhat in the future. References 50, 51, 51a and 52 contain engineering analyses and comparisons of the two types of extruders.

RECIPROCATING SCREW TIP ASSEMBLIES

The reciprocating screw machine uses the screw as a plunger. As the plunger comes forward the material could flow past the screw head and back into the flights. For more viscous materials, such as PVC, a plain tapered tip (Figure 1-21) on the front of the screw is sufficient to permit the screw to act as a

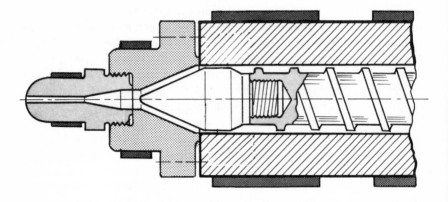

Figure 1-21 Plain screw tip used on reciprocating screws for viscous heat sensitive materials.

plunger. The rapid forward motion of the plunger does not let too much material leak back. Moreover, the plain tip is also good for molding heat sensitive material such as PVC, because this type of screw front provides the least opportunity for hangup and material degradation.

The less viscous materials require a check valve to prevent back flow over the screw tip. Screw tips, in either case, are a source of frictional heat, material hangups, intermittent malfunctioning, and high maintenance costs. The only way to eliminate them is to convert the machine into a two stage screw-pot, a situation which has other problems. Different valve types are listed below:

1. *The ring type nonreturn valve.* This valve (Figure 1-22) is a three part assembly. The check ring and seat are slipped on the main body which contains the tip. The assembly is then screwed into the reciprocating screw. The sliding ring fits snugly in the barrel. When the screw rotates the force of the plastic pushes the ring toward the nozzle end and permits the plasticized material to flow under it through flutes or grooves on the main assembly. The screw slides back until the amount of material necessary for the shot is plasticized. On the forward or injection stroke, the ring slides toward the seat and seals the rear of the screw from the front, so that material cannot leak by as the plunger comes forward.

If the fit of the ring is too tight it will cause excessive barrel wear. It will also increase the resistance of the screw to retraction, which is equivalent to raising the back pressure. This may not be desirable as extra stress is put on the threaded stud. A ring that is too loose allows excess leakage, and because of the high shear rate, degradation of the plastic. Plastics are not lubricants, which compounds the fitting problem. The abrasive action of some fillers accelerates wear in this type valve. Additionally, tramp metal can lodge between the seat

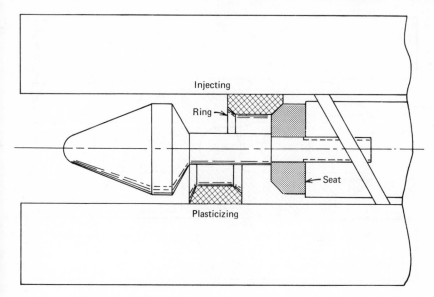

Figure 1-22   Ring type nonreturn valve for reciprocating screws.

and the ring preventing seal off. It may be flushed through later but can leave a dent. When there is decomposition of the plastic in this area the trapped gas may etch the valve seat.

Wear on the ring can cause the ring to cock, resulting in severe erosion of the barrel. Wear also increases the pressure loss which increases the variation in shot size with all its resulting difficulties. If the ring valve does not close immediately, part of the material will flow back over the flights causing the screw to turn during injection. In any type of check valve, such rotation means the valve is not seating properly.

2. *Ball shutoff valve.* Some of the limitations of a sliding ring type shutoff are overcome with a ball shutoff, Figure 1-23. Such a valve should be guided so that it is not affected by gravity or centrifugal forces. It should have a short stroke and enough passage area so that the plasticized material can move without generating excessive frictional heat. It is advantageous to have a one piece design with a removable pin for ease of maintenance. Two piece units welded together are difficult to maintain and may cause problems at the joint. Ball-type nonreturn valves are easy to maintain, usually requiring only a relapping of the seat and a new ball. Their design is inherently more streamline than the sliding ring type. An excellent analysis of both types is found in Ref. 53.

3. *Other shutoff valves.* Neither shutoff valves are completely satisfactory, and new valves are being developed (54). One of the most interesting uses a body

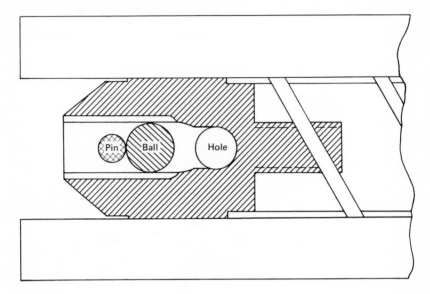

Figure 1-23   Ball type none return valve for reciprocating screws in its plasticizing position.

which has a hole drilled in it at right angles to the screw. A large hole is drilled from the nozzle end and through the body to meet this hole. On the nozzle side of the body is a round disk covering this hole and floating on two stripper bolts. When the material is being plasticized its force coming out of the hole separates the disk from the body. When the plunger comes forward, the disk is forced back covering the hole. One of the advantages of this type of valve, aside from its low cost, is its ability to pass large pieces of foreign material without damage. It can be readily converted into a normally closed type of valve by putting springs on the stripper pins. A review of European nonreturn valves is found in Ref. 55.

SCREW DRIVES

One method of applying the driving force to the screw is to attach it to an electric motor through a speed-reducing, gear train with different speed ranges (56). A second method is to connect the screw to a nonvariable speed-reducing coupling which is driven by an hydraulic motor.

The Hydraulic Motor

The hydraulic motor is driven by an hydraulic pump (57, 58). Hydraulic motors and pumps have either fixed or variable displacements. Fixed displacement pumps and motors are used for screw drives and have a constant torque output therefore, their horsepower varies with speed. This combination is the least expensive and simplest form of hyrdostatic transmissions. The maximum torque depends on the pressure setting of the controlling relief valve. Their running torque depends on the load imposed by the screw.

One of the outstanding features of the hydraulic system is the stepless control function. This characteristic is very important because of the importance of precise control over screw speed. Precise control is important because melt temperature and the time required to plasticize a given shot weight vary with screw speed. On the one hand, high screw speed is desirable, and on the other hand, slower screw rotation produces a more uniform melt temperature and degrades the material less. Good balance and control are more easily obtained with a stepless system.

Thus for many reasons it is desirable to plastify at the lowest possible screw speeds. The drive must supply enough torque to accomplish this, but not enough to mechanically break the screw. Changes in torque are needed because of the different processing characteristics of plastics (much higher torque is required to plasticize polycarbonate than polystyrene). It is desirable to have two torque-speed ranges for handling the different materials. If a material is being molded with a minimum torque, at a given speed, increasing the speed will increase the horsepower requirements.

The highest torque requirement is needed at the start of the rotation. This is primarily due to the inertia of the plastic material, that is, the viscosity. In addition, there is the inertial mass of the screw, coupling gears, and drive system. As the screw turns, the material is heated and the viscosity decreases, reducing the torque requirements.

Combinations of variable displacement pumps and motors will give different torque, horsepower, and speed operating characteristics. And the new high torque, low speed hydraulic motors, allow the elimination of the speed reducing unit found in many hydraulic systems.

The availability of high torque at low speeds by the electrical drive system is very useful in molding viscous materials. This can be achieved in hydraulic systems by having one pair of the gears in the speed reducing unit reversible. Another method is to use two motors attached to the same screw drive mechanism. They can be operated singly or together. When one motor operates it runs at full speed and lower torque. When both are used they share the available supply and run at half speed. This doubles the torque at the same horsepower input. For more on hydraulic drives, see Chapter 6.

The Electric Motor

The available torque of an electric motor drive is similar to the pattern described above. An overload factor as high as 100% is available. An hydraulic drive, on the contrary, builds up its starting torque slowly and never can exceed its maximum operating torque. The maximum torque of the hydraulic system must be at least as large as the peak torque required to start turning the screw.

Thus a smaller electric motor can be specified for an electric drive as against an hydraulically driven screw of equal starting torque capacities. This, coupled with the higher efficiency of the electrical drive, results in significant operating economies.

This effect can be a mixed blessing. Suppose during operations that a heater band failed, or the pyrometer was set too low, or start-up was attempted at too low a temperature. Under these conditions, the availability of 100% torque overload may be excessive for the screw. At the worst, this could break the screw. There are a number of safety devices of varying degrees of effectiveness. None of these devices have the reliability and ease of operation of the pressure relief valve in an hydraulic drive system. This valve will limit the torque (which is controlled by the pressure supplied to the motor) to a safe value. Oil is by-passed through this valve to tank and the stalled screw and hydraulic motor will suffer no damage.

Electric drives do not have independent speed and torque controls. The speed is changed by gear trains. Since the input and output power is constant, the change in either speed or torque will inversely affect each other. This restricts finding the optimum molding conditions. The hydraulic system, on the other hand, has stepless speed control of the screw. It would be difficult to overstress this difference.

The rotating screw pushes the material in front of it, forcing the screw and injection carriage to the rear. The electric motor weighs considerably more than an hydraulic motor for comparable output. Therefore, the hydraulic system results in considerably lower back pressure at the "zero" setting of the back pressure valve.

Screws can be driven by special low-torque hydraulic motors which are either mounted outside directly on the screw or internally (Figure 1-6:27). A direct drive to the screw has no gears, speed reducers, thrust bearings, and develops minimum noise and vibration. No lubrication is required and maintenance is at a minimum. If the motor is inside the cylinder the maintenance is even less, but more difficult when it occurs.

Drive Characteristics and Speed Reducers

The majority of screws are driven by electric motors or hydraulic motors through a gear type speed reducer. Speed reducers are rated on a constant torque basis. One must be careful in selecting speed reducers. A speed reducer rated at 50 hp at 200 rpm, is good for only 12½ hp at 50 rpm. A speed reducer attached to an electric motor should not have a capacity larger than the strength of the screw. It is much less expensive to strip gears than break screws. Of course, a properly designed and operated system will do neither. Hydraulic motors, if properly sized, cannot break the screw. Table 1-1 compares some of the important characteristics of the hydraulic and electric drives for injection molding screws.

Torque. The work done in a screw (melting the material) is done by rotating a screw in a stationary barrel. Under appropriate conditions, the polymer molecules slide over each other (shearing) creating heat. A second source of heat, but of much lower magnitude, occurs when the molecular chain is broken. The rotational force is called torque. It is the product of the

Table 1-1 Comparison of some characteristics of hydraulic and electrical drives for screws

Characteristics	Hydraulic	Electrical
Efficiency	Low 60–75%	High 95%
Screw safety	Relief valve will prevent screw damage.	Overload protection difficult, particularly with small diameter screws.
Size	Small and light weight. Good shot weight control.	Heavy. Poorer shot weight control. More stresses in the system.
Torque	Constant output. Infinitely adjustable. Must have maximum torque requirements built into the system even though it might be used only at startup.	Varies with screw speed. Excellent starting characteristics and at low screw speeds.
Speed	Stepless control easily adjusted. Gives best control of molding conditions	Limited number of speeds. More difficult to adjust and maintain.
Melt quality	Best	Acceptable

tangential force and the distance from the center of the rotating member. For example, if a 1 lb weight is placed at the end of a 1 ft bar attached to the center of the screw, the torque would be 1 ft X 1 lb or 1 ft-lb. Torque is related to horsepower:

$$hp = \frac{\text{torque (ft-lb) rpm}}{5252} \tag{1-11}$$

$$hp = \frac{\text{torque (in.-lb) rpm}}{63,024} \tag{1-12}$$

It is clear that the torque output of an electric motor of given horsepower will depend on its speed. A 30-hp motor will have the following torque at different speeds.

rpm	Torque (ft-lb)
1800	87 1/2
1200	133
900	175

The speed of a given horsepower motor is built into that motor. Obviously, the higher torque unit (lower speed) motor will have a larger frame than the lower torque unit. The change in speed and torque can also be accomplished by changing the output speed of the motor by using a gear train. The change in torque will vary inversely with the speed.

Screw Strength and Speed.   The strength of the screw limits the input horsepower. As input horsepower is increased, a point will be reached where the torque that can be provided will be above the yield strength of the metal   screw. The strength of the screw varies with the cube of the root diameter.

Overpowering the screw will either degrade the plastic at excess speeds or shear off the screw at too low speeds. For a given horsepower, the slower the speed the higher the torque as shown in (1-11). For a 2 1/2 In. screw at 200 rpm the maximum permissible drive input is about 40 hp. For a 3 1/2 In. screw at 200 rpm it is 120 hp, and for a 4 1/2 In. screw at 150 rpm it is 180 hp. Table   1-2 shows the maximum horsepower per screw diameter that should be used on injection molding machines.

The shear rate, (1-3) is directly dependent upon the screw speed. Excess shear rate will degrade the material. The shear rate is highest near the barrel wall. The maximum surface speed with present screw technology is about 150 ft/min. Some of the more shear sensitive materials limit the surface speed to 100 ft/min.

Table 1-2

Screw in.	Diameter mm	Max. hp for Injection Molding	rpm for 100 ft/min (30.5 ms/min)	rpm for 150 ft/min (46 ms/min)
1 3/4	44.5	7 1/2	220	330
2	50.8	15	190	290
2 1/2	63.5	40	155	230
3 1/2	88.9	75	110	165
4 1/2	114.3	150	85	130
6	152.4	200	65	95

Screw Temperature. The temperature increase of the material caused by shearing action is

$$\Delta T = \frac{\pi D N \eta}{C h}$$

where ΔT = temperature increase of material   (1-13)
        D = diameter of screw
        N = rate of screw rotation
        C = specific heat of plastic
        h = screw channel depth
        η = non-Newtonian viscosity.

This relationship is not easy to use quantitatively because of the difficulty in accurately determining viscosity but is of value in explaining the parameters that affect ΔT because of the following reasons:

1. The viscosity of polymeric materials is very sensitive to the rate of shear (discussed later). The shear rate is also dependent upon the rate of screw rotation, Eq. 1-3.
2. The viscosity is temperature dependent.
3. The mathematics are further complicated by the lack of knowledge of the mixing patterns in the channel.
4. Lastly, this equation only measures the frictional energy and not the heat received from the barrel.

Nonetheless, increasing speed of screw rotation increases temperature if everything else remains constant in an extruder. In a molding machine, the screw output is not continuous and the temperature rise is influenced by screw speed and also time of screw turning. Using a 1 3/4 in screw to injection mold a 0.96 density, 3.5 MI, polyethylene bowl weighing 130 g the following results were obtained (28):

Screw Speed (RPM)	Back Pressure (PSI) (Kg/cm²)		Barrel Temp °F	°C	Average Plastic Temp °F	°C	Time of Screw Rotation (sec)
24.5	0	0	420	216	472	244	33
66.	0	0	420	216	464	240	14
129	0	0	420	216	464	240	10
66	5100	359	420	216	491	255	38
129	5100	359	420	216	491	255	20

In going from 24.5 to 66 rpm the required time of screw rotation falls in approximately that proportion. Going to 129 rpm reduces the time of screw rotation but nowhere near the extent expected. Among the reasons for this is the small time to receive conducted heat from the barrel.

Horsepower Requirements.   If the back pressure is increased to 5100 psi, the time required increases from 14 to 38 sec at 66 rpm, and 10 to 20 sec at 129 rpm. The increased energy needed to raise the pressure came from the extra time of screw rotation.

The work done in screw plasticizing raises the temperature of the material to the molding temperature. The total energy required to do this would be the product of the specific heat, temperature change, and output. The screw also acts like a pump and the energy required for pumping is the product of the increase in pressure and the output. This can be expressed as follows:

$$Z = C (T_p - T_f) Q + \Delta P Q \qquad (1\text{-}14)$$

where Z = power
C = average specific heat (Btu/lb, °F)
T_p = temperature plasticized material (°F)
T_f = temperature feed material °F)
Q = through-put (lb/hr)
ΔP = back pressure (psi)

If we disregard the energy required for pumping $(\Delta P \times Q)$, use a 30% loss factor for efficiency and heat provided from the barrel, and convert the expression into consistent units (1-15) becomes a good approximation of the amount of shaft horsepower required for a given output.

$$\begin{aligned} hp &= C (T_p - T_f) Q \\ &= 0.00056 \, C (T_p - T_f) Q \end{aligned} \qquad (1\text{-}15)$$

For example, high impact polystyrene has an average specific heat of 0.42.

How much horsepower is required to plasticize 1 lb/hr ($Q = 1$) when the room temperature is 70°F?

$$hp = (0.00056)(0.42)(400 - 70) 1 = 0.078 \ hp$$

This is equivalent to 13 lb/hr for each horsepower input.
Molding materials range from 6 to 14 lb/hr for each horsepower input.

The enthalpy is the same as the product of the average specific heat and temperature rise. Enthalpy tables for plastic are available (59), (1-15) might also be written as

$$hp = 0.00056 \ (h_p - H_f) \ Q \tag{1-16}$$

h_p = enthalpy of plasticized material

h_f = enthalpy of feed material

Q = output (lb/hr)

For the previous example the enthalpies are 150 and 15, respectively; thus,
$h_p = (0.00056)(150 - 15) 1 = 0.076 \ hp = 13 \ lb/hr$

Using a 30-hp motor at a room temperature of 80°F, what is the maximum output of low density polyethylene ($C = 0.8$) at 380°F?

$$30 = (0.00056)(0.8)(380 - 80) Q$$
$$Q = 223 \ lb/hr$$

Supposing the material temperature were raised to 450°F. Would the output increase?

$$30 = (0.00056)(0.8)(450 - 80) Q$$
$$Q = 181 \ lb/hr$$

Thus raising the material temperature *lowers* the maximum output. The molder therefore molds at the lowest possible melt temperature. This gives maximum screw output and reduces the time needed for reversing the process, that is, polymer cooling in the mold.

It is interesting to note that the output of the machine as defined in (1-14) is completely independent of the screw diameter. If, for example, a 2 1/2 In. and 3 1/2 In. screw, each having the same L/D ratio and the same input drive horsepower, were operated at the maximum capacity of the input drive, they will both deliver the same output (lb/hr). The plastic will remain longer in the 3 1/2 In. screw. Why then have large screws? The answer is found in (1-3), (1-11), and Table 1-2. Screw speeds must be kept low to prevent degradation. With a constant horsepower, the slower speeds can develop a torque high enough to shear the screw. Therefore, the higher horsepower required for higher output needs larger diameter screws to prevent screw breakage.

Thus it is obvious that in a screw machine, the horsepower rating available for screw rotation is a very important specification. Assuming similar efficiency for different screw designs, the maximum output, which is a primary concern of injection molders, is largely determined by the horsepower rating of the screw. With this criteria, analyzing machine specifications of different manufacturers becomes more productive and interesting.

Finally, the amount of power consumed in shearing per unit screw length equal to one turn of a flight is

$$Z = \pi^3 D^3 \left[\frac{L}{h} \eta_c + \frac{S}{\delta} \eta_L \right] N^2 \qquad (1\text{-}17)$$

where Z = power
        D = diameter
        L = lead (pitch length)
        h = channel depth
        S = flight land width
        N = screw speed
        η_c = viscosity in channel
        π_L = viscosity at land clearance
        δ = clearance between flight and barrel

Again, since the viscosities are difficult to obtain, the purpose of the equation is to show the factors which affect power output.

Advantages of Screw Plasticizing.   In a screw, the melting of the plastic is caused by the shearing action on the polymer between the barrel and root of the screw. As the polymer molecules slide over each other they convert the mechanical energy of the screw drive into heat energy. The heat is applied directly to the material. This process and the mixing action of the screw contribute to its major advantages as a plasticizing method. These advantages follow:

1. High shearing rates are obtained. As we see later these high rates lower the viscosity of the melt making the material flow easier.
2. Good mixing is developed resulting in an homogeneous melt. This usually means lower injection and hence clamp pressures.
3. Flow is nonlaminar.
4. Residence time in the cylinder is much less than in a plunger machine.
5. Most of the heat is supplied directly to the material.
6. Since relatively little heat is supplied from the heating bands compared to a plunger machine the cycle can be delayed for a longer period before purging, since the screw is not turning and little heat is being generated.
7. The action of the screw reduces chances of material hold-up and subsequent degradation.
8. Machine can be used with heat sensitive materials, such as PVC.
9. The screw is easier to clean than the plunger.

10. The screw is easier to purge than the plunger.

There is considerable literature comparing screw preplasticizing with plunger plasticizing and describing the operation of the in-line screw (60-63).

The reciprocating screw is the most popular injection end in the United States.

In the reciprocating machine, the material is fed from the hopper, plasticized in the screw, and forced past the one-way valve (if there is one) at the injection end of the screw. The material accumulates in front of the screw, forcing back the screw, the screw drive, and the motor. When the screw reaches a position, determined by the amount of feed required, a limit switch is contacted stopping the screw rotation. This happens while the previous shot cools in the mold. After that shot has been ejected and the mold closed, hydraulic injection cylinders bring the screw assembly forward and use the screw as an injection ram. The advantages of an in-line screw have been amply reported in the literature. Some of the advantages and disadvantages of the in-line screw can be deduced from the discussion of a screw-pot machine.

Screw-Pot (Screw Plunger). A screw-pot injection end is shown in Figure 1-9. It is essentially similar to the plunger-plunger machine (Figure 1-8) except that a fixed screw is used for plasticizing. The fixed screw need not be mounted vertically above the machine. It can be mounted at an angle or next to it on the same level. A ball type check valve is usually used between the screw and the front of the shooting cylinder.

A reciprocating screw could be used in place of a fixed screw. This permits continual operation of the screw throughout the whole cycle. It is primarily used in rapid cycling machines. For most applications it is less expensive to use larger cylinders and motors.

The screw pot and the reciprocating screw are both preplasticizing systems. Conceptually the difference is limited to the location of the pot which is in front of the reciprocating screw and a separate cylinder in the two stage machine. Although most machines sold today are of the reciprocating type, many molders have found significant advantages in screw-pot equipment. Some of them follow:

1. Because the screw does not act as the injection ram, lighter bearings can be used. There is no need for the heavy thrust assemblies found on reciprocating screws. Higher thrust assemblies could mean reduced maintenance costs.

2. The extruder barrel need only be strong enough to maintain the pressure of the material during plastification which is rarely over 10,000 psi. In contrast, the barrel for the reciprocating screw must contain the 20,000 psi used for injection.

3. There is less wear because the screw does not move.

4. A simple ball check valve can be used as the connection between the two stages. This is a trouble free and easy to maintain system. Furthermore, it

presents minimum flow resistance. The nonreturn valve at the tip of a reciprocating screw wears; sometimes it does not seat properly (preventing consistent molding), or sometimes causes wear in the barrel. In this valve material may hang up and degrade; it is also much more expensive than a ball check valve.

5. The connection between the two stages result in better mixing of the melt.

6. A very important advantage of a two-stage machine is that all the material goes over the full flights of the screw, receiving the same heat history. In a reciprocating screw only the first material in goes over the full length of the screw.

7. In a two stage machine, the screw pumps only against the injection ram which is floating in oil in the hydraulic cylinder. The reciprocating screw must push back the whole weight of the carriage and all the equipment on it. For this reason the shot size control is usually considerably more accurate in the two stage machine.

8. Extremely high injection pressures are available.

9. The size of a pot in front of a reciprocating screw is limited by the length of the feed stroke. If the screw goes too far back the material will not plasticize correctly. In a two stage machine there is no theoretical limitation. Thus a 2-in. reciprocating screw normally has a maximum shooting capacity of 13 oz while the same diameter screw can be readily designed to shoot 60 oz in a two stage machine.

There are a number of disadvantages of the two stage machine:

1. It requires two cylinders and two sets of heat controls.

2. It is slightly more difficult to clean.

3. It is slightly more difficult to set up.

4. It does not process high heat sensitive materials as well as a reciprocating screw.

5. Cylinders for molding thermosets and rubber are designed only for reciprocating screws.

6. It takes up more space than a reciprocating screw.

Injection End Specifications

A number of specifications for the injection end of the molding machine are common to all plasticizing equipment. The *injection capacity* (in.^{3}) is the maximum volume of material that can be injected in one shot. It is a measure of the geometry of the cylinder, plunger, and plunger stroke. It is sometimes given in ounces of polystyrene. To convert cubic inches to ounces of 1.00 density material multiply by 0.578. Multiplying by the density of the material in question will give the injection capacity in ounces of that material. For example, polystyrene with a density of 1.06 is equivalent to 0.61 oz/in.^{3} (0.578 x 1.06). This is a major machine specification. A 16-oz machine means that it will inject 16 oz (usually with polystyrene) per shot.

The *injection rate* (in.^{3}/min; oz/min) is the maximum rate at which the injection cylinder can eject fully plasticized material into the air, on a single shot basis. This is somewhat different than the speed achieved during molding which may be limited by the resistance to flow of the mold. The availability of a high injection rate is desirable since you can always adjust for lower rates if need be, but it is good to have the high rates when necessary. Injection rate should not be confused with the plasticizing capacity of the system, which is an indication of the maximum amount of moldable material produced in a given time.

An important quality of the machine is the amount of pressure that is placed directly upon the plasticized material. This *injection pressure* is easy to determine in a two stage machine or an in-line screw. It depends on the diameter of the screw or plunger, the diameter of the piston of the hydraulic injection cylinder, and the oil pressure. For example, a 2 1/2 in.-diameter reciprocating screw uses two 6-in.-diameter injection cylinders in a 2000-psi hydraulic system. The area of the injection cylinders is 2 x 28.3 in.^{2} = 56.6 in.^{2}. The force on the injection plunger (Eq. 1-18) is 56 in.^{2} x 2000 psi=113,200 lb. This is applied by the 2 1/2-in.-diameter (4.91 in.^{2}) screw upon the plastic material. The pressure applied on the material is 113,200/4.91 = 23,000 psi.

In a straight plunger machine, the injection pressure on the material is usually given based upon the same factors. This is erroneous as there is at least a 30% pressure loss on the material from the feed end to the nozzle. A minimum pressure on the material of 20,000 psi should be available on general purpose machines.

Screw Recovery Rate.   The major specification of the injection end is the screw recovery rate. This tells how many ounces per second are plasticized while the screw is running. The limitations of the manufacturer's specification and interesting data on the effect of screw speed and back pressure are presented in Ref. 66.

Through the efforts of the Injection Molding Professional Activity Group (PAG) of the Society of Plastic Engineers, the Society of the Plastic Industry adopted a standard, as of January 1, 1968 (67). In summary, a thermocouple, placed in the nozzle, is attached to a fast response temperature recorder. The temperature variations shall not exceed 10°F. The temperature of extrusion for polystyrene is 420°F, polyethylene 460°F, and nylon 540°F. For a given test, the shot size is 25, 50, 75, and 100% of rated injection capacity. The ejection rate for machines up to 30 oz is 1 1/2 to 2 oz/sec; for 30 to 150 oz, 3 to 4 oz/sec; and for machines over 150 oz 6 to 8 oz/sec. The screw running time is 50% of the total cycle.

When the temperature is stablized, ten successive shots are taken recording both their weight and the screw rotating time. The recovery rate is calculated by taking the average weight in ounces of the 10 shots and dividing it by the average screw running time in seconds. Before buying a machine one should ascertain the rating under this standard and make comparisons.

Table 1-3 gives typical injection end specifications for plunger, reciprocating screw, and screw plunger type machines.

Table 1-3   Injection end specifications of three different types of molding machines

Injection unit		#1-10 Ounce plunger type 275 ton toggle clamp	#2-28 Ounce 2½" reciprocating screw 375 ton toggle clamp	#3-60 Ounce 3½" screw-plunger type 425 ton hydraulic clamp
Material injected	oz/shot	10	28.5	75
	g/shot	284	809	2130
Material injected	in.^{3}/shot	–	52.6	134.0
	cc/shot	–	862.	2196.
Maximum inection rate	in.^{3}/sec	–	15	35
	cc/sec		246	574
Plunger displacement	in.^{3}/min	34	–	–
	cc/stroke	557	–	–
Injection ram forward	in./min	360	–	–
	mm/min	9144	–	–
Plunger diameter	in.	2 5/8	–	–
	mm	67	–	–
Screw diameter	in.	–	2 1/2	3 1/2
	mm	–	64	89
Screw L/D ratio		–	20:1	20:1
Screw drive			Hydraulic	Hydraulic
Screw speed	rpm	–	0–200	0–200
Injection stroke (maximum)	in.	11	8 1/2	–
	mm	279	216	–

Table 1-3 *(Continued)*

Injection unit		#1-10 Ounce plunger type 275 ton toggle clamp		#2.28 Ounce 2½" reciprocating screw 375 ton toggle clamp	#3-60 Ounce 3½" screw-plunger type 425 ton hydraulic clamp
Injection pressure on material	psi kg/cm²	20,000 1406		20,000 1406	20,000 1406
Recovery rate	oz/sec g/sec	– –		2.9 82.	3.7 105.
Plasticizing capacity (poly-styrene)	lb/hr kg/hr	120 54		– –	– –
Hopper capacity	lb kg	150	68	110 50	200 91
Heater load	kW	16		25	32

Other Injection End Auxiliaries

Below are some items found on many molding machines, some of which are essential and others helpful.

1. All molding machines should have a tachometer, so that the screw speed can be ascertained.

2. For those machines requiring it, a permanently mounted ladder at the hopper is useful. Plants using blended material can have a material blender which mixes in reinforcements with the plastic mounted directly over the hopper.

3. Some machines ride the injection carriage on chrome plated tie bars instead of on ways.

4. Some machines use rotary joints instead of hoses for connecting the movable carriage to the hydraulic oil supply.

5. The injection rate is a variable which should be controlled by a hydraulic flow control valve.

6. Two injection pressures, with at least 20,000 psi in the material available.

Flow (Intrusion) Molding

In flow molding, the material is extruded directly into the mold until it is filled. In practice, reciprocating machines are converted to flow molding machines by changing the hydraulic and electric circuits. The circuits are changes to allow the screw to come forward after extrusion and apply pressure to pack the material into the mold, as opposed to collecting the material in front of the screw and injecting an exact shot volume.

This technique is primarily used for molding heavy sections. Flow molding effectively extends the capacity of the machine.

Void elimination is the major problem encountered in conventional molding of heavy sections. The walls and the gate freeze too quickly to permit enough material to enter to compensate for shrinkage. In clear lenses, for example, the different densities cause distortion. Flow molding, combined with hot molds, keeps the plastic molten long enough to mold an homogeneous part. The author has molded a 30 1/2 oz large heavy section clear acrylic lens by flow molding using a 2 1/2-in. reciprocating screw on a machine whose normal capacity is 20 oz. The part required an exceptionally slow, controlled injection rate which could not have been done with conventional molding equipment.

Flow molding cannot be used effectively for thin walls. Because of the slow plasticizing rate of in-line screws, the material sets up too quickly to receive the benefit of the injection pressure at the end of the cycle.

Different Ways to Flow Mold.   There are three ways of flow molding. The first is to start with the screw in the full forward position. The screw starts to rotate after the mold is closed and fills the mold. It continues extruding, producing a small cushion. Pressure is then applied to the injection ram to apply the final injection pressure.

The second method starts with a full barrel of previously plasticized material. The screw acting as a ram comes forward slowly and when it reaches its full forward point it starts to turn until the mold is filled. Then a small cushion is maintained and the hydraulic pressure is applied to complete the cycle.

The third method is to start with a full barrel and extrude until the point is reached where when the screw comes forward a small cushion will be left.

The last two methods allow for a more rapid filling cycle. In all instances, unless molding heat sensitive materials, the nonreturn valve at the front of the screw is desirable. Flow molding can be used with all materials, including crosslinked polyethylene, and FEP fluorocarbons and polyurethanes.

Other Methods of Melting Polymers

Because of the large volume of plastic manufactured by extruding devices, there is a continuing search for new methods. The plunger machine is in a sense an extruder. It has a number of disadvantages, the most important of which are discontinuous output, long heating time, and air entrapment. Its advantages are high extrusion pressures which are independent of the temperature, and the low shear level. The disadvantages have been overcome by using two rams and a shuttle valve feeding into two separate reservoirs, each with their plunger (64). At the moment it is primarily a laboratory tool, but has potentials for injection molding.

When a viscose elastic material is introduced between a rotating and stationary disk, the material moves spirally and centripetally, heating up and plasticizing. It is extruded through an opening in the center of the stationary disk. This was first described in 1947 by Weissenberg. The advantages of this elastodynamic pump are its good heat transfer, degassing, mixing, nonpulsing flow, and short residence time. It is also a simple extruder to build. A Belgian machine using this principle appeared on the market in 1968. However, this machine has no commercial significance today.

A combination of the Weissenberg effect and the discovery of a lubrication phenomenon, in which an oil wedge capable of developing high pressures was developed between a journal and its bearing-resulted in a hydrodynamic screwless extruder (65, 65a, 65b). The advantages claimed for this extruder are its simplicity, good heat transfer, low residence time, low shear, minimum thermal degradation, uniform pressure, high pressure capability, unlimited size, and rapid adjustments for operating conditions. The use of rapid impact to melt plastic is described in Ref. 74, 74a.

CLAMPING MECHANISMS

The clamping mechanism is a major factor in the cost of a machine and clamping force is a major machine specification (68). The two most common methods of clamping are to use an hydraulically operated toggle system or a fully hydraulic clamp. A toggle is a mechanical device used to amplify force. Figure 1-24 shows a common double acting toggle. The mechanical advantage can be as high as 50:1. In a molding machine it may consist of two bars joined together, end to end, with a pivot. The other end of one bar is attached to a stationary platen and the other end of the second bar to the movable platen. When the mold is open, the toggle is in the shape of a V. When pressure is applied to the pivot the two bars come into a straight line. The force to straighten the toggle is applied by an hydraulic cylinder. In a fully hydraulic system the force is supplied by an hydraulic cylinder alone.

Clamp Force Rating

The clamping force is rated in tons. For a hydraulic mechanism force is related to pressure and area by the following equation:

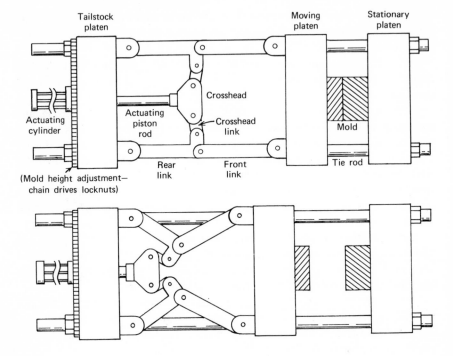

Figure 1-24   Double toggle clamping system.

$$F = PA \qquad \qquad (1\text{-}18)$$

where F = force (lb)

P = pressure (lb/in.^{2}) (psi)

A = area (in.^{2})

For example, a press with an hydraulic clamp has a 20-in.-diameter clamping cylinder. Assume the maximum working line pressure is 2,000 psi. The clamping force is

$$F = 2000 \; \pi(10)^2 = 628,000 \; \text{lb}$$
$$= 314 \; \text{tons}$$

This press would be called a 300-ton press. Machines available today range from a 5.5-ton clamp for a machine whose maximum shot is 10 g, to a 5500-ton unit (69) that can mold 54 lb of polystyrene per shot. Machines with 10,000-ton clamps are being built (70).

When the molten plastic is injected into the mold, the pressure generated upon the material by the injection ram is preset by the operator; 20,000 psi is a common maximum. This is opposed by the clamping force. If there were full pressure transmission through the material, a clamp force of over 10 tons/in.^{2} of projected area of molded surface would be needed to prevent the mold from opening. The projected area is the area of molded parts (including runners) parallel to the clamp-platen. Because of the hydraulic action of the machine, and the changes in viscosity and pressure transmission as the plastic cools in the mold, normal production can be conducted with 2 1/2 tons/in.^{2} of projected area. It is affected by the material, design of the part, speed of injection, and controls of the machine.

In both hydraulic and toggle systems, an hydraulic cylinder provides the force which causes the clamping action. This force is equally opposed by a stress in the tie rods. The amount of stretch is equal to the product of the load and the length of the rod, divided by the product of the cross sectional area of the rods, and the modulus of elasticity of steel (approximately 30×10^6 psi). A 3-in. tie rod on a 400-ton press will stretch approximately 1/16 in. Amplifying and measuring the deflection will give an accurate reading of the clamping force. This would only be necessary in a toggle machine. Clamping force in hydraulic machines can be read directly from the oil pressure gauge.

Toggle Clamps

In the mold open position of double acting toggle in Figure 1-24 the hydraulic actuating cylinder has retracted pulling the cross head close to the tail stock platen. This pulls the moving platen away from the stationary platen and opens the mold. It is difficult to stop the moving platen before the completion of the full stroke. Where this is important, nylon bumpers can be used as a mechanical stop. To close the mold the hydraulic cylinder is extended. The moving platen moves, starts slowly, reaches maximum speed at midstroke, and automatically decelerates as the crosshead extends and straightens out the links. A very small motion of the crosshead develops a large mechanical advantage causing the locking. (If the mechanical advantage were 50:1, a 300-ton press would only require a 6-ton hydraulic locking cylinder.) When the mold opens the full breaking force is developed as it utilizes the same leverage. This might be useful for molding.

There are a number of important advantages in a toggle system:

1. Because a much smaller hydraulic cylinder is required, there is significant economy in the size of the hydraulic equipment and, more important, in its operating cost. The savings are very significant over the life of the machine.

2. The second major advantage is the inherent speed of the design. Fully hydraulic clamps are available that move as fast as 2000 in./min. This is faster than most toggle machines. However, the cost of a fast moving hydraulic unit is considerably higher than the equivalent toggle system.

3. Another advantage of the toggle system is that it can be self-locking if it is designed to go "over center." Once the links have reached their extended position they will remain there until retracted. Most machines are live hydraulic or nonlocking. An hydraulic system requires maintenance of the line pressure at all times (7.1).

The toggle system used on molding machines has several disadvantages:

1. The toggle system doesn't provide for any simple positive indication of the clamping force. It is therefore more difficult to adjust and monitor. Therefore the possibility of overclamping and damaging molds is real.

2. The clamping force may not remain constant. As the temperature of the mold and tie bars change, their lengths change, thus affecting the clamping force. In a hydraulic system this is automatically compensated by the compressibility of the oil.

3. It is difficult to control the speeds and force of the toggle mechanism as well as stopping and starting at different points.

4. The toggle system requires much more maintenance than the hydraulic system and is comparatively more susceptible to wear.

Operating the Toggle Clamp.   To clamp properly the toggles must be fully extended. Therefore, the distance of the tail stock platen has to be changed to accommodate different mold heights. One way to do this is to have a chain which simultaneously moves the four locking nuts on the tail stock platen. A centralized die height adjustment with the addition of a moving back platen can also be used (Figure 1-6;6). The mold height adjustment screw (Figure 1-6;8) is moved by a crank (Figure 1-6;14), and the four tie bar nuts are not changed. The mechanism can be operated manually, electrically, or hydraulically.

Figure 1-6 shows the toggle system of Figure 1-24. The tail stock platen (Figure 1-6;2) is attached to the crosshead links (Figure 1-6;5) whose other ends are attached to the moving back platen (Figure 1-6;6). The hydraulic cylinder (Figure 1-6;1) has its piston (Figure 1-6;3) attached to the crosshead (Figure 1-6;4) which is supported on guide bars (Figure 1-6;13). As the cylinder extends the crosshead moves toward the injection end, forcing the toggles into a straight line and closing the mold. Some machines have four toggles, one on each corner of the platen.

The stroke of the moving platen is limited because the movement of the toggles is limited by the width or height of the platen. This can be overcome by attaching the hydraulic cylinder to the base with a pivot, the cylinder being parallel to the platens (Figure 1-25).

A further variation in the Monotoggle* (Figure 1-26) where the cylinder is attached to the toggles only. This permits longer strokes without loss of leverage.

*Copyrighted and patented (2969818); Improved Machinery, Inc.

The single toggle (Figure 1-25) may be activated mechanically instead of hydraulically. This is done by using a motor with a variable speed pulley attached to a fly wheel. On the fly wheel there is a spiroid pinion which is attached to a spiroid gear. The gear is connected to the toggle mechanism so that the rotation of the gear causes the toggle to extend or retract. The motion of the spiroid gear is controlled with a clutch and brake mechanism.

Hydraulic Clamps

Many of the smaller machines are available with either hydraulic or mechanical clamps. Most of the larger machines have hydraulic clamp systems. Some of the advantages of hydraulic clamping follow:

1. The clamping force is infinitely variable to its maximum.
2. The clamping force can be continually monitored by a pressure gauge on the hydraulic cylinder.
3. The clamping force can be changed at any time with hydraulic controls.

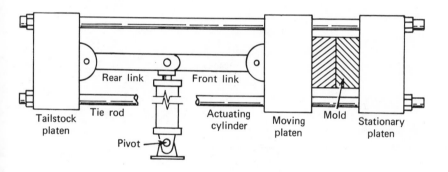

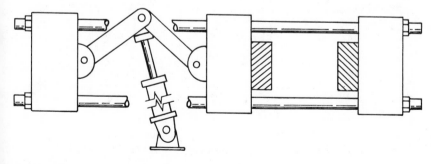

Figure 1-25 Single toggle clamping system.

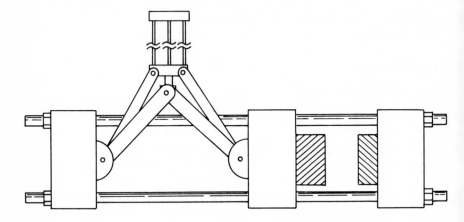

Monotoggle © system, clamp opened

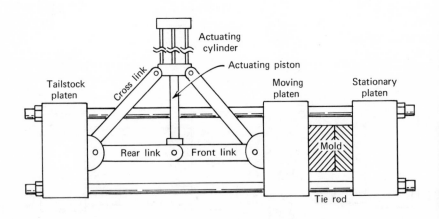

Figure 1-26   Monotoggle© system, clamp locked (Improved Machinery Inc.).

4. The speed of the ram in either direction can be infinitely varied below its maximum by hydraulic controls.

5. The stroke of the ram can be set by using a limit switch.

6. The machine stroke is not fixed and can be kept at a minimum.

7. The platen accelerates and decelerates smoothly.

8. The clamp can be stopped and started at will. This is useful in cam action and unscrewing molds.

9. Low pressure protection is easier to obtain hydraulically than with a toggle system.

10. The breakaway opening force and speed is adjustable.

11. Mold setup is quick. The stroke is adjusted by setting a limit switch and the pressure by the pressure control valve.

12. Since the moving parts of the hydraulic clamp float in oil, wear and maintenance are at a minimum. With proper care the only maintenance is the replacement of the chevron packing around the ram. By contrast toggles require forced lubrication, continual inspection and eventual replacement of the toggle pins and bushings.

Operating the Hydraulic Clamp.   The major disadvantages of an hydraulic system are the higher initial cost, the higher operating costs because of the larger pump and motor, and the increased possibility of oil leakage (72). Most clamping systems use 2000 psi oil supplied from the pump. There are some systems where pressure intensifiers are used for pressures up to 8000 psi, which are held by check valves. Other intensifier systems use their own small pump. They are fed through flanges into the clamping ram to eliminate piping and leaks. Figure 1-27 shows the clamp and piping of a standard molding machine which uses a manifold to minimize interconnected piping. All valves, piping, and motors are readily accessible for adjustment and maintenance.

Hydraulic versus Toggle.   Both toggle and hydraulic clamps work well. A comparison of their features, based on equivalent performance (clamp force and speed) is summarized below.

Hydraulic	Toggle
1. Much higher original cost	1. Lower original cost
2. Higher horsepower needed therefore more expensive to run	2. Lower horsepower needed, more economical to run
3. Nonpositive clamp	3. Positive clamp
4. Unlimited stroke potential	4. Limited stroke
5. Direct readout of clamp force	5. No direct readout
6. Easy adjustment of clamp force	6. More difficult
7. Easy mold setup	7. More involved mold setup

8. Varies stroke to mold height

9. Clamp speed easily controlled or stopped at any point

10. Low maintenance as parts are self lubricated

8. Constant mold stroke

9. Clamp speed more difficult to control and stop

10. Higher maintenance costs

Other Clamp Systems

Neither the hydraulic clamp nor the toggle system were originally designed for molding machines. In the larger sizes it became impossible to reconcile a toggle

Figure 1-27 Clamp-end piping. The simplified piping arrangement employs a manifold to minimize interconnected piping. All valves, piping and motors are readily accessible for adjustment and maintenance. Full-time filtration of the hydraulic oil is provided (Farrel Corp.).

mechanism with the need for high clamp force, long stroke, platen size, and floor space. The simple hydraulic system, with its large power always available (even though not needed until the final clamp), was too costly and cumbersome. A number of systems were developed which overcame these limitations in varying degrees.

1. One of the earliest solutions was the use of a small diameter cylinder to move the clamping ram rapidly until the mold faces touched. This uses a relatively small amount of oil and, with large pumps, will give rapid motion. The main locking force is established by a large diameter cylinder which only need move a short distance. If the small cylinder is bored inside the large cylinder it is called a jack ram (Figure 6-19). If oil had to be pumped behind the large ram, obviously, the benefit of the jack ram would be lost. Instead of pumping, oil is supplied by gravity from the tank above the clamp or by suction if the tank is below the ram. Another way to do it is to have a large hole in the clamping ram. As the jack ram pushes the clamp ram forward the oil flows from front to back through the hole. Near the very end of the stroke a rod inside the clamp housing seals the hole allowing full clamp pressure behind the piston (73).

2. As the clamp capacity became larger and the stroke requirement longer, hydraulic clamps of this size became very expensive and unwieldly. A system was designed for injection molding machines which is commonly called "lock and block" (Figure 1-28). A small high speed rapid traverse cylinder is used to move the movable platen. A spacer which may be a hollow tube or rod is attached to the movable platen. At the end of the stroke of the rapid traverse cylinder the locking mechanism, an hydraulically operated movable plate, is inserted between the spacer and a large diameter short stroke hydraulic cylinder. The large clamp cylinder moves forward approximately an inch to provide the locking force. This type of "lock and block" mechanism requires a mold height adjustment.

All systems of this design, while losing a very slight speed advantage because of the three motions, gain in economy because of smaller hydraulic cylinder sizes, lower power requirements, and no huge toggle links requirement.

One of the disadvantages of this system is its long nonadjustable stroke with a corresponding waste of floor space and long stressed tie bars. To overcome this clamping units have been designed where the moving tailstock platen is not firmly anchored to the tie bars at all times (Figure 1-29). The rapid traverse cylinder moves the moving tailstock platen and clamping plate until the mold closes. The half nuts are closed over the tie bars, anchoring the moving tailstock platen. The large diameter, limited stroke main clamp cylinder extends giving the clamping force. To open the mold, the clamp cylinder is vented, the half nuts are released, and the rapid traverse cylinder retracted. This has the advantage of adjustable stroke, minimum stress length of the tie bars, increased rigidity, and smaller floor space. Careful engineering and maintenance is required to lock the

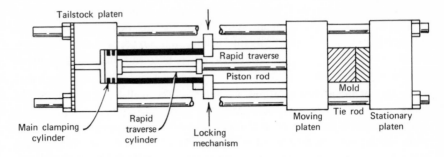

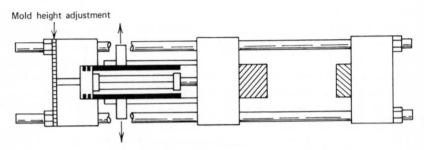

Figure 1-28 "Lock and block" hydraulic clamp.

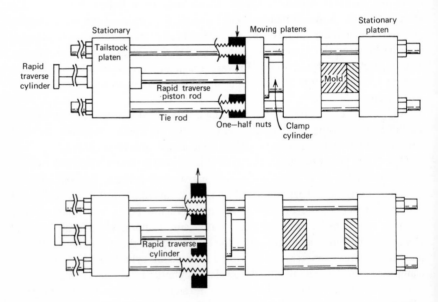

Figure 1-29 Long stroke, rapid acting clamping system using one-half nuts to anchor the moving platen.

half nuts consistently in place.

3. Another system moves the movable platen with a small long stroke cylinder. The tie bars are completely out of the movable platen and are inserted at the very end of the stroke. They are locked by half nuts, hydraulically operated. Hydraulic cylinders at the end of each tie bar are located between the platen and lock. When they are extended they stretch the tie bars for clamping action.

Small single toggles can be used for rapid traverse of the platen. The clamp pressure is supplied by a large diameter short stroke cylinder. Since the toggle is only acting as a spacer it does not require the massive proportions that would be needed to generate large blocking pressures.

4. A very ingenious method (Figure 1-30) has the large diameter small stroke clamping cylinder moving on ways. A small cylinder lifts this main clamp vertically over the movable platens. The movable platens are activated by small diameter long stroke hydraulic cylinders. This system permits long strokes in a minimum tie bar length and uses minimum floor space.

Vertical or Horizontal Clamp?

Users of large presses must consider whether to buy a vertical or horizontal

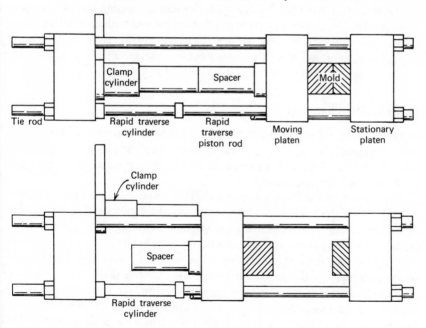

Figure 1-30 Long stroke, rapid acting clamping system with movable clamp cylinder.

press. One of the considerations is the space required. A 5000-ton horizontal machine occupies 720 ft^2. A 4000-ton vertical press requires 190 ft^2. Twelve feet are usually below floor level and 20 ft above it.

The size of the mold that can fit into a horizontal machine is limited by the spacing of the tie rods. There is a practical limit to this dimension. Vertical machines can be designed for much larger molds. Long strokes and large daylight are difficult to obtain in horizontal machines, but pose little problems for vertical ones. Parts can be too heavy to be removed by the operator. In horizontal machines the parts usually fail on a conveyor and are removed. (This procedure may sometimes damage the part). Vertical machines can also be unloaded automatically with such devices as suction cups, clamp assemblies, and air cushions to float the part out of the press. One of the disadvantages of the vertical machine is that the injection unit feeds from the side. If a part cannot be edge gated, a long hot runner system is used, a situation which can pose problems when using heat sensitive materials.

MACHINE SPECIFICATIONS

As noted before, a machine is classified by its clamping and injection ends. The clamping end is specified by its type (hydraulic, toggle) and its clamping tonnage; the injection end by its type (plunger, two-stage plunger, reciprocating screw, two-stage screw) and its shot capacity usually given in ounces of polystyrene per shot. If it is a screw machine, the specification includes the nominal diameter of the screw.

Clamp End Specifications

Tables 1-3 and 1-4 shows the specification for three machines, a plunger type with a toggle clamp, a reciprocating screw with a toggle clamp, and a screw-plunger machine with an hydraulic clamp. These specifications were taken from manufacturers' literature.

The Society of the Plastics Industry (SPI) has established patterns for tapped holes on the platens. They also have standardized the knock-out pattern so that molds are readily interchangeable from machine to machine and plant to plant. A discussion of the clamp end specifications may be helpful.

The *clamping force* (given in tons) has been discussed. The *clamp stroke* (given in inches), is the maximum distance the moving platen will move. It is desirable that this motion be fully adjustable to permit the minimum stroke required by the mold. On hydraulic equipment, the travel is stopped by a limit switch. Mechanical stops are helpful when mold stroke control is needed. The *maximum daylight* (inches) is the furtherest distance the platens can be separated from each other. The *minimum mold thickness* (inches) is the closest

Table 1-4 Clamp end and general specifications for three different types of molding machines

		#1-10-Oz plunger type 275-ton toggle clamp	#2-28-Oz 2½ In. reciprocating screw, 375-ton toggle clamp	#3-60-Oz 3½ In. screw-plunger-type, 425-ton hydraulic clamp
Clamping Force	Short tons	275	375	425
	Metric tons	250	340	387
Break-away	Short tons	40	45	55
force	Metric tons	36.4	41	50
Tie rod	in.	3 1/2	4 3/4	5
diameter	mm	89	121	127
maximum	in.	42	55	34
daylight	mm	1067	1397	864
Clamp stroke	in.	8–20	6–24	22
(min.-max.)	mm	203–508	152–609	560
Minimum mold	in.	6	8	12
thickness	mm	152	203	305
Maximum mold	in.	22	31	–
thickness	mm	560	787	–
Distance between	in.	16½ X 18	24 X 20	24 X 23
tie rods hori-	mm	419 X 457	610 X 508	610 X 584
zontal X ver-				
tical				
Platen size,	in.	27 X 27	36 X 32	39 X 36
height X	mm	686 X 686	914 X 813	991 X 914
width				
Ejector type		Mechanical	Hydraulic	Mechanical
Ejector stroke	in.	4	7	5
maximum	mm	102	178	127
Ejector force	Short tons	–	8	–
	Metric tons	–	7.3	–
Speeds				
Clamp ram	in./min	925	1380	600
forward	m/min	23.5	35.	15.2

Table 1-4 *(Continued)*

		#1-10-Oz plunger type 275-ton toggle clamp	#2-28-Oz 2½ In. reciprocating screw, 375-ton toggle clamp	#3-60-Oz 3½ In. screw-plunger-type, 425-ton hydraulic clamp
Clamp ram return	in./min	1185	1680	650
	m/min	30.	42.7	16.5
General				
Motor	hp	30	50	75
Hydraulic pump capacity at 2000 psi (141 kg/cm^2)	gal/min	44	90	130
	l/min	166	340	492
Oil reservoir	gal	155	170	250
	liters	585	640	945
Hydraulic pressure	psi	2,000	2,000	2000 injection 2800 clamp
	kg/cm^2	141	141	141 injection 197 clamp
Machine dimensions $(L \times W \times H)$	ft	19 X 6 X 7	21 X 6 X 9	17 X 5½ X 10½
	cm	579 X 183 X 213	640 X 183 X 274	518 X 168 X 320
Weight	Short tons	8.5	15	18
	Metric tons	7.7	13.7	16.4

distance that the two platens can come to each other while still maintaining clamp pressure. This can be further reduced by using a spacer or "bolster" plates. The minimum mold thickness is the difference between the maximum stroke and daylight. These are important specifications in that they tell how deep a piece may be molded and whether a mold of a given depth will fit in a machine.

The *clearance* between the tie rods (inches) is the determining factor whether a mold of a given length or width will fit into the press. For example, a press has 20 in. clearance vertically and 18 in. horizontally. Therefore, a mold less than 20 in. wide, but over 20 in. long will fit vertically. A mold less than 18 in. high but over 18 in. long will fit horizontally. The length and width dimensions of a mold are often determined by the side which is parallel to the knock-out plate.

The *clamp speed* (inches per minute) is an important specification. Slow clamp speed as well as excessive clamp motion means wasted productive time. Losing 0.5 sec/shot on a machine producing 120 shots/hr will waste 110 productive hours per year.

The *knock-out stroke* (inches) determines the maximum knock-out movement available. When molding deep draw pieces or using the knock-out system for camming, long knock-out stroke is desirable.

Controls, hydraulics, knock-out systems, and safety, are all necessary components of any injection molding machine installation. A brief description of what is required is given below.

Controls

The electrical controls can be mounted in a separate enclosure (Figure 1-1) or on the machine (Figure 1-6). The latter are less expensive to install in a molding plant, take up less floor space, and are easier to move in the event the plant is relocated. Separate panels remove the components from machine shock and vibration, are roomier permitting easier maintenance and modification, and can be located for maximum visibility to operating personnel. The area housing the controls should be well lighted.

There are a number of conveniences found in some molding machines. The heating and control circuits should be separate. This permits the molding temperature to be reached at approximately the same time a repair is finished. Symbols as well as labels should be put on the controls as some operators have difficulty reading. There should be sequential signal lights for electrical and hydraulic circuits for rapid location of malfunctions. Plug in components (relays, timers, and limit switches) are available and desirable.

There should also be provision for extra instrumentation including heat control units, timers, and relays. A nonresettable cycle counter can be attached to the automatic ram forward solenoid control. A volt meter with long leads can be permanently mounted in the control cabinet.

The manual control panel (Figure 1-6;20) is mounted on the machine for the operator's convenience. It should be located so that he cannot possibly get hurt when purging material. It should also be designed so that the controls cannot be accidently moved. Hydraulic gauges and controls (Figure 1-6;28; 29; 36) are mounted for convenience in piping. It is very desirable to have multiple hydraulic plug-in testing points to facilitate hydraulic maintenance. Additional core pulling circuits are also desirable.

Hydraulics

Location of the hydraulic components and electric motors depend on the individual machine. Most oil reservoirs (Figure 1-6;16) are located in the base of the machine. Fully hydraulic large capacity clamps may have them above the clamping end. This helps the oil flow in by gravity. There should be an oil filtering system for continually filtering the oil. A magnet in the reservoir will collect the iron resulting from the normal wear and tear of the components. An oil temperature gauge should be installed on the machine preferably with a warning light if the oil goes above the predetermined temperature. The machine should be designed for easy maintenance, with as many components as possible readily accessible. Unfortunately this is not aways the case.

Cooling water is needed because the hydraulic system generates heat. For example, a 16-oz capacity press running approximately 100 shots/hr required the removal of 25,000 Btu/hr through a heat exchanger (cooler). This is in addition to a substantial amount of radiation loss from the machine body. Provision for bringing water to and from the machine is therefore required. Water for cooling the hydraulic oil is usually supplied by a water tower or a well. Automatic throttling valves will control the amount of water going through the machine heat exchanger. (Provision is also required for mold cooling water. This can be tower water or more preferably, a mechanically refrigerated coolant). A fully hydraulic clamp usually requires lubrication only on the tie rods and the ways. Machines with toggle clamps require good lubrication. The trend is to automatic lubrication (Figure 1-6;12).

Knock-out Systems

The knock-out (K.O.) system is needed to eject the part from the mold. A separate knock-out plate in the mold changes its location relative to the rest of the mold. Attached to this plate are knock-out (ejector) pins or other devices which push against the molded parts as the mold completes its opening. This ejects the parts from the mold. On most machines there is an adjustable stationary ejector plate (Figure 1-6;7) to which are attached the ejector bars (Figure 1-6;17) which go through the movable platen. As the mold returns, the ejector bars go through holes in the back plate of the mold, contact, and stop the knock-out plate. The rest of the mold continues moving a predetermined distance and stops.

There are pins attached to the knock-out plate which project to the mold parting surface. They are pushed by the injection side of the mold as the mold closes. This forces the knock-out plate back to its normal position. Such pins are called push back or return pins. A much superior but more costly way to activate the knock-out system is by the use of hydraulic cylinders (Figure 1-6;18). This is an advantage since the knock-out system can be advanced or retracted at any time with any speed or force. Hydraulic cylinders are particularly useful in cam action molds and insert molding.

Safety

The unprotected molding machine can be dangerous. People have lost limbs, fingers, and lives in the moving parts of the machine. A safety gate in front of and in the rear of the mold area is a necessary part of a safety system. The gate should be large enough and high enough so that it is impossible for anyone to get any part of his body in the platen area when the platen is closing. The safety gate should be permanently mounted on slides with enough clearance for hydraulic cylinders or other mold attachments. A mechanical safety operated by the gate drops a bar of steel between the platens so that the machine cannot close with the gate open even if there is an hydraulic or electric failure.

Guards should also be placed around the machine to prevent anyone from putting any part of his body in the toggle area or other moving parts of the machine. The gate should also be electrically interlocked. A machine must have electrical and hydraulic safety systems too.

The open feed section of the screw can also amputate fingers. A safety screen is difficult to design for this section. Hoppers that slide away from above the machine and which have material emptying chutes tend to minimize this danger and at the same time make the cleaning of the hopper easier. All the safety switches should have fail safe circuits and have provisions for testing.

In addition observe the following safety procedures:

1. At no time should any repairs between the platens be made with the motor running.

2. The heating cylinder should have a cover to prevent direct contact with the heating bands.

3. The purging controls should be located so that the operator is protected from spattering by overheated material. Clear plastic purging guides can also be used here.

4. Electrical locks should be provided so that no one can start the machine when mold or machine repairs are being made.

Occupational Safety and Health Act of 1970 (OSHA)

OSHA, which became effective on April 27, 1971, has the declared Congressional purpose to "assure so far as possible every working man and woman in the Nation safe and healthful working conditions and to preserve our human resources."

Each *employer* under the act has the general duty to furnish each of his employees employment and places of employment, free from recognized hazards causing, or likely to cause, death or serious physical harm; and the employer has the specific duty of complying with safety and health standards, promulgated under this act.

Each *employee* has the duty to comply with these safety and health standards, and all rules, regulations, and orders issued pursuant to the act which are applicable to his own actions and conduct.

The standards can be roughly divided into two parts. One deals with standards applicable to almost all manufacturing operations. Examples would be maintaining aisles and passageways in good repair and clear of obstructions, and maintaining recognized fire prevention standards, such as fire extinguishers and keeping all material at least 36-in. below sprinklers. The second type would pertain to the particular industry.

Any employee (or representative, such as a union) who believe that a violation of a job safety or health standard exists which threaten physical harm, or that imminent danger exists, may request an inspection by sending a signed written notice to the Department of Labor.

Complete records of all injuries must be kept in a specific place and notice of this law must be posted throughout the plant.

The law puts the obligation for a safe plant directly upon the employer. Inspections, which are unannounced, see that the law is complied with; they do not tell the employer what to do. Therefore, penalties and fines are issued at the first inspection. This is contrary to most types of inspection, where a time is given to correct faults without any penalty.

Penalties of up to $10,000 may be assessed for each violation. A willful violation by an employer which results in the death of any employee is punishable by a fine of up to $10,000 or imprisonment of up to 6 months. A second conviction doubles these criminal penalties.

The plastics industry has one of the highest injury rates in industry and has been selected by OSHA as one of their early priorities. One would be well advised to become familiar with the requirements of this act. Providing a safe place has always been morally imperative. It now also becomes financially desirable. (76, 77, 78)

SPECIAL TYPES OF MOLDING MACHINES

Specific needs of plastics processors have led to the development of special types of injection machines modifications, four of which are described here.

1. Off-center molding modifications. Many times it is necessary to gate a molded part off center. Normally this would require an off center mold with the sprue bushing placed directly in the center of the machine. Another way of overcoming this off-center problem is to build either a three plate, hot runner, or insulated runner mold. The mold is placed centrally in the machine and the runner system takes care of the gate location. This method overcomes the objection of the first method where the off-center mold puts unequal stress on the tie bars. A molding machine has been developed which has the injection carriage move laterally on the bed as well as axially. The platen is correspondingly cored so that the injection cylinder can be brought to meet an off-center sprue.

2. Insert molding modifications. The need for insert molding, particularly molding vinyl electrical cord sets, led to the development of a shuttle table machine. This is a vertical machine with one core plate attached to the movable platen. While one set of cavities (A) is molding, the operator removes the previously molded parts and inserts new inserts into the other identical cavity section (B). When the cycle is done A slides out and B slides under the force and the cycle is repeated.

3. Two color molding modifications. The need for two color molding, such as the numerals in typewriter keys, led to the development of machines with more than one injection cylinder. Many ingenious mold designs are used to transfer the very first molding and use it as a matrix for the second molding.

4. Rotary and multistation machines. A rotary type machine with a turret system was developed for rapid production of dissimilar parts with relatively low mold costs. As the turret moves from station to station, the material is injected, cured, and ejected. While one cavity is being molded the other parts are cooling. The individual mold stations can be equipped with different cavities of different weights, thicknesses, and dimensions. This sytem is easier to control than a conventional multicavity mold. There is also more time available for unscrewing threaded parts. In the event a cavity is damaged, the machine can skip that part. Turret machines are considerably more complicated mechanically than the standard machines (75).

REFERENCES

1. "Machinery in Place," L. I. Naturman, *PT,* February 1970, p. 37.

2. "Predicting Mold Flow by Electronic Computer," H. L. Toor, R. L. Ballman, and L. Cooper, *MP,* December 1960, p. 117.

3. "Predrying of Molding Powders," F. I. McCosh, *MP,* August 1957, p. 120.

3a. "You Can Vent Injection Molding Machines," I. Ronzoni, A. Casale, and G. DeMarosi, *SPE-J,* October 1971, p. 74.

3b. "Improvement of Screw Plasticizing in Injection Molding." H. Nakagawa, Japan Plastics, October 1971, p. 20.

4. "The Injection Machine Heating Cylinder, C. E. Beyer, R. B. Dahl, and R. B. McKee, *MP,* "Temperature and Pressure Measurements," April 1955, p. 127; "Temperature Variations, May 1955, p. 110; "Effect of Design Factors," June 1955, p. 127.

5. "Induction Heating for Injection Molding Machines," *BP,* August 1960, p. 380.

6. "Copper Clad Cylinders for Induction Heating," K. J. Cleereman, and D. S. Chisholm, *MP,* June 1962, p. 131.

7. "Rotating Spreader," N. Keiser, L. H. Cirker, and P. D. Kohl, *SPE-J*, April 1961, p. 340.

8. "Melt Extractor," A. Spaak, *MP*, October 1960, p. 107.

8a. "A New Approach to the Heating Cylinder Problem," A. R. Morse, *Plastics Industry*, February 1959, p. 16.

9. "Preplasticizing in Injection Moulding: Part 1," E. G. Fisher, and W. A. Maslen, *BP*, September 1959, p. 417; Part 2, R. Wood, October 1959, p. 468; Part 3, E. G. Fisher, and W. A. Maslen, November 1959, p. 516.

10. "Converting to Screw Plastication," C. J. Waechter, and L. J. Kovach, *MP*, March 1963, p. 125.

11. "Mechanical Aspects of Extruder Cylinders," L. J. Mattek, *SPE-J*, January 1965, p. 80.

12. "Hard Surfacing of Extruder Screws," L. I. Naturman, *PT*, September 1963, p. 945.

13. "Purging Injection Cylinders," J. H. DuBois, *Plastics World*, August 1964, p. 26.

14. "How Glass-fiber Fillers Affect Injection Machines," B. A. Olmsted, *SPE-J*, February 1970, p. 42.

14a. "How to Avoid Damage to Injection Barrels and Screws When Molding RP," W. B. Evans, *MP*, March 1972, p. 68.

14b. "Injection Machine Screw and Barrel Wear," R. A. Butler, *B. P.*, June 1970, p. 139.

15. "Selecting the Right Injection Molding Machine Nozzle," A. R. Morse, *PDP;* Part 1, June 1968, p. 23; Part 2, July 1968, p. 23.

16. "Effects of Radial Screw Clearance on Extruder Performance," R. A. Barr, and C. I. Chung, *SPE-J*, June 1966, p. 71.

17. "How Friction Affects Extruder Operation," R. B. Gregory, *SPE-J*, October 1969, p. 55.

18. "Feeding the Single Screw Extruder," R. L. Miller, *SPE-J*, November 1964, p. 1183.

19. "Effect of Feed Condition on Screw Conveying and Plastifying," R. M. Griffith, *SPE-J*, November 1967, p. 65.

20. "How Particle Size Influences Molding and Extrusion," L. A. Landers, and R. A. Tiley, *MP*, October 1963, p. 213.

21. "Plastifying Extrusion," L. F. Street, *IPE*, July 1961, p. 289.

21a. "Does the Internally Heated Screw Help?" F. Cumings, *MP*, June 1972, p. 78.

22. "Evaluation of Screws for Nylon," R. M. Bonner, *SPE-J*, October 1963, p. 1069.

23. "An Improved Mixing-Screw Design," B. H. Maddock, *SPE-J*, July 1967, p. 23.

24. "Calculating Extruder Performance," E. C. Bernhardt, *MP*, February 1955, p. 125.

25. "Metering Screw Performance with Temperature Gradients," *SPE-J;* Part 1, D. I. Marshall, I. Klein, and R. H. Uhl, October 1965, p. 1192; Part 2, I. Klein, D. I. Marshall, and C. A. Friehe, November 1965, p. 1299; Part 3, I. Klein and D. I. Marshall, December 1965, p. 1376.

26. "Metering Screws in a Screw Injection Molding Machine," P. N. Richardson, and J. C. Houston, *SPE-J*, January 1965, p. 44.

27. "Flow Patterns in a Single Screw Extruder," *SPE-J*, September 1960, p. 1015; W. D. Mohr, P. H. Squires, and F. C. Starr. (This is a condensation of "Flow Patterns in a Non-Newtonian Fluid in a Single-Screw Extruder," W. D. Mohr, J. B. Clapp, and F. C. Starr, *PES*, July 1961, p. 113.

28. "How to Get Best Performance in Screw Injection Molding," R. B. Staub, *SPE-J*, November 1963, p. 1182.

29. "Processing of Thermoplastic Materials," E. C. Bernhardt, (Ed.), Reinhold Publishing Co., New York, Chap. 4, "Extrusion," 1959.

30. "Fundamentals of Plasticating Extrusion," Z. Tadmor, D. I. Marshall, and I. Klein, *PES*, July 1966, pp. 185-212.

31. "Melting in Plasticating Extruders, Theory and Experiment," Z. Tadmor, I. J. Duvdevani, and I. Klein, *PES*, July 1967, pp. 198-217.

32. "The Effect of Design and Operating Conditions on Melting in Plasticating Extruders," Z. Tadmor and I. Klein, *PES*, January 1969, pp. 1-10.

33. "Simulation of the Plasticating Screw Extrusion Process with a Computer Programmed Theoretical Model," I. Klein and Z. Tadmor, *PES*, January 1969, pp. 11-21.

33a. "Pressure Profiles in Plasticating Extruders." R.C. Donovan, PES, November 1971, p. 484.

34. "Computer Design of Plasticating Extruder Screws," I. Klein and Z. Tadmor, *MP*, September 1969, p. 166.

34a. "Extruder Design by Computer Printout," B. H. Maddock and D. J. Smith, *SPE-J*, May 1972, p. 12.

35. "A Visual Analysis of Flow and Mixing in Extruder Screws," B. H. Maddock, *Technical Papers SPE-ANTEC V*, 1959, 45-1.

36. "Effects of Recent Fundamental Investigations on Extruder Design," G. P. M. Schenkel, *IPE;* Part 1, July 1961, p. 315; Part 2, August 1961, p. 364.

37. "New Ideas About Solids Conveying in Screw Extruders," C. I. Chung, *SPE-J*, May 1970, p. 33.

37a. "A New Theory of Solids Conveying in Single-Screw Extruders," W. Tedder, *SPE-J*, October 1971, p. 68.

37b. "Solids Conveying in Screw Extruders,"   E. Broyer, and Z. Tadmor, PES, January 1972, p. 12.

38. *Computer Programs for Plastics Engineers,* I. Klein and D. H. Marshall, Reinhold Publishing Co., New York, 1968.

39. *Principles of Plasticating Extrusion,* Z. Tadmor and I. Klein, Reinhold Publishing Co., New York, 1970.

40. "Extrusion Screw Design for Polysulfone," A. M. Fazzari, B. H. Maddock, and R. B. Staub, *SPE-J*, January 1967, p. 31.

41. "The Performance of Single Screw Extruders in the Processing of Polyolefines and PVC," O. Schiedrum, O. Domininghaus; Plastics; Part 1, February 1961, p. 83, Part 2,, March 1961, p. 81.

42. "Extruder Heating & Cooling Systems," P. H. Squires, *SPE-J*, August 1962, p. 863.

43. "Screw Extrusion of Thermoplastics," B. S. Glyde and W. A. Holmes-Walker, *IPE;* Part 1, August 1962, p. 338; Part 2, September 1962, p. 396.

44. "Single-Screw Extruder Design and Operation," W. F. O. Pollett, *IPE;* Part 1, May 1964, p. 142; Part 2, June 1964, p. 187.

45. "Screw Injection Machine Design," M. Jury, *MP*, April 1965, p. 119.

46. "A New Theory for Single-Screw Extrusion," C. I. Chung, *MP*, Part 1, September 1968, p. 178; Part 2, December 1968, p. 110.

47.   "A Modified Melting Model for Plastifying Extruders," D. L. Hinrichs and L. U. Lilleleht, *PES,* September 1970, p. 268.

47a. "A Theoretical Melting Model for Plasticating Extruders," R.C. Donovan, PES, May 1971, p. 247.

47b. "An Experimental Study of Plasticating in a Reciprocating Screw Injection Molding Machine," R.C. Donovan, D.E. Thomas and L.D. Leversen, PES, September 1971, p. 353.

47c. "A Theoretical Melting Model for a Reciprocating Screw Injection Molding Machine," R.C. Donovan, PES, September 1971, p. 361.

48.   "Mixing and Residence Time Distribution in Melt Extruders," G. Pinto and Z. Tadmor, *PES,* September 1970, p. 279.

49.   "How Plastics Melt in an Extruder," R. B. Gregory, *SPE-J,* June 1971, p. 49.

50.   "Single and Twin Screw Extruders – A Technical Comparison," F. Martelli, *SPE-J,* June 1967, p. 53.

51.   "Twin-Screw Extruders," J. J. Prause, *PT,* November 1967, p. 41; February 1968, p. 29; March 1968, p. 52.

51a. "Twin Screw Extruders – A Separate Breed," F. Martelli, *SPE-J,* January 1971, p. 25.

52.   "Pressure Measurements in Twin Screw Extruders," H. Marhenkel, *Plastics,* August 1965, p. 57.

53.   "Reciprocating Screw-Tip Shutoffs," A. R. Morse, *PT;* Part 1, July 1967, p. 46; Part 2, August 1967, p. 46.

54.   "Non-return Valves in Screw Injection Moulding Machines," E. Bauer, *BP,* February 1968, p. 83.

55.   "Reciprocating Screw Tip Designs," A. R. Morse, *PDP,* February 1969, p. 18.

56.   "Electric Screw Drives for Injection Machines," R. M. Norman, *PT,* April 1966, p. 31.

57.   "Fluid Screw Drives for Injection Molding Machines," R. J. Lindsey, *PT,* April 1966, p. 33.

58.   "Hydraulic Injection Screw Drives," J. Newlove, *MP,* May 1966, p. 237.

59.   *Modern Plastics Encyclopedia,* October 1969, p. 1026.

60.   "Graphic Comparison of Screw and Plunger Machine Performance," W. G. Kriner, *MP,* May 1962, p. 121.

61.   "Ram vs. Screw Injection," C. L. Weir and P. T. Zimmerman, *MP;* Part I, November 1962, p. 122; Part II, December 1962, p. 125.

62.   "Single Screw Injection Molding," H. Frimberger and J. G. Fuller, *PT,* May 1961, p. 53.

63.   "Ram vs. Plunger Screw in Injection Molding," L. W. Meyer and J. W. Mighton, *PT,* July 1962, p. 39.

64.   "Continuous Flow Ram Type Extruder," R. F. Westover, *MP,* March 1963, p. 130.

65.   "A Hydrodynamic Screwless Extruder," R. F. Westover, *SPE-J,* December 1962, p. 1473.

65a. "Scaling Up the Elastic Melt Extruder," B. Maxwell, *SPE-J,* June 1970, p. 48.

65b. "Miniature Injection Molder Minimizes Residence Time," B. Maxwell, SPE-J, February 1972, p. 24.

66.   "How to Evaluate Screw Injection Machine Plasticating Performance," V. R. Grundmann, *MP,* November 1966, p. 117.

67. "New Standards for Plasticating Performance of Screw Injection Machines," *MP*, March 1968, p. 107.

68. "Hydraulic & Mechanical Presses," C. J. Olowin, *SPE-J,* February 1970, p. 21.

69. "5500 Ton Molding Machine," J. Hauk, *MP,* September 1969, p. 108.

70. "Giant Machines for the Big Jobs," J. E. Hauk, *MP,* January 1969, p. 101.

71. "Why Choose a Toggle Clamp," T. Debreceni, *PT,* January 1968, p. 39.

72. "Why Choose a Hydraulic Clamp," T. Erwin, *PT,* January 1968, p. 35.

73. Patent 3,416,415 (Sohmers).

74. "Injection Machine of Tomorrow," MP, July 1970, P. 71.

74a. "Heat Sensitive Materials? Turn on the Pressure," G. Menges, W. Dalhoff, and P. Mohren, SPE-J, April 1972, p. 72.

75. "Multi-Station Injection Machines With Single or Dual Injection Units," R. H. Schlueter, *SPE-J,* April 1970, p. 95.

76. "Making Machinery Safer," *MP,* July 1971, p. 39.

77. "Controlling Noise in the Plant," L. H. Bell and A. J. Raymond, *MP,* January 1972, p. 84.

78. "Solving the Noise Problem," L.H. Bell, and A.J. Raymond, SPE-J, January 1972, p. 16.

CHAPTER 2

Molds

The injection mold is the mechanism into which the hot plasticized material is injected and maintained under pressure while it cools into a commercially acceptable shape. When the plastic material has sufficiently solidified, the machine opens, separating the two halves of the mold. The plastic pieces are ejected (Figure 2-1). The quality of the part and its cost of manufacture are strongly influenced by mold design, construction, and excellence of workmanship.

As machine capacity increases, the molds become larger and more expensive. It is not uncommon, for example, for a mold to produce a garbage can and cover that costs over $25,000.00. Molds for 12-oz machines usually range from $3000 to $9000 (1). This is only a small part of the investment. The original idea, market testing, samples, prototypes, advertising, selling, and commitment for the initial order are many times the cost of the tool.

The two most critical steps in the production of a plastic part are the piece part design and the mold design. A failure of the first naturally results in an unacceptable part. Mold design failure does not necessarily result in piece part failure, although well it may. It does cause low productivity, high mold maintenance, and the probability of reduced part quality. An improperly functioning mold will take excessive supervisory time in the plant and tool maintenance department. This can cause neglect of other operations, compounding the normal problems of running a molding plant.

It is for this reason that the molder, the moldmaker, and on occasion, the purchaser, should give maximum attention to the design. Unfortunately this is not always done.

The last opportunity to change the part easily is when the mold is designed. In many instances the molder or purchaser will consult the moldmaker prior to ordering the mold. If not, suggestions by the moldmaker for changing the part to simplify the mold are to be given full consideration. The moldmakers are a versatile, ingenious, knowledgeable, and highly experienced group and have contributed much to the success of the industry. Moldmaking establishments

Figure 2-1   Operator removing molded part from machine (Crucible Steel Co.).

vary from small shops with the owner and two or more employees, to plants with several hundred employees and a full range of production equipment. There is no correlation between the size of the tool shop and the quality of its molds.

Limits of Responsibility

Let us consider the relationship between the molder, moldmaker, and user. The limits of responsibility and allocation of costs should be determined initially. The following are the current trade practices which have worked quite successfully. If the end user is the molder, he will be fully aware of his costs and the relationship with the particular moldmaker. If the moldmaker has a machine available for testing, he will submit molded parts for approval. The molder will then be reasonably assured of a mold that he can put into his machine and start producing immediately without any significant additional costs. This will be particularly true if he is able to watch the final molding test. If the molder tests the mold himself, he must include the costs of material, machine time, changes that might be made in the molder's machine shop, and the cost of transporting the mold back and forth to the mold maker. These costs can be substantial. The molder will evaluate them based on his estimate of the mold maker's ability, his experience with him, and the complexity of the mold.

If there is a third party, the end user, he can order the mold in two ways. He can order the mold through the molder, who will assume full responsibility for delivering a workable mold and the acceptable molded part. In this instance the molder will include the costs of testing, either in the mold price that he quotes to the end user or in his piece part price.

If the end user buys the mold directly from the mold maker, he assumes the responsibiuty of delivering a working mold in the condition required to produce the part at the quoted price. He should assume full responsibility for the testing and other costs needed to make the mold operate.

Mold Design Procedure

The first step in designing a mold is to have an accurate fully dimensioned drawing that notes tapers and where they start, tolerances, shrinkage specifications, surface finish specifications, material in which it is to be molded, part identification, and any other pertinent information. It is desirable to have a model of the part to be molded. If the drawing or the part is complicated, this is a necessity. Mold design is significantly more reliable, and molding problems are more readily anticipated. Most models need not be made exactly to the sizes on the drawing. They should always be full scale and may be made of plastic, metal, plaster, clay, wood, cardboard, paper, or any other material.

There should be representatives of four plant functions during the mold designing stage — engineering, molding, mold maintenance, and quality control. While they need not meet together all the time, they should approve the final mold design. If any one of them feels there might be problems, these problems should be noted with the suggested remedies. Obviously the mold maker must agree to the design. At this time the metal for the cavity and the core will be selected. The cavity and core are those parts of the mold which, when held together by the closing of the mold, will provide the air space or cavity into which the molten plastic is injected. The following will now be decided: the number of cavities (2); the parting line of the piece (where the faces of the cavities and cores touch); the type and location of the gates (the entry point of the hot plastic into the cavity); the runner system (which brings the hot plastic from the plasticizing chamber to the gate); the method of ejection (which removes the molded plastic parts from the mold); the location of the ejecting devices; the location and size of the temperature control channels; and the type and location of the venting system (which removes the air that is displaced by the incoming plastic).

Any additional information required by the molder or moldmaker is discussed and drawings of the mold are submitted. These must be reviewed very carefully. A list of 73 items to take into account is found on p. 169 which should be checked against each mold drawing.

Studying a mold design and drawing is a time consuming and often ignored task. Actually, few activities are more productive for the molder and moldmaker. It is much easier to change a drawing than a completed mold. A mold takes several months to complete so there is great urgency to produce molded parts immediately after the mold is delivered. Because of this, and the difficulty, cost, and time to make substantial mold changes, the minimum amount of work is done "just to get the mold running." Too often the mold remains in this condition as a monument to poor or careless mold design.

One cannot overemphasize the importance of getting a mold to function correctly. If this is not done, the foreman will be occupied with the problem for an excessive amount of time, preventing him from caring for the other machines; unacceptable parts will be passed by the operator and burden quality control and a poorer part will be produced. There is no more expensive error than bad tooling.

The state of the art has not advanced to the point where there is a complete guarantee that all parts can be molded, molded in the material selected, or molded at a cost consonant with the quoted price. The time to minimize these possibilities is when the mold is designed (3, 4).

In-House Moldmaking

Injection molders may have their own tool making facilities. This gives them control of the design, quality, and delivery of the tools. It permits them to expedite really urgent projects.

There are a number of disadvantages to in-house moldmaking. An outside source may be selected which has particular equipment or experience best suited for that particular job. He is less susceptible to pressures to put aside someone else's mold and concentrate on that of a good customer. He has a smaller capital investment and does not have the problem of managing and staffing a mold shop. Many molders compromise by having a small shop for some of their requirements and contracting out for the balance.

THE MOLD BASE

The steel parts which contain the cavities and cores are called a mold base, mold frame, mold set, die base, die set, or shoe. Figure 2-2 shows an exploded view of the parts of a standard mold base which are described below.

Seating Ring and Sprue Bushing

The locating or seating ring centers the sprue bushing on the stationary or injection platen of the machine directly in line with the nozzle of the injection cylinder. The sprue has an opening in the center of a concave spherical surface, whose counterpart is an equivalent convex surface on the nozzle of the injection

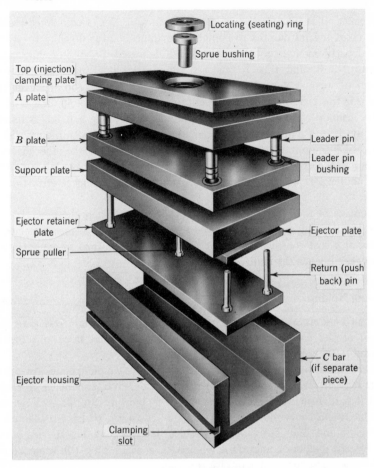

Figure 2-2   Exploded view of standard mold base (D-M-E Corporation).

heating cylinder. The opening of the sprue bushing must always be larger than the opening in the nozzle, so when the plastic hardens it will not form an obstruction larger than the sprue opening and cause the sprue to stick. The sprue has a generous taper to facilitate removal of the plastic. The standard radii for sprue bushings and nozzles are either 1/2 or 3/4 in.

Clamping Plates and Leader Pins

The injection (top) clamping or another backup plate supports the cavity or A plate. Usually a pocket is cut through the A plate to hold the cavity. The cavities are attached to the backup or clamping plate or preferably in milled pockets in the A plate. The A plate has four leader pins which are guided into four corresponding bushings in the B plate. The B plate has the cores attached to it or in pockets cut through it. Sometimes the location of the pins and bushings are reversed. The leader pins and bushings align the A and B plates and hence the cavities and cores. One pin is offset so the mold cannot be put together in the wrong position. The leader pins should be long enough so that if part of the mold slips during molding the leader pins will hit the A or B plates first, rather than a cavity or a core. When the leader pins or bushings wear, they can cause the A and B plates to shift several thousands of an inch with corresponding misalignment of the cavities and cores. If a container with an 0.018 wall was being molded and the leader pins and bushings wore and misaligned the cores and cavities by 0.004, then one wall of the container would be 0.014 and the other 0.022. This would cause severe filling, and ejection problems, and provide insufficient strength for the intended use. Leader pin wear problems are compounded by wear on the tie bars and bushings of the molding machine.

Wedge Locking Device

In critical applications a wedge locking device on the cavities and cores or the A and B plates is required. This will eliminate the effect of leader pin and bushing wear during molding. It will not cure ejection problems or the scratching of the part when the mold opens. Replacement of the pins or bushings will be required. If the machine was fully clamped at all times the angle of the wedge lock would not be important. Even with machines of adequate clamping capacity there will be times when the machine will open and the core and cavity separate during molding. Assuming a 0.006-in. opening, a $20°$ taper would permit the core to shift 0.002, a $45°$ 0.006, and a $60°$ taper 0.010. Other reasons for core shift are discussed later.

A stripper ring should not be used to align the cavity and core. The compression of the stripper ring when used for alignment may permanently deform it and will lead to excessive wear and possible galling on the core.

Back-Up Plates

The back-up or supporting plate under the B plate serves the same function as the one behind the A plate.

The B support plate rests on the ejector housing. Some molds have the ejector housing consisting of a back plate and two C bar spacers. Now these parts are fabricated as one unit. Holes are drilled in the back of the ejector housing to permit the ejector bars of the machine to contact the ejector plate of the mold before the machine stops opening. In mechanical systems the ejector plate is stopped while the balance of the mold moves further back. This establishes a motion between the ejector plate and the B plate on which are the molded plastic parts, the common expression notwithstanding, "the ejector plate moves forward." In machines with hydraulic knockouts, the K.O. plate does move forward.

Ejector Pins and Plates

Ejector pins are made either from H-11 (a high chrome vanadium steel) or a nitriding steel. They are hardened to a surface hardness of 70 to 80 Rc to a depth of 0.004 to 0.007. The inside core is tough. The heads are forged for maximum strength. They are honed to a super finish. The K.O. pins come in fractional and letter size diameters. Each size can be had 0.005 oversized. These are used when K.O. holes are worn and flash occurs around the pins.

The ejector pins are countersunk into the ejector plate. The ejector retaining plate holds them in position when the two plates are screwed together. The ejector plate also contains four push-back pins. These pins are contacted by the A plate when the mold closes and pushes the knockout plate to its back position, so that the knockout pins and/or the cavities are not damaged.

Knock-Out (K.O.) Mechanisms

The proper way to open a mold is to use a bar to drive the K.O. plate forward. The push back pins will separate the mold. This prevents cocking the plates and possible damage to the leader pins, and is much superior to driving a wedge between the A and B plates.

In the center of the plate there is a sprue puller (pin) attached to the knockout plate, directly opposite the sprue. If a Z shaped undercut is machined on the sprue puller, plastic will mold around it. When the knockout plate moves forward, the pin is raised above the mold surface so that the runner can slide off the undercut. Other ways of pulling the sprue are to machine the top of the hole in which the sprue puller moves with a reverse taper (about $5°$ per side) or to machine a ring undercut (0.010 to 0.020 radius $\times$ 1/16 in. wide).

Sometimes the knockout arrangement of the machine falls outside of the mold base. Then the knockout plate is extended further than the rest of the mold or steel wings are bolted or welded to its sides. See "Ejection Systems" for additional data on ejection (p. 128).

Support Pillars

Most molds require an additional support in the area between the parallels of the ejector housing. These support pillars go through the ejection plate and have to be located without interfering with the ejection system.

The D-M-E Corporation, which pioneered standard mold bases and parts suggest the following calculations for support pillars:

$$W = \frac{8ZS}{M} \qquad Z = \frac{B^2 L}{6} \qquad A = \frac{W}{P}$$

where W = permissable load (pounds on support plate)
Z = section modulus (in.^{3})
A = permissable cavity area (in.^{2})
S = permissable working stress of steel (12,000 psi)
M = distance between supports (in.) (based on row of support pillars on 3 in. centers.)
B = thickness of support plate (in.)
L = length of support plate (in.)
P = effective polymer pressure on support plate (10,000 psi)
Combining these equations:

$$A = \frac{1.6 B^2 L}{M} \qquad\qquad (2\text{-}1)$$

For a 15-in. plate, 1.875 in. thick, and 8½ in. between supports, the maximum projected molding area is 10 in.^{2}. Obviously for a mold plate of this size, the projected area is too low. If a row of support pillars were added down the middle, reducing M to 4¼ in., then each side could contain double the area and the plate four times or 40 in.^{2}. If two rows were added then nine times the area is available or 90 in.^{2}. One should be careful to have enough support pillar area so that the clamp force of the machine will not hob the pillars into the backup plate.

The obvious result of insufficient support is flashing. As the plate flexes other things may occur. The K.O. pins and pushback pins may bind. Cavities and cores may become angular in relation to each other causing ejection problems, parts scoring, and possible cracking. If the machine clamping tonnage is sufficient, the first place to check in case of trouble is the support pillar location, calculating the allowable projected molding area.

Types of Bases Available

A number of companies manufacture standard mold bases and parts. Because of their high volume, they have equipment which usually makes their mold bases less expensive and superior to those manufactured by the moldmaker. In addition, replacement parts are standard and readily available to the molder at a low cost.

There are cavities and cores of such size or shape where a mold base is best built around them. There are several types of steel available for the mold base. It is strongly urged that the best quality be used for any mold which requires high quality parts or long production runs.

Many mold bases are made from a medium carbon, silicon-killed forging quality steel AISI-C type 1030 with a hardness range of 165-185 Brinell. A better grade steel, AISI type 4140 which is slightly more expensive and a little more difficult to machine is recommended. It comes normalized and drawn to relieve internal stresses and preheat treated to 252-302 Brinell. This steel can be hardened to Rc-45. Mold sets are also available in prehardened steel into which cavities and cores can be cut.

A rigid mold is essential and is the result of mold design, correct material selection of the mold frame, sufficient supporting pillars, and proper core, cavity, and gating. If the frame does not remain rigid the best cavities and cores will not stay in line. (A general discussion of molds is found in Ref. 75, 76, and 77.)

DESCRIPTION OF BASIC MOLD TYPES

The injection mold is identified descriptively by a combination of some of the following terms. They will be described in the text following.

Mold Parting lines
    Regular
    Irregular
    Two plate mold
    Three plate mold
Runner system
    Hot runner
    Insulated runner
Material
    Steel
    Stainless steel
    Prehardened steel
    Hardened steel
    Beryllium copper
    Chrome plated
    Aluminum
    Epoxy
Number of cavities
Methods of manufacture
    Machined
    Hobbed
    Cast
    Pressure cast
    Electroplated
    EDM (spark erosion)

Gating

Edge	Diaphragm
Restricted (Pin Pointed)	Tab
Submarine	Flash
Sprue	Fan
Ring	Multiple

Ejection

Stripper ring	Removable insert
Stripper plate	Hydraulic core pull
Unscrewing	Pneumatic core pull
Cam	Knock out pins

PARTING LINE

When a mold closes, the core and cavity meet, producing an air space into which the plastic is injected. If one were inside of this air space and looking toward the outside, this mating junction would appear as a line. It so appears on the molded piece and is called the parting line. A piece might have several parting lines if it has cam or side actions. The expression "parting line" is usually restricted to that line which is related to the primary opening of the mold. The mold separates at the parting line in a two plate mold and at the parting line, and runner plate in a three plate mold.

The selection of the parting line is largely influenced by the shape of the piece, method of fabrication, tapers, tolerances, method of ejection, type of mold, aesthetic considerations, post molding operations, inserts, venting, wall thickness, the number of cavities, and the location and type of gating.

Figure 2-3a shows a simple parting line on a flat plane. It is simple to make. The surfaces are ground flat. Figure 2-3b shows an irregular parting line. In this instance the shape consists of flat plane and arcs of circles which can be mechanically machined. These shapes are more difficult to make and maintain. Parting lines on figures or animals, for example, are completely irregular and usually require hand fitting.

Figure 2-4 shows a slab of plastic 3/16 in. thick, 1/2 in. wide, and 1 in. long. The shape of the molded piece will depend on the parting line chosen. The tapers have been exaggerated in the drawing. On a molded piece the eye readily picks up very small tapers. If the parting line is selected as in A, the two bottom edges would be sharp while the top could be sharp or radiused. This might not be acceptable for a handle as the sharp edge could be uncomfortable. The B has the parting line in the middle. The two cavities are shallow but have to be accurately made. If not, there will be a mismatch between them. This is not possible in A because the plastic is molded in one cavity and sealed off with a flat

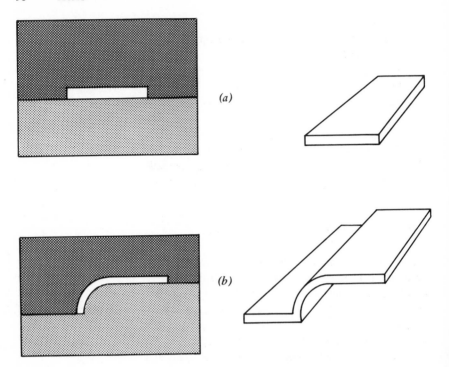

Figure 2-3 (a) Straight parting line; (b) curved parting line.

back plate. In B you would be concerned about having the parts stick on the ejection side. Since both sides are identical a slightly smaller taper would be put on the knockout side. This would not be necessary in A.

Moreover, C and D show a similar selection of parting lines except that the piece is turned on the 3/16 in. end. This gives a deep cavity harder to make and maintain. If a $1°$ per side taper was needed for ejection, configuration C would have a difference of 0.035-in. from maximum to minimum on the 3/16-in. dimension. This contrasts with a 0.003 difference for this dimension using B. The part could also be molded on the 1-in. side as shown in E and F. A, C, and E could be parted on the diagonals.

The projected area is the net area of molded surfaces parallel to the platens. Normally a minimum of two tons per square inch of projected area is required in the clamping system to prevent the mold from opening. Many times the limitations of a machine is its clamping capacity rather than its plasticizing capacity. Parting lines A and B have a projected area of 0.5 in.^{2}, E and F have a projected area of 3/16 in.^{2}, and C and D of 3/32 in.^{2}. Thus if the projected area (clamping capacity of the machine) was the limiting factor, five times more parts could be

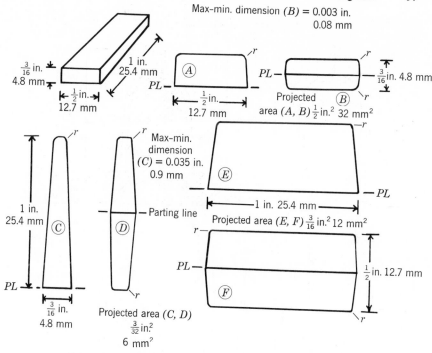

Figure 2-4   Effect of parting line location. 1° taper/side

molded using parting line C and D than by using those of A and B.

It is always more difficult to hold a dimension that crosses the parting line. Under the severe conditions of injection there is always the possibility that the clamping mechanism will move several thousands of an inch. This will increase the dimension controlled by the parting line accordingly. Therefore, if the 1-in. dimension is critical it would not be wise to use C or D. Any of the other four would contain the 1-in. dimension within the cavity and produce parts to a closer tolerance.

Sometimes the location of the parting line is determined by the characteristics of the molding machine available. Consider molding a 10-in. deep cylindrical container (Figure 2-5a). Constructing the mold with the parting line shown in Figure 2-5b with the cavity attached to the stationary platen and the core opening on the movable platen, the following mold sizes are required. The cavity half would be 12-in. long, the core side 16-in. long, and the closed mold height 18-in. To eject the piece after molding the machine would first have to open 10-in. to get the cavity out of the core and another 10-in. to clear the piece and 2-in. so that the piece can be ejected. This requires a minimum machine daylight of

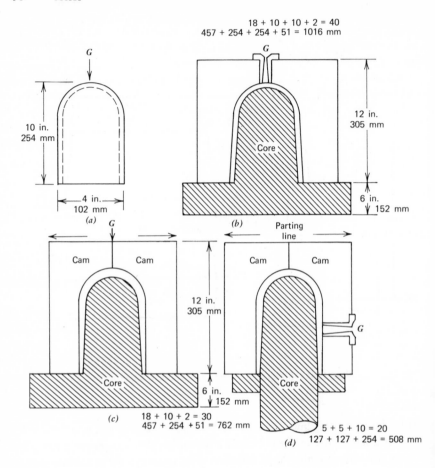

Figure 2-5   Effect of parting line on machine stroke.

40-in. If the container was 0.100 thick and molded in polystyrene it would weigh approximately 8 oz. Most machines with this injection size do not have a minimum 40-in. daylight.

If the parting line was changed so that there were two cams (Figure 2-5c) the maximum daylight could be reduced to 30-in. If the mold was split as shown in Figure 2-5d with a removable core, the daylight requirement would be less than 20 in.

Changing the parting line may cause or permit other changes. Configuration B results in a tapered container, usually with walls of even thickness. In configurations C or D the option is added of having parallel outside walls (or any other taper).

The parting line selection would also affect the gate location. Mold B must be gated at point G. If it were gated on the parting line entrapped air would prevent molding a complete piece. Mold C can only be gated at point G. Mold D can be gated anywhere along the 10-in. section. As we see later there will be distinctly different mechanical properties when the gating is changed from the side to point G. The projected area in molds B and C is about 13-in.^{2}. In mold D it is 38.3 in.^{2}. Asthetic requirements might not permit mold D, because of the gate mark location.

Two Plate Molds

Figure 2-6 shows a schematic representation of the cross section of part of a regular two plate injection mold. The part being molded is a shallow dish, gated on the edge. There are temperature control channels in both backup plates and in the cores and cavities. Since there is significant insulation between two pieces of metal, the use of channels directly in cores and cavities gives better and more efficient temperature control than just cooling the A and B plates.

Note the support pillars which are anchored to the ejector housing (back plate) and support the backup (support) plate underneath the cores in the B plate. A machine knockout bar is shown. As mentioned, they remain stationary, and as the moving platen returns, they stop the ejector plate. The mold opens on the parting line and the sprue puller, which in this instance is shaped like a Z, pulls the molded sprue and runner with it. The part design and molding conditions keep the plastic on the core. As the ejector mechanism works, the parts are pushed off the core and the sprue puller moves out if its hole, allowing the parts and runner to be removed or fall.

Three Plate Molds.   Suppose the dish was deeper and could only be gated in the top center section. The mold could be constructed as a one cavity mold feeding directly from the sprue. There are other alternatives, one of which is shown in Figure 2-7 and 2-8. There are six cavities located in two parallel rows of three. One cavity and core is shown with other significant parts of the mold.

The difference between this type mold and the one illustrated in Figure 2-6 is that it separates between the runner plate and the pin plate (PL-1) as well as at the regular parting line (PL-2). This is called a three plate mold even though a third plate is not always used or a fourth plate may be added. The plastic is injected through the sprue bushing into the runner channel, which is trapezoidal, and tapered and cut into the runner plate with the wider face of the trapezoidal cross section facing the injection side. The plastic flows into the part through an auxiliary sprue bushing. While this can be machined directly into the runner plate and cavity, it is good practice to have a separate bushing so that it can be replaced or changed.

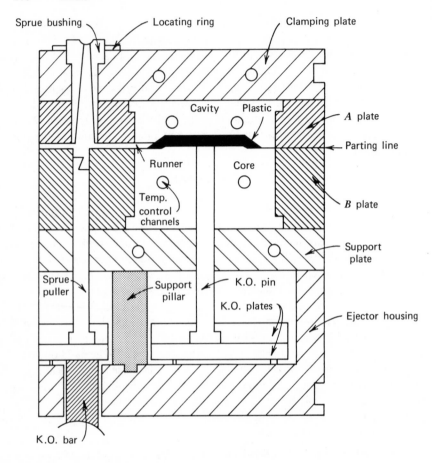

Figure 2-6 Schematic drawing of a two plate mold. (Robinson Plastics Corp.)

When the mold opens, the *A* and *B* plate moves together. Sometimes this will occur normally. Other times latching mechanisms are needed. The mold opens initially on parting line 1 (PL-1). This breaks the gate and leaves the runner attached to the pin plate because of the undercut pins (*A*) attached to the injection backup plate and extending into the runner. After the separation has occurred at PL-1, the mold continues to open, separating at PL-2. The molded pieces remain on the core. They are then ejected in the conventional manner by a stripper plate. The tie rod (*C*) pulls the cavity plate, which in turn pulls the pin plate through stripper bolt (*D*). The pin plate is limited in its travel by stripper bolts *B*. When it is moved forward opening at PL-3 (it could also be moved by

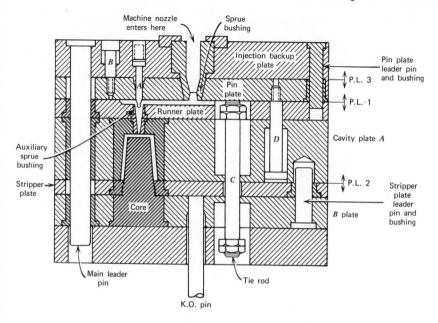

Figure 2-7 Schematic drawing of a three plate mold in closed position.

latches, chains, ejector bars or cylinders), the runner is stripped off the undercut pins, and the plastic sprue is moved forward out of the sprue bushing. The runner fall down, can be removed by hand, an air blast or mechanical wiper. The pin plate always stays on its leader pins. They must be long enough so that the plates can separate far enough to remove the runner. It is good practice to support the pin plate on its own leader pins, attached to the injection backup plate. This will prevent it from binding on the main leader pins.

This type mold works very well provided the workmanship is of good quality and the components fully sized and adequately designed. If not, cocking and binding will occur relatively quickly on heavy molds. It is sometimes necessary to put an extra set of leader pins and bushings to support the A plate. These should not be used to line up the A and B plates. In other instances small leader pins and bushings are put into the A and B plates to assure good line-up and compensate for wear on the longer leader pins.

Another use for three plate molds is to mold parts on the plates that open on PL-1, Figure 7. This is an old technique used for molding records; it is not new in injection molding either (5a). Today they are called "stack" or "stacking" molds. Suppose one were to injection mold three 8 x 10 in. plaques directly over each other in a mold with the prerequisite number of plates in a stack mold. What clamping force would be required, based on a 3 ton/in.^{2} of projected area

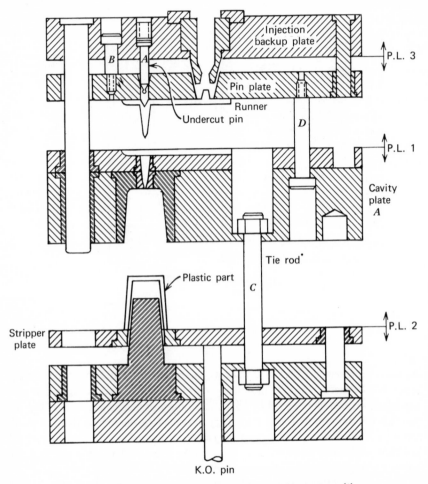

Figure 2-8 Schematic drawing of a three plate mold in open position.

design figure? The answer is 3(8 x 10) or 240 tons, not 3(8 x 10 + 8 x 10 + 8 x 10) or 720 tons. For each force there is an equal and opposing force, therefore, the force in the rear of the first and the front of the second plaque cancel each other out, as do the rear of the second and front of the third.

RUNNER SYSTEMS

The runner is the connection between the sprue and the gate. It is a necessary evil. It should be large enough to allow rapid filling and minimum pressure loss, but not so large as to require the cooling cycle to be extended for the runner to harden enough for ejection. Most jobs will permit the runner to be reground and reused. Regrinding is expensive, wastes material, is a source of contamination, is a place for foreign material such as screw drivers and other metal parts to enter, and causes a probable lowering of the physical properties of the plastic.

There are various types of runners. The full round runner is preferable, because it has the smallest surface to volume ratio. When a runner has to be on one side only, the best compromise is a trapezoidal shape. Half round and rectangular runners should not be used. The runner should have no undercuts and be polished. This gives less turbulence in the flow and slightly faster filling rates.

The shape of the runner depends to some degree on the location of the cavities. It must be remembered that the initial surge of material will cool as it goes through the sprue and runner. Injection of this cooler material into the cavity tends to set up stresses. This should be avoided where possible by cold slug wells and runoffs. The design of the runner is an important factor in keeping the cooler material out of the cavity. Figure 2-9a shows a radially runnered mold, with

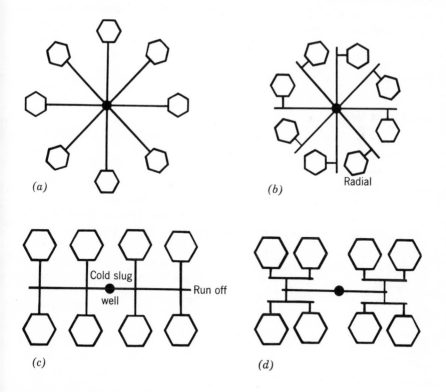

Figure 2-9 Runner systems: (a) radial, poor design; (b) radial, good design; (c) standard design; (d) "H" design.

the runner running directly into the piece; *b* shows the preferred method, allowing the cooler material to continue to the end of the runner and warmer material to branch off into the cavities. Sharp corners should be avoided to reduce turbulent flow.

In multicavity molds it is essential for all cavities to fill at the same rate. If not, there will be packing and incomplete shots. Some people attempt to do this by a "balanced" or *H* runner system (Figure 2-9*d*). The theory is that if the plastic flows exactly the same length to all cavities, all will be uniform. Unfortunately this is not always so, as the mold is not usually the same temperature throughout. The heat history of the molding material is also not consistent. The gates still have to be balanced; therefore, the theoretical gains compared to a standard runner system (Figure 2-9*c*) may not outweigh the considerably larger runner system which has to be reground.

The size of the runners depend on the type of material and the size and thickness of the parts. As a general rule, the less viscous materials, such as the styrenes, approximately 5-in. or less from the sprue bushing, use from 1/8 in. to 1/4 in. full round runners. Runners occasionally go to 5/16 in. and very rarely over 3/8 in. Runners for viscous materials like acrylics and polycarbonates are usually about 3/8-in. full round. It is better to start with smaller runners as they are much easier to increase than decrease. If much experimentation is expected the runners should be cut into removable blocks.

Hot Runner Molds

The runners must be reground and reused in most operations, if possible. A logical extension of the three plate mold overcomes this and is called a hot runner mold (Figure 2-10). This mold has a hot runner plate, which is a block of steel heated with electric cartridges, usually thermostatically controlled. This keeps the plastic fluid. The material is received from the injection cylinder and is forced through the hot runner blocks into the cavities. Theoretically this is fine. Practically, hot runner molds take considerably longer to become operational initially when the mold is first tried. There are problems of temperature control, gating, balanced flow, and drooling. The greatest difficulty is to prevent the nozzle leading into each cavity from freezing. If the nozzle is too hot, the material will drool. If it is too cold, it will freeze. In multicavity molds, it is difficult to balance the gates (essential for proper molding) and still prevent either freezing or drooling.

The hot runner mold is especially susceptible to tramp metal, paper, wood, and other contaminants which will quickly clog the nozzle. Start up for each run is more difficult as freezing of one or two nozzles will pack or flash the other cavities.

Various gate controlling valves have been tried with varying success in the

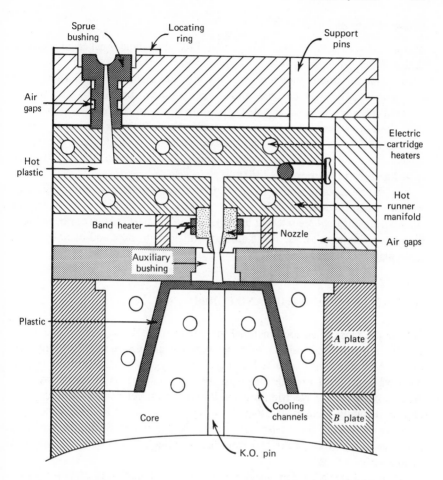

Figure 2-10 Schematic drawing of a hot runner mold. (Robinson Plastics Corp.)

nozzle of the hot runner mold (6). The most successful require temperature control on the heater around the nozzle and in the blocks surrounding them. They are comparatively expensive although they often can be used on other molds. The consideration for their use is usually economic.

Unless the job is long running or there are technical considerations, three plate molds are preferable. This is not to be construed a condemnation of hot runner molds. The author has and has seen many such molds that run and run well. When they do they are usually run automatically, are a pleasure to watch, and highly efficient (7, 7a, 11).

Insulated Runner Molds

A cross between a hot runner mold and a three plate mold is called an insulated runner mold. The gating system is very similar to that of a three plate mold except that the runners are very thick, at least 3/4 in. in diameter. There is no runner plate. The back-up plate and A plate are held together by latches. The material, usually olefins, is injected. The outside of the runner freezes but insulates the center permitting the core to remain fluid. This acts as a hot runner. If the runner freezes during start up the two plates are separated and the runner system removed. As soon as the runner reaches equilibrium, the latch is kept closed and the mold is operated that way.

The nozzles present similar problems to those of a hot runner mold. Usually a heated torpedo is inserted in each gate and kept on continually. The wattage of the cartridge heaters are selected by experience. The heat output can be reduced by using resistors in series. These are more difficult to start and operate than a three plate mold and are usually restricted to the olefins (8, 12).

MATERIALS FOR CAVITIES AND CORES

Steel is the most often used material for injection mold sets, cavities and cores. Beryllium copper and other nonferrous materials are also usable. Iron ore is reduced by burning with coke and on occasion, oxygen, in blast furnaces to produce pig iron. Steel is made from pig iron by removing its impurities in open hearth, Bessemer, or electric furnaces. Impurities are removed by various techniques including vacuum degasing. It is beyond our scope to discuss the technique of producing the quality steels required for molds. Such information is available from the literature and from the pamphlets and booklets of the steel companies.

When tool steels are deoxidized by silicone and aluminum to prevent gas evolution upon solidification, they are called "killed steel". These pockets of gas leave voids which are not desirable in mold steels.

The steel is cast into ingots which require additional mechanical work. If this work is done above the critical temperature, it is called hot worked steel, and below is called cold worked steel. Cold worked steel is usually done at room temperatures and increases the tensile and yield strength and reduces the ductility. The ductility is a property which allows permanent deformation of the steel without facturing by stress in tension. Hot working increases the mechanical properties but is particularly valuable in increasing the ductility.

Steel is converted by either rolling or forging. In rolling, the steel is continuously passed through rolls reducing it to the desired size. Forging is done with a hammer forge or with more slowly applied hydraulic pressure in a forging press, the latter method being preferred.

Steel is converted either singly, such as hot worked steel which is converted to the final size in one continuous operation, or in multiple conversions, where the steel undergoes a series of heating, forging, and cooling operations. The harder alloy steels are multiple converted.

Alloys are made by melting other elements into steel to affect its properties. A very brief summary of what each alloy does follows.

Manganese (Mn) is found in almost all tool and alloy steels. It is an excellent deoxidizer and hardens and strengthens the steel though less than carbon. It decreases the cooling rate, increasing the hardenability.

Silicon (Si) in the range of 0.2 to 0.35% is specified in all alloy steels for deoxidation. Increasing silicon content raises the critical temperature, increases the susceptibility to decarburization and graphitization, and when combined with other alloys promotes increasing resistance to high temperature oxidation.

Nickel (Ni) is used to improve the low temperature toughness of steel. It lowers the critical temperature lessening distortion in heat treatment. Nickel permits lowering the carbon content to achieve a given strength level thereby increasing toughness and fatigue resistance.

Chromium (Cr) improves the hardenability best of all the alloying elements and is used in air hardening steels. It improves the surface resistance to wear, promotes carburization and in high percentages reduces corrosion.

Molybdenum (Mo) increases the strength toughness, hardness, and hardenability of steel. It increases the machinability. It has a strong tendency to form stable carbides inhibiting grain growth and making the steel fine grained. This adds toughness. It also intensifies the effects of other alloys.

Vanadium (V), while expensive, increases strength, hardness, and impact resistance. It inhibits the growth of grains during heating, permitting a higher tempering temperature. It intensifies the effect of other alloys.

Tungsten (T) increases hardness, strength, and toughness.

Aluminum (Al) is a deoxidizer and a degassifier. It slows down grain growth and is used to produce fine austenitic grain size. It is used to the extent of 1% in steels to be nitrided.

Steels are designated by the American Iron & Steel Institute (AISI) and the Society of Automotive Engineers (SAE), whose designations for carbon and tool steels are the same. Four numbers are used to designate steels other than tool and die steels. A letter prefix designates the method of manufacture: *C* designates open hearth steel; *B*, Bessemer steel; and *E*, electric furnace steel. The first two numbers designate the type of steel. The last two numbers indicate the average carbon content in one one-hundredth of a percent. Hence a C-1045 steel would be an open hearth steel containing a 0.43 to 0.50% carbon (average 0.45%). Table 2-1 shows the series designation and the approximate alloy con-

Table 2-1   AISI Numbers and the corresponding alloying elements which cause their designation. The xx indicate the average carbon content of the steel in one one-hundredth of a percent.

Series Number	Approximate % of Alloying Elements Designated by the Identifying Number
13xx	Mn 1.75
40xx	Mo 0.20, 0.25
41xx	Cr 0.50, 0.80, 0.95; Mo 0.12, 0.20, 0.30
43xx	Ni 1.83; Cr 0.50, 0.80; Mo 0.25
44xx	Mo 0.53
46xx	Ni 0.85, 1.83; Mo 0.20, 0.25
47xx	Ni 1.05; Cr 0.45; Mo 0.20, 0.35
48xx	Ni 3.50; Mo 0.25
50xx	Cr. 0.40
51xx	Cr. 0.80, 0.88, 0.93, 0.95, 1.00
52xxx	C 1.04; Cr. 1.03, 1.45
61xx	Cr 0.60, 0.95; V 0.13, 0.15 (minimum)
86xx	Ni 0.55; Cr 0.50; Mo 0.20
87xx	Ni 0.55; Cr 0.50; Mo 0.25
88xx	Ni 0.55; Cr 0.50; Mo 0.35
92xx	Si 2.00

tent of carbon and alloy steels.

The designation of special tool and die steels is found in Table 2-2. The P-1 steels are easily hobbed because of their low alloy and carbon content. Since they are water hardening steels they have a high tendency to distort during heat treatment. The P-4 and P-5 are a little more difficult to hob because of their chromium content. They are used because the chromium permits oil hardening and less distortion, as well as increased cores strength and resistance to wear. The P-2 and P-6 steels are more difficult to hob but give a tougher finished cavity than the other steels.

Steels for hobs are usually of the S-1, A-6, or 0-2 types. The H-13 and H-23 steels are used for hot hobbing beryllium. H-13 steel is commonly used for cores because of its low dimensional change in hardening and its good toughness. The P-20 steels are very tough and generally supplied in a prehardened condition of approximately R_c30. It can be carborized and hardened later. Other types of prehardened steels are supplied with hardnesses up to R_c42. The advantages of a prehardened steel evolves from the fact they do not have to be hardened. Hardening can cause distortion and on rare occasions cracking. It also increases the polishing time (14-16).

The 18% Nickel maraging steel is relatively new. It is delivered in a R_c30

Table 2-2   Tool and die steels (AISI) classification.

Type	Prefix	Description
Water hardening	W	
Shock resisting	S	
Cold work	O	Oil hardening
	A	Air hardening — medium alloy
	D	High carbon — high chrome
Hot work	H 1 — H-19	Chromium base
	H 20 — H-39	Tungsten base
	H 40 — H-59	Molybdenum
High speed	T	Tungsten base
	M	Molybdenum base
Special purpose	L	Low alloy
	F	Carbon tungsten
Mold steels	P 1 — P 19	Mold steels — low carbon
	P 20 — P 39	Mold steels — other types

hardness which results from air cooling from 1500°F. It can be precipitation hardened by ageing at 900°F for 3 hr. This gives an R_c of 52 to 54. Because of the low hardening temperature there is practically no distortion (17).

Stainless steels are used in molds to eliminate the effects of corrosion. Corrosion can be caused by water or molding compounds, particularly PVC. By commercial usage they are defined as steels containing 11 1/2 to 20% chromium. Type 420 is most often used. It is magnetic and can be hardened by oil quenching to approximately R_c52. Inserts of stainless steel are valuable as they have lower thermal conductivity than tool steel, hence can reduce warpage due to nonuniform cooling of plastic.

Hardness

Hardness seems to be an easy concept but has not yet been adequately defined. It is probably not a fundamental property but a combination of work hardening, elasticity, yield strength, and tensile strength. Nonetheless it is an important measurement in mold building. It is commonplace in ordering a mold to specify the hardness of the cavities, cores, pins, and other components. The method most used in America is the Rockwell test method. There are ten scales. The one used for steel is the C scale. The Brinell system is also used here and in Europe. Conversion tables are available in the Appendix. The Rockwell system is a measure of the difference of the depth of penetration of a ball into steel between an initial small load and a final large load. If the steel is too hard it becomes brittle. If too soft, it does not provide enough protection against damage and wear. A Rockwell reading of R_c50-55 will give good results. Steel this hard is difficult to machine even with carbide. It is easily worked by grinding or electrical (EDM) or chemical removal equipment. The cavity or core is readily machined from steel in the soft condition as it comes from the steel mill. It is then hardened. Pre-hardened steel is a compromise between machinability, hardness, distortion, and wear (18).

The harder the steel, the greater its compressive strength. This should not be confused with the modulus of elasticity of steel which is 30,000,000 psi and is the same for all steels. This is a measure of the deflection of the steel by bending. Hardness does not significantly effect this at all. The choice of materials will. For example, beryllium used in molds has a modulus of elasticity of approximately 20,000,000 psi and will therefore deform elasticly about one and a half times as much as steel under the same stress conditions.

Hardening

The hardening and hardness of molds, cores, and cavities should not be left to chance. The selection of steel, design of the piece, heat treating before and during the machining operations, and the type and specifications for heat treating have much to do with the longevity, usefulness, and trouble-free operation of a mold. A brief review of the hardening of steel is in order.

Iron has two crystalline structures. Alpha iron has the iron atoms at each corner of the cube and one in the middle. Gamma iron has iron atoms at each corner of the cube and one each centered on each of the six sides. Heating pure iron to $1670°F$ changes its structure from alpha to gamma. At $2535°F$ it goes back to the alpha form but is called Delta iron. At $2802°F$ iron melts.

Carbon is the important alloy in iron relating to hardening. Alpha iron contains very little carbon in solution. The solution of carbon or other alloying elements in alpha iron is called *ferrite*. Gamma iron holds considerably more carbon in solution. When it contains carbon or other elements in solid solution it is called *austenite*. When the carbon is not in true solution with the iron it forms iron carbide (Fe_3C), which is an extremely hard and brittle compound known as *cementite*. When the eutectic mixture of iron and carbon is cooled, alternate layers of ferrite and cementite precipitate out. This structure is called *pearlite*. *Banite* structure contains cementite needles in a ferrite matrix.

If the steel is cooled slowly enough to maintain the equilibrium, pearlite structures result which are soft structures. As the rate of cooling increases the cementite and ferrite precipitate out more quickly and the distance between the layers becomes smaller. The carbide is more dispersed, making it more difficult for the layers to slide over each other, which causes the steel to become harder. When austenitic steel is cooled very quickly (quenched) the carbide does not have time to separate out. This supersaturated carbon structure is called *martensite*. This is very hard and brittle and must be tempered before it can be used.

The minimum rate of cooling to form a fully martensitic structure is called the *critical cooling rate*. Cooling the steel more slowly will give a combination of pearlite and martensite. When, as mentioned, the steel is cooled slowly enough to reach equilibrium the composition is completely pearlite.

Steel is hardened only when martensite is formed. Therefore, the steel must be cooled or quenched fast enough to prevent the formation of pearlite or banite. The hardening behavior of steel is shown by TTT curves (derived from T for temperature, T for transformation, and T for time). Pure steel of the thicknesses used in molds does not have a high enough thermal conductivity to cool the inside quickly enough. To overcome this, alloying elements are added which in effect shift the TTT curves to the right.

The steels are quenched either in water (brine), oil which quenches more slowly than water, and air which is the slowest cooling medium. They are commonly classified as water hardening, oil hardening, or air hardening steel.

In large pieces, the surface can be converted to martensite while the core is still warm enough to be 100% austenite. Martensite formation increases the volume and cooling decreases the volume. Therefore, the outside surface is cool and contracting while the inside surface is expanding while converting from austenite to martensite. This causes stress in the steel which may result in warping or cracking. Obviously, the slower the cooling (air hardening) the less probability of either occurrence.

Because of this, the steel should be tempered immediately upon reaching approximately 150°F. To temper, it is heated to a given temperature below the critical range depending upon the grade of steel and the hardness desired. It is held there for 1 to 6 hr. It is then allowed to cool in air. Tempering removes the brittleness and reduces the hardness. The resultant degree of hardness depends on the temperature at which it is tempered.

Annealing is a term used for the heating and slow cooling of steel. It can remove stresses, change the toughness, hardness, and other physical properties.

The simplest form of annealing is stress relieving. The purpose is to remove the stresses generated in previous heat treatment, cutting, forming, or other operations performed on the steel. The steel is heated and soaked at about 1200°F and then allowed to cool in the furnace to about 930°F. It is then cooled in air. Steel should be stress relieved before hardening. If full annealing is required (to permit machining, for example) it is done by heating the steel to about 100°F over the upper limit of the critical temperature allowing it to soak fully and then cooling it in a furnace or other slow cooling medium. The cooling rate is shown by the TTT curve.

Case Hardening

Alloy steels will not form or hob as readily as a low carbon steel. Therefore, most hobbings are done in low carbon steels. To obtain a hardness suitable for use in plastic molds, elements have to be alloyed on the surface. This is called case hardening. A case hardened steel will have a very hard surface and a tough ductile interior. These hardened surfaces are required to give the high polish and long wear needed for injection molds. The most common method of case hardening is called carburizing. In pack carburizing, a steel container is lined with a commercial compound usually containing about 20% of metallic carbonate and charcoal. The steel to be hardened is packed into this compound and a cover put on the box. It is heated to 1550 to 1750°F, depending on the steel and the results required. The length of heating time is determined by the temper and the depth of hardness desired. For example, in a particular steel carburizing at $1600°$ for 28 hr gave a depth of the case of 0.070 in. This is approximately the depth of case used in mold making. It permits the grinding and fitting operations to be done without going below the hardened steel. The parts are then quenched in oil and tempered for the desired hardness.

Nitriding

Nitriding is a process in which nitrogen is introduced to the outside surface of the steel. It gives an extremely hard, wear-resistant surface which is retained at elevated temperatures. A nitrided case is as hard at $750°F$ as it is at room temperature. It can be heated to a $1000°F$ without losing its hardness when cooled. This would immediately suggest its use for heating cylinders, plungers and bushings.

Nitriding is accomplished by heating the steel in a controlled ammonia atmosphere between 950 to $1000°F$. The processing time is measured in days. Since the part is not quenched distortion and warpage are very low. There is a slight dimensional growth for which allowances can be made. Parts which might be difficult to harden by other methods are sometimes nitrided. Nitriding is best done with a special alloy steel, Nitralloy®, containing 0.85 to 1.20% aluminum. The case hardness is in the area of $R_c 70$. The case depth can vary from 0.007 to 0.080 in.

Design for Heat Treating

A mold and its parts should be designed with heat treating kept in mind. Nearly all serious failures of hardened steel parts are caused by internal stresses. Since molding develops repeated thermal and mechanical stress special care is needed.

There are a number of simple principles and practises which improve the design for heat treating. The major cause of cracking caused by heat treating is sharp corners. Sharp corners are exceedingly dangerous to molds and MUST be avoided. Fillets, radiuses, and tapers are preferred.

Avoid sudden changes of thickness. This prevents even heating and cooling. Holes should be at least 1 1/2 diameters away from the edge. Water lines should be kept 3/4 in. away from the mold surfaces. Ample lands should be provided. Holes should be placed with as much metal around it as possible. It is sometimes better to grind a slot after hardening, such as a key way. Assuming there is a 7X 9-in. plate with a 6 X 3-in. cutout on one side. It might be better to make two sections which will be hardened and fitted separately. Deep drilled holes should be avoided as they are difficult to quench. If there is any question it would be well to consult the heat treater.

Steel Requirements for Cavities and Cores

Uniform structure and freedom from defects are obviously essential for a good cavity. It is more than distressing to invest weeks of labor on a cavity or core and find that there is a defect or flaw, making it unusable. It is, therefore, advisable to only use steels specified for molds by the steel manufacturer.

Machinability, which directly affects cost, is an important factor. It is very hard to rate. High machinability is sometimes produced by adding sulphur and other nonmetallic elements. This reduces the polishibility as does large amounts of nickel. Most moldmakers have "pet" steels which they rate as very machinable.

The steel should be able to take such surface finishing or polishing as required for the application. It should also have the suitable corrosion resistance.

If the steel is to be hardened, its heat treatability and distortion must be considered. Since there is always a possibility of repair, either during or after moldmaking, it should be readily weldable.

The strength and toughness requirements of the steel depends on the nature of the cavity and core. It is also related to the method of gating and venting. When molding abrasive materials, such as glass filled thermoplastics, erosion resistance should be considered. References about the selection of steels are 14, 15, 16, 19, 20 and 20a.

Surface Finish

The surface finish of the cavities and cores is an important specification. It will affect appearance, ejectability, and cost. In many instances, the difference in mold quotes is the different conception the mold makers have of the polish required. Heretofore in the molding industry surface finish specifications have been merely descriptive, such as good polish, medium polish, and sandblast. Mold surface specifications are available as well as strips of metal which are finished to these specifications. A list of terms commonly used to describe surface finish follows.

Lay	The direction of the predominant surface pattern caused by the method of fabricating the metal. It might be tool marks or stoning marks.
Flaw	An irregular or infrequent surface mark such as a hole, scratch, and ridge.
Roughness	Relatively fine spaced surface irregularity in the direction of the predominant surface pattern.
Roughness width	The maximum width in inches, of the surface irregularity included in the measurement of roughness height.

Roughness height   A major specification, measured in microinches*
and specified in one of the following ways; maxi-
mum peak to valley height, average peak to valley
height, average deviation from the mean surface
(either Root mean square, rms or arithmetical).

When one number is used to specify the height or width of an irregularity, it
indicates the maximum value. Any lesser will be satisfactory. When two numbers
are used, they specify the maximum and minimum permissible values.

Waviness is a surface irregularity which has greater spacing than roughness.
This might develop from work deflections, vibrations, heat treatment, or warping strains. The height and width are specified. It can be seen then that "polish"
should not be the only specification. A roughness height of 2, 4, or 8 μ in. would
give a very high polish, but if the waviness height is 0.008 or 0.010 the piece
might well stick in the cavity or on the core.

Drilling and milling give surface finishes of 63 to 250 μ in., finish turning,
broaching, boring, and reaming 32 to 125 μ in., grinding 32 to 63 μ in., and
polishing down to 2 μ in., all measurements rms. Roughness heights are usually
specified as 2, 4, 8, 16, 32, 50, 125, 250, and 500 μ in.

If the cost of finishing a 120-μ in. surface is 1, then 50 μ in. would be double,
16 μ in. triple and 8 μ in. quadruple.

Mold Polishing

The need for a properly contoured, polished mold surface is self-evident (20b,
20c). The penalties for neglect are sticking (which makes consistent mold cycles
and automation impossible), scratching, and surface blemishes. A stuck piece
wastes molding time in removal and reestablishment of the cycle, materials in
purging, impairment of physical properties, and possible damage to the mold
while the part is being removed.

Attention to the ultimate polishing throughout the manufacture of the cavity
and core is profitable. If they are hobbed, the higher the finish on the hob, the
easier the polishing on the final piece. If parts are to be hardened they should
not be highly polished, as this changes the mold surface and causes flaws that
will not come out in subsequent polishing. Heat treating must be done carefully
to minimize scale (21, 21a).

After hardening, the scale (which is a deposit of ferric oxide) must be removed. This is done by alkaline reverse current methods, vapor blasting or immersion in ultrasonic pickling solution.

Molds are polished by using abrasives starting with the coarser grits for more
rapid material removal. The abrasives that are used are stones, abrasive papers,

*Dimensions are often given in microinches (1 μin. = 0.000001 in.).

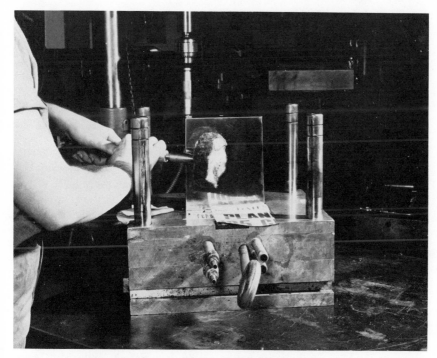

Figure 2-11 Mold Polishing. (Robinson Plastics Corp.)

polishing compounds suspended in a liquid medium (carborundum or diamonds), and polishing compounds suspended in solid media. Each moldmaker has his own combination of polishing compounds and techniques.

Stoning and abrasive papers are usually done by hand, although reciprocating units are available. Compounds suspended in liquid or solid are applied by cotton or felt buffs, or bristle brushes. Figure 2-11 shows the core of a fish tank aerator being polished with a cloth buff attached to a drill press. Bar compounds used for metal polishing in other industries are useful. A fast cutting stainless steel compound, red rouge, and a chrome oxide green rouge are used. A hard wheel, using a stainless steel compound, will bring up a good polish. A soft wheel with chrome oxide green rouge will give a high luster polish.

The direction of polishing is important. If the polish is done in the direction of ejection, it is called draw polishing. Many times this will eliminate a severe ejection problem. As noted in the section on ejection (p. 128), high polish is not only not always desirable but can be detrimental. The selection and manufacture of a proper mold polish is a highly skilled art.

Steel is highly susceptible to rust, especially after polishing, and should be protected by a suitable rust preventative when not in use. Molds that are run cold should be brought to room temperature before being coated with a rust preventative. If not, in humid air particularly, water will condense on the mold and cause rust damage. Using stainless steel or chrome plating the mold prevents water damage. Maintaining a proper polish is the responsibility of the molder (22-24).

Mold surfaces can be textured and patterned by photoengraving and etching (25).

Chrome Plating

The use of hard (industrial) chrome plating on forces and cavities is commonplace. When properly used, it offers significant advantages to the molder. The chromium will not rust, oxidize, or tarnish. It is comparatively inert to the chemicals and by-products of molding material except for those giving off hydrochloric acid. Even under those conditions it is superior to regular steel.

Its surface will not be affected by water or dampness. This means that polishing is no longer a large part of mold maintenance and that care to prevent rusting or corrosion of cavities and cores is eliminated.

Chrome plated surfaces are hard. At a thickness of about 0.0015 in. their hardness is about 1000 Brinnel. It has a low co-efficient of friction. In most cases chrome plating will help mold release.

Chrome plating will not improve the surface underneath. If anything, it will accentuate blemishes, scratches, dullness, and pit marks. Therefore the preparation and polishing of the mold before plating is extremely important. Many platers specializing for the plastic industry have polishing departments and it is recommended that they do the final polishing on the steel.

There are a number of disadvantages to chrome plating. Once the plating is partially removed, either deliberately, or by damage or wear, it is virtually impossible to blend the remaining surface so that a mark will not show on the molded part. It will continue to peel and must be stripped and rechromed. The mold should be in good working condition with all vents ground in and changes made before chroming. The cost, too, of chroming is not insignificant. There is the possibility of hydrogen embrittlement particularly in heavier coatings. Stress relieving in hot oil at 350°F will help the condition.

Small hardened parts are plated approximately 0.0003 to 0.0005 in. thick. As the cavities get larger the thicknesses increase to a maximum of 0.005 in. In prehardened steel the thicknesses are from 0.001 to 0.003 in., in beryllium cavities 0.0005 to 0.003 in., in steel castings to 0.001 to 0.003 in. and in electro formed molds 0.0002 to 0.003 in. Within these limits chrome plating can be used to build-up the surface for making dimensional changes.

Beryllium Copper — Pressure Casting

The second major material used in cavities and cores is beryllium copper. It is an alloy of copper containing approximately 2 3/4% beryllium and 1/2% cobalt. It is usually supplied to the mold maker as a pressure casting from a hob. It is also available in thin rods and bars. The material is excellent for molds and will give long wear and quality parts (26).

Beryllium copper is chosen because of its thermal conductivity, its resistance to rusting, and its method of fabrication.

An important function of the mold is to remove heat from the plastic. This is done by circulating a cooling fluid through the mold. Assuming identical molds with adequate cooling capacity the rate at which heat is removed from the mold will depend on the thermal conductivity of the metal. The thermal conductivity of beryllium is approximately two and a half times that of tool steel and four times that of stainless steel. Since the time required for cooling a plastic in the mold is a function of heat removal, a beryllium mold should give faster cycles if cooling is the limiting factor in the molding cycle. There are times when rapid cooling is not desirable or different cooling rates for different parts of the mold are required. If these factors are critical, they will form a major consideration in selecting the mold materials (27).

The surface of beryllium copper does not rust. Very often the refrigerated cooling media are circulated causing condensation on the molding surfaces if the cycle is interrupted. This will cause no mold problem in beryllium copper as compared to steel which will rust severely. A 0.0003 to 0.0005-in. flash chrome will appreciably increase the life of the cavity.

A major reason for using beryllium copper is because of the relative ease of fabrication using pressure casting, or more accurately hot hobbing (28, 29).

A hob is made proportional in size, but larger than the finished plastic part. (After hobbing the beryllium copper shrinks as it cools, just as plastic shrinks after molding.) The hob is placed in the bottom of an insulated cylindrical container. (For more on hobbing see the hobbing section, p. 119.) The melted beryllium copper is poured over it and a plunger comes down exerting pressure on the beryllium and hob. When the beryllium is cooled, the hob is separated. When casting cavities, use a shrinkage of 0.004 in./in. and on cores 0.008 in./in. is suggested. A steel hob is used made of a good hot working die steel such as H-13 or H-23 which will not deform under the temperature of casting. For one or two cavities, beryllium copper itself can be used for the hob, but as the number of cavities cast increases, the beryllium copper hob will tend to deform. A draft of 1½° per side is desirable though one-half of a degree can be used. The pressure cast cavities are reproducible with a tolerance of about ±0.002 in./in.

The nature of the process permits hot hobbing of thin sections and projections, intricate carvings, fluting, serrations, and other delicate shapes not hobbable in steel. The tensile strength of steel is not enough to prevent breaking of delicate sections under the high pressures of cold hobbing.

Hobs for pressure casting can be cast from an original sample model by a number of casting processes. If there is an irregular parting line the cavity can be cast from the hob and the wall thicknesses machined off the hob and then used as a core. This gives a good parting line match and saves most of the machining of the core.

Pressure cast beryllium copper cavities are usually supplied to the mold maker annealed and hardened. The hardening range R_c35 to 40 is soft enough to machine but hard enough to perform well in long production molding runs. They have an ultimate tensile strength of 140 to 165,000 psi. The polishing technique is similar to that used for steel. The cavities can be silver soldered or electric arc welded.

The cost of beryllium copper and steel cavities are similar. Cast beryllium copper cavities are usually sold on a poundage basis. Its specific gravity is about 8.09 which is equivalent to 0.292 lb/in.^{3}.

Other Mold-making Materials. Other cavity and core materials used for sampling or very low production molds (30) are brass, aluminum, steel filled epoxy (31), and metal sprayed concrete shells (32). These are not materials of choice for production runs or quality parts.

METHODS OF FABRICATING CAVITIES AND CORES

Most fabrication of metals for molds is done with tool room equipment. This equipment is used for machining mold bases, cores, cavities, pins, blocks, and such. More advanced equipment permits running of tool room equipment electronically by punched tape, which can be computer generated. For those not familiar with tool room equipment, a very brief description follows.

1. *A drill press* is a tool that has a stationary table above which is a rotating motor driven shaft. The shaft contains a chuck to which the drill or other tool is attached. The shaft moves up and down. It can be hand or automatically fed.

2. *A milling machine* is a drill press with a table which can be moved left and right, backwards and forwards, and up and down. The rotating shaft or spindle in the head moves up and down. This is an indispensable tool. These movements can be automatically or manually controlled. A separate attachment has a stylus which moves to trace a three-dimensional replica of the part to be cut in steel. As the tracer (stylus) moves in one direction the cutting tool moves in the same direction with a proportional movement. This is now called a *duplicator.*

3. *A lathe* has a rotating head to which is attached the material to be cut. The material rotates and the carriage, which contains the cutting tools, moves along the length of the bed or across it. The tail stock is equipped with a chuck for drilling and reaming.

4. *A grinder* has a rotating head to which is attached a grinding wheel. The

table reciprocates and can move in and out at a predetermined distance per reciprocation of the table. The height of the grinding wheel above the work is accurately adjusted. The grinder is used to obtain accurate dimensions and a good surface finish. It will readily grind hardened steel. Cylindrical grinders are used for grinding materials that can be held on centers and rotated. It is similar in concept to a lathe. External or internal grinding of round shapes is done.

5. *Band saws* consist of two wheels, one of which is motor-driven, upon which rotates an endless belt of saw material. The material to be cut is fed into the blade. *Cut-off* saws are movable while the material to be cut is held in a vise. There are two types; one is a band saw on a pivot and the other is a reciprocating saw which is mounted on a pivitong flame.

6. *A shaper* is used to cut and square blocks of steel. The steel is stationary and the head with the cutting tool reciprocates back and forth cutting on the forward stroke.

7. *A planer* does the same thing as a shaper except that the cutting head tool is stationary and the steel reciprocates.

8. *A jig borer* is a precision drill press with a table that moves similarly to a miller. Its purpose is to locate holes with extreme accuracy.

9. *Welding equipment* is of two types, electric and gas. In electric welding the heat required for melting is derived from an electric arc. In gas, by burning a combination of oxygen and acetylene. It is primarily used for repairing or changing molds and requires the services of a skilled technician. Welding is a common method of joining metal and is used on machinery and equipment throughout the plant.

Hobbing

Hobbing is the cold forming of metal. The term is used in plastics to designate the cold displacement of one material by another, caused by high pressures. For example, if a piece of plastic is left on a mold and the clamping pressure of the machine forces the plastic into the steel, the plastic is said to have hobbed itself into the mold. The term is used in mold making for the process which takes a hardened steel replica (hob) of the plastic part and by means of high pressure forces it into a soft iron block. Iron is very ductile. It flows around the hob giving an identical, but reversed impression. This is much the same as forcing a coin into a piece of clay. The steel selected for the hob must have good strength, a tough core and take a good polish. The hobbing blank, usually of soft Swedish iron, is carefully polished and fitted into the hobbing ring. This ring is a large piece of steel to contain the hobbing blank as the pressure of the hobbing press is applied. The iron of the hobbing blank slowly yields. In many instances the hobbing has to be annealed to relieve stresses and pushed again.

The design of the hob, the steel selected, its hardening, the amount of draft, and the preparation of the hobbing blank are critical. If they are not correct, or the hobbing is not done properly the hob can crack.

Hobbing is a fast economical way to produce multiple cavities. All the cavities are the same size compared to each other and the hob. A high polish on the hob will be transferred to the cavity. Since the cavities are iron they must be carburized after they are machined to size. Figure 2-12 shows the hob for a plastic column. The hob is on the upper right. Beneath it is the molded plastic part which is identical in shape. The size of the molded parts will be smaller than the hob because the plastic shrinks in cooling. The upper left-hand piece is a hobbed cavity, and beneath it is the finished hardened, polished cavity, ground to size and ready to be put into the mold base (33, 34).

Casting

Recent technique for casting is so improved that this method is readily adaptable for injection molds (35, 36, 37, and 38). Any metal can be successfully cast, particularly with the Shaw process. In this patented process, a sample (or a plaster reversal) of the part in plastic, wood, metal, or other material is cast against a ceramic slurry. The slurry is fired and gives a reverse ceramic reproduction with a micrograin structure filled with small air gaps. The gaps act as vents so that the molten metal can achieve a good reproduction of the surface. The resultant cavity is not as dense as those produced by other methods, and there is the possibil-

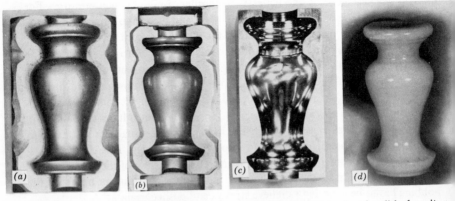

Figure 2-12 (a) Hardened metal hob; (b) hobbed cavity; (c) hardened polished cavity finished to size; and (d) molded part. (Robinson Plastics Corp.)

ity of small pits. The appropriate shrinkage factors for the slurry, metal, and molding must be calculated. A new slurry casting must be made for each new cavity. The major advantages of casting are its speed and cost. A cavity can be made in less than a week. The economics depend on the size and the nature of the part.

Electroforming

Electroplating is old. It was tried unsuccessfully for mold cavities and failed primarily because the stress in the tool caused by the plating, caused the tool to deform in the molding process. This has now been overcome. A master, sometimes called a mandrel, is the exact reverse of the cavity. It can be made of any nonporous substance which will stand mild heat and upon which a conductive metal can be deposited. The deposition is usually a silver nitrate-silver system or vacuum deposition. The mandrel acts as the cathode.

A nickel cobalt compound is deposited at approximately 0.0005 in./hr to a total depth of about 0.150 in. Behind this is plated pure electrolytic copper at the rate of 0.003 in./hr through the thickness required for the cavity. The copper is harder than mild steel being approximately 220 Brinell. The deposition takes approximately 2 weeks for the nickel and 2 weeks for the copper. This is not excessive in terms of cavity production by other means. The polish of the mandrel can be reproduced with finishes in the neighborhood of 5 μin., so that cavity polishing is eliminated. For areas impossible to reach with polishing equipment, or where polishing will effect delicate lettering, this is invaluable.

Slots as deep as 3 in. with a opening of 1/4 in. on top are formable. For projections a hardened steel blade can be set in and the electro forming be done around it, sealing it permanently into the cavity.

If a number of cavities are made at once on the same plate they can be electroformed very closely together with the runner. This can increase the number of cavities in a given area as no side wall is required for support between the cavities.

The regular parting lines can be made to match perfectly. Since there is no hardening there is no heat distortion. The mold surface is noncorrosive. The cavity has exceptionally high thermal conductivity. Very high precision is possible.

Electroforming is used for such qualities as reproducing of fine surfaces, very close tolerances, and shapes difficult to fabricate by other means such as gears.

The process adapts itself well where accuracy is required. It is used, for example, for gear cavities, slide rule cavities, and intricate electrical connectors. Making cavities for external threads on plastic pieces is difficult by conventional machining. An electrode can be turned in a lathe and can readily form two matched cavities. Even though the process is entirely different, it can be considered to have the advantages of cold hobbing in hardened steel.

The surface finish is determined by the electrode and current density and frequency. The usual finish is approximately 10 to 20 μin., and can easily be polished by conventional methods. The quality of the finish is determined by economics and the physical shape of the parts (39, 39a).

Duplicating

Duplicating is mechanical reproduction by cutting tools which are guided by a master, proportional in size to the desired finished parts. Duplicating is mostly used for large parts, as hobbing and casting will usually reproduce a smaller one more economically. Large automatic duplicators are basically powerful horizontal millers with hydraulically controlled feeds. Using feedback and electronic techniques maximum cutting speed is obtained. Such processes as producing mirror images are easy. They can be automatically run and tape controlled. Small duplicators are often used in making hobs or engraving small designs, letters, and numerals on cavities. A major disadvantage of duplicating is its comparatively poor surface finish (40, 41).

Erosion

A new and useful method for removing steel is by electrical discharge machining (EDM) (42, 43 and 43a). An electrode, usually made of carbon although it can be made of any conducting material, is made in the reverse shape of the part to be produced. The steel and electrode are immersed in a circulating solution, which serves to flush away the eroded material and cool the work. AC power is rectified and charged into a capacitor system. This discharge between the electrode and the cavity creates a spark which erodes the steel. The electrode is eroded about one-eighth as fast as the steel. Roughing electrodes are used to bring the cavity to its approximate shape and a finishing electrode brings it to size.

The process is accurate, produces good detail, can be used with hardened steel so that no heat distortion takes place, and can be used for cutting thin slots. By eroding on one plate the distance between cavities can be reduced. Cutting is relatively slow. The preparation of the electrodes and the operation of the equipment requires an excellent toolmaker. Spark erosion is widely used in changing and correcting hardened steel cavities.

Chemical removal of steel is slowly being accepted in the mold-making industry (44).

GATES

The gate is the connection between the runner and the molded part (45-47a). It must permit enough material to enter and fill the cavity, plus the extra amount required to prevent excess shrinkage. The literature is full of articles relating to the size, type, and location of gates, and their effect on the molding process and the physical properties of the molded part. This section considers gate types and sizes. The effect of gates are discussed in the section on molding theory.

Gates can be classified as large or restricted (pin pointed). Restricted gates are circular in cross section and for most materials do not exceed 0.060 in. in diameter. The more viscous materials may have restricted gates as large as 0.115 in. diameter. An example of a large gate, which is usually square or rectangular, is 1/4 in. wide by 3/16 in. high. They are used for molding heavy sections and where the restricted gates give a surface blemish problem.

Restricted Gates

The restricted gate is successful because the apparent viscosity of the plastic is a function of the shear rate. The faster it moves, the less viscous it becomes. As the material is forced through the small opening its velocity increases. The shear rate is directly related to the velocity. In addition, some of the kinetic energy is transformed into heat, raising the local gate area temperature. Once the gate is opened to the point where it loses this shear rate effect (viscosity improvement), a much larger opening is required to get any reasonable flow. This is why there is a jump in size from a restricted to a large gate.

The size of the restricted gate is small enough so that when the flow ceases, the gate material will cool, sealing off the hydraulic pressure of the runner system from the cavity very quickly. This means that the flow of plastic must continue, once started, until the cavity is filled. If the plastic freezes at the gate there will not be enough pressure in the runner system to force the plugged gate into the cavity and resume the flow of material. Therefore, in multicavity, restricted gate molds, the gates must be balanced so that all cavities fill together.

Since the gate freezes quickly when the flow of material stops, there is little chance of packing the gate area. Gate packing is a common cause of overstressing plastic causing environmental failure. Since it will also prevent adding material to compensate for shrinkage, pieces molded with restricted gates show measurably higher shrinkages than the same piece molded with a large gate. When properly designed restricted gating will minimize packing and sticking in the mold.

One of the obvious advantages of small gates is the ease of degating. In most instances the parts are acceptable if cleanly broken from the runner. It also lends itself to automatically shearing the gates during the ejection part of the cycle. This is helpful in automatic molding. It is important that restricted gates come from full round runners. This permits the hottest material to enter the cavity and not have the cold material, which would be dragged along the half runner, restricting the flow into the mold.

Before opening gates to correct molding conditions such as insufficient filling, sink marks, or bubbles, one should be sure the runner system is adequate. If there is insufficient transmission of hydraulic pressure due to the runner, the flow might slow down enough so that the gate freezes before complete filling. Also, low hydraulic pressure will slow down the rate of filling, reducing the velocity-viscosity effect. If a radical increase of injection pressure has little results on the piece, it is an indication of insufficient runner size.

When material from a restricted gate shoots into the mold it is in the form of a thin stream or jet. It shoots across the cavity, hits a wall, is cooled and folds over and over, creating a mass of cool material in the mold which is not always reheated and absorbed into the plastic. This can result in internal stresses, flow marks and surface blemishes. One way to overcome this is to mold against a tab or a pin or wall. Another way is to heat the mold section where the plastic hits. This can be checked by locally heating it with a torch. If effective a small cartridge heater under the gate area or a beyrillium heat sink may do the job. If not the gate can be opened to a large size. Control of the injection rate so that a slow initial fill builds back pressure in the cavity before the major high speed fill is initiated will minimize this effect.

When molding heavy sections it is necessary to add a considerable amount of material to the place the volume lost as the plastic decreases in temperature. Small gates freeze quickly preventing material from entering. Large gates stay open much longer. Unless the machine conditions are correctly set large gates will permit the material to flow backwards out of the cavities and into the runner.

Restricted gates have the benefit of better mixing. It is virtually impossible to mold a good variegated pattern (mottle) without going through a large gate. Dispersion or mixing nozzles on the machine use the principle of the restricted gate. Many small restrictions are placed on a plate inserted between the nozzle and the cylinder. This restricts the flow as the effective cross section is much smaller than in an unrestricted nozzle.

Location of Gates

Gates are also described by location, such as edge gated, back gated, submarine gated, tab gated, and nozzle gated. Figure 2-13 shows examples of various types of gates. A sprue gate feeds directly into the piece from the nozzle of the machine or a runner. It has the advantage of a short direct flow, with minimal pressure loss. Its disadvantages include the lack of a cold slug, the possibilities of sinking around the gate, the high stress concentration around the gate, and the need for gate removal. Most single cavity molds of any size are gated this way.

Edge gating is most common. It can be the large type or restricted. If the

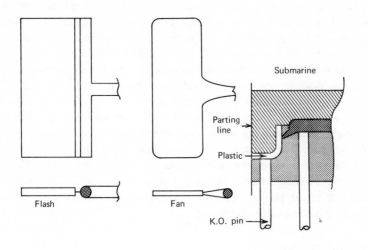

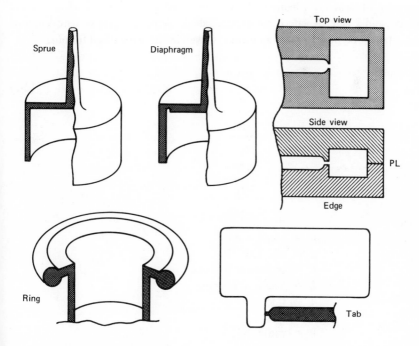

Figure 2-13 Different gating designs. (Robinson Plastics Corp.)

large gate is spread out, it is called a fan gate. If the gate is extended for a considerable length of the piece and connected by a thin section of plastic, it is called a flash gate (47a). Sometimes it is necessary to have the gate impinge upon a wall. This distributes the material more evenly and gives improved surface conditions. Walls are not always available. To overcome this a rectangular tab is milled into the piece and the gate is attached there. This is called a tab gate.

In gating into hollow tubes, flow consideration can require an even injection flow pattern. A single gate will not be sufficient. Four gates 90° from each other will often give four flow lines down the side of the piece which can be objectionable. To overcome this a diaphragm gate is used. The inside of the hole is filled with plastic directly from the sprue, and acts as a gate. It must be machined out later. A ring gate accomplishes the same thing from the outside.

A submarine gate is one that goes through the steel of the cavity. When the mold opens the plastic shears at the gate. A properly placed knockout pin, using the flexibility of the plastic, ejects the runner and pulls out the gate. This type gate is usually used in automatic molds.

Flow Through the Gates

Figure 2-14 shows a schematic drawing of a cavity block with half of a round runner and a gate. A gate is sometimes called a land; the gate or land length is important. The flow through an orifice is given in the generalized formula:

$$Q = \frac{K\Delta P}{\mu} \qquad (2\text{-}2)$$

$$\text{Circle } K = \frac{\pi R^4}{8L}$$

$$\text{Slit } K = \frac{Wh^3}{12L}$$

where
Q = flow rate
ΔP = pressure drop
μ = viscosity
K = geometric constant
L = length of opening
h = thickness
R = radius
W = width.

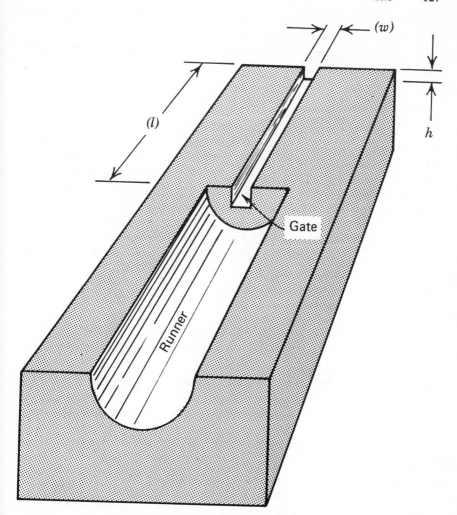

Figure 2-14 Schematic cross section of one-half runner and gate: (*h*) land depth, (*w*) land width, and (*l*) land length.

Increasing the size of the opening (R^4 or Wh^3) has a major effect in the flow rate. However, this is counteracted by reducing the velocity which will raise effectively the viscosity. A point will be reached where small increments of R will no longer increase the flow, but because of viscosity changes, will decrease the flow. This is the effect described previously comparing large and small gates. With very large increments of R^4 the effect of viscosity will be less important. Reducing the land length (L) will increase Q and since it does not change the cross sectional area, it will increase the velocity. This will have the added effect of reducing the viscosity. It also increases ΔP. For these reasons land lengths are kept to a minimum. Experience has shown that the shorter landed gates remain open for a longer period.

Ejection Systems

After the part is molded it must be ejected from the mold. Parts are ejected by K.O. (knockout) pins, K.O. sleeves, stripper rings, or stripper plates, either singly or in combination. The considerations for ejection are similar to those for parting lines. Additionally the quality of the molded piece is affected. We will not consider undercuts at this point. (An undercut is an interference by the mold to delay or prevent mechanical ejection of the plastic parts.)

The geometry of the parts and the plastic material are the major factors in selecting the knockout system. Most parts eject readily with a 1° per side taper. They can be ejected with smaller tapers but this should be done only if required. A high polish is not always required for easy ejection. The direction of the polish is more important. Draw polishing (stoning and polishing in the direction of ejection rather than randomly) is important in difficult cases. With some materials, such as the olefins and nylon, fine sand blasting may help. Normally a moderately polished surface will not present ejection problems.

The cross-sectional area of the knockout pins or rings must be large enough so that the knockout does not damage the piece. Aside from the obvious, when the knockout pins go through the molded parts, serious stressing can be caused in the knockout area. Birefringence studies of transparent molded parts show this clearly. It is desirable, although not always possible, to use large diameter knockout pins.

The molder and moldmaker usually are able to predict knockout problems. Often aesthetic considerations or lack of room for knockouts prevent using the number and size of pins desired. It is poor practise to build a mold under these conditions. If satisfactory ejection cannot be designed initially on paper, it will be difficult, if not impossible, to install same on the completed mold. Sometimes the parts have to be redesigned or made in two parts and joined later. Unless a mold can eject consistently, an even cycle cannot be maintained, and this part cannot be produced on a production basis.

The location of the cooling channel directly affects the location of the knockout pins. While it is possible, using "0" rings, to send a knockout pin through a cooling channel, it is certainly not desirable. Cooling channels should be designed away from knockout locations.

Figure 2-15 shows a pen barrel 6 in. long, 5/8 in. in diameter and closed at one end. The mold has sixteen cavities and was gated in the back. The core is slender and unsupported. An initial attempt was made to mold one cavity with the gate in the center (Figure 2-15a). This was unsuccessful because the pressure

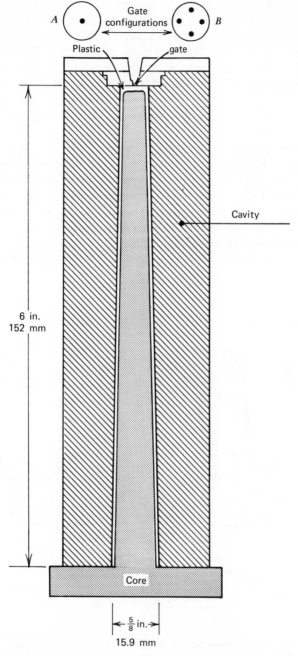

Figure 2-15 Effect of gating on molded pen barrel.

of the molten plastic bent the core. When the mold opened the steel straightened out and scratched the plastic piece on the edge of the cavity as the platen moved back. This was overcome by gating each cavity in four spots (Figure 2-15b). This equalized the pressure around the core to the extent that it prevented noticeable bending and ejection problems. As an experiment all 16 cavities were singly gated at a later date. At the first full shot the mold froze closed. As the machine opened it stripped the threads of the clamps holding the mold to the platen. While this is an extreme case, it illuminates the problems that can be caused by the flexing of steel.

Figure 2-16 illustrates a condition known as entrapped material also caused by the elasticity of steel and its deformation under molding pressure. The item molded is a cup with an innercircular compartment. When the mold opens the molded piece remains on the core. The annular plastic ring A is entrapped between the core C and the annular steel ring B. This can be very difficult if not impossible to eject, particularly if A is thin. The proper way to build the mold is to have C retract in relation to the rest of the core before ejection.

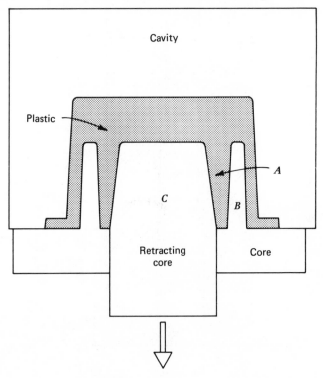

Figure 2-16 Use of retractable core to prevent ejection difficulties caused by entrapped material.

In molding deep parts it is sometimes necessary to vent the core. If not, the vacuum will prevent ejection. Figure 2-17 shows such a system. A vent pin which seats on a taper is held in place by a light spring. When the material injects, the

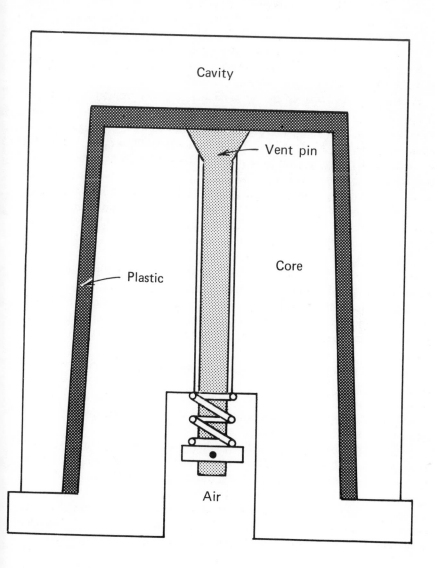

Figure 2-17   Use of venting pin to break vacuum on core. Pin is held closed by spring and sealed by the pressure of the molded material. (Robinson Plastics Corp.)

force of the material will seal the pin. As the part ejects, the pin will move up, against the spring pressure, venting the core. Air pressure can be used to help eject the part.

A cause of part sticking related to the elasticity of steel is a condition called packing. This results from excessive injection pressures on the material. Aside from elastically deforming the steel, sticking is caused by deeper penetration of the plastic into the pores of the mold. Packing is most common in multicavity molds where the filling rate is not balanced. Some cavities fill first, their gate seals off and the full injection pressure is concentrated on the remaining cavities. Balanced filling will cure this condition.

Once a mold is completed and the mold does not eject properly, a number of things can be done. The mold surfaces should be carefully inspected to remove burrs, scratches, pits, and undercuts. Possibilities of core and cavities shifting or misalignment should be eliminated. The part should be filled evenly. Increasing the curing time should be tried to be sure that the part is fully hardened. Packing should be reduced to a minimum. The plastic should be inspected to see if it has been degraded in the processing. Lubricants should be tried. The most common is a silicone aerosol spray. Different formulations are available for different materials (48). The stearates of the heavy metals are also excellent lubricants. Graphite impregnation of the steel is reported to greatly reduce ejection problems (48a).

If all the above fail, attention must be paid to mechanically changing the mold. Tapers should be evaluated. Locations for additional knockouts should be considered. Increasing the diameter of existing knockouts might help. The molder and moldmaker have many techniques to overcome ejection problems. Notwithstanding there are a few parts which cannot be ejected the way they are designed.

From a properly designed and built mold, parts should eject cleanly. If they do not, mechanical modifications should be made during the initial run to achieve this end. Lubricants, manual operation, unusual cycles, air, and such should be considered only as temporary solutions pending permanent mechanical solution of the ejection problem.

Figure 2-18 shows a stripper plate ejection system. The cores are stationary. Around them are hardened stripper bushings which are mounted in the stripper plate or plates. There is clearance in the lower part of the stripper bushings to minimize wear. The knockout bars cause the stripper plate to move in relation to the core pin leaving the part either on the plate or free to fall off. Figure 2-19 shows a stripper ring ejection system. The stripper bushings or rings are attached to the ejector plate and act as ejector pins as they move in relation to the core pin.

Sometimes it is necessary to eject in two stages. Figure 2-20 shows a double acting K.O. system. When the K.O. bar is activated the main K.O. plate, K.O.-1,

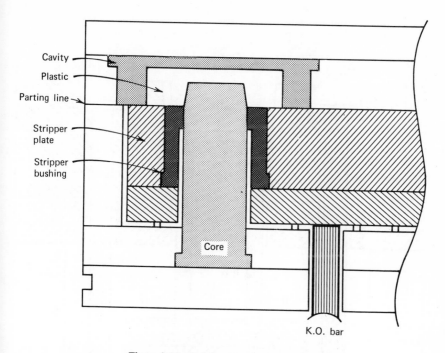

Figure 2-18 Stripper plate ejection system.

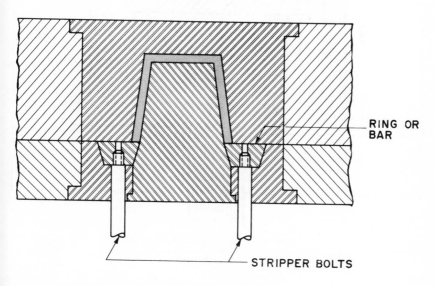

Figure 2-19 Stripper ring ejection system (Drawing reproduced with the permission of the National Tool, Die and Precision Machinery Association).

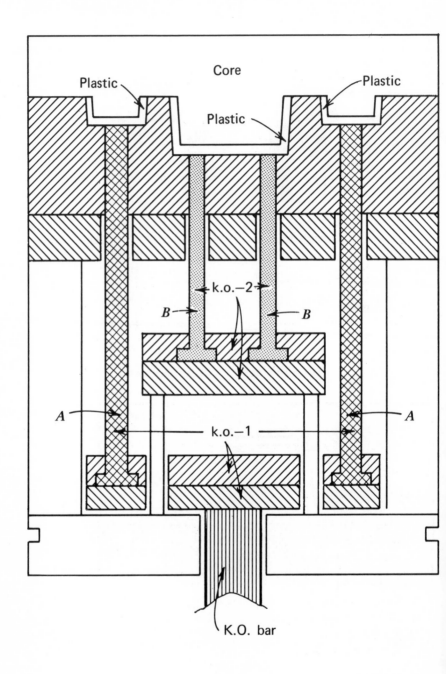

Figure 2-20 Double acting ejection system.

moves forward ejecting by means of ejector pins *A*. As the ejector plate continues to move it hits the second ejector plate, K.O.-2, which actuates ejector pins *B*. A separate push-back system is needed for returning the second K.O. plate.

Double ejection motion of a stripper plate followed by knockout pins is obtainable by mechanical methods utilizing a modified return pin (48b).

If hydraulic ejection is not available from the machine, a simple substitute can be used. The knockout plate should be extended beyond both sides of the mold. Two hydraulic cylinders should be attached to the extended plate, one on each side, with their rods attached to the moving platen. Activating the cylinders will be, in effect, equivalent to the hydraulic knockout action built into the molding machine.

In trying out new molds, parts often stick in cavities and cores. They should be carefully removed by a competent tool maker or molding expert using wood and soft copper. Occasionally the only way to remove the plastic without disassembling the mold is to heat a piece of metal with an undercut, such as a screw or part of a saw blade, and carefully insert it hot into the plastic. When it is cool, pulling the metal part may also remove the stuck plastic. On deep parts, blowing a jet of compressed air across the top, after the part has cooled slightly, may affect a release. Many times short shots do not eject because they do not mold over the knockout pins. If this tends to be a problem, extra pins for the short shot can be installed.

Cam Actions

Cam acting molds are common in the injection field. The cams are primarily used for molding parts with undercuts and holes whose cores, if left in place, would prevent ejection in the direction of the machine movement (49, 50). They are also used for engineering consideration relating to ejection, venting, and gate locations.

Cam action molds are more costly to build, more expensive to maintain, and require closer supervision during molding. It is sometimes possible to eliminate the cams. It might be possible to drill a hole or machine a slot after molding. Frequently this operation can be done at the machine without additional cost. Even if the work has to be done away from the machine it might be less expensive than a cammed mold. This is particularly true of short runs.

Assembling of two or more parts can avoid cam actions. Again, this might be done at the machine. In some instances assembly can provide other benefits. Internal threads are easily formed by molding in two parts and assembling. Internal threads can be made by molding over removable inserts. They are unscrewed on the bench. Extra inserts are made so that the cycle can continue normally. This eliminates tapping plastic which should best be avoided. Tapping creates sharp edges which are focal points for breakage. This is called the "notch" effect. A metallic hexagonal nut can be readily located and held firmly by assembling it between two plastic parts, eliminating the necessity of an insert or tapping.

Figure 2-21 shows a sliding cam actuated by an air or hydraulic cylinder. It could also be run electrically through gear chains and sprockets although such mechanisms are mainly used in rotating unscrewing devices. This type of cam action is independent of the mold motion and may be actuated at any time in the cycle. This contrasts with mechanical cam action (Figures 2-22 and 2-23). The cam motions can be caused by the machine travel or the knockout system. Regardless of the activating mechanism, cams are held in place by some mechanical lock to prevent the injection pressure from moving the cam and putting excess stress on the cam bar. Sometimes it is necessary to have an ejector pin move in the path vacated by the cam. Unless the ejector pins are returned before the cam moves they will collide causing damage. To prevent this the

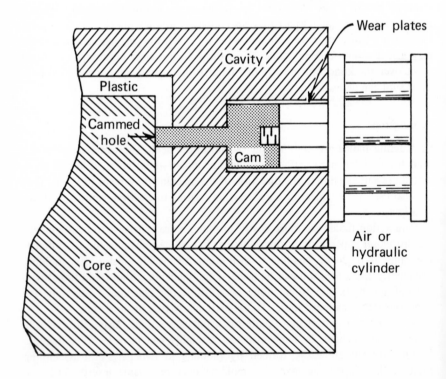

Figure 2-21 Externally operated cam for molding hole in plastic. (Locking device not shown.)

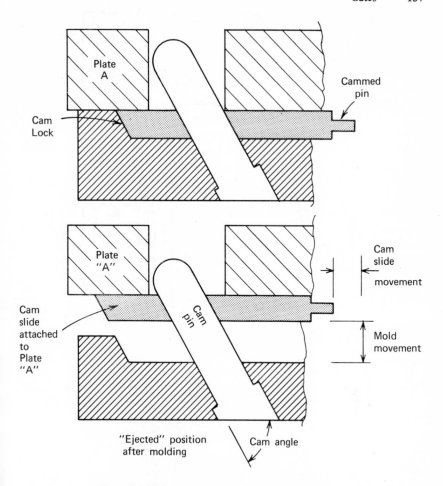

Figure 2-22 Mechanical cam action using cam pins.

knockout plate should activate a switch which is tied into the machine circuit so that the cam cannot move unless the knockout pins are in place. If cams move so that the cam pin does not meet the hole, damage may result. It is sometimes necessary to use switches activated by the cam to prevent the machine closing if this happens.

Springs can be used to release undercuts. Figure 2-24 shows a cross section of an oval container with undercuts on its side. As the mold opens up the springs force the cavities up the angled slide releasing the parts. When the mold closes it forces the cavity down and compresses the springs. A mold of this design, producing a part 10 in. long, 4 in. wide and 4 in. deep, molder over 500,000 parts without any maintenance on the cam mechanism.

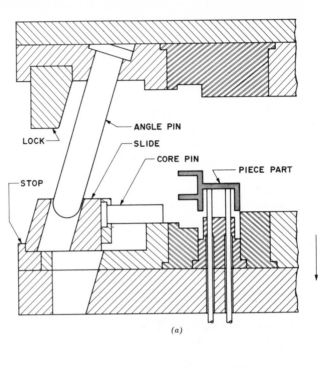

(a)

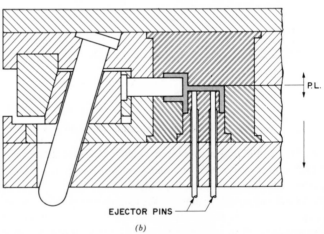

(b)

Figure 2-23 (a) Mechanical cam in ejection position (b) Mechanical cam in molding position. (Drawing reproduced with the permission of the National Tool Die and Precision Machinery Association).

138

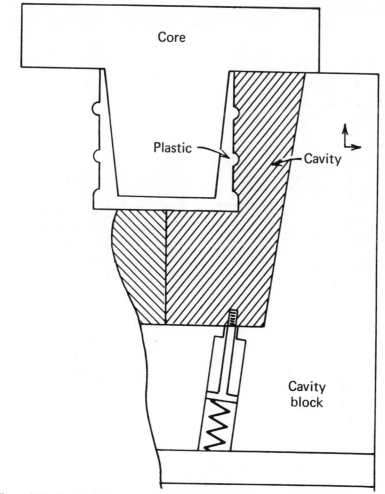

Figure 2-24 Cavity slides up and out in the cavity block propelled by springs, until the undercuts are cleared. The plastic part stays on the core and is ejected normally.

Figure 2-25 shows a method of using K.O. pins for molding undercuts. They are called jiggler pins. The left-hand side shows a pin in the molding position. As the knockout plate advances the pin is pushed to the inside of the plastic part because of its configuration. It is anchored into the K.O. plate loosely so that it can slide laterally.

There are numerous other cam action systems. These actions are limited only by the ingenuity of the mold designer. Even though they are more expensive to build than conventional molds, one should not hesitate to use them if it

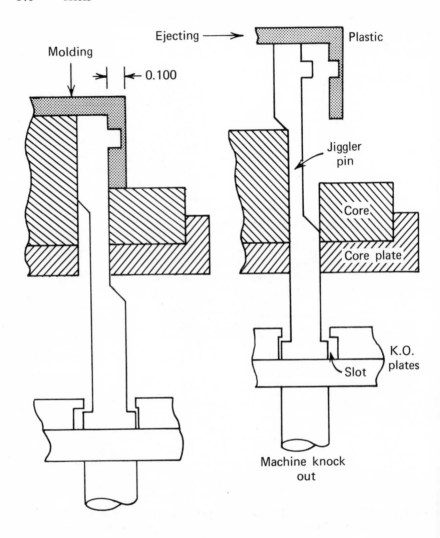

Figure 2-25 "Jiggler" pins for molding undercots.

simplifies engineering of the part or the mold.

Internal or external threads can be molded automatically, or by using inserts in the mold which are removed with the piece and unscrewed on the bench. Internal threads can sometimes be made by molding one half of the thread (without cams or unscrewing actions) and letting the screw cut its own threads on the unthreaded side of the hole (53a). Extra inserts are used to save machine time. Automatic unscrewing driving mechanisms include racks and pinions, gears, sprockets, electric motors, and hydraulic motors. Automatic unscrewing molds are considerably more expensive to build and maintain (51-54a). Thought should be given to using inserts in the mold (55-55a), adding metal inserts in a post molding operation, tapping the hole in the plastic, and using self-tapping screws.

VENTING

When the hot plastic is injected into the mold it displaces air. In a well-built properly clamped mold without vents, the injecting material may compress the air to such an extent as to prevent proper molding. The heat of compression in a nonvented mold might actually ignite the mixture of air and plastic, leaving a characteristic burn mark. Inadequate venting may also decrease the filling rate of the cavity, and in some instances prevent the cavity from completely filling.

The location and size of vents are still governed mainly by experience. Vents are first put in the obvious places before testing the mold. Additional vents are added as required.

Vents are usually ground on the parting line. The size of the vents depends on the nature of the material and the size of the cavity. A typical vent would be 0.001 in. deep and 1 in. wide. After the vent extended for a half inch the depth would be increased to 0.005 in.

Clearance between knockout pins and their holes provide venting. Sometimes special pins are placed in the mold just for venting purposes (Figure 2-17).

The gate location has a lot to do with venting and one is often restricted in gating because of the inability to completely vent the mold.

One way to decide whether venting is a problem in mold filling is to put enough brass shim washers over each leader pin so that the mold remains open approximately 0.005 in. If this does not cure the filling problem, its cause is not in venting. The ends of runners should always be vented.

Some parts cannot be adequately vented without vents so large that they cause objectionable molded edges. This can be overcome by machining a small tab in the desired location and connecting it to the molded part with a restricted gate.

Weld lines result from two hot masses of plastic joined in the part. Sometimes the only way to eliminate it is by molding a similar tab to remove the cold welded material.

Venting technique has been extended to the point where molds are evacuated by a vacuum to promote rapid filling and minimum degradation (56, 57).

TOLERANCE

A tolerance is the total permissible variation of size, form, or location. Tolerance should relate to the dimensional and performance characteristics which have to be met and kept for the effective and acceptable performance of the part. Tolerances should be specified for no other reasons.

Tolerance terms and dimensions are the method for describing the size of the finished piece. A definition of some common terms follow:

A *dimension* is a numerical value indicated on drawings with lines, symbols, and notes to define the geometrical characteristics of an object.

A *basic dimension* is a theoretical value used to describe exact size, shape, or location of a feature. It is a standard from which permissible variations are established by tolerances on other dimensions or notes.

A *form dimension* is one that specifies a feature which cannot be properly defined by dimensions of size or location, such as the angle of a thread or the angle of the frustrum of a cone.

Location dimension is one that specifies the position or distance relationship of one feature of an object with respect to another.

Size dimension is a specific value of a diameter, width, length, or other geometrical characteristics directly related to the size of an object.

A *reference dimension* is one without tolerances used for informational purposes and does not govern manufacturing or inspection operations.

Datum, parts, lines, and surfaces are features assumed to be exact for purposes of reference or computation and from which the location of the features may be established.

Size is a designation of magnitude. When a value is assigned to a dimension, it is thereafter referred to as the size of that dimension. The *actual size* is the measured size. *Basic size* is the one from which limits of size are derived by means of allowances and tolerances. The *design size* is that size from which the limits of size are derived by only the application of tolerances. If there is no allowance the design size and basic size are the same. The *nominal sizes* is the one used for general identification. For example, a rod may be referred to as a - 1/4 - in. rod but the actual dimension is 0.248. In this instance 1/4 in. is the nominal size.

An *allowance* is the prescribed difference between the maximum material condition (MMC) of mating parts. It is the minimum clearance (positive allowance) or maximum interference (negative allowance) between such parts.

Unilateral tolerance is a tolerance in which the variation is permitted only in one direction from the design size, (i.e. +0.000, −0.010). In *bilateral tolerance* the variation is permitted in both directions from the design size, (i.e., + 0.002, −0.006). In *limit dimensioning methods* the largest and smallest permissible dimensions are indicated. The tolerance is the difference between the two.

Fit is a term used to signify the range of tightness which may result from the application of a specific combination of allowances and tolerances in the design of mating parts.

The definitions above are taken from military standards (58) which is recommended for additional information.

In making drawings, fundamental rules, although obvious, are unfortunately not always followed. Enough dimensions should be shown so that the intended sizes and shapes can be determined without calculating or assuming any distances. At no time should one scale a print for dimensions. Each dimension should be stated clearly and be interpreted in only one way.

The purpose of a drawing is to elucidate. Not every one who uses a drawing is skilled in its interpretation. In complicated parts considerable thought and effort are required by even a competent draftsman. There is always a possibility of missing a vital spatial relationship. Since the plastic parts have drafts, parting lines, and perhaps undercuts, any method to simplify the understanding of a drawing is desirable. A simple isometric drawing, a perspective sketch or a brief description, can be helpful. A Polaroid® photograph of the part, enclosed with the drawing, is greatly appreciated.

When a part is molded initially all dimensions are inspected. For production purposes a limited number are checked. It is assumed that if the mold is not changed and certain specified dimensions are correct the remainder will be acceptable. It is important to determine which dimensions are to be inspected, the conditions of inspection, and the method of inspection.

The tolerances in mold making are relatively easy to define. Figure 2-26 shows the tolerance range of various types of machining. The relative cost of turning and finishing a steel casting is shown in Table 2-3.

It is evident that the higher the tolerance, the more expensive the mold. Costs rise geometrically. It is thus well to review mold dimensions and specify maximum tolerances.

Table 2-3 Tolerance versus cost for machining a steel casting

Operation	Tolerance (±) in.	mm	Cost ($)
Rough turning	0.015	0.381	1
Rough turning	0.005	0.127	2
Semifine turning	0.003	0.0762	3
Fine turning	0.001	0.0254	5
Grinding	0.0005	0.0127	10
Honing	0.00025	0.00635	16

Tolerances

Range of sizes		.00015	.0002	.0003	.0005	.0008	.0012	.002	.003	.005
From	To and incl.									
.000	.599	.00015	.0002	.0003	.0005	.0008	.0012	.002	.003	.005
.600	.999	.00015	.00025	.0004	.0006	.001	.0015	.0025	.004	.006
1.000	1.499	.0002	.0003	.0005	.0008	.0012	.002	.003	.005	.008
1.500	2.799	.00025	.0004	.0006	.001	.0015	.0025	.004	.006	.010
2.800	4.499	.0003	.0005	.0008	.0012	.002	.003	.005	.008	.012
4.500	7.799	.0004	.0006	.001	.0015	.0025	.004	.006	.010	.015
7.800	13.599	.0005	.0008	.0012	.002	.003	.005	.008	.012	.020
13.600	20.999	.0006	.001	.0015	.0025	.004	.006	.010	.015	.025

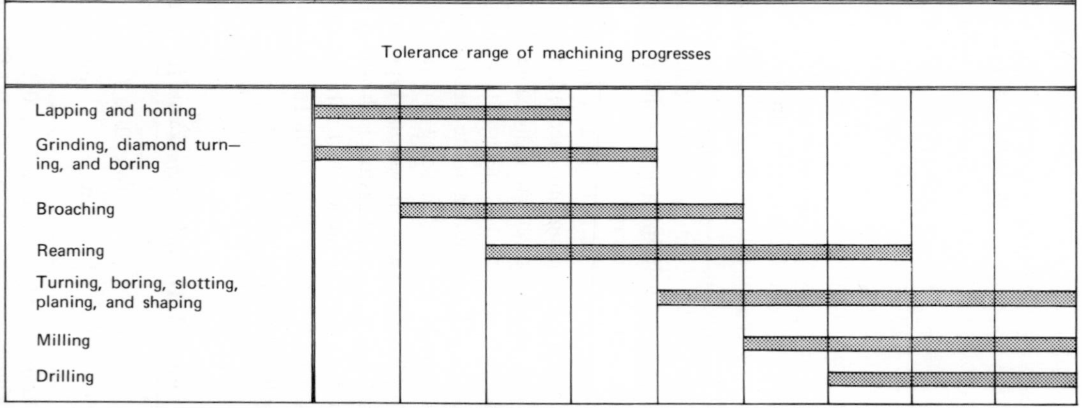

Tolerance range of machining progresses

Lapping and honing

Grinding, diamond turning, and boring

Broaching

Reaming

Turning, boring, slotting, planing, and shaping

Milling

Drilling

Figure 2-26 Machining tolerances for steel (MIL-STD-8B; 11-16-59).

Specifying mold dimensions is a function of the moldmaker and molder. One must estimate the shrinkage of the molded material, table 3-11. To some extent this is controlled by the molding conditions and the type and condition of the equipment. It is best to design to that side of the tolerance which, if violated, would require removal of metal from the mold. This is usually easy as contrasted to adding metal to the mold which at its worst might be impossible and at its best difficult. If hole or pin locations may be troublesome, they can be made by using larger pins in the mold turned down to size. They can later be removed, and new ones turned off center and relocated at a minimum cost. Consideration of molding and dimensional problems are mandatory and must be done in the design stage of the mold.

Plastic Tolerances

Tolerances are important because they prescribe the limits for the part. Ideally, these limits should be controlled by function and aesthetics. Many times they are controlled by ignorance or the desire to "play it safe." These practises are costly in dollars, time, and customer satisfaction. Excessive tolerances are indicative of extremely poor engineering and should warn the molder and mold maker. They should go further into the application and see the functional and aesthetic requirements. Engineering drawings customer supplied notwithstanding, the part must function correctly. Ultimately the molder and the moldmaker will be faced with producing such a part. Even if they are paid for their extra work, the loss of their time and of customer satisfaction is serious. If the parts function properly and are aesthetically pleasing, they will almost invariably be accepted, print tolerances notwithstanding (59-61).

Suggested tolerances on molded parts in different plastics have been published by the Society of the Plastics Industry, Inc. (SPI). Figure 2-27 is for polystyrene. For a discussion of plastic tolerances see p. 275. If finer tolerances are required, the rejection rate will rise significantly and the part should be priced accordingly (62).

It is not too difficult to build a mold to tolerance. The tolerance limits of the mold-making techniques are accurately known. They can be used to produce a mold meeting the specifications of the drawing.

Selecting the right cavity size is essentially an educated guess. As the material cools, the volume shrinks. The shrinkage range which is specific for each material is given in inches per inch. These are what normally would be expected in molding. For example, polycarbonate has an expected shrinkage of 0.006 in./in. If a finished dimension of a part was to be 6.000 in., the cavity size would have to be 6 × 0.006 in. plus the shrinkage of the 0.036 in. (0.002) or 6.038 in. (63) (see p. 284).

The reproducibility of the molded part or the tolerance to which a part can be held will depend, among other things, on the shrinkage variation from shot to

STANDARDS AND PRACTICES OF PLASTICS CUSTOM MOLDERS	Engineering and Technical Standards POLYSTYRENE

NOTE: The Commercial values shown below represent common production tolerances at the most economical level. The Fine values represent closer tolerances that can be held but at a greater cost.

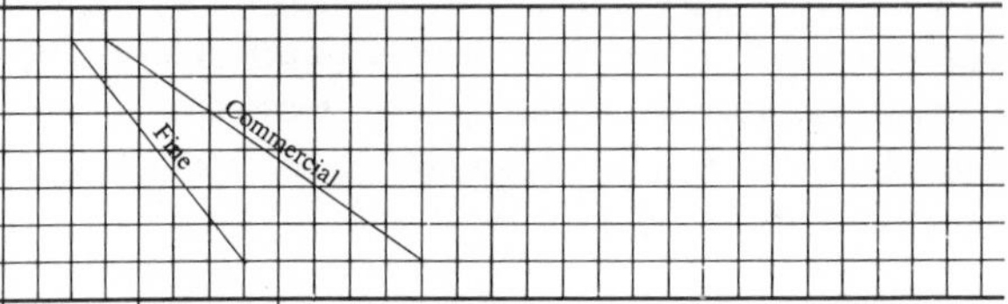

Drawing Code	Dimensions (Inches)	Plus or Minus in Thousands of an Inch	
A = Diameter (see Note #1) B = Depth (see Note #3) C = Height (see Note #3)	0.000 / 0.500 / 1.000 / 2.000 / 3.000 / 4.000 / 5.000 / 6.000	1 2 3 4 5 6 7 8 9 10 11 12 13 14 15 16 17 18 19 20 21 22 23 24 25 26 27 28	
		Comm. ±	Fine ±
6.000 to 12.000 for each additional inch add (inches)		.004	.002
D=Bottom Wall (see Note #3)		.0055	.003
E = Side Wall (see Note #4)		.007	.0035
F = Hole Size Diameter (see Note #1)	0.000 to 0.125	.002	.001
	0.125 to 0.250	.002	.001
	0.250 to 0.500	.002	.0015
	0.500 & Over	.0035	.002
G = Hole Size Depth (see Note#5)	0.000 to 0.250	.0035	.002
	0.250 to 0.500	.004	.002
	0.500 to 1.000	.005	.003
Draft Allowance per side (see Note #5)		1½°	½°
Flatness (see Note #4)	0.000 to 3.000	.007	.004
	3.000 to 6.000	.013	.005
Thread Size (class)	Internal	1	2
	External	1	2
Concentricity (see Note #4)	(T.I.R.)	.010	.008
Fillets, Ribs, Corners (see Note #6)		.015	.010
Surface Finish	(see Note #7)		
Color Stability	(see Note #7)		

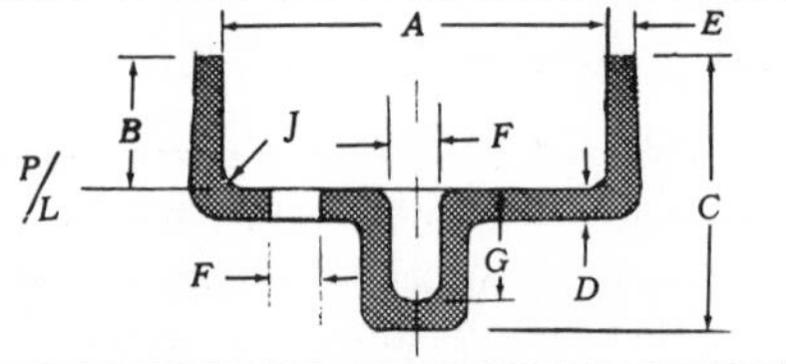

REFERENCE NOTES

1 – These tolerances do not include allowance for aging characteristics of material.

2 – Tolerances based on ⅛″ wall section.

3 – Parting line must be taken into consideration.

4 – Part design should maintain a wall thickness as nearly constant as possible. Complete uniformity in this dimension is impossible to achieve.

5 – Care must be taken that the ratio of the depth of a cored hole to its diameter does not reach a point that will result in excessive pin damage.

6 – These values should be increased whenever compatible with desired design and good molding technique.

7 – Customer-Molder understanding necessary prior to tooling.

Figure 2-27 Molding tolerances for polystyrene (The Society of the Plastics Industry Inc.).

shot. In polycarbonate, for instance, which has very high reproducibility, a close molding tolerance can be held. In this instance 6.000 in. plus or minus 0.010 in. is realistic. In polypropylene 6.000 in. plus or minus 0.025 in. is a realistic tolerance. The reproductability or the amount of tolerance to which a material can be held is an inherent function of the molecular structure, molecular size and molecular weight distribution of the polymer, which affects the volume-dimensional change in cooling from a hot fluid to a cold solid.

Given a cavity of a specific size there are many things which will cause variation in dimensions. The gate size, the gate location, the number of gates, and the size of the runners will change dimensions. By the time the mold is ready for production they will have been optimized. The other variables are inherent to the molding process. These include mold temperature, clamping pressure, material temperature, injection speed, injection time, injection pressure, overall cycle, and variations in materials. We shall subsequently see during the discussions of molding theory that these variations can only be predicted in general terms. The selection of proper mold dimensions is one of the more difficult phases of mold design.

Since these molding variables are compounded by uneven cycles caused by the operator, the tolerances should be ideally set so that these variations would still mold acceptable parts. When this is not the case, some method of control is required. The molded parts shrink for about 1 day with the major shrinkage occurring within the first hour of molding. Gages can be developed or measurements designed so that inspection information obtained soon after molding can be translated into acceptable or rejectable parts. This is one reason why close tolerances are particularly expensive in injection molding.

Sometimes shrink fixtures are used to maintain dimensions. In other instances rapid chilling of the molded parts, usually in water, will help maintain sizes. Consideration should be given to salvaging out of tolerance parts. Sometimes a simple machining or a post-forming operation would be economical.

WATER AND MOLD TEMPERATURE CONTROL

Reduced to its ultimate fundamental, injection molding is really the controlled heating and cooling of plastic. This section is concerned with the cooling or removing of heat from the plasticized material. Cooling of the molding machines is also briefly discussed.

Water is the universally used heat transfer medium in its temperature range. The equipment manufacturers will recommend chemical systems for use outside of that range which are compatible with their product. Although we generally consider that cooling water is readily available, ecologists warn that the present supply and distribution of potable water is not sufficient for the world's or even the community's predictable needs. Consequently many municipalities require treatment and reuse of waste water particularly in such applications as molding.

Furthermore, an efficient molding system uses a great deal of water so that recirculation often makes good economic sense. Recirculating can also yield additional benefits in water quality and productivity.

A typical 16-oz molding machine uses approximately 300 gal or 40 ft³/hr. Assuming 6000 hr of operation a year, it will use 240,000 ft³. Based on a typical water and sewerage cost of $2.00/1000 cubic feet, cooling the machine alone (not including the mold) costs $480. If the water was recirculated using a cooling tower the cost would be less than $100. With water rates spiraling (New York City's is $4.00/1000 ft³), economics and good (industrial) citizenship, therefore, dictate recirculating systems for molding plants.

Basic Water Cooling Theory

The standard unit for measuring heat is the British Thermal Unit (Btu); 1 Btu is the amount of heat necessary to change the temperature of 1 lb of water 1°F. The unit of refrigeration is a commercial ton of refrigeration which is defined as the removal of heat at the rate of 200 Btu/min, or 12,000 Btu/hr. The standard ton of refrigeration is 288,000 Btu, or the amount of Btus removed by a commercial ton of refrigeration in 1 day. It should be noted that the standard ton has the dimensions of heat while a commercial ton has the dimensions of heat divided by time. Table 2-4 gives some useful data for cooling calculations.

The branch of physics devoted to measuring the thermodynamic properties of moist air is known as psychrometry. Cooling towers, which are used in most plants, operate on psychrometric principles.

When a liquid changes its state to a gas (evaporation), it requires energy to loosen the molecular bonds. It receives this energy in the form of heat, taking the heat from the surrounding substances, thus lowering the temperature. This heat is called the *latent heat of vaporization* or the latent heat. Thus when water evaporates, it will remove heat from the surrounding water and air, lowering their temperature. This is the theoretical basis for cooling tower action. Conversely when gas is condensed, as in refrigerating systems, energy in the form of heat is liberated and the surrounding substance (the condenser) heats up.

Sensible heat is the heat required to change the temperature of the air or water without changing its state. It is a measure of the internal kinetic energy and changes with the absolute temperature of the body. The *latent heat* is potential energy and shows itself in changes of the physical state of the body (evaporation, condensation) and is not accompanied by any changes of temperature. The latent heat of evaporation of water (Btu/lb) at 60°F is 1059.3, at 80°F is 1048.1 and at 100°F is 1036.7.

The sources of cooling water are

1. Rivers and lakes
2. Cooling ponds and wells
3. Spray ponds
4. Government water supplies
5. Cooling or evaporative towers
6. Mechanical refrigeration

Table 2-4   Useful data for cooling calculations

One standard ton 288,000 Btu
One commercial ton
   removes   200 Btu/min
              12,000 Btu/hr
              288,000 Btu/day

   cools   20 lb of water $10°F/min$
             200 lb of water $1°F/min$
               6 gal $4°F/min$
              12 gal $2°F/min$
              24 gal $1 F/min$

Cooling 1000 gal of water $25°F$ requires removal of 208,000 Btu

1 gal	=	231 in.^{3}
1 gal	=	0.1337 ft^3
1 gal/min	=	8.0208 ft^3/hr
1 ft^3	=	7.42 gal
1 ft^3/min	=	0.1247 gal/sec
1 ft^3 of water	=	62.43 lb
1 lb of water	=	0.016 ft^3
1 lb of water	=	0.1198 gal
1 gal of water	=	8.345 lb
1 Btu	=	778.2 ft lb
1 hp	=	0.7068 Btu/sec
1 kw	=	0.9478 Btu/sec
1 Btu/sec	=	1.055 kw

Rivers and Lakes.   These sources are rarely available to molding plants. They probably will require filtering, settling, or chemical treatment. A constant check is required because of the possibility of upstream pollution and other changes. The use of this water usually requires the permission of local authorities.

Cooling Ponds and Wells.   Aside from the sources just mentioned the cheapest method of cooling water is by means of a cooling pond. This is a pond of water where hot water enters on one side and cold water is removed on the other side. While it is inexpensive it is also inefficient and often unsatisfactory for molding plants. It has a low heat transfer rate and needs a large size. The cooling depends on air temperature, relative humidity, wind speed, and heat gain from the sun. During summer in middle northern latitudes the minimum water temperature expected would be about $86°$.

If the pond is used for cooling and hydraulic system of machines and refrigerating equipment, the size of a pond to handle 1000 gal/min would be approximately 60,000 ft^2. The pond has to be protected from children, algea, bacteria, and such other contaminators that may fall on an open body of water. Needless to say the water must be chemically treated and filtered. Well water can be used if the supply is adequate and the temperature low enough.

Spray Ponds   An improvement over cooling ponds is a spray pond, which is a body of water over which a spray system is installed. The nozzles are approximately 8 ft above the water level. By presenting a much larger area for evaporization more cooling will occur. The disadvantages are those of an open water system, dependence on wind velocity, plumbing the spray, relatively high water losses, and possibility of the spray being a public nuisance.

Cooling Towers.   An important source of water cooling in the molding plant is the cooling tower. There are two types, one that depends on prevailing winds (atmospheric) and the other that depends on a forced air feed by fans (mechanical) which is used in molding plants. When the fan is placed on the bottom of the tower, it is called a forced draft tower and when on top it is called an induced draft tower.

This type of tower has the advantage of a small ground area per unit cooling. It can be located anywhere including inside loft buildings; it requires low pumping head; it can control the temperature of the water more closely than of any of the previously mentioned systems; and it is economical in terms of water consumption. Its main disadvantages are that it has a high operating cost primarily for the air circulating fan. Its maintenance costs are comparatively high; it is subject to mechanical failure, and can present problems in removing the hot, moisture-filled exhaust air.

The hot water is pumped into the top of the cooling tower from where it falls by gravity over a grid, cooling itself during its fall. It is collected in the bottom or basin from which it is pumped out as the cooled water for the processing system. The cooling is done primarily by evaporization. It is estimated that between 10 and 20% of the cooling is done by convection heat transfer between the cool air and warm water. The water drops on slats to break its fall. The amount of cooling depends on the length of time the drop of water is exposed to evaporization. Since falling bodies are accelerated by gravity the slats effectively decelerate the droplets. Additionally it breaks the "Thomson effect," which hinders evaporization because of a difference in electrical potential at different points on the sphere of water, caused by surface tension effects. The slats act to break up the flow of water and form new drops, thus giving a larger surface area and breaking this thermal barrier at the surface. They are made either of redwood or polyethylene.

The *cooling range* of a tower is the difference in temperature between the water intake and outlet. The *heat* load of a tower is the number of Btu per minute removed by the tower. The *circulation rate* is the amount of water going through the tower per unit time. The heat load is the product of the circulation and the range. The amount of heat removed by the tower can be increased by the increasing of the area over which the water flows, increasing the amount of air flowing per unit time (velocity), raising the inlet water temperature and reducing the humidity of the air.

The principle involved in a cooling tower is the removal of sensible heat because of the difference in air and water temperature and the removal of latent heat by the change of state from water to water vapor of a small amount of the fluid. It takes approximately 1000 Btu to evaporate 1 lb of water. This is the amount of heat required to cool 100 lb of water 10°. Therefore for each 10° of cooling approximately 1% of the water circulated must be evaporated. This water plus the spray loss (tenths of a per cent) has to be replaced and is called the *makeup* rate.

All water contains dissolved chemicals. These are brought into the tower during makeup while pure distilled water departs during evaporization. Therefore in time the chemical composition of the water will cause scale and other problems in the molds and coolers. Water must be treated and filtered. The large velocity of air will cause dust and other particles to collect in the tower requiring cleaning. In northern climates during winter there is a possibility of the tower icing. This is overcome by simply bypassing some of the hot inlet water into the basin.

Mechanical Refrigeration

The mechanical refrigerator, like the cooling tower, removes heat from the system. The practical difference is that the heat removed by the cooling tower cools the water to temperatures which are controlled by atmospheric conditions. The heat removed in a mechanical refrigerator is removed at a temperature based on the design of the machine. For example, heat can be removed to cool water to 40°F while having the removed heat dissipated at temperatures of 85 to 100° Mold temperature control is a basic requirement for accuracy and economy in molding. The temperature of the mold should be determined by the optimum molding conditions, not by the available temperature of the cooling water. For this reason mechanical refrigeration is required in a molding plant.

The simplest mechanical refrigeration system would be a closed box containing an open dish of a low boiling chemical (refrigerant), a fan, and a vent. If the refrigerant were liquid ammonia it would evaporate at minus 28°F at atmospheric pressure. One pound would absorb 589.3 Btu in evaporating (latent heat of evaporation). If the temperature surrounding the box is above minus 28°F, the heat absorbed by the ammonia in evaporating would come from the surrounding media. This is the same theory as water evaporating in a water tower. This method is not practical for many reasons. If the refrigerant were ammonia the odor and toxicity would make it unusable. The cost of the refrigerant would make it uneconomical. The technique would not allow for adequate temperature control. To overcome this the refrigerant is evaporated and mechanically condensed in a closed system and continually reused.

If the pressure is increased, the temperature at which the ammonia (refrigerant) will evaporate and condense is raised. At 47.6 psig, the temperature of vaporization is 32°F, at 92.9 psig it is 60°, and at 197.2 psig is 100°F. Thus by changing the system's pressure the temperatures at which the change from liquid to vapor or vapor to liquid can be controlled. This means, in effect, that heat can be removed from the system at any convenient cooling temperature without depending on the atmospheric temperature as is required in a cooling tower. Machines designed for air-cooled operation are about 15% less efficient. For units up to 5 tons the convenience is worth the extra cost. These are mainly on portable units. For larger sizes water cooling is preferred.

In mechanical refrigerators the liquid refrigerant is charged into the receiver. The compressor is started. When the system operates, the gas is compressed. The condenser section removes heat from the vapor (either by air or water cooling) causing it to condense into liquid which is stored in the receiver, still under pressure. In large units the heat from this section can be used to help heat the plant. The gas is now expanded by an automatically controlled throttle valve which reduces the pressure so that the liquid refrigerant will evaporate. It does so in the evaporator absorbing heat from the surrounding environment. This causes the cooling. It then goes to the compressor where it is compressed again and the cycle repeated. In water cooled mechanical refrigerators, 2.4 gal/min per ton of refrigeration of cooling water are required for each 10° of cooling. This is a good approximation for preliminary estimates, as many cooling towers used for injection molding operate at about that range during the summer months. The most common refrigerant is Freon F-12® (dichlorodifluromethane).

A mechanical refrigerator or a chiller is designed for outgoing water of a specific temperature. The most common is 50°F. Any deviations from this will change the efficiency of the unit. For example, water leaving at 60° would raise the efficiency to 120%, 40° would reduce it to 80%, and 30° to 60%. Therefore a 10-ton chiller designed for 50° would deliver 12 tons at 60° and 6 tons of refrigeration at 30°.

The refrigerated water can be supplied either from a central system or portable coolers for each machine. The main advantages of the central system is low initial cost and freeing floor space around the molding machine. It has a number of disadvantages. It provides water at one temperature requiring elaborate mixing systems for mold temperature control. It is relatively inflexible in terms of capacity. At the initial installation one has to guess the cooling requirement for the future. Individual chillers can be bought as required. Molding can be scheduled for their maximum utilization. For most custom molding plants portable chillers are preferable (64).

Cooling Requirements

Two convenient equations for determining cooling loads follow:

$$\text{ton} = \frac{\text{Btu}}{(12,000)} \ (\text{hr}) \tag{2-3}$$

$$\text{gal/min} = \frac{(\text{tons})\,(12,000)}{(\Delta t)\,(60)\,(8.3)} = \frac{(24)\,(\text{tons})}{\Delta t} \tag{2-4}$$

Molding machines are usually cooled with tower water. If the temperature-humidity conditions are too severe mechanical refrigeration is added as required. Tower water is much more economical and should be used when possible.

The heat exchanger of a 16-oz 400-ton hydraulic clamp molding machine with 45 connected horsepower was instrumented to determine the heat removed. This averaged 25,000 Btu/hr. It was relatively independent of the cycle time and ambient temperature. This is not indicative of the total heat loss as the machine radiates a considerable amount of heat energy. The cooler would require approximately two tons of refrigeration (Eq. 2-3). A convenient approximation of machine cooling requirement is 1 ton/20 connected hp. If a water tower with a $6°$ approach was used the machine would require approximately 8 gal/min of water (Eq. 2-4).

Mold cooling requirements are relatively easy to estimate. The enthalpy of plastics is a measure of their heat content and given in Btu per pound. Graphs are available of enthalpy versus temperature. In crystalline material they include the heat of fusion. By subtracting the enthalpy at room temperature from the enthalpy of the material at the cylinder temperature the number of Btu to be removed is obtained. Table 2-5 shows this for some thermoplastics.

These figures do not actually describe what occurs. When the molded part is removed from the mold, a considerable amount of heat is still in the part, which cools in the air. There is a significant radiation loss from the mold itself. The author molded a plaque of general purpose styrene 7 in. $\times$ 3 in. $\times$ 0.150. The heat loss through the mold water cooling was measured. The molded part was put in a calorimeter and the residual heat measured. The enthalpy graph of this particular material showed a heat content of 140 Btu/lb between molding and room temperature. There was 38 Btu/lb removed by the cooling water, 57 Btu/lb remained in the molded part and the balance of 45 Btu/lb was radiated from the mold. About 80 lb/hr were molded. The amount of refrigeration required is 38 $\times$ 80 or 3040 Btu/hr. This is approximately one fourth of a ton. Using the enthalpy from the graph, 140 Btu/lb, one would expect that a ton of refrigeration would be needed. Practically the amount would vary with the geometry and thickness of the part and the size of the mold. Using 50 to 75% of the figures in Table 2-5 will give a good approximation of the required cooling (65).

Table 2-5   Enthalpy difference or heat content (Btu/lb) of some thermoplastics between approximate molding temperatures and room temperature

Polystyrene	155
Acetate	180
Acetal	180
Polypropylene	210
Low density polyethelyne	260
Nylon 6	270
High density polyethelyne	310
Nylon 6/6	340

Mold "Heating" Units

The function of the fluid circulation through the mold is to control the rate of heat transfer, hence the cooling rate at the plastic. Elevated temperatures are used when slow cooling is required.

The temperature of the cooling medium will depend on the molding requirements. When the cooling medium is above room temperature, requiring the addition of heat, it is commonly called a mold heater, even though it is in effect cooling the mold. A mold heater is, in essence, a tank with a motor driven centrifugal pump recirculating a fixed amount of fluid from the tank through the mold. Adding heat to the fluid is done by electrical resistance heaters. When the molding conditions are on the border line of adding or removing heat from the circulating fluid, a coil attached to a cooling medium is inserted in the tank. A temperature sensing element activates the heating or cooling circuit for the temperature at which it is set. For temperatures above the boiling point of water nonaqueous fluids are used. It is essential to keep the fluid clean as rust scale and other contaminants seriously reduce the efficiency of the heat removal. When operated at high temperatures extreme care must be used in the selection and maintenance of the connecting hoses. A ruptured connector may result in serious burns (66).

Mold temperature control units will accurately control mold temperature. A unit attached to a mold running a half pound shot of general purpose styrene at 83 cycles/hr was instrumented. The inlet and outlet mold temperatures were read every 6 sec. They were charted with the heat on-off and water on-off controls of the unit, the cycle time of the machine and the mold temperature. In a typical case with the cooling water and heating elements each cycling alternately every three shots, the inlet water temperature varied from 88.5 to 90.3°F, and the outlet water temperature from 89.7 to 91.5°F. The temperature difference between the outlet and inlet water was plotted for each 6-sec reading. It varied from 2.0 to 2.9°F with a mean of .9°F. The cycle of the curve followed that of the units heating cooling cycle. The mold temperature as read by a dial thermometer and pyrometer showed no readable change. By changing the molding conditions slightly so that cooling water was used all the time in the mold temperature unit the outlet temperature was 77.5° and the inlet temperature varied between 73.4 and 73.6°F. The variation between the difference of the inlet and outlet water never exceeded 0.2°F. The dial thermometer and pyrometer in the mold showed no change. In the first instance 4590 Btu/hr were removed and in the latter 4740. These figures show that commercial units can produce accurate and consistent mold temperature control which is required for proper molding.

Heat Transfer

The three methods for exchanging heat are radiation, convection, and conduction. We are primarily concerned with convection and conduction. In the coolers for the molding machines the heat from the hot oil is exchanged into the tube walls, and from the tube walls into the circulating water. In the mold the heat from the plastic is transferred to the cavities and cores, which in turn transfers the heat to the mold temperature circulating medium. Some of the factors which affect the rate and amount of heat transfer are material of the container, size and shape of the container, rate of flow of both materials, temperature, viscosity, specific heat, thermal conductivity, density, and surface conditions of both sides of the container. The mathematics of these processes have not been quantitatively completed. Notwithstanding, a qualitative discussion of some of the factors affecting heat transfer is valuable.

The rate of heat removal equals the overall heat transfer coefficient times the area of exposed surface, times the difference in temperature between the two fluids (plastic and water).

$$Q = UA\,\Delta t \qquad (2\text{-}5)$$

Q = rate of heat removal (Btu/hr)
U = overall heat transfer coefficient Btu/(hr) (ft^2) (°F)   (2-6)
A = area (ft^2)
Δt = difference in temperature of the two fluids (°F).

This equation shows, as one would expect, that the lower the temperature of the cooling medium, the faster the heat removal. The rate could also be increased by increasing the material temperature. This would be self-defeating because the higher removal rate would not compensate for the additional amount of heat to be removed, thus lengthening the cycle. The lower limit of the cooling temperature is the molding condition. Molds that are too cold may not fill, may develop surface blemishes and lower some physical properties.

The area of the cooling surface is limited by the geometry of the mold. Table 2-6 shows the effect of different size cooling channels. Using a 3/8-in. pipe instead of a 1/8-in. one will increase the cooling rate by a factor of 1.8. Large cooling channels are one of the easiest ways to reduce cycle time. Unfortunately this is often overlooked in mold design.

By use of electrical analogies the overall heat transfer coefficient is described as:

$$\frac{1}{U} = \frac{1}{h_1} + \frac{1}{h_2} + \frac{1}{h_3} ...+ \frac{X}{k}$$

X = thickness of wall (ft)
k = thermal conductivity of wall
   Btu/(hr)(ft^2) ÷ °F/ft = Btu/(hr)(°F)(ft)
h = individual heat transfer coefficients Btu/(hr)(ft^2)(°F)

This equation leads to some very interesting conclusions. It is important to notice that the heat transfer rate is controlled by the coefficient at the point of maximum resistance. For example ignoring X/k, if there are only two coefficients, $h_1 = 20$ and $h_2 = 1000$, U would equal 19.6. Suppose h_2 were changed from 1000 to 500, then U would equal 19.23. Therefore, even though one coefficient were changed by 50% it would only change the total coefficient by 2%. The film coefficient for water in the cooling system is approximately 1500. However, if scales, sludge and dirt enter the system this can drop to as low as 200 introducing serious resistance to heat transfer and probable increase in mold cycles. Therefore, clean circulating water and cooling channels are very important (67).

For molds the X/K factor is important. The rate of heat removal will vary directly with the thermal conductivity of the mold material. Therefore if the K for beryllium is 70 and steel 24 the beryllium will remove or add heat to the plastic approximately three times as fast as steel. This is an important factor in mold material selection (27). It is also obvious that the closer the cooling channel is to the plastic (a minimum X) the higher the rate of heat removal. Cooling channel location should be designed so that there is even cooling of the mold surface. Since heat removal varies directly with the distance between the cooling channel and the mold, equally spaced circles from the cooling channel will be roughly the same temperature. They can be drawn on a mold layout and a good indication of the temperature profile of the cavity or core obtained (68).

It is also evident that the highest heat transfer coefficient will occur when the

Table 2-6 Physical characteristics of drilled mold cooling holes related to mold temperature control

Nominal Pipe Size	Tap Drill Used In Mold	(in.) I.D.	I.D. Area (in²)	I.D. Circumf. (in)	Surface Area ft² per ft of length	Ratio of Cooling Area to 1/8 in pipe	Capacity at 1 Ft/sec	
							gal/min	lb/hr
1/8	5/16	0.3125	0.0767	0.982	0.0818	1.0	0.24	120
1/4	7/16	0.4375	0.1503	1.374	0.115	1.4	0.47	234
3/8	9/16	0.5625	0.2485	1.767	0.147	1.8	0.77	387
1/2	11/16	0.6875	0.3712	2.160	0.180	2.2	1.15	580

157

cooling channels are directly in the cavity or core. If put in the surrounding mold base the controlling coefficient will be between the cavity and the mold base. Interface losses are very significant and should be avoided if possible. This is one of the advantages of EDMing cavities in one block.

The ability of the plastic to change temperature is a factor in cooling the part. This is called the thermal diffusivity and is defined as the thermal conductivity divided by the product of the specific heat and the density. There is nothing the molder can do to change this as it is an inherent property of the material. It will explain why some materials cool more readily than others (69).

It stands to reason that the velocity of the cooling media would affect the heat transfer rate. From dimensional analysis and experimental work the heat transfer coefficient is affected by, among other things, the Reynolds number (p. 214). There is a velocity factor in the Reynolds number. When the number is below 2100 there is laminar flow and the heat transfer coefficient (inside a tube) varies as the 1/3 power of the velocity. Above 2100, turbulent flow, it varies as the 0.8 power of the velocity. The probable reason for this is that the turbulent flow provides better mixing and the metal-water interface is broken more often. In turbulent flow, for example, if water flowing through the tube had a film coefficient of heat transfer of 300 $Btu/(h_r)(ft^2)(^{\circ}F)$ at a given velocity, and its velocity were doubled, the new film coefficient would be 300 $(2^{0.8})$ or 522. This means that increasing the velocity of the cooling fluid will increase the rate of heat transfer. This should not be overlooked. It may be necessary to increase the pumping capacity of the mold temperature control unit. The amount of fluid circulating can be easily determined with a water meter. With this information and the cooling channel dimensions the Reynolds number can be calculated.

When mold cooling seems inadequate the first things to be done are to clean the cooling system and mold channels, increase the velocity of the cooling medium and lower its temperature. If these methods do not work, consideration must be given to enlarging or adding to the cooling channels. Similarly, the heat exchanger for cooling the oil in the molding machine should be kept clean and periodically examined. If the machine overheats and the water temperature is normal, the cooler should be cleaned. If that does not help there probably is a malfunction in the hydraulic system permitting oil to bypass and generate heat.

The heat exchangers for cooling oil consists of tubes through which the cooling water flows, and a shell through which the hot oil, to be cooled, flows. They are mainly single pass exchanges; that is, the liquids flow in one direction. They can be connected in two ways. In parallel, the hot oil and the cold water enter at the same end so that the cooler oil and heated water will emerge at the other end. In counter flow the hot oil will enter at one end and the cold water will enter at the opposite end. Molding machines are connected counter flow, since it will remove approximately 10% more heat from the oil.

This has relevance in mold cooling, as a mold is a heat exchanger with the hot plastic as a heat source and water as the cooling medium. Most molds have the hottest section at the sprue, primarily because of the radiation effect of the outside of the mold base. Attaching cold water to the outside of the mold (analogous to counter flow) will remove more heat than putting the cooling water directly into the sprue section first. Properly designed molds will permit the plastics engineer to adjust mold temperature accordingly.

Mold Temperature Control

The two reasons for providing for good mold temperature control are (a) economic and (b) part quality. The temperature control system includes the cooling fluid, means for its circulation, method of temperature control, and cooling channels in the mold. Its purpose is to remove heat from the plastic part at a controlled rate. The goal is the removal of heat as rapidly as possible so that the part can be removed from the mold in a condition which will result in acceptable pieces. This cannot be done without a good temperature control system which will permit the molding conditions to establish the mold temperature rather than the adequacy of the equipment.

An incorrect and inconsistent mold temperature will create serious difficulties. We shall assume an adequate system and discuss some of the problems caused by incorrect mold temperature. It is not always possible or necessary to predict the best temperature for a given mold and material. With thermostatically controlled temperatures, trial and error is not difficult. In many instances there will be several different temperatures maintained for different parts of the mold.

To cool any given part a specific number of BTU will have to be removed. Equation 2-5 shows that the greater the temperature difference between the plastic and the cooling fluid the higher the heat removal rate. Therefore, a lower mold temperature will permit the part to be removed more quickly.

Material or Mold Temperature Too High.   If the temperature is too high, cycle time increases. This is not the only disadvantage. Some of the others are described below.

Because the plastic will remain fluid longer with higher temperatures, there is a greater tendency for the material to flash. The gate will remain open longer permitting more material to be packed into the cavity. This excessive packing particularly at the gate may lead to an over stressed part and difficulties in ejection. The packed material either deforms the steel and/or adheres to it more strongly than usual, increasing the possibility of sticking.

Since the part will probably be softer on ejection there is a greater chance for the ejectors to force their way into the material. Furthermore, an overheated mold may (*a*) not allow the sprue to solidify resulting in it sticking, and (*b*) cause excessive shrinkages, sink marks, and burning, although these are less frequent. If one side of the mold overheats the thermal expansion of the mold may cause one side or plate to seize. This can pull the mold off the platen.

When hot thermoplastic hits a cold mold, part of the polymer freezes against the wall and some is stretched or "oriented" in the direction of flow. This orientation is greatest near the surface of the mold and decreases towards the center of the part. The amount will depend in some measure upon the mold temperature. It can be helpful or troublesome. Orientation leads to molded-in stress, and highly oriented parts which are usually not desirable. The part has a higher tensile strength in the direction of flow because it has more carbon-carbon linkages than perpendicular to flow, where the main cohesive forces are the weaker electrostatic bonds. This is discussed in detail in Chapter 3.

Molds Too Cold. Molds that are too cold will also cause considerable difficulty. If the mold is filled the plastic most distant from the gate will have a lot less material than that close to the gate, which will be warm enough to receive some packing during injection. This will leave an uneven density distribution causing severe molded-in stresses.

Moreover, the material might freeze before it fills the cavity giving incomplete shots. There may not be enough material forced into the mold before the material stops flowing. This can be caused either by premature freezing of the plastic in the mold or premature freezing of the gate. In these instances there will not be enough material in the part, which can mean voids, excessive shrinkage, and severe reduction of the mechanical properties of the part, the insufficiency of material will emphasize sink marks.

The cold mold surface may also cause tails, tears, and surface smears which are caused by the skidding of cold material that does not remelt into the polymer, as well as emphasizing the negative aspect of weld lines, which are the junction of two fronts of molten polymer.

A characteristic of cold molds are tiny ripples on the surface, which are in the form of wavefront perpendicular to the direction of flow.

These are just some of the major problems caused by incorrect mold temperature setting. These are all compounded if the cooling system lacks the capacity or instrumentation to maintain a consistent temperature. Varying mold temperatures are even worse than incorrect temperatures and make quality molding impossible.

Mold Cooling Channels.   Before discussing mold cooling channels, we reemphasize that materials for cavities and cores must be evaluated in terms of their thermal properties. This concept extends even to the use of a combination of metals. It might be desirable, for example, to hollow out the core of a tumbler mold and put in a beryllium insert to improve its thermal conductivity.

In evaluating the economics of a mold, it is rare that extra money spent for cooling is not quickly recouped. Also, the cooling system (excluding the mold) must have enough cooling capacity and pumps able to deliver water at high velocities. Dropping the mold temperature and increasing the velocity of the water to at least Reynolds 3000 will increase the rate of heat removal, and the amount of heat removed from the mold. Attention should be paid to the way the cooling water is attached to the mold cooling channels. For example, if there are a number of cooling channels drilled parallel, they should be hooked up in parallel from the water supply and not in series. The outlets should be at least as large as the inlets.

Figure 2-28 shows an adequate cooling channel pattern for a mold plate. The pattern on the top is inferior because the left-hand side of the plate will be at a lower temperate than the right-hand side. Depending on what is molded, a significant difference in the plastic part might be observed. The lower design shows the same holes but with a different baffle arrangement. The cooling could be attached either in parallel or counter flow. By rearranging the baffles and plugs any type of local mold temperature control could be achieved.

Figure 2-29 shows what can be done to increase the cooling of a core. The core is hollowed out. A stainless steel insert is turned to the same taper with a spiral groove running from top to bottom. The cooling fluid enters from the bottom, fills up the spiral, and drops down a hole drilled in the center to the out port. This will give tremendously superior temperature control when compared to conventional bubblers. While this, too, is more expensive to build it is the most economical mold design.

Cores, cavities, and pins are sometimes cooled with bubblers. Figure 2-30 shows the series cooling of a pin. The water enters the first pin, flows over a blade into the second pin, and so on. This is a very poor design. The first pin would be much much cooler than the last. Even a slight bit of corrosion or dirt can clog or seriously restrict the flow. Because of the high resistance there will be a minimum velocity. The proper way for such cooling is shown in Figure 2-31. Two holes are drilled in the plate, one under the other. The metal between them is tapped and a bubbler pipe of a noncorrosive metal is screwed in. The water flow is equal in all channels. The clogging of one pin will not effect the others. Maximum velocity is obtainable. The "in" channel must be directed to the top channel so that the tube will fill up with water before it overflows. If the water entered the outside and the in-channel had a larger capacity the pin might never fill with water. The references on mold cooling contain many ingenious ways of increasing the cooling capacity of molds (70-75A).

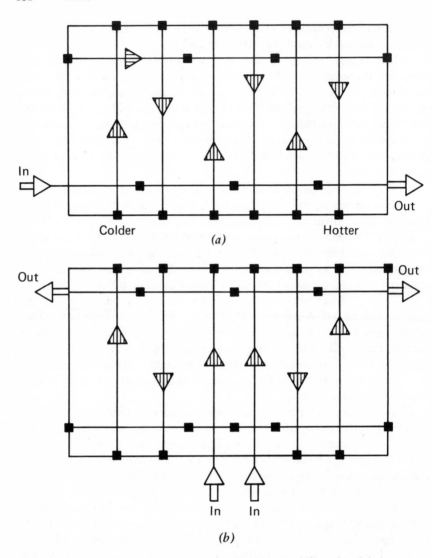

Figure 2-28 Cooling channels: (*a*) adequate design, and (*b*) preferred design.

Automation

Many times one hears the expression, "automatic machines" when referring to automatic molding. This is a misnomer; all machines today are automatic. What makes automatic molding automatic is the mold. There are a number of requirements for automatic molding:

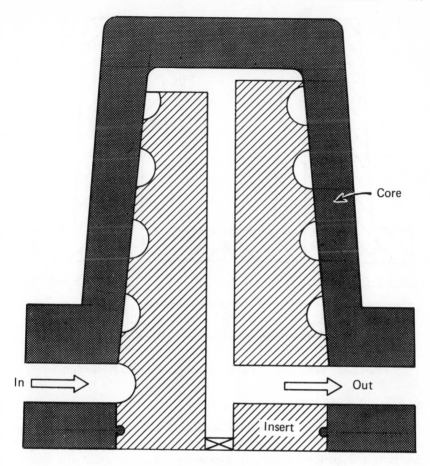

In

Out

Core

Insert

Figure 2-29   Core with insert to provide maximum cooling.

1. The machine must be capable of consistent, repetitive action.

2. The mold must clear itself automatically. This means that all the parts have to be ejected using a runnerless mold, or that the gate and parts have to be ejected in a conventional manner and fall free of the mold. There usually is some method for assisting in the removal of the pieces and gates, in the form of a wiper mechanism or an air blast. Some systems weigh the shot after ejection and stop the machine or sound an alarm if the shot is too light.

3. Indicating that a part is stuck is necessary. All machines used automatically must have a low pressure closing system which prevents the machine from closing under full pressure if there is any obstruction between the dies. The

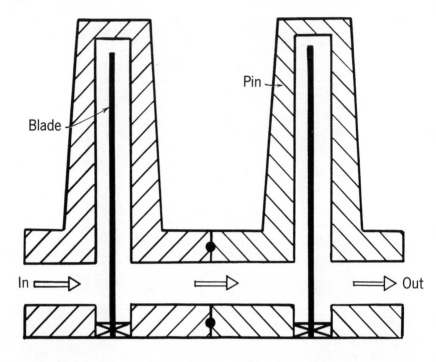

Figure 2-30   Series cooling of pins, poor design. (Robinson Plastics Corp.)

machine is shut off and/or an alarm is sounded.

Automatic molding, which usually produces better parts more rapidly, does not necessarily eliminate the operator. Many times an operator is present to pack the parts and perform secondary operations. However, some new systems automate this function as well. Usually in automatic molding an experienced person attends several machines. Unless the powder feed and part removal are automated he will take care of them.

Automation means replacing high labor costs with high capital costs for molds and parts handling equipment. Excellent machinery, good molds, trained employees and managerial skill are all required. When the quantity of a part permits, it is a very satisfactory and economical operation (79, 80, and 81).

Mold Maintenance

Mold maintenance can be done either on an emergency basis or between runs. Management policy will determine whether a mold should be fixed during the run. It will depend on how badly the customer needs the part, the length of time for a temporary repair, the time for a permanent repair, the availability of

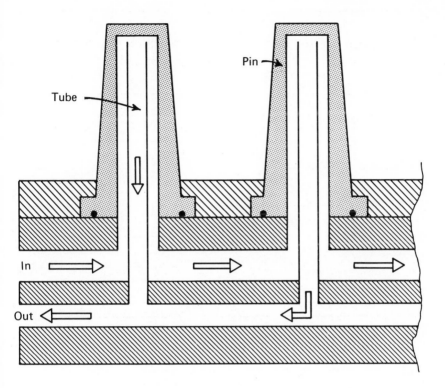

Figure 2-31   Parallel cooling of pins, correct design. (Robinson Plastics Corp.)

manpower, and the difficulty of setting up the machine again for the same job as well as economic factors.

Mold repairs should be carefully evaluated. Since mold repairs may effect the customer and the economies of the job, they should be approved by management rather than left to the discretion of operating personnel.

Mold records are essential. Figure 2-32 shows the front of a typical mold record card. The other side of the card is left blank for specialized information.

MOLD RECORD CARD   ROBINSON PLASTICS CORP.

DATE   FITS MACH. #   MOLD

BUILT BY ~   MOLD #

H		L		W		WT
#	CAVITY	GM.	PC/LB	LBS/M	MATERIAL	

SHOT   GM   GATE   GM

100 LBS. ~   PCS.   1000 PCS. ~   LBS.

SPRUE BUSHING   DME#   O   RADIUS

A   K   L

DESCRIPTION OF ACTION

PARTS INDENTIFICATION

Figure 2-32 Typical mold record card.

Figure 2-33 shows a "mold out" form. This is filled out every time a mold is removed after a molding run. It is countersigned by the molding foreman and head of the tool shop. This permits them to discuss what has to be done before

Mold Out Form

Mold _____ Machine_____ Date _____
Shift _____ Removed by_____ Foreman_____
Authority for Removal _____

Removal Procedure—Initial Each Operation as Performed:

1.___ Save last four shots prior to shutting down. If mold becomes inoperative, submit shots taken last for inspection purposes. Final shots are to be complete to include runners and sprue.

2.___ Close mold, remove all water lines, and blow out all lines until no evidence . of moisture is present.

3.___ Open mold. Allow to come to room temperature if cooled. Dry completely if moisture is present and use antirust on all interior surfaces—no grease.

4.___ Close mold, coat with antirust, place dust caps or plugs into all water lines, and remove mold from press.

5.___ Put special parts and fixtures into box and label. List parts on the reverse of this sheet. Parts stored in location _____ .

6.___ Note on reverse side of this sheet any descrepancies not listed on Molding Control Sheet attachment

7.___ Return Molding Requirements Form and Inspection Sheet to file.

8.___ Note on reverse side all repairs that have to be made on the mold.

Figure 2-33   Mold out reminder form. (Robinson Plastics Corp.)

REPAIR TICKET   ROBINSON PLASTICS CORP.

MOLD #   (IN MACH. #   )

MACHINE #   OTHER

DATE   SHIFT   DATE   SHIFT

DATE   SHIFT   DATE   SHIFT

TROUBLE & REPAIRS DONE	TIME	BY	COST

PRODUCTION HOURS LOST

PARTS USED	COST

CAUSE

RECOMMENDATIONS:

Figure 2-34 Mold repair ticket.

the mold is run again. A copy of this form and samples of the molded shot are given to the tool room when the maintenance is scheduled. They fill out the "repair ticket," Figure 2-34, which is evaluated and filed with the original copy of the "mold out" form. Periodic inspection of these files will reveal molds causing trouble and suggest remedies. It is interesting to observe how often the same repair is required on a mold.

CHECK LIST FOR MOLDS

The mold design is the final stage in the engineering and planning of a plastic part. A good mold is prerequisite to the economic production of quality parts. If the mold is able to consistently produce parts that are dimensionally within tolerance, flash free, and mechanically and esthetically acceptable, the need for supervision and quality control is reduced greatly though not eliminated. The nature of operating molding machines is such that it is almost impossible to have every operator, on every shift, able enough to make adjustments, sort parts, and do whatever else is required to nurse a poor mold. It is for this reason that mold design and its implications are extremely important. "Savings" in mold engineering and mold costs are illusory.

The purpose of this check list, which has been successfully used and expanded for many years, is to force systematically, consideration of the important parts of mold design. Certainly an experienced mold engineer will automatically consider all of these without the formalized need of going through them one by one. In practice the author has found that the more experienced the designer, the happier he is about a list of this kind. It has been an indispensable training device and is of significant help to the sales engineer. His involvement with the customer or user must be directed to eliminate as many difficulties as possible before the mold is designed.

Brief comment has been made on certain questions. Other questions are covered more fully in the text. Still others are amplified in the bibliographical material.

Piece Parts

1. Is this piece part drawing approved? Often there is more than one drawing made for the part. Drawings might be changed. Samples might have been made and the drawing not changed. The final drawing might have been verbally modified by the parties concerned. It is always desirable to have the piece part drawing from which the mold is being built approved, initialed, and dated. Often the toolmaker redraws the part. It is good practise to have this drawing approved also. The final print should be signed by the moldmaker, molder, and customer.

2. Have you read all the notes pertaining to the job? Notes are just as important and binding as dimensions. If a note reads, for example, "this part must fit --- part," it is incumbent upon the molder and toolmaker to understand enough of the other part so that he can meet the specification. If he cannot then the limitations should be approved or the note removed.

3. Is the type of plastic materials indicated? Many times there is a possibility of running the part in another material. If so, it might be necessary to have runner bars and gates removable. Modifications in the cooling and knockout mechanisms might have to be planned.

4. Is the function, location and use of the piece understood? This is one of the most important items. It is part of the plastic fabricator's responsibility to interpret plastic properties for the specific application. This is true even if a piece part drawing for the plastic is submitted. While the moldmaker may not be economically responsible, the difficulties and delays must eventually harm his relationship with his customer. Knowledge of the parts use influences the mold design. The end user might not be fully aware of the significance of parting lines selection, gate location, draft, surface finish, tolerances, weld lines, and so on. The plastics engineer should have these explained to the end user, who should agree to the final decision.

5. Can any changes be recommended to make a simpler or better piece? Such suggestions are always in order. Sometimes the people involved in the design of the part and/or system have lost a certain amount of perspective. The mold designer may have a fresh viewpoint which can be most helpful.

6. Are the number of cavities correct?

7. Are tolerances indicated on all critical dimensions?

8. Can these tolerances be maintained? Tolerances are discussed on p. 145, 275. The tolerances referred to in steps 7 and 8 are those of the molded plastic part. The mold drawing will give the mold dimension whose tolerances are very different. One cannot overemphasize the importance of establishing moldable tolerances. The mold must be designed so that it can be adjusted. It is almost always easier to remove metal than to add to it. Many times it is desirable to leave too much metal on the mold, try the part, and then remove enough metal to bring the piece into tolerance. While maintaining the tolerances during molding it is not strictly a moldmaker's problem, it should be reviewed here. If they cannot be maintained, and the part will not function unless they are, provision must be made for a postmolding operation. If this cannot be done and there is no other solution, then there is no point to building the mold.

9. Are the dimensions given including or excluding shrinkage?

10. What shrinkage factor is to be used? This should be specified by the molder. See p. 280, 284.

11. Has adequate draft (taper) been specified? The purpose of draft is to ensure that the part can be ejected from the mold. This is a function of many other things. See p. 114, 128. Many times drawings do not have any taper specifications at all. These should be selected by the molder and mold maker and approved by the end user.

12. Has the parting line been approved? See p. 95.

13. Has the gate location been approved? See p. 123.

14. Is the gate location in the best possible place for maximum physical properties?

15. Is the gate location in the best possible place for finishing?

16. In designing gate location, will anticipated weld lines prove objectionable esthetically or mechanically? In considering steps 13, 14, 15, and 16, do not lose sight of the ultimate consideration whether the gate is in a position which will permit molding.

17. Will the piece hang on the injection side? It is not always safe to assume that the part will remain on the ejection side. The mold should be designed so that tapers can be changed and undercuts provided to hold the part on the right side of the mold. It might be necessary to design an ejector mechanism on the injection side of the mold. For a discussion on ejection, see p. 114, 128.

18. Has the ejector mechanism(s) been decided?

19. Have the location of the ejector mechanism(s) been approved?

20. Is the ejection mechanism(s) sufficient?

21. Has polish been specified? This should have been specified before the mold was quoted as it strongly influences its cost. See p. 114.

Machine.   While steps 22 to 26 seem obvious, they are often overlooked. If the mold is being designed to be operated at a specific molding plant, the machines in which the mold will fit should be indicated on the blue-print. This information should be permanently recorded and passed on to the molding room.

22. Will the molds physically fit into the presses being used?

23. Is the mold thicker than the minimum thickness required of the presses? If the mold is too thin, bolster plates can be added. It is usually easier to increase the spacer bars of the mold.

24. Is the stroke of the machine long enough to allow for part removal? This should be critcally reviewed. When there is a minimum amount of room the overall cycle may be longer and this should be considered when quoting the piece part.

25. Is the ejection stroke of the machine long enough to allow for part removal? If the stroke is insufficient it may be overcome by extending the knockout plates and attaching pneumatic or hydraulic cylinders to it on both sides, with the rod end toward the movable platen. When pressure is applied to the head end of the cylinder the knockout plate will move. If the rod is attached to the platen the knockout system will be hydraulically controlled and can be made to move in either direction at any time.

26. Can the mold be clamped into the press? Sometimes additional holes might have to be drilled in the platen. If cylinders or other equipment are to be attached to the mold they should be mounted so that they will not interfere with the clamping of the mold.

27. Is the clamping capacity of the machine enough for the parts?

28. Is the injection capacity of the machine enough for the parts?

29. Do the ejector holes correspond to the ejection mechanism of the presses to be used? Molds may be run in several presses. Not all presses have the standard SPI ejector hole pattern. After the types of presses upon which the mold may run have been listed, it should be determined whether the mold can be made so that its ejector holes will accommodate them all.

30. Are knockout mechanisms needed on the injection side? It is strongly suggested that holes be drilled through the injection platen and hydraulic cylinders be mounted on each side. When such hydraulic ejection is available, mold design may often be simplified, particularly when three plate molds and ejection on the stationary side is used.

31. Are water lines located so that they will not be in the way of the operator, or the removal of the part and gate? When water lines are externally connected they can be in the way of the operator or catch the gate or piece as it falls. When the mold is heated, the connection should be out of the operator's reach so that he will not be burned. When refrigerated water is used the mold should be designed so that condensation does not fall where it can cause damage.

32. Do water lines interfere with tie bars or other mechanisms?

33. In the event of requirements for heating the mold, are the heating elements and control units placed safely to be out of the operators way?

34. Have the dimension of the locating ring been shown?

Mold Design

35. Have the materials for cavities, cores and other parts been specified? There are other things beside economics which dictate the material choice. For example, if dimensional changes caused by hardening could cause trouble, prehardened steel might be used. Another method might be to use hardened steel and finish the cavity with EDM. Beryllium copper might be chosen because of its high thermal conductivity. Stainless steel might be chosen because of its low thermal conductivity, or its resistance to rusting. Parts moving against each other should be of different hardness or material to prevent galling.

36. Are the mold plates and component parts strong enough for the piece?

37. Is there sufficient steel surrounding the cavities and cores? This is particularly important in multicavity molds. They are usually inserted in milled out sections of the retainer plate. Unless there is sufficient steel retaining them on the outside there will be a tendency for the mold to spread. There is no practical way to calculate the requirement. It is better to have more steel than less.

38. Are there sufficient support pillars? See p. 92.

39. Is one leader pin and bushing unsymmetrical? The reason for this is to prevent the mold from being assembled in the press or on the table incorrectly.

40. Will the leader pin enter before any other part of the mold? This is very important. In the event a mold slips during assembly or on the press, the leader pin will hit the other side of the mold rather than the core. If the leader pin hits a noncritical area no damage will result. If it is the force that hits any part of the other side of the mold there is good possibility of damage.

41. Is there ample clearance for leader pins in the other side of the mold? In addition the hole into which the leader pin goes should be vented.

42. Is there sufficient travel for the ejector plate?

43. Is the ejector plate strong enough?

44. If a stripper mold, is the stripper plate properly supported?

45. Have push back pins been provided?

46. Does the sprue bushing fit the machine? This is a good opportunity to reexamine the size and length of the sprue, and cold-slug. The mold designer should know the maximum depth into which the machine nozzle can penetrate into the mold. The length of the sprue bushing should be kept to a minimum. The diameter of the sprue bushing should be large enough to provide free flow but not that large where the cycle will be held up because of its cooling.

47. Have the dimensions of the sprue bushings been recorded?

48. Are there sufficient cooling channels in the mold and cavities? This is exceptionally important. It should be reviewed again after the mold has been completely designed. See p. 160.

49. Do the knockouts clear the water holes? While knockout pins can be sent through water holes using "0" rings, it is not desirable. If extra knockouts might have to be added in the future water lines should be kept clear of the sites.

50. Are runners specified? See p. 102.

51. Have gates been specified? See p. 123, 279.

52. Have run offs been provided when required? See p. 141.

53. Has venting been specified? See p. 131, 141.

54. In cam acting molds, have provisions been made for hardening moving parts? See p. 135.

55. In cam action molds, have provisions been made for replacing worn parts and tightening cams?

56. In cam action molds, can cam pins be replaced without removing the mold from the machine? The purpose of steps 55 and 56 is to minimize mold down time. This should be a constant thought throughout the mold design stages.

57. If there are electrical heaters or devices on the mold, have they been made safe? Since molds are subject to vibrations, shock, water, and mechanical actions, the requirements for electrical safety are rigorous. One should attempt to visualize what could go wrong and design to prevent it. Electrical shocks are dangerous as well as unpleasant.

58. Can the electrical parts be replaced without removing the mold?

59. Have provisions been made for closing opening and depressions which might be filled up by flashed shots? This is particularly important on three plate molds. Often screw holes are filled when a shot flashes and are never cleaned out properly. This causes depressions in the mold and flashing. It is usually easy to keep openings away from the plastic surfaces.

60. Are all steel and metal specifications shown? Items 60, 61, 62, 65, and 66 are extremely helpful for the molding and mold maintenance department. It is rather frustrating to try to properly weld or reheat treat an unknown steel.

61. Have the heat treating specifications been shown?

62. Have the surface specifications been shown (including chrome plating)?

63. If the mold has to be heated, have provisions for difference in expansion been made? In hot runner molds provisions will automatically be made for expansion and contraction of the hot runner block. If one expects to run one side of the mold at a significantly different temperature than the other, provisions might have to be made to prevent galling of leader pins and other malfunctions.

64. Have eye bolts been provided on both halves of the mold? This permits each half of the mold to be safely handled by means of a crane. Many plants drop molds in from the top rather than slide them in on the side. It permits each side of the mold to be handled separately, not only at the machine but also during repairs and maintenance in the mold shop.

65. Where there are expendable parts such as springs, "0" rings, and switches, has a specification chart been provided? This is exceptionally valuable for maintenance in the plant, as anyone trying to measure a badly damaged "0" ring can attest. It also permits ordering spare parts without having to disassemble the mold to find out what these specifications are.

66. Are bolt sizes specified?

67. Are mold parts (sprues, etc.) standard? It is much more expensive for a molder to make a sprue bushing, leader pins, knockout pins, and other standard parts, than it is to purchase them from companies who specialize in their manufacture. Therefore, standard parts should always be specified. This is particularly important when purchasing molds built in other countries. Most countries have these standard American parts available. If not, they can be purchased here and sent to the moldmaker.

68. Have any spare parts to be furnished with the mold designated?

69. Can mold and cavities be disassembled within a minimum of time?

70. Are all the component parts numbered so as to allow for proper reassembly?

71. Has the mold been properly marked for identification?

72. Are the dimensions on the prints the same as the dimensions on the mold? This happens so rarely that the mold drawings are seldom used without rechecking dimensions. However, when a dimension has been checked it should be so marked on the print.

73. Is there a schedule of completion dates for stages of the mold work? If one is concerned about prompt mold deliveries, it is too late to expedite when the mold is scheduled for completion. The only way to find delays in time for remedial action is to have a schedule from the mold maker. Periodically this should be checked visually. With good reasons, mold makers are notoriously optimistic about the time it takes to perform a task.

BIBLIOGRAPHY

Injection Mould Design R. G. W. Pye, Iliffe Books Ltd., London, 1968.

Plastics Mold Engineering, DuBois and Pribble, sponsored by SPE, Reinhold Publishing Co., New York, 1965.

Moldmaking and Die Cast Dies for Apprentice Training, J. Kluz, National Tool Die and Precision Machining Association, 1411 K. St. N.W., Washington, D.C.

Runnerless Molding, E. P. Moslo, Reinhold Publishing Co., New York, 1960.

Heat Transmission, W. H. McAdams, McGraw-Hill Book Co., New York, 1954.

REFERENCES

1. "Estimating Mould Manufacturing Costs," S. T. Dawson, *IPE* April 1964, p. 112.

2. "Determining the Economic Optimum Number of Cavities,"J. C. Goettal, *SPE-J* July 1958, p. 32.

3. "Designing Large Injection and Compression Moulds," J. D. Robinson *IPE* May 1963, p. 173.

4. "The Injection Moulding of HD Polyethelyne - Principles of Mould Design," D. G. Briers and D. Burgess, *B.P.* March 1960, p. 100.

5. "A Molder's Guide to Mold Design," H. A. Perras, *PT* January 1968, p. 43.

5a. "Impact Styrene Out-Perform Metal," *MP*, December 1963, p. 88.

6. "Mold Design for Automatic Runnerless Molding," M. I. Ross, *SPE-J*, June 1965, p. 559.

7. "Hot Runner Systems," *Plastics*, November 1967, p. 1311.

7a. "The Design of Hot Runner Moulds," V.T. Gardner, *Plastics*, May 1969, p.515.

8. "Mold Design for High Speed Production of Disposables," L. Temesvary, *SPE-J*, February 1968, p. 25.

9. "Processing of Fiber Glass Reinforced Thermoplastics Improved by Hot-Runner Molding," T.P. Murphy, *PDP*, May 1964, p. 12.

10. "New Standards for Hot-Runner Molds," G.B. Thayer, *MP*, March 1969, p. 92.

11. "Hot Runner Molding," A. Seres, and R. Horvath, *PDP*, April 1970, p. 20.

12. "Runnerless Molding Without Hangups," E. J. Csaszar, *SPE-J*, February 1972, p. 20.

13. "What Does Non-Ferrous Tooling Cost?" M. Savla, *SPE-J*, March 1971, p. 24.

14. "Steel for Plastics Mold Cavities and Cores," J. M. McArthur, *SPE-J*, June 1966, p. 65.

15. "How to Select Mold Steels," J. R. Schettig, *PT*, November 1964, p. 36.

16. "Tool Steel for Plastic Molds," H. G. Becker, *SPE-J*, October 1954, p. 19.

17. "Vacuum Melting and Maraging Steels," C. N. Younkin, *SPE-J*, June 1967, p. 65.

18. "New Ways to Strengthen Metals," D. Pekner, *Mat. Des. Eng.*, October 1961, p. 11.

19. "The Choice of Tool Steels," A. G. Shaw, *BP*, August 1970, p. 86.

20. "Steels for Plastic Molds," E. E. Lull, *PDP*, November 1969, p. 20.

20a. "Consider Shock Resisting Tool Steels for Tough, Wear-Resistant Molds," J. W. Sullivan, *PDP*, May 1972, p. 18.

20b. "Mold Finishing," R. D. Balint, *SPE-J*, March 1966, p. 31.

20c. "Hints for Better Mold Performance," W. Young, *SPE-J*, December 1965, p. 1362.

21. "The Effect of Heat Treatment on the Polishability of Mold Steels," E. E. Lull, *SPE-J*, April 1956, p. 30.

21a. "Orange Peel and Pitting-Their Causes, Their Cures," W. Young, *PT*, May 1967, p. 41.

22. "Mold Polishing," A. W. Logozzo, *SPE Tech Pap.*, 1958, p. 322.

23. "Diamond Compounds for Mold Finishing," *IPE*, May 1961, p. 171.

24. "What Designers Should Know About Mold Finishes," L. Gabriel, *PDP*, April 1965, p. 10.

25. "Mold Texturing by Photo Etching," W.M. Schumacher, *SPE-J*, September 1967, p. 22.

26. "Beryllium Copper Molds," H. Mast, *PT*, October 1969, p. 52.

27. "Thermal Consideration in Mold Design," W. J. B. Stokes, *SPE-J*, April 1960, p. 417.

28. "How to Hot Hob Beryllium Copper," I. Thomas, *MP*, July 1961, p. 101.

29. "Beryllium Copper Molds-Make or Buy," *MP*, December 1969, p. 76.

30. "How to Use Short-run Tooling," M. Austin, *SPE-J*, March 1971, p. 29.

31. "Epoxy Molds," *PDP*, July 1965, p. 12.

32. "New Mold Making Technique," P. J. Garner, *SPE-J*, May 1971, p. 18.

33. "How & When to Hob," I. Thomas, and E. W. Spitzig, *MP*, February 1955, p. 117.

34. "Hobbing for Raised and Undercut Sections," N. S. Metrocavich, *SPE-J*, November 1964, p. 1191.

35. "Things to Look for in Selecting Metals for Cast Molds," R. E. Schoeller, *SPE-J*, March 1971, p. 35.

36. "Cast Moulds-Ferrous and Non-Ferrous," M.J. Butler, *BP*, June 1970, p. 144.

37. "Cast Mold Cavities," I. Lubalin, *MP*, October 1957, p. 147.

38. "Ceramic Castings for Large Injection Molds," I. Lubalin, *SPE-J*, January 1963. p. 61.

39. "Templectroforming of Mold Cavities," D. H. Wright, *SPE-J*, October 1968, p. 113.

39a. "Electroforming: An Easy Way To Low-Cost Tooling," J. J. Pawlak, *PT*, June 1968, p. 45.

40. "Duplication of Plastics Moulds by Die Sinking," I. Thomas, *IPE*, March 1962, p. 102; April 1962, p. 175.

41. "The Art of Engraving," I. Thomas, *IPE*, November 1961, p. 496, I. Thomas and R. Koegl (also *MP* May 1945).

42. "Tool Making by Spark Erosion," P. J. C. Gough, *IPE*, September 1961, p. 399.

43. "Spark Erosion Machining," *IPE*, March 1961, p. 67.

43a. "Breakthrough In Mold-Making-Electro-Erosion," F. Jacques and J. Schmidt, *MP*, December 1960, p. 109.

44. "Electro-Chemical Machining," T. E. Aaron and R. Wolosewicz, *Mach. Des.*, December 11, 1969, p. 160.

45. "Feeding (Gating) Techniques for Injection Moulds," D. G. Briers, *IPE*, April 1961, p. 102; May 1961, p. 166.

46. "Some Problems of Feeding and Gating Technique for Cellulose Acetate and Cellulose Acetate Butyrate Compounds," L. Hille, *IPE*, September 1961, p. 421.

47. "Gating and Cooling Techniques for Polypropylene," J. D. Robinson, *Plastics*, August 1965, p. 47.

47a. "Injection Moulding With Film (Flash) Gating," *IPE*, November 1964, p. 347.

48. "Progress Report on Mold Release Agents," *PDP*, March 1965, p. 10.

48a. "Self-lubricating Molds Boost Production," *Plast. World*, December 1970, p. 54.

48b. "New Ejector Doubles The Action," J. R. Byrne, *SPE-J*, January 1972, p. 24.

49. "Undercuts on Injection Mouldings," G. Ward, *IPE*, July 1961, p. 274.

50. "Cam Actions of Injection Molds," W. Lewi, *PT*, February 1955, p. 27.

51. "Some Aspects of Injection Mould Design," F. J. Lupton, *Plastics*, December 1961, p. 80.

52. "Injection Mold with Two Threads," *IPE*, July 1962, p. 325.

53. "Economics of Unscrewing Molds," I. Thomas, and E. J. Csaszar, *MP*, November 1958, p. 181.

53a. "The Half-Threaded Boss for Screw Holes," J. Andras, *SPE-J*, August 1968, p. 57.

54. "Collapsible Core," J. Andras, *SPE-J*, May 1967, p. 35.

54a. "New Collapsible-Core Tooling System," *BP*, September 1971, p. 195.

55. *"Inserts: Plastics Engineering Handbook,"* 3rd ed., Reinhold Publishing Co., New York, pp. 347-366.

55a. "Inserting Molding With High Density Polyethylene," W. L. Price, and W. A. Hunter, *SPE-J*, July 1960, p. 697.

56. "Continuous Mold Venting," S. E. Giragosian, *M.P.*, November 1966, p. 122.

57. "Vacuum Venting of Molds," G. S. Bohannen, *MP*, December 1956, p. 162.

58. "Dimensioning and Tolerancing," *MIL-STD-8B*, November 16, 1959.

59. "Cost and Value of Small Tolerances and Smooth Finishes," W. W. Gilbert, ASME Paper 61-MD-12.

60. "Is This Tolerance Necessary?" M. A. Sanders, *SPE-J*, January 1961, p. 51.

61. "Writing Meaningful Specifications for Plastic Parts," H. S. Byrne, R. L. Miller, and R. N. Peterson, *SPE-J*, May 1961, p. 469.

62. "Tolerances: Plastics Engineering Handbook," 3rd ed., Reinhold Publishing Co., New York, pp. 310-346.

63. "Estimating Mold Cavity Size-Factors Affecting Shrinkage of Acetal Resin Homopolymer," J. D. Bruton and W. C. Filbert, Jr. *PDP* November 1961, p. 12.

64. "Cooling It, But How?" *MP*, April 1969, p. 142.

65. "Sizing Chiller to Mold," C. E. Waters, *MP*, April 1969, p. 147.

66. "A Practical Improvement in Mold Temperature Control," J. D. Robertson, *SPE-J*, April 1969, p. 72.

67. "Water Treatment Pays Off," C. E. Waters, *MP*, March 1968, p. 114.

68. "What You Should Know About Mold Cooling," H. A. Meyrick, *MP*, October 1963, p. 219.

69. "Easy Way to Calculate Injection Molding Set-up Time," R. L. Ballman, and T. Shusman, *MP*, November 1959, p. 126.

70. "Water Cooling and Temperature Control of Injection Molds," G. Ward, *IPE*, March 1961, p. 48.

71. "Mold Cooling; Key to Fast Molding," L. Temesvary, *MP*, December 1966, p. 125.

72. "Improved Surface of Acetal Molding," C. W. Filbert, and T. M. Roder, *SPE-J*, February 1964, p. 149.

73. "Mold Cooling System Design," O. Doubek, *SPE-J*, June 1969, p. 47.

74. "Good Planning Needed for Effective Mold Cooling," C. E. Waters, *PT*, November 1970, p. 43.

75. "Injection Moulds," F. J. Lupton, *BP* 1970, Part I, September p. 124: Part II, October, p. 142.

76. "New Techniques in Mold Temperature Control," *PDP*, February 1971, p. 21.

77. "Automatic Molds," H. G. Brown, *Plastics,* May 1967, p. 539.

78. "Are Small Hoses Wasting Your Chilling Dollars?" A. Prasad, *MP*, September 1971, p. 80.

79. "How and Why Automation Cuts Molding Costs," D. Cook, *PT*, August 1971, p. 33.

80. "Automated Parts Removal," L. L. Scheiner, *PT*, April 1971, p. 37.

81. "The Challenge of Automation," *MP*, October 1971, p. 108.

CHAPTER 3

Theory and Practice of Injection Molding

This chapter develops a theory of injection molding and is concerned primarily with the qualitative aspects of plastic flow after the material leaves the injection cylinder. Practical applications that logically follow the theory are presented, where possible, immediately thereafter. Therefore, the presentation of theory is not limited to this chapter, but given in logical places in the book. For example, the theory of plastification can be found with the discussion of the screw in the first chapter.

A mathematical expression of injection molding process has not yet been accomplished. The reasons become clear when we analyze the machinery, materials, and process. Most factors that effect the process—material temperature, temperature profile, pressure, material velocities in the cylinder and mold, and mold temperature and flow patterns—are not measured at all or noncontinually at isolated points. There is no feedback of the processing variables to the molding machine to compensate automatically for changing conditions.

Plastic material is never the same. It has different heat histories, molecular weights, molecular weight distributions, degrees of polymerization, and impurities. In processing, the material is exposed to moist air and compressed with it, in the heating cylinder, with varying amounts of oxidation. It is heated by convection, conduction, and shearing. The heat content of the plastic changes as it moves through the cylinder and mold. Pressure changes from 0 to possibly 30,000 psi.

The physical properties of the material are generally not linear with respect to temperature and velocity. The plastic is compressible, stretchable, elastic, and subject to changing properties after removal from the mold. It changes its dimensions and properties after processing, and these properties may differ,

179

depending on the direction of the material flow in the mold. It has time-dependent properties which are strongly altered by its environment. In some materials varying degrees of crystallinity are neither predictable nor reproducible.

Therefore, one should not be surprised if injection molding has not yet yielded to mathematical analysis with the same accuracy attained in bridge building or motor design. However, the qualitative understanding of plastic behavior and the mathematical treatments (based on certain simplifying assumptions) are sufficiently advanced to permit a reasonably accurate understanding of the process.

BASIC THEORETICAL CONCEPTS AND THEIR RELATIONSHIP TO PROCESSING

Rheological data, concepts of energy levels, molecular structure, molecular forces, theory of heat transfer, and the theory of flow can be combined to develop a cohesive picture of what happens during the injection molding process. This conceptual catalogue can be of great value in preventing problems and solving difficulties. Furthermore, it is extremely helpful in understanding the literature.

A picture of this kind will contain certain generalizations. While existing theory and experimental evidence may slightly limit or modify these concepts, they do not limit their usefulness for our purposes. The picture is essentially a simple one and can best be explained by starting with certain fundamental concepts.

Phase Changes

Materials exist in three forms—solid, liquid, and gas. As heat is applied to a solid a point is reached where the solid is in equilibrium with the liquid. This is the melting point. Continual application of heat will not raise the temperature of the mixture until the heat energy required to convert the solid to the liquid is absorbed. This amount of energy is called the latent heat of fusion. After all the solid is melted, the application of heat will raise the temperature of the liquid until it reaches its boiling point and begins to vaporize. Similarly the liquid will remain at the same temperature until enough heat is supplied to completely vaporize the liquid. The amount of energy to convert the liquid to vapor at constant temperature is called the latent heat of vaporization.

Melting Points

In plastics processing we are concerned with the liquid and solid phases. In crystalline materials the change from solid to liquid is abrupt and easily discernible. In an amorphous (noncrystalline) polymer the change is not abrupt or readily apparent. The material softens over a wide temperature range and there is no dramatic visible change in its flow properties at any given point, such as found at the conversion of ice to water. If we plot certain properties of the plastic such as specific volume or heat capacity against the temperature, we notice at a point an abrupt change of the slope of the line. In the thermodynamic sense it is called a second-order transition, although there is evidence to suggest it might be a first-order transition (1). In polymer science, this point is called the glass transition point (T_g). Below the glass transition temperature the polymer is stiff and dimensionally stable behaving like a solid. It has brittle characteristics with little elasticity and its properties are relatively time independent. Above T_g the polymer will behave as a viscous liquid. It will evidence elastomeric properties depending on the chemical structure, cross linking, and degree of crystallinity. Its properties are now highly time dependent. (The glass transition mechanism will be discussed subsequently.) Additional heat applied to the polymer brings it to its melting range.

In a crystalline polymer the upper limits of the melting range is the melting point. This is the temperature above which crystals cannot exist. The melting point of an amorphous polymer is more difficult to determine. Since melting is accompanied by considerable changes in properties, a comparison of them with temperature is used for determining the melting point.

How Solids Behave

The differences between the three forms of matter can be explained in terms of molecular attraction. In a solid, the closeness of the molecules to each other permits the strong cohesive force of molecular attraction to limit their motion relative to each other. While solids can be deformed, it takes a comparatively large amount of energy to do so. If the solid is stressed below its elastic limit, it will be deformed. An ideal solid obeys Hooke's law which states that the amount of strain (movement) is directly proportional to the stress (force). The constant of proportionality, E (stress/strain) is called Young's modulus or the modulus of elasticity. When the stress is removed the molecular bonds which have been stretched contract, bringing the solid back to its original position.

Plastics are not ideal solids. They exhibit both Hookean elastic properties and delayed elastic responses, which are combined with viscous or flow properties; hence they are called viscoelastic materials.

There are two energy systems to consider within a material. One is the potential energy of a Newtonian gravitational nature and is a measure of the forces between the molecules. The other is kinetic energy which is the energy of motion and is related to the thermal or heat energy of the system.

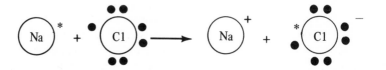

Ionic bond. Electron given from one to another. Not usually found in polymers.

Coordinate bond. Electron to be shared donated by one donor. Not usually found in polymers

x = Hydrogen electron: ● = First carbon electron: △ = second carbon electron

Covalent bond. Electrons shared. Primary bond in polymers. Bond distance is 1.54 A°; dissociation energy is 83 Kcal/mole.

Hydrogen bond. Between Hydrogen attached to an acidic group (-NH₂, -COOH, etc.) and to an oxygen, halogen, or a nitrogen in a basic group. Bond distance is 2.4-3.1 A°; dissociation energy is 3-7 Kcal/mole.

Van der Waals Forces. Result of interaction of electrostatic fields. Varies inversely as the sixth power of the distance. Dissociation energy is 2-5 Kcal/mole.

Figure 3-1 Nature of bonding forces.

Heat is energy in transit. When it flows over into the boundaries of the system it is called internal energy which is related to the random motion of the molecules—Brownian movement. The higher the internal energy (heat) of the system, the more the random motion. If we heat a solid of a given number of molecules, it is safe to assume that the increased internal energy has been used to increase the intermolecular distances.

In a solid, the potential energy (forces of attraction between the molecules) is larger than the kinetic energy (energy of movement tending to separate the molecules). Hence a solid has an ordered structure, with the molecular attraction strong enough to limit their motion relative to each other.

As more energy is put into the system the solid turns into a liquid where the potential and kinetic energy are equal. The molecules can move relative to each other but the cohesive forces are large enough to maintain a contiguous medium. Additional thermal energy results in the kinetic energy becoming larger than the potential energy. This separates the molecules to the extent they repel each other and will fill the container with equal density throughout. This is called the gaseous state.

Bonding Forces. It is appropriate here to review briefly the nature of the of the bonding forces in the polymer (Figure 3-1). The atom consists of a relatively small nucleus which contains most of the atom's mass and the positive charges. If the atom were the size of a house, the nucleus would be a pinhead at its center. Atoms are a few angstroms (10^{-8} cm) in diameter. The nucleus is some five orders of magnitude smaller, a few ten-trillionths of a centimeter in diameter. The orbits of the electrons form outer concentric shells each of which have certain stable configurations. The outermost, most loosely bound electrons (valence electrons) are involved in chemical reactions and primary bond formation.

When one element donates an electron to another to complete the stable configuration we have a strong electrostatic bond called an *ionic bond*. This is characteristic of compounds of metallic elements, such as sodium chloride (NaC1).

The *covalent bond* is of a different nature. Here the electron in the outer shells of two atoms are shared between them. This is typical of carbon-carbon bonds and is the primary bond found in polymers used commercially. The C-C bond is 1.54 Å long with a disassociation energy of 83 Kcal/mole.

A *coordinate bond* is one in which the electrons that are shared are donated by only one of the atoms. The strength of the coordinate bond is between that of the ionic and covalent ones.

There is another bond of importance in polymers which is not fully understood. Hydrogen should theoretically only be able to form one covalent bond. However, there is another weaker covalent type bond found primarily on hydrogen atoms attached to acidic groups (COOH, NH_2) which bond with oxygen, nitrogen, chlorine, and fluorine atoms attached to basic groups in the same or different molecules. Typical *hydrogen bonds* have a bond length of 2.4 to 3.1 Å and disassociation energies of 3 to 7 Kcal/mole.

There are secondary bonding forces in polymers which are sometimes called *Van der Waals* forces. Their disassociation energy is 2 to 5 Kcal/mole. They are between molecules and molecular segments and vary as the sixth power of the distance. They are much weaker than the primary bonds, and are part of the resistance to flow. The energy attracting molecules is sometimes called the "cohesive energy," and is that energy required to move a molecule a large distance from its neighbor.

If we take a cubic inch of a plastic and raise its temperature, its volume will increase. Since we are not adding any molecules to the cube, it is reasonable to believe that the distance between the molecules has increased. The Van der Waals forces decrease with the sixth power of the distance. Therefore, the molecules and their segments become much more mobile. Since these forces are decreasing as the sixth power of the distance, there will be a relatively narrow range in which the polymer properties change from "solid" to a "liquid." This is the glass transition point (T_g). Also since these cohesive forces form a major portion of the strength of the polymer, we can expect polymer properties to be very temperature dependent. This is the case.

Structure of a Plastics Molecule

It is important to have a physical concept of a plastic molecule. This will make it easier to understand its flow properties and characteristics. As an example, let us consider the polyethylene polymer. It is made by linking ethylene molecules. The double bond between the carbons in ethylene are less stable than the single bond in polyethylene. We can, with appropriate temperatures, pressures, and catalysts, cause the ethylene molecules to react with each other to form polyethylene. The idealized reaction is

$$
\begin{array}{ccccccccc}
H & H & H & H & H & H & & H & H & H & H & H & H \\
C = C & + & C = C & + & C = C & + & \cdots \rightarrow & - C - C - C - C - C - C - \\
H & H & H & H & H & H & & H & H & H & H & H & H
\end{array}
$$

If the number of molecules of ethylene in a reaction was one hundred, the polymer would have an approximate molecular weight of 3200. A typical Ziegler-type polyethylene polymer might have 7000 ethylene molecules with a molecular weight of approximately 200,000. The polymerization does not proceed as simply as the above indicates. Each polymer molecule will not be the same length or configuration.

When we discuss molecular weight of polymers we mean the average molecular weight. The molecular weight distribution is an important characterization of the polymer. If the molecular weight is spread over a large range (broad spectrum material) its properties will differ from those whose molecular weight distribution is narrow. For example, a wide spectrum material will show more elastic effects and extreme pressure sensitivity. Viscosity, solubility, and stress crack resistance are some of the other properties effected. Molecular weight and molecular weight distribution are discussed in greater detail later.

To get some idea of the size of a polyethylene molecule imagine that the methyl$\left(-\overset{\text{H}}{\underset{\text{H}}{\text{C}}}-\right)$group is 0.25 in. in diameter. A typical polyethylene molecule would be one city block long. A molecule of water would be about the size of the methyl group. When one considers the possibilities of entanglement, kinking, and partial crystallization of the huge polyethylene molecule, compared to the small size and simplicity of the water molecule it would not be unexpected to find considerable differences in flow properties. Flow of water is relatively simple (Newtonian). Viscoelastic flow (of polymers) has been qualitatively described but not yet mathematically defined in a quantitative manner.

When ethylene molecules polymerize they could theoretically do so and produce a straight line of carbon linkages, as shown in the top of Figure 3-2. It is also possible for the polymerization to take place so that the carbon atoms attached to each other in a nonlinear fashion, branching out to form chains, as shown in the bottom section of Figure 3-2. The amount of branching will depend on the method of manufacture. Polyethylene made with high pressure processes has more branching, with a typical polymer having 20 to 25 methyl side groups per 1000 carbon atoms. Polyethylene made under low pressure conditions (Ziegler, Phillips) might have from 1.5 to 4 methyl side groups per 1000 carbon atoms.

To understand further the nature of the polymer molecule it should be noted that the carbon atoms are free to rotate around their bonds and can bend at angles less than 180°. This swiveling and twisting permits the molecules and segments of the molecules to twist and entangle each with the other. This cohesive force consists of Van der Waals type attraction. The other type of force in the polymer is, of course, the carbon-carbon (C-C) linkages.

With our simple concepts of molecular structure we should be able to predict the different properties of the linear and branched materials.

Density. The linear structure should permit the polymer segments to get closer to each other than the branched structure so that it is denser. Obviously the branched side chains will keep the main polymer backbones further apart from each other in those areas where the chains exist. The longer the chains the greater this effect. There are three density ranges in commercial polyethylene, low density 0.910 to 0.925 g/cc, medium density 0.926 to 0.940 g/cc, and high density 0.941 to 0.965 g/cc. The high density polyethylene is commonly called linear polyethylene, recognizing the linear quality of the polymer chain. Some of the properties that we may predict from our model (Figure 3-2) follow.

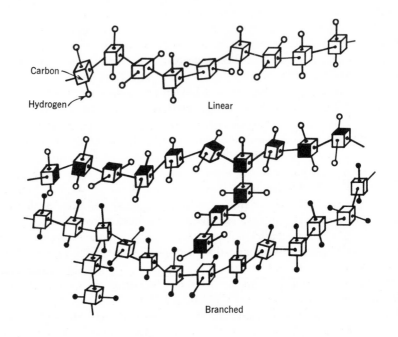

EFFECT OF BRANCHING

Figure 3-2 Branching of polyethylene causes chains to be kept further apart, reducing density and rigidity and affecting other properties. (Robinson Plastics Corp.)

Product Yield. The higher the density the fewer pieces of molded parts per pound of polyethylene can be produced. This is not an unimportant consideration in material selection.

Permeability to Gases and Solvents. Since the branching effect of a low density polymer provides more space between the atoms, one would expect that molecules of gas or solvent would diffuse more readily through the low density material. This is the fact. For example, the permeability of 0.92 density polyethylene film to oxygen or carbon dioxide is five times that of 0.96 density. The percentage weight gain tested for 7 days at 73°F in 5% acetic acid is 0.20% for low density, 0.16% for medium density, and 0.06% for high density. Similarly, for turpentine it is 11% for low density, 9% for medium density and 1.9% for high density.

Tensile Strength. One would expect that the linear material being closer together would have higher intermolecular forces and therefore greater tensile strength. A 0.915 resin has a typical strength of 1400 psi as compared to a 0.96 resin with a tensile strength of 4300 psi.

Percentage Elongation to Failure. Since the linear molecules can entwine and kink more than branched molecules, one would expect that it would be more difficult to separate them, so that applying a strong tensile force would rupture the molecule rather than causing it to flow and elongate. The branched material having lower intermolecular forces would slide considerably more before rupturing. High density polyethylene (HDPE) breaks at 30% elongation. Low density PE breaks at 450% elongation.

Stiffness. Linear PE being closer together would allow less room for segmental motion of the chains and bending of the backbone. Therefore it should be stiffer; 0.915 density PE has a stiffness or flexural modulus of 19,000 psi; 0.960 density has a modulus of 150,000 psi.

Heat Distortion. The heat deflection temperature under load is that temperature at which a specified bar under given conditions loaded to produce an outer fiber stress of 66 psi or 264 psi will deflect, (ASTM D-648). At a given temperature the molecular forces binding a high density material are higher than a low density material because the segments are closer together. Therefore, it will take a certain amount of heat energy (temperature increase) to separate the linear configuration so that it will have the same binding forces as the branched configuration. The more dense the material the more heat will be required to separate the molecules for equivalent strength. A low density PE will have a heat distortion temperature of approximately 110°F, a medium 130°F, and a high density of 160°F.

Softening Temperature. For similar reasons the softening temperature of the high density material is higher than that of the low density.

Hardness. Since the linear molecules are closer together, one would expect them to be more resistant to penetration. Hardness is measured by a penetration test. Hardness as measured on the Shore D scale is 67 for linear PE and 48 for branched PE.

Resistance to Creep. Creep is the amount of flow (strain) caused by a given force (stress). One would expect the higher intermolecular forces of the linear material to be more resistant to strain. The 100-sec tensile creep modulus at 68°F for a 1% strain is 20,000 psi for a 0.918 density and 125,000 psi for a 0.957 density. The stress required to produce a 1% strain is 170 psi for the 0.918 density PE material, 620 psi for a 0.939 density, and 1020 psi for the 0.957 density (2).

Flowability. Again, because of the stronger molecular attraction, the linear should be more difficult to flow. The standard test of flowability is the melt index (MI). This melt index and its limitations are discussed shortly. One would expect that the branched material, having less molecular attraction, would flow much more readily. As a practical example low and high density PE of similar MI were used to mold a rectangular dishpan. The minimum injection pressure to fill the mold, using low density PE, was 7600 psi, and for high density 11,500 psi (3).

Compressibility. Since there is a lot more open space in the branched PE it should compress more easily. At 15,000 psi the linear material compressed 1% and the branched material 3%. At 30,000 psi the linear compressed 3% and the branched 6 1/2% (4).

Impact Strength–Stress Cracking. One would expect from the molecular structure that the linear material would have a higher impact strength. This is not the case. Polyethylene is a crystalline material. Because linear material can get closer together there is an increased likelihood of crystallization. It can have so few branches that the crystallinity may reach the 90% level. On the other hand low density material can be so branched that the level of crystallinity can be as low as 15%. The formation of crystal structures brings the CH_2 units closer together. In addition, the formation of the crystal structure itself requires energy. The closeness of the CH_2 units and the energy of formation of the crystal structure should make the impact strength of the linear polymer much higher than the branched polymer whose CH_2 groups are much further apart. This is not so because of the characteristic of crystalline materials which permits the propagation of a crack very readily along the crystal structure. The energy of impact in a branched material will be absorbed by elongation in the amorphous phase. In a crystalline material the energy which won't be absorbed as readily follows the crystal and break it. For this reason, too, the resistance to stress cracking is higher in a branched material. Crystallinity can be controlled to some degree by the molding conditions. We shall therefore consider the effects of crystallinity although in some instances they are parallel to the effects of density.

The effects of density on some polymer properties are summarized in Table 3-1.

CRYSTALLINITY

The properties of crystalline materials such as polyethylene, polypropylene, nylon, and acetal are affected by the amount of amorphous material, the amount and nature of the crystalline phase, and the orientation. In a given

Table 3-10   Effect of density on polymer properties — Increasing
density causes the indicated change in the property
listed in the first column

Increasing density changes the	Increase	Decrease
Yield (pcs/lb)		X
Permeability to gases and solvents		X
Tensile Strength	X	
Percentage elongationtto failure	X	
Stiffness	X	
Heat distortion	X	
Softening temperature	X	
Hardness	X	
Resistance to creep		X
Flowability		X
Compressibility		X
Impact strength		X

polymer these properties will be strongly affected by the molding conditions and post molding treatment.

When a polymer is melted, the molecules are separated to the extent that there no longer is an ordered structure. Large molecular segments vibrate and rotate to give a totally disordered structure. When the plastic is cooled a point is reached where the forces of attraction are strong enough to prevent this free movement and lock part of the polymer into an ordered or latticed position. The segments can now rotate and oscillate only in a small fixed position. Since the molecular configuration is the same throughout, the intermolecular distance for this phenomena should be the same throughout the polymer. This distance is controlled by the temperature. At a given temperature the effect is extreme and the material starts to crystallize. This is known as the crystallization temperature. This is usually 10° or more below the equilibrium melting point.

Crystalline polymers have a latent heat of fusion which is a measure of the energy required to form or melt the crystal. The heat of fusion of high density PE is 104 Btu/lb, acetal 70, low density PE, 56, and nylon 66, 56. For example, it takes approximately 180 Btu/lb to bring acetal to its molding temperature. Of this, 70 Btu (about 40%) are for melting the crystalline structure. In contrast polystyrene, which has no latent heat of fusion, requires only 160 Btu/lb, because there are no crystals to melt (4a). It should be noted that cylinder output ratings are usually given in polystyrene, and produce considerably less output for crystalline materials unless additional heat can be delivered to the polymer.

As one would expect the crystalline or ordered structure occupies much less space. The sharp decrease in volume is indicative of the onset and amount of crystallization. Amorphous materials do not have any volume change caused by crystallization. Therefore, all crystalline polymers show greater shrinkage than amorphous ones in going from a hot liquid to a room temp solid.

Since the amount of crystallinity varies with molding conditions, it is usually more difficult to maintain tolerances in highly crystalline materials.

The measurement of the amount of crystallinity has theoretical and practical significance. It is done by measuring a particular intensive property of the polymer. The following properties have been used: specific volume, specific heat, specific enthalpy, specific enthalpy of fusion, infrared extinction coefficients, specific line widths in nuclear magnetic resonance, and specific X-ray diffraction intensities (5). The properties of the polymer with different degrees of crystallization can be measured and correlated with the processing conditions (5a).

Crystalline materials, on the other hand, have a considerable degree of rigidity between T_g and the melting point, ranging up to a relatively rigid solid. The amount of "rigidity" depends on the amount of crystallinity.

For example, amorphous polystyrene, which is commercial, has a modulus (stiffness) of about 100,000 psi at 175°F. Just above its T_g, 190°F, it drops to close to 0. Crystalline polystyrene which is a laboratory product follows the value of amorphous polystyrene until T_g, but instead of dropping precipitously, continues decreasing slowly so that it has a modulus of about 1000 psi at 385°F. As it approaches its melting temperature, its remaining modulus disappears rapidly. Polyethylene at 75°F has a modulus of 15,000 psi when 40% crystallized and 110,000 psi when 70% crystallized.

These results are not unexpected in terms of our concept of molecular structure. The crystalline state has a more compact structure so that the Van der Waal's forces are stronger. Thus more force is required to move the polymer segments, that is, the polymer is "stiffer." This is another way of saying that the crystalline structure is at a higher energy level (the heat of fusion) and that more energy is required to move the segments than at the lower amorphous level.

Structure of the Crystalline Polymer

We now consider the structure of the crystalline polymer. There are several excellent summaries of polymer morphology. Reference 6 is primarily concerned with the structure of the polymer crystal. Reference 7 considers the structure of the crystallite, orientation, and the effect on polymer properties. Reference 8 (47 pages) is a fine presentation of the theory of polymer crystallization including nucleation, and Reference 9 questions the formation of regularly folded interfaces. A full presentation of polymer single crystals will be found in Ref. 10, and a discussion of crystallinity is found in Chapter 5 of Ref. 11 and Chapter 10 of Ref. 12. Discussion of the structure of crystalline polymers is also found in Ref. 13 and 14. Measurement of crystallinity is reviewed in Ref. 15.

The initial concept of crystalline structure was a series of almost perfectly formed crystals surrounded by an amorphous region. A single molecule might be part of several crystalline structures. The part of the molecule not in the crystal consisted of the amorphous region. This "fringed micelle" concept was not questioned until 1957 when single crystals of polyethylene made by slow crystallization from dilute solutions showed a folded chain structure. This was the only structure that could fit the X-ray diffraction patterns and the known size of the crystals and molecules.

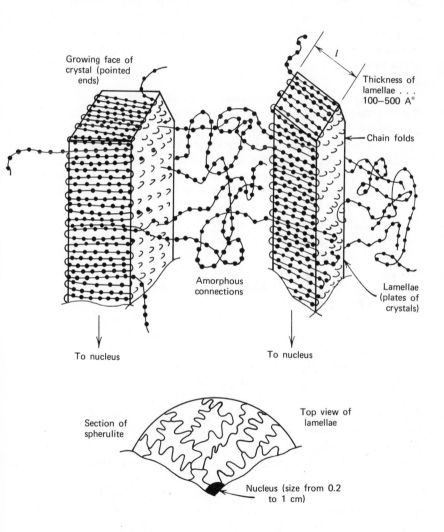

Figure 3-3 Schematic view of crystal structure, showing plate-like lamella, amorphous connections and section of spherulite.

The chain-folding theory of crystalline structure is illustrated in Figure 3-3. The crystalline part of the polymer consists of lamellae (platelets) which are plates of crystal material. They consist of chains of molecules folded into the flat plates (lamellae) which are between 100 and 500 A° units thick (l). It is generally agreed that the chain segments reenter in adjacent positions rather than randomly jumping to other parts of the crystal. The crystals grow on the pointed front end of the lamellae. The lamellae are held together by polymer chains going from one to another. These are randomly located and are the amorphous part of the plastic. When the polymer is oriented, either by processing conditions or stretching, the amorphous part tends to become parallel and to straighten out the lamellae.

Crystal growth requires measurable time and high mobility of the chain so that the molecular segments can get close together. Therefore, the slower the cooling (high mold temperatures) particularly as the polymer passes thru its melting temperature, the higher the degree of crystallization. Anything that affects the polymer mobility should affect its crystallinity. High pressure on the material decreases crystallinity and reduces density. Using a high density Phillips type PE at a mold temperature of 230°F, the density with 5000 psi injection pressure was 0.962, and with 20,000 psi 0.955 (16).

Most polymers under the conditions of processing form spherulites (Figure 3-3). As the polymer cools, crystallization starts at a nucleus which may vary from several tenths of a micron to several millimeters in diameter. The crystalline structure spreads out more or less in a spherical manner. The crystalline part (lamellae) consists of stacks of thin blades radiating from the nucleus. They branch and rebranch at small angles to the radius to fill the volume of the spherulite. They may twist in a regular manner. The polymer chain axis are approximately perpendicular to the spherulite radius. Unless quenched rapidly enough the spherulites will grow until they touch each other filling the volume of the plastic.

The spherulites may fill the whole structure. There is no reason to believe that the amorphous content between the lamellae in the spherulite is any different than that between the spherulites. The properties of the plastic will depend in part on the number and quality of the spherulites.

The type and size of the crystallite also have a strong influence on polymer properties. As the polymer cools from above its melting point, crystallites form around the nucleus. The spherulitic growth continues until the temperature drops to the point where the lack of chain mobility prevents crystallization. The rate of crystallization affects the size of the spherulite, and this rate is highly temperature dependent. It is a combination of the rate of nucleus formation and of crystal growth. They are each affected differently by temperature. The rate of nucleation increases continually as the temperature falls. The rate of crystal growth increases as the temperature falls and then decreases. When the crystallization rate is low, such as in isotactic* polystyrene, rapid cooling, such as quenching in water, will prevent crystallization. High density polyethylene, on the other hand, has such a high crystallization rate that it is not possible to cool it quickly enough to prevent some crystallization.

Nucleation. Therefore, at higher temperatures the nucleation is relatively slow and the crystallization relatively rapid, so that large spherulites form. The rapid quenching of the material gives much smaller spherulitic structure. The importance of this is evidenced, for example, in that the Gardener impact strength of a linear polyethylene with a 4.2 - μ - diameter spherulite is 104 ft/lb. While the same material with a 6.8 - μ - diameter spherulite drops to an impact strength of 13 ft/lb (17). As we shall see, other properties are affected by the character and number of the spherulites.

Spherulite formation can start in three ways:

1. The random motion of the polymer chains produces a crystal of sufficient size so that the surface energy thereof will propagate continued crystalline growth. This is called spontaneous or homogeneous nucleation and is the vastly predominant mechanism.

2. The polymer is not completely melted and the unmelted portions form nuclei.

3. Foreign substances are deliberately added which are called nucleating agents. A nucleating agent for nylon, for example, may be colloidal silica at a concentration of about 0.1%. Nucleating agents for polypropylene may be finely ground higher melting polymers, silica, and some metal salts of organic acids.

In a molded part it is desirable to have as homogeneous a structure as possible. Rapid cooling tends to produce more and smaller spherulites. Nucleating agents normally produce more uniform development of the spherulite in terms of growth, size, and distribution. It usually will produce more crystallinity, and some of the effect of nucleating agents are in reality the result of higher crystallinity. Some of the effects of nucleating agents follow:

1. Nucleation raises the temperature at which the crystallization rate is maximum. A nonnucleated polypropylene has a crystallization rate peak at about 240°F. With nucleating agents it can be raised as high as 287°F (18). The

---

*The term "isotactic" refers to the spatial or stereo configuration of a regularly repeating unit in a polymer chain such as the methyl group in polypropylene. *Isotatic* polypropylene has the methyl groups all on one side of the polymer chain. *Syndiotactic* polypropylene has the methyl groups alternating from one side to the other. *Atactic* polypropylene has the methyl groups located randomly. Isotatic polypropylene crystallizes readily, while atactic cannot crystallize at all and is a soft rubbery material which can be extracted with hydrocarbon solvents to leave the desired percentage of isotactic material.

amount of crystallinity determines the stiffness of the material which determines the length of time a part must cool in the mold before it is stiff enough to eject. Since the crystallization temperature is higher, the setup temperature (stiffness) is higher, and the part can be removed at a higher temperature, significantly lowering cycles. Depending on the shape of the piece cycle reductions of 3 to 30% have been reported (19).

2. The fine spherulitic structure of nucleated polypropylene significantly reduces the dispersion of transmitted and reflected light giving a marked increase in transparency and surface gloss (18).

3. Because of the small spherulite size nucleated materials tend to shrink more uniformly. As the spherulitic size is less temperature dependent the polymer tends to be more uniform. A nonnucleated material will have larger spherulites in the center where the temperature remains higher longer. It has been found that nucleation permits molding thicker sections without voids. Similarly the tendency for sinks to form on the surface is reduced.

4. Nucleation tends to reduce warpage particularly in olefins. Probably this is caused by a more uniform material and the higher tensile modulus. The latter effect might be due to more crystallinity rather than the size of the crystals.

5. Nucleation increases the hardness slightly, and also the abrasion resistance.* Tensile strength, rigidity, and the tensile yield strength are moderately increased. Conversely the elongation decreases and the impact strength drops. These effects are probably predominantly caused by the increase in crystallinity. It is interesting to note the effect of nucleation, spherulite size, and crystallinity on the impact properties of compression molded polypropylene (Table 3-2).

6. Nucleation of the polymer raises the heat distortion temperature of the molded part. For example, a nonnucleated nylon 6 has its heat distortion temperature raised from 152 to 180°F by nucleation.

7. Water absorption is slightly lower for nucleated materials, probably due to the increased crystallization.

8. Nucleated polymers seem to shrink more (higher crystallinity) than nonnucleated materials. This can be used to advantage if a part is molded in a nonnucleated material and is sticking in the cavity. Sometimes the additional shrinkage of the nucleated material will keep it on the force.

Summarizing, nucleation generally results in the following:

Shorter cycle time
Increased cost of material
Less voids
Minimum sink marks
Reduced flow lines
Increased surface smoothness

---

*A description of plastic properties is found in Chapter 4.

Table 3-2 Effect of nucleation, spherulite size, and crystallinity upon the impact strength of compression molded polypropylese. (7)

Crystalization Conditions	Nucleation	Spherulite Diameter (μ)	Crystallinity Measured from Density (%)	Impact Data at 23°C	
				Elongation (%)	Energy $(kg-cm-cm^{-2})$
Water quenched	No	~ 2	52	11.34	940–2600
Cooled slowly	No	~ 20	67	4.2	540
Water quenched	Yes	<2	62	4.4	510
Cooled slowly	Yes	<2	73	3.6	470

Improvement in see through clarity
Reduction of warping
Increased shrinking
Increased flexual modulus
Increased tensile modulus
Increased tensile yield strength
Increased hardness
Increased abrasion resistance
Increased heat distortion temperature
Decreased elongation
Decreased impact strength

How to Regulate Crystallinity.   The molder has a limited ability to regulate the percentage and type of crystallinity in polymers. This is done primarily by using the following variables:

1. Mold temperature.
2. Cooling method after removal from the mold.
3. Cycle time.
4. Injection and holding pressure.

As the polymer in the mold is well below its melting temperature, we would expect the higher the mold temperature the more molecular and segmental motion of the chains. This condition increases the probability of bringing the molecules close enough together to be frozen into a crystalline state. Therefore, the higher the mold temperature the more the crystallinity. Similarly, the quicker the part is cooled outside the mold the less the crystallinity. Pressure increases the rate of crystallinity because it brings the segments closer together. A 1/2-in.-diameter high density polyethylene rod was molded with a melt and mold temperature of 400°F. The mold was cooled at a rate of 10°F/min. The cooling time to produce 50% crystallinity in the rod was measured. At 30,000 psi it occurred after 6 min. At 1000 psi the time more than doubled, requiring 15 min (20). The effects of the cycle time can be to increase or decrease crystallization. This will depend on the interaction of the mold temperature, subsequent temperature treatment of the piece, and the injection pressure.

For a given material the amount of shrinkage should depend on the crystallinity. The crystalline structure has the molecules closer together than the amorphous structure. Consider a cube completely filled initially with amorphous material. As the material crystallizes it requires less volume, decreasing the size of the original cube (shrinkage). It is easily seen, therefore, the more crystallinity the smaller the cube (higher shrinkage). This is shown in Figure 3-4 where by raising the mold temperature from 100 to 175°F the shrinkage of polypropylene in the direction of flow increased from 0.014 to 0.024 in./in. Following the same reasoning it is evident that the more crystals in the material, the higher its

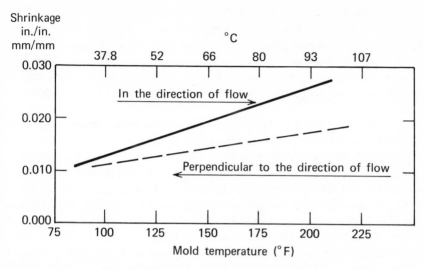

Figure 3-4 Shrinkage of polypropylene as related to mold temperature, both in the direction of flow and perpendicular to the direction of flow (Avisun Corporation).

density.

The Effect of Crystallinity on Properties

We shall briefly discuss the effect of crystallinity on physical properties.

There are many references in the literature relating processing conditions to properties of crystalline polymers. References 16 and 21 refer to high density polyethylene, Ref. 22 to polypropylene, and Ref. 23 to nylon.

1. *Density*.   Since crystallinity gives a more compact structure, the density increases with crystallinity. A compression molded sample of polypropylene with 70% crystallinity had a density of 0.896. Increasing the crystallinity to 95% gave a density of 0.903 (24). As expected increasing the mold temperature increases the crystallinity and density. Nylon-6 measured immediately after molding showed a density of 1.094 with a 68°F mold temperature and 1.122 with the mold temperature at 212°F (23). High density polyethylene at a mold temperature of 75°F gave a density of 0.944. Raising the mold temperature to 250°F increased the density to 0.950 (16).

2. *Stiffness*.   The flexibility of a plastic depends on the ability of its segments to rotate. Crystalline structures hinder such rotations; therefore, a crystalline material is significantly stiffer than the equivalent plastic in its amorphous condition. The tensile modulus (stiffness) of a 70% crystallized

polypropylene is 65,000 psi and at 95% crystallization it is 150,000 psi (24).* Amorphous polymers lose their stiffness about 50°F above the glass transition point. The effect of crystallization is to extend the upper temperature limit for a usable modulus. For example, high density polyethylene maintains a fairly high modulus to within approximately 20°F of its melting point (275°F). Additionally, because the crystalline structure is stiffer there will be less tendency to creep.

3. *Tensile strength.*   The tensile strength increases significantly with increasing crystallization. One would expect to need more force to break the closer bonds of a compact crystalline structure than that of an amorphous material. The tensile yield strength (force required to start the polymer flowing) of a 70% crystalline polypropylene is 4000 psi. Increasing the crystallinity to 95% gives a tensile yield strength of 6100 psi (24). Since the flow is primarily in the amorphous section, it would be expected that the percent elongation at break would be higher when there is more amorphous material, that is lower crystallinity. A typical slowly cooled polypropylene will have a yield strength of 3750 psi and a 50% elongation at break. Cooling the same part rapidly will decrease its crystallinity lowering its yield strength to 3000 psi but increasing its percentage elongation at break to 800%.

4. *Impact strength.*   From the strength of the crystalline structure one would expect the impact strength to increase with increasing crystallinity. This is not the case. Crystalline structures in all materials tend to rapidly propagate impact energy along the faces of the crystals where they break. This is used to advantage in some industries such as diamond cutting, but is a distinct disadvantage in molding. Seventy percent crystalline polypropylene has a notched Izod impact strength of 2.8 ft-lb/in. notch. Increasing the crystallinity to 95% reduces the impact strength to 0.9 (24).

5. *Shrinkage.*   As mentioned previously, increasing crystallinity decreases the volume, thus increasing shrinkage.

6. *Hardness and Abrasion Resistance.*   Hardness, which is a measure of depth of penetration, increases with crystallinity. The increase is relatively neglible until the crystallinity is over 80%. The wear and abraision resistance are enhanced by crystallinity (23). For practical purposes, the changes in hardness and abrasion resistance caused by molding conditions is very small.

7. *Heat properties.*   Increasing crystallinity has an important effect in raising the softening and heat distortion temperatures. Seventy percent crystalline polypropylene has a deformation temperature under load (ASTM D648) of 257°F, while at 95% it is 304°F (24). The higher softening point or heat distortion temperature causes the polymer to remain stiffer at higher temperatures. Since stiffness is the criteria for ejection from the mold, higher

---

*The tensile modulus is the slope of the initial portion of the stress-strain curve. It is a good measure of stiffness. See page 339.

crystallinity reduces cycle time. Since higher crystallinity is caused by higher mold temperatures one would expect that cold mold temperatures would increase the cycle, particularly for thin walled objects (23). Crystallinity will increase the brittleness at low temperatures. The brittle temperatures of 55%, 83%, and 95% isotactic polypropylene (isotacticity being a measure of crystallinity) are 32°, 50°, and 68°F, respectively.

8. *Permeability.* Because of the closeness of the crystalline structure it is more difficult for a gas or liquid molecule to find its way through the plastic. For this reason increased crystallinity increases the resistance to permeability to gases and vapors.

9. *Stress Cracking Resistance.* The higher the crystallinity, the lower the resistance to stress cracking. This is probably due to a mechanism similar to the lowering of the impact strength with increasing crystallinity. Therefore, higher mold temperatures (which promote higher crystallinity) increase the effect of stress cracking. A polyethylene dishpan molded with mold temperatures of 118°, 135°, 150°, and 175°F had the following percent failures after 24 hr when subjected to a stress crack resistance test — 30%, 60%, 80%, and 100%.

10. *Optical Properties.* There is no simple explanation of the effect of crystallinity on optical properties. Higher crystallinity will give a higher surface gloss. A distinction is made between transparency or light transmittance and freedom from haze. Usually higher crystallinity will cause less transparency and more haze. However, higher crystallinity in nylon usually increases transparency. Nucleation of polypropylene has been reported to increase transparency. A good review of the optical properties of crystalline materials can be found in Ref. 25.

11. *Warpage.* Crystalline materials exhibit more warpage than the amorphous ones. It is probably related to the effect of temperature on crystallinity. When an amorphous material cools, the difference in density throughout the piece is primarily a function of how much material is forced into the cavity after the mold is initially filled. To this, a crystalline material adds the effect of varying amounts of crystallinity caused by differences in polymer temperature in the mold. (The amount of crystallinity is dependent on the polymer temperatur.). The resulting differences in density causes differential shrinkage which sets up high internal stresses. When they exceed the strength of the polymer, strain or warpage occurs.

How Annealing Affects Properties

There is a great deal of uncertainty concerning the details of the effects of annealing crystalline polymers. Theoretical discussions not directly applicable to molding practices can be found in Chapter 5 of Ref. 10, and Ref. 26 and 27.

As one might expect annealing a crystalline material should increase its crystallinity. A 0.952 density PE annealed at 149° for 1 hr increased its density to 0.953. Annealing the same material at 212° increased its density to 0.955. Increasing the annealing time to 1 1/2 hr did not change the density when the annealing temperature was 149°. At 212° it increased the density to 0.956 (21). For practical purposes the main advantage of annealing is to relieve internal stresses which is true for both crystalline and amorphous polymers and is discussed subsequently.

Orientation Also Affects Properties

Orientation strongly affects the properties of polymers. This is fully discussed later on in this chapter. The mechanism of orientation of a crystalline polymer involves at least three steps. There is an instantaneous deformation of the spherulitic structure. There is a slower slipping of the lamellae within the spherulite. Finally there is a still slower viscoelastic flow of molecules within the crystal (28).

Molding Crystalline Plastics

The processing of thermoplastic materials essentially involves applying heat to melt the material in the cylinder and removing this heat while the plastic part is in the mold and after removal of the part from the mold. The rate of heat transfer, which is important in determining the cycle time, will also affect the degree of crystallinity with its profound effect on physical properties. In molding, the size of the gate is also important; it affects rate of fill, as well as length of time to freeze.

Mathematical and graphical formulations of freezing time are of great value to the molder. Reference 29 shows such calculations with polyethylene, polypropylene, and polyacetal. Figure 3-5a, taken therefrom, shows the freeze-off time for 0.945 PE starting at a cylinder stock temperature of 450°F. The curve shows the time to reach a freeze-off temperature of 266°F at the center of the part. For example, a 2-in. slab molded with cooling water at 70°F would require about 52 min for the last material in the center of the slab to reach 266°F. A 1/16 diameter (1/32 radius) gate under the same conditions would have a center freeze-off in 1.2 sec. While the 2-in. slab need not be kept the full 52 min in the mold for ejection, such information permits the molder to correlate his practice and the graph. It will then enable him to make accurate predictions. Such graphs are very useful in computing gate freeze-off times. Figure 3-5b gives the same information for polypropylene. Reference 30 shows the calculations for heating, cooling, and gate freeze-off temperatures for acetals. Reference 29a describes the effects of crystal size, orientation, and density on the thermal conductivity of polyethylene. Reference 31 is concerned with 66 nylon. Reference 20 has information on polyethylene though in less useful form. Reference 32 gives a general numerical procedure for calculating temperature

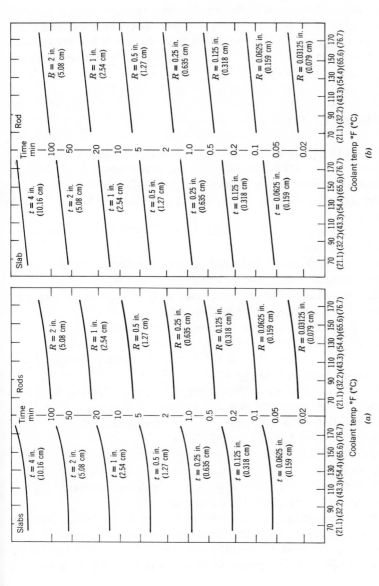

Figure 3-5 Freeze-off time of rods and slabs of polyethylene and polypropylene (Ref. 29). (a) Freeze-off time of polyethylene (0.945 density) melt at 450°F (232.2°C): coolant as shown. (b) Freeze-off times of polypropylene melt at 500°F (260°C): coolant as shown.

201

profiles during the cooling of molded parts. Good agreement is obtained between theoretical and actual results for a 10 MI, 0.935 PE. Temperature, pressure, and cooling rate curves for different conditions are shown. References 32a and 32b describe, mathematically, heat transfer to molten polymers.

If a crystalline polymer is cooled rapidly, its crystal structure is fine with many small spherulites. If it is crystallized slowly, the crystal structure is coarse and the spherulites are fewer but considerably larger. Crystalline texture has a significant effect on the properties of the molded part so that the rate of cooling is an important variable in molding.

CROSS-LINKED PLASTICS

The properties of plastics that are associated with crystallinity are derived from the closeness of the crystalline structure. A similar result is obtained by cross-linking the carbon atoms in polymer systems. To date cross-linking in thermoplastics has been mostly performed on polyolefins used in wire and cable applications.

After cross-linking the material loses its thermoplastic characteristics and becomes thermosetting.

Cross-linking significantly improves the heat resistance, tensile strength, tensile elongation, and the stiffness. The tensile creep behavior of cross-linked polyethylene is similar to that of acetal, which is one of the most creep resistant thermoplastics. It is considerably better than nylon.

Cross-linking also improves the resistance to environmental stress cracking. Solvent resistance is strongly improved. For example, uncross-linked polyethylene in toluene for seven days at 155°C broke into pieces. The cross-linked material only increased its weight by 45%.

The two methods used are irradiation and chemical cross-linking. Irradiation techniques use high-speed electrons and neutrons, gamma rays, X-rays, and ultraviolet radiation. The properties of most commercial plastics (except cellulosics) are improved by radiation (32c, 32d).

Chemical cross linking systems use organic peroxides which decompose when heated, to yield highly reactive free radicals. These catalyze the chemical cross linking. A filler which may be a silica, carbon, calcium carbonate, clay, or alumina is generally required with these compounds.

Cross-linked polyolefin materials are molded with oil heated injection cylinders. The material is injected below the temperature required for the decomposition of the peroxide. Mold temperatures of 370 to 450°F catalyze the cross-linking. Cycle times are similar to those for molding regular polyolefins (33 to 36).

MOLECULAR WEIGHT: AFFECTS PROCESSING AND PROPERTIES

The distinguishing characteristic of polymers is their high molecular weight. Polymeric behavior is a function of molecular weight (MW). At room temperature ethylene, with a molecular weight of 30, is a gas. When ethylene is polymerized to a molecular weight of 170 it becomes a liquid. At 1000 it is a grease, at 4000 a wax, and at 12000 a plastic.

It is not possible to polymerize ethylene (or any polymer) so that every polyethylene molecule will have the same molecular weight and configuration. The molecular weight distribution may be controlled within relatively large limits. However, the range can be extremely broad, such as in polystyrene, which may have molecules with MW's of 5000 and 2,000,000 in the same resin.

To determine the molecular weight distribution (MWD) it is necessary to fractionate, or separate the polymer into fractions of similar molecular weights. This is accomplished by ultracentrifuging the polymer or using its solubility characteristics. A high molecular weight material is less soluble than a low molecular weight. The fractionation is done either by precipitation or solution methods. None of these procedures are particularly precise, but they yield useful information which has been correlated with processing and properties.

The molder controls the molecular weight and the molecular weight distribution of a material only in that he is involved in its specifications. This can be done in consultation with the material supplier (37, 37a, 38). Since they do affect his processing the product a general statement about them is desirable. More detailed information will be found in Chapter 4 of Ref. 39, Chapters 1 and 2 of Ref. 12, and Chapter 3 of Ref. 11.

The concept in measurement of molecular weight and molecular weight averages in polymers presents some difficulties. It is distinctly different than in a monodisperse system which consists of identical molecules. In ice, for example, all the molecules are identical. Therefore the molecular weight and any average molecular weight will be the same, that is, 18. Polymers have different chain lengths, different branches, and different molecular arrangements. In polydisperse systems the term average molecular weight alone has no meaning. Consider the case of a fatal disease striking people of the average age of 46 without discrimination regarding sex. The significance of this average lessens when one finds out there are only two cases reported, a girl 1 year old and a man 91 years old. In this instance averages are totally misleading. Therefore, in polymer science average molecular weights are only used when related to a specific property whose functions are directly related in some manner to the change in that molecular weight average.

Molecular Weight Determinations

There are two major molecular weight determinations. One is based on the number of polymer molecules (M_n) and the other on the molecular weight of the polymer molecules (M_w).

The number average molecular weight (M_n) is determined by taking the sum of this average for each fraction. Obviously the more fractions taken the more accurate the determination. The extreme difficulty and costliness of these determinations, as well as their relative inaccuracy, usually keep the number of fractions at a minimum.

Number average molecular weight (M_n) is the sum of

$$
\begin{array}{l}
\text{Number of monomer units} \\
\text{in one polymer molecule} \\
\text{in fraction } (A)
\end{array}
\quad \times \quad
\frac{
\begin{array}{l}
\text{number of polymer molecules} \\
\text{in fraction } (A)
\end{array}
}{
\begin{array}{l}
\text{total number of polymer} \\
\text{molecules in all} \\
\text{fractions}
\end{array}
}
\qquad (3\text{-}1)
$$

(Similarly for fractions B, C, etc.)

The weight average molecular weight (M_w) is

$$
\begin{array}{l}
\text{Number of monomer units} \\
\text{in one of polymer molecule} \\
\text{in fraction } (A)
\end{array}
\quad \times \quad
\frac{
\begin{array}{l}
\text{molecular weight of polymer} \\
\text{molecules in fraction } (A)
\end{array}
}{
\begin{array}{l}
\text{total molecular weight} \\
\text{of all fractions}
\end{array}
}
\qquad (3\text{-}2)
$$

(Similarly for fractions B, C, etc.)

Two examples will clarify this concept. Consider a polymer of two polymer molecules, one with 10,000 monomer units and a molecular weight of 10,000 and the other with 100,000 monomer units and a molecular weight of 100,000. Let us mix an equal number of each polymer molecule. For simplicity let us use a two molecule plastic with fraction A have the one molecule with Mw equal 10,000 and fraction B one molecule with MW equal 100,000. Therefore,

$$
\begin{array}{llll}
1 & \text{polymer molecule } Mw \text{ (molecular weight)} & = & 10,000 \\
\underline{1} & \text{polymer molecule } Mw \text{ (molecular weight)} & = & \underline{100,000} \\
2 & \text{total polymer molecules} \quad \text{Total } Mw & = & 110,000
\end{array}
$$

The M_n from (3-1) is

$$10,000 \quad \times \quad \tfrac{1}{2} = 5,000$$

$$100,000 \quad \times \quad \tfrac{1}{2} \quad = \quad \frac{50,000}{55,000} \quad = M_n$$

The M_W from (3-2) is

$$10,000 \quad \times \quad \frac{10,000}{110,000} \quad = \quad 909$$

$$100,000 \quad \times \quad \frac{100,000}{110,000} \quad = \quad \frac{90,909}{91,818} \quad = \quad 92,000 \; = \; M_w$$

Let us now construct a polymer with the same molecules but using equal molecular weights instead of equal molecular numbers. Therefore,

10 polymer molecules of M_W (10,000)	=	100,000 M_W	
1 polymer molecule of M_W (100,000)	=	100,000 M_W	
11 Total number of total M_W	=	200,000	
polymer molecules			

The M_n from (3-1) is

$$10,000 \quad \times \quad \frac{10}{11} \quad = \quad 9,090$$

$$100,000 \quad \times \quad \frac{1}{11} \quad = \quad \frac{9,090}{18,180} \quad = \quad 18,000 \; = \; M_n$$

The M_W from (3-2) is

$$10,000 \quad \times \quad \frac{100,000}{200,000} \quad = \quad 5,000$$

$$100,000 \quad \times \quad \frac{100,000}{200,000} \quad = \quad \frac{50,000}{55,000} \quad = \quad M_w$$

It should be noted that M_W is always greater than M_n., M_W is more sensitive to high molecular weight molecules while M_n is affected more by the lower molecular weight fractions.

The ratio of the weight average (M_W) over the number average (M_n) molecular weight is usually used as a measure of the broadness of a particular polymer.

$$\text{Degree of hetrogeneity or polydispersity} \quad = \quad \frac{M_W}{M_n} \tag{3-3}$$

Using our examples with equal number of polymer molecules

$$\frac{M_W}{M_n} \quad \frac{92,000}{55,000} \quad = \quad 1.7$$

With equal weight of polymer molecules

$$\frac{M_w}{M_n} = \frac{55,000}{18,000} = 3$$

This ratio does not show the distribution curve and is particularly sensitive to errors in the number average molecular weight. It is useful as a comparative value of the dispersity of a polymer rather than its absolute value.

Molecular Weight Measurements

Molecular weights are measured indirectly by such methods as osmotic pressure, light scattering, and sedimentation equilibrium. A second way is to measure properties which are connected to the molecular weight by a known function, such as viscosity. The number average molecular weight for polymers with weights below 20,000 are usually determined from freezing point, boiling point, and vapor pressure measurements. Above 20,000 and up to a 1,000,000 it is measured by the osmotic pressure of solutions of the polymer. When molecular weights are very high, over a million, an electron microscope is used for analysis of molecular weight and molecular weight distribution. Under certain conditions low molecular weight polymers are weighed by analyzing the end groups. Tagging end groups radioactively is also used.

The number average molecular weight is used in glass transition calculations, diffusion coefficients, sendimentation coefficients and analysis of polymerization kinetics.

The weight average is determined by light scattering techniques and ultracentrifuging. The weight average is used for viscosities at low shear rates, tensile impact strength, and turbidity. Some properties are best described by a molecular weight between M_n and M_w. This is occasionally called M_z and is used for viscosity at high shear rates, percentage elongation at failure, and tensile strength.

Melt Index Viscosity and Molecular Weight

All the preceding techniques are difficult and time consuming. A relationship has been found between the viscosity and the molecular weight. In dilute solutions the emperical relationship is

$$\eta = K M_w^a \tag{3-4}$$

where K and a are constants depending on the nature of the polymer, the solvent, and the temperature but independent of molecular weight.

Flory found an emperical relationship for molten polymers,

$$\log \eta = A + B M_w^{1/2} \tag{3-5}$$

where the constants A and B depend on the nature of the polymer and the temperature. The relationship holds for a number of linear condensation polymers over wide ranges of molecular weights. Although Eq. 3-5 has subsequently been modified, its form is intact and it is the basis for using the melt index (MI). The MI measures the viscosity of a polymer at low (Newtonian) shear rates, which, in effect, is a measure of the molecular weight. The viscosity of polymers are strongly influenced by the M_W. (39a).

Melt Index (MI) and Molecular Weight.   Since MI is a measure of molecular weight, the change in properties with changing MI is best described in terms of molecular weight. High molecular weight materials would be expected to entangle more than lower molecular weight ones. Therefore they should be more resistant to flow and extrude less material (weight) under equal conditions, that is, a lower MI.

In general the physical properties of a polymer increase with increasing molecular weight. Above a certain molecular weight, specific for each polymer, the rate of increase rapidly diminishes. The processability (viscosity) of a polymer decreases exponentially with increasing molecular weight (Eq. 3-5). A point will be reached where the polymer is so viscous that it cannot be processed without degradation. A commercial polymer is a compromise between physical properties and processability.

MI Measurement   The melt index is measured with an extrusion rheometer as described in the ASTM Test D-1238 (Figure 3-6). It consists of an insulated, heated chamber 3/8 in. in diameter with an orifice of 0.0825 in. The polymer is put into a cylinder under a plunger on which is a weight of 2160 g (equal to 43¼ psi on the polymer). The material extrudes from the orifice. When equilibrium conditions have been reached the extrudate is cut off and the weight, in grams, extruded in 10 min is the melt index.

The MI is primarily used for olefins. For polyethylene the chamber is heated to 374°F. The melt index of polyethylene normally used for injection molding varies from 0.2 to 30. The melt indexing of polypropylene and an analysis of some of the problems is given in Ref. 40. The MI is occasionally used in evaluating polystyrene, ABS, acrylics, nylon, and acetals. The MI measures the flowability or viscosity of the polymer at very slow flow rates. This is not necessarily characteristic of the flowability during injection molding, which is at much higher shear rates (41 – 46). Other methods have been developed for studying the flow rates during molding. They include plots of the shear stress against the shear rate, the apparent viscosity versus the shear rate, and the length of flow under specified conditions in spiral molds.

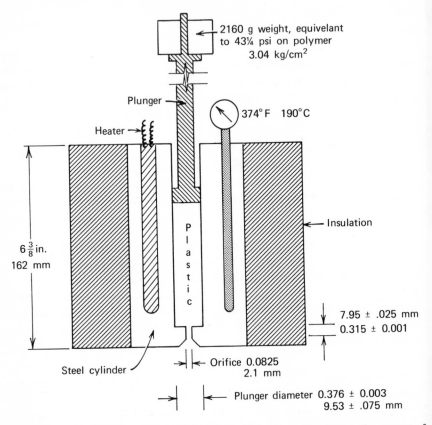

Figure 3-6 Melt indexer used for olefins. The melt index number (MI) is the amount of plastic extruded in 10 min reported as grams.

Molecular Weight versus Density

The material becomes less dense as the molecular weight increases (MI decrease). It can be shown, for example, that polypropylenes of equivalent isostatic content (a measure of the maximum amount of crystallinity) processed under identical conditions will yield lower crystallinity as the molecular weight increases (MI decreases). It would seem that the smaller molecules have a greater number of chain ends per volume and can move more freely to positions favoring crystallization. The lower crystallinity explains some of the property changes caused by the molecular weight.

Molecular Weight versus Notched Izod Impact Strength

The notched Izod impact strength increases materially with increasing MW (lower MI). For example, a 5 MI polyethylene has an Izod notched impact strength of 3-ft-lb/in. notch. Decreasing the MI to 0.5 (increasing MW) raises the izod to 21. A 1.7 MI polypropylene has a notched impact strength of 1.2. Reducing the MI to 0.3 raises the Izod to 3.8. In polypropylene an MI of 0.3 corresponds roughly to a molecular weight of 250,000, while an MI of 1.7 corresponds to 120,000. One would suspect that the increasing resistance to failure by impact, with increased molecular weight, is due both to the lowered crystallinity and the increased strength of molecular entanglements.

Tensile Yield versus Molecular Weight

Tensile yield strength increases slightly with increasing MW (decreasing MI). A 15 MI, 0.96 density PE has a tensile yield strength of 4300 psi. A 0.42 MI has a tensile yield of 4600 psi. The increasing tensile yield strength reflects the longer molecules and the increased degree of entanglements.

Tensile Strength versus Molecular Weight

For the same reason tensile strength increases with increasing MW (lowering MI). A 0.95 density polyethylene has a 3200-psi tensile strength at a MI of 8 and a 4000-psi tensile strength at a MI of 2.

Stiffness versus Molecular Weight

As the molecular weight increases (MI decreases) the stiffness decreases. This is probably the net result of the effects of crystallinity and chain length. A 0.4 MI polypropylene has a stiffness in flexure of 160,000 psi, while a 4 MI is at 220,000 psi.

Percentage Elongation to Failure versus Molecular Weight

The percent elongation to failure increases sharply with increasing MW (decreasing MI). The larger molecules have more entwining and more C-C linkages to stretch. A 2 MI polypropylene has a 40% elongation as compared 0.3 MI with a 600% elongation.

Brittleness Temperatures and Impact Strength versus Molecular Weight

The brittleness temperature decreases with increasing MW (decreasing MI). A linear polyethylene with a 5 MI has a brittleness temperature of -100°F. At 1.5

MI it is -180°F. Since brittleness temperature is a measure of the cohesive strength of the molecule, one would expect the higher molecular weight material to behave better at lower temperatures. For the same reason the impact strength at a given temperature is higher with increased MW (lower MI). At 120°F a 3 MI polypropylene has an Izod impact strength of 3, while a 0.3 melt is 11.

Environmental Stress Cracking versus Molecular Weight

The higher the MW (lower MI) the greater resistance to stress cracking. The greater entanglement of the polymer makes it stronger and probably leaves less molecular space for the "invading" molecules. For example, the time for a 0.96 density molded PE part to show stress cracking in a chemical environment was 15 hr. for a 1 MI and 270 hr for a 0.1 MI.

Shrinkage and Warpage versus Molecular Weight

The lower molecular weight (higher MI) materials flow much more rapidly into the mold. They also have lower melting points, requiring less time for cooling. Both these factors reduce the internal stress which is the main cause for warping and shrinking.

Creep versus Molecular Weight

Creep decreases with increasing molecular weight (lower MI). It is the permanent deformation resulting from a prolonged application of stress below the elastic limit. At room temperature creep is sometimes called cold flow. Below the glass transition temperature, molecular weight has little effect on creep. At these temperatures the molecular movement is basically that of segments or chains rather than the whole molecule. Therefore the total molecular motion will be fairly constant for a given amount of the polymer, regardless of the molecular weight.

Above T_g the polymer acts as a viscous liquid, with the whole molecule moving. The rate of creep is almost entirely dependent upon the viscosity which is directly related to the molecular weight (Eq. 3-5). (See Table 3-3).

MOLECULAR WEIGHT DISTRIBUTION AND PROPERTIES

There are properties of plastic which cannot be fully described by the density, weight average, or number average molecular weight of the polymer. They can be correlated to the molecular weight distribution (MWD) (47). The MWD of a given grade of polymer is determined by the process of manufacture and is not altered by normal processing. The polymer consists of short, medium, and long

Table 3-3   Effect of molecular weight (MI) on polymer
properties — increasing MW (lowering MI) causes a
change in the property listed in column A

A	Increase	Decrease
Crystallinity		X
Density		X
Notched impact strength	X	
Tensile strength	X	
Stiffness		X
Percentage elongation to failure	X	
Brittleness temperature		X
Shrinkage and warpage	X	
Creep		X
Environmental stress cracking	X	

chain molecules. The percentage of each determines the molecular weight distribution. If it is made up mostly of chains of the same length it is called a narrow molecular weight distributions (NMWD). If the distribution contains a relatively broad range of molecular weights it is called a broad molecular weight distribution (BMWD). This is approximately what is measured in the ratio of the molecular weight average and molecular number average (Eq. 3-3). A more accurate presentation is a graph showing the molecular weight distribution. Two high density polyethylenes with the same M_w/M_n distribution but with different distribution curves produced polymers having separate flow, mechanical, and morphological properties (48). This method can be used to construct polymers with a combination of properties which might otherwise be unavailable.

Increasing the broadness of the molecular weight distribution makes the polymer more sensitive to the rate of flow (shear rate). Adding lower molecular weight fractions causes a rapid increase in flow (decrease in viscosity) at high shear rates. The reasons are discussed more fully in the sections on rheology and orientation. In effect, the small chains separate the large chains from each other. This reduces the Van der Waal's forces. Additionally they are acting as lubricants for the larger molecules. Because of this, flow at very low shear rates (the melt index range) is not necessarily comparable to those at very high shear rates. A broader MWD will flow more readily. It is therefore possible to have a higher MI material flow more slowly under molding conditions than a lower MI material.

In general the broad MWD materials are more sensitive to molding conditions. A modified spiral flow mold was used to measure the flow length of a 20.5-MI 0.95-density PE at a cylinder temperature of 425°F, a cylinder pressure of 10,000 psi, and a mold temperature of 32°F. The same material with a narrow molecular weight distribution flowed 21 in. and a broad molecular weight distribution flowed 27 in. (44). Increasing the mold temperature increased the flow length linearly. This relationship did not change with the MWD.

The increased pressure sensitivity was shown with the same material, using the same cylinder temperature and mold temperature, but varying the injection pressure.

Cylinder Pressure		5000 psi; 352 kg/cm^2	15,000 psi; 1,055 kg/cm^2	Increase in Flow Length
Narrow MWD	in.	15	27	12
Flow length	mm	381	686	305
Wide MWD	in.	19	35	16
Flow length	mm	483	889	406

Increasing the pressure 10,000 psi increased the flow length of a narrow MWD 12 in. while that of the broad MWD 16 in.

The use of M_w/M_n ratios and MI for describing the effects of different MWD is discussed in Ref. 45.

The effects of broad MWD on the viscosity and shear rate are shown in Refs. 43 to 46.

The effects of MWD on some polymer properties are shown in Table 3-4.

Table 3-4   Effects of molecular weight distribution on properties and processing characteristics

To obtain	Make the MDW Narrower	Broader
Higher impact strength	X	
Lower shrinkage	X	
Lower warpage	X	
Better dimensional stability	X	
Higher tensile strength	X	
Higher resistance to creep	X	
Better low temperature brittleness	X	
Improved flow		X
Higher sensitivity to pressure		X
Higher sensitivity to temperature		X

RHEOLOGY AND WHAT IT MEANS TO THE MOLDER

Rheology is a study of liquid flow and deformation properties in terms of stress, strain, and time. Obviously this is of major importance for the injection molder.

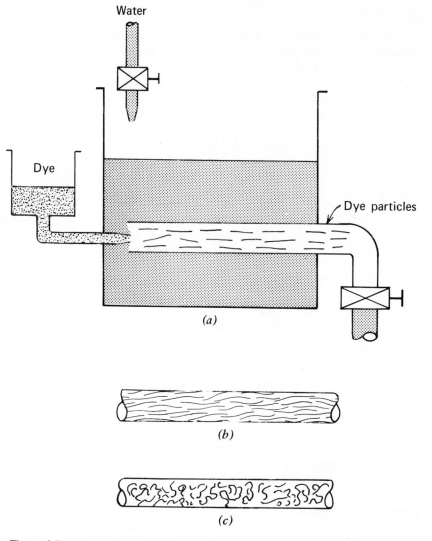

Figure 3-7 Osborne Reynold's flow experiment: (a) laminar – to 2000; (b) transition – 2000 to 3000; and (c) turbulent – over 3000.

The principles of rheology are concerned with the flow of hydraulic fluids in the molding machine, the fluids in the heat transfer of the mold, and in the flow of polymers in the heating cylinder and mold.

Reynolds Number

In the eighteenth century, Osborne Reynolds discovered some of the basic principles of flow. He took a container (Figure 3-7) to which he attached a tube with an outlet valve which could control the amount and rate of flow in the tube. There was a source of inlet water to maintain the level in the container. Feeding externally, directly into the tube, was a pot of dye with a controllable flow rate. When the flow rate in the tube was low, the dye assumed a straight path parallel to the tube. This Reynolds called a laminar flow. As the flow increased there came a range where turbulence began to occur. When flow became very rapid there was complete mixing of the dye in the tube. He derived a formula to describe this phenomena for a circular tube using a dimensionless number known as the Reynolds number.

$$N_R = \frac{\rho U D}{\mu} \tag{3-6}$$

where N_R = Reynolds number
ρ = mass density $[(lb)(sec^2)/ft^4]$
U = velocity (ft/sec)
D = diameter (ft)
μ = viscosity $[(lb)(sec)/ft^2]$.

When the Reynolds number is below 2000, the flow is laminar. Between 2000 and 3000 there is a transition period and above 3000 there is turbulent flow. It is obvious that the laminar flow is the most efficient. Imagine a molecule flowing the length of the tube. In laminar flow it will be least likely to collide with any other molecules or deviate largely from the direct straight line path. In turbulent flow considerable energy is lost in collisions with other molecules and in the much longer distance the molecule has to travel from one end of the tube to the other. In an oil hydraulic system, turbulent flow generates more heat, which has to be removed. Inspecting Reynolds' equation we know that the mass density and viscosity of the oil will not change significantly. The velocity and pipe diameter are the variables.

Reynolds' formula was developed for Newtonian fluids (p. 223). Although thermoplastics are non-Newtonian, experience shows the parameters described by Reynolds can apply. An outstanding example of this is in the molding of variegated colors. When two colors are molded simultaneously through a large gate, poor mixing is obtained and the desired mottled patterns ensue. If the gate is restricted (pin pointed), it will be impossible to obtain the same patterns. The turbulence generated by decreasing the diameter (increasing the velocity) will give excellent mixing. This principle is used in dispersion disks inserted in nozzles for improving the mixing and homogeneity of plastics alone or with colorants.

Newtonian Flow Rates

Two men, Hagan and Poiseuille, independently derived the volumetric flow rate for a Newtonian liquid.

$$Q = \frac{\pi}{8} \frac{R^4}{L} \Delta P \ \frac{1}{\mu} \tag{3-7}$$

where Q = volumetric flow rate
R = radius of tube
L = length of tube
ΔP = pressure drop
μ = viscosity.

This formula, for a cylindrical tube, shows that the volumetric flow rate depends on three parameters. The first is the physical constants of the tube, R^4/L. A Newtonian liquid is extremely sensitive to the radius. This is less so with plastics since, as we shall shortly see, viscosity varies with the shear rate (velocity). Increasing R will increase Q, but also will decrease the velocity. This will increase the viscosity because of the polymer's sensitivity to shear rate and partially reduce the full flow effects of the increase of R. As we would expect the more pressure (ΔP) the higher flow rate. Finally, the more viscous (μ) the material, the less the flow rate. The flow rate is influenced by the length of the tube (L). This corresponds to the land length of a gate. When we wish maximum flow into a cavity, which is the usual situation, we keep the land length to a minimum. The land length can be used to help balance the flow of material into multicavity molds.

In rheology the word stress is not used in the sense of a force acting on a body. It is the internal resistance of a body to an applied force. This resistance consists of the attraction of molecular bonds and forces. Thus when we say that we increase the shear stress to increase the shear rate, we really mean that we have to overcome increasing molecular resistance to achieve a faster flow rate.

Shearing stress is the measure of the resistance to flow of sliding layers of molecules. It is reported in pounds per square inch. Force is measured in the same units but is different in two respects. Force acts perpendicular to the body while shear stress acts parallel to the containing surface. Pressure is force per unit area while shearing stress is a resistance to force. Newton developed the concept of this doctrine by using concentric cylinders. He stated that the shear stress caused by the viscosity of the liquid is proportional to the shear rate. We can understand this concept by imagining a stationary plate (Figure 3-8) with a movable plate of area A, moving at a velocity U, with a force F, stationed at a distance X, above the stationary plate. Neglecting "slip" we assume the velocity of the liquid at the stationary plate as zero, and the maximum velocity U at the moving plate. The rate of change of velocity across the liquid is the slope of the

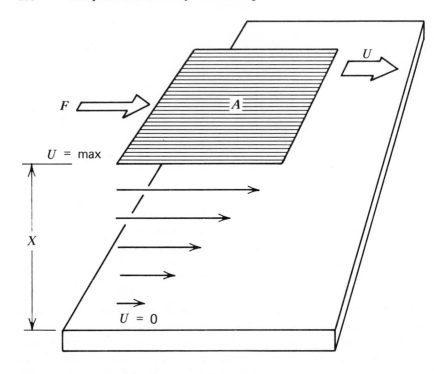

Figure 3-8 Concept of viscosity.
   F is proportional to A and U (velocity).
   $f = \mu$(Newtonian viscosity) or η (non-Newtonian viscosity)

$$F = (f)\, A\, \frac{du}{dx}$$

$$\frac{F}{A} = \text{shear force } \frac{lb.}{in.^2} = \frac{\text{dynes}}{\text{cm}^2} = \tau$$

$$\frac{du}{dx} = \text{shear rate} \left(\frac{in.}{sec.}\right)\left(\frac{1}{in.}\right) = \sec^{-1} = \gamma$$

$$\mu = \frac{\tau}{\gamma} = \frac{(lb)\,(sec)}{in^2} = 68{,}948\,\frac{(dyne)\,(sec)}{cm^2}\ \ (\text{poise})$$

line connecting the velocity vectors or du/dx.
   The force is proportional to the area and velocity.

$$F = (f)\,(A)\,\frac{du}{dx}$$

The proportionality constant (f) is called the viscosity and designated μ for Newtonian liquids or η for non-Newtonian liquids. Shear force or stress is represented by the Greek letter τ, and shear rate by γ.

Rearranging the terms we have the following:

$$\mu = \frac{\dfrac{F}{A}\,(\text{shear stress})}{\dfrac{du}{dx}\,(\text{shear rate})} = \frac{\tau}{\gamma} \tag{3-8}$$

In a Newtonian liquid, therefore, the shear force is directly proportional to the shear rate; double the unit force and you double the unit rate. In thermoplastic material this does not turn out to be the case. In the processing range a unit increase in the shear force may quadruple the shear rate. The viscosity is dependent on the shear rate and drops exponentially with increasing shear rate.

Rheometers

Rheometers were developed to measure the shear stress and shear rate during polymer flow. Viscosity is then calculated from Eq. 3-8.

Let us consider flow in a tube (Figure 3-9). Imagine the tube to be constructed by bending the flat plate of Figure 3-8 so that it becomes a cylinder. At the cylinder wall the velocity is zero (assuming that there is no slip of the polymer along the wall). Since one end of the polymer is anchored to the wall the shearing stress will be maximum at the wall. The shear rate (du/dx) is given by the slope of the velocity profile at any point. This would vary from zero at the center of the tube to its maximum at the wall. In the center of the tube the velocity will be maximum and the shear stress and shear rate will be minimum. To characterize the viscosity we must compare shear rate and shear stress at the same point. They are both maximum at the wall and rheometric calculations refer to this point.

The wave front of the fluid in a tube is a parabola. The volume under a parabola is

Volume $= \dfrac{\pi}{2}\, r^2 h \; > \; h = \tfrac{1}{2}r \qquad$ (Figure 3-9)

Volume $= \dfrac{\pi r^3}{4}$

shear rate $(\gamma) = \dfrac{\text{Volumetric flow rate}}{\text{Volume}} = \dfrac{Q}{\pi r^3/4} = \dfrac{4Q}{\pi r^3} \tag{3-9}$

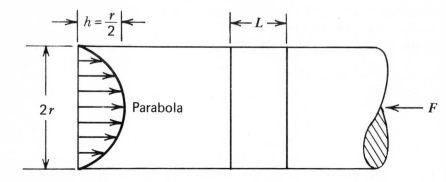

Figure 3-9   Flow in a capillary tube

At the wall of the tube
Velocity = 0
Shear rate = maximum (slope du/dr)
shear stress = maximum

At the center of the tube
Velocity = maximum
Shear rate = minimum
Shear stress = minimum

Shear Rate
Volume of parabola $= \frac{1}{2}\,\pi\,r^2\,h = \frac{1}{2}\,\pi\,r^2\,\frac{1}{2}\,r = \frac{\pi r^3}{4}$

Shear rate $(\gamma) = \dfrac{\text{Volumetric flow rate }(Q)}{\text{Volume}} = \dfrac{Q}{\dfrac{\pi r^3}{4}} = \dfrac{4Q}{\pi r^3}$

Shear Stress
$F = P \times A$
Force (F) acting parallel to tube $= P\,\pi\,r^2$
Stress opposing (F) = shearing stress (τ) $\times$ shearing area $|2\,\pi\,r\,L|$
The two forces are equal, thus:
$P\,\pi\,r^2 = \tau\,2\,\pi\,r\,L$
Shearing stress $\tau = \dfrac{Pr}{2L}$

Analyzing this equation dimensionally yields.

Shear rate $=\left(\dfrac{\text{in.}^3}{\text{sec}}\right)\left(\dfrac{1}{\text{in.}^3}\right) = \text{sec}^{-1}$

Therefore, shear rates in rheometry are reported in units of reciprocal seconds.
The shearing stress (τ) can be computed by considering what happens to a

plug of liquid L inches long. The force (perpendicular to it) is equal to the pressure times the area ($P\pi r^2$). The force opposing it would be the shearing stresses at the wall multiplied by the area of the wall (circumference × length = $2\pi rL$) or $\tau(2\pi rL)$. Since the force required to move the plug is that required to overcome the shearing stress we can equate the two and solve for τ

$$P\pi r^2 = \tau(2\pi rL)$$

$$\tau = \frac{Pr}{2L} \tag{3-10}$$

Analyzing this equation dimensionally yields

Shear stress $= \left(\dfrac{lb}{in.^2}\right)\left(\dfrac{in.}{in.}\right) = \dfrac{lb}{in.^2}$

Therefore shear stress in rheometery are reported as $lb/in.^2$.

The viscosity is the shear force over the shear rate. Substituting dimensions

Viscosity $= \dfrac{lb}{in.^2}\ \dfrac{1}{1/sec}\ \dfrac{(lb)(sec)}{in.^2}$

The metric system uses dyne-sec/cm^2 which is called a poise named after Poiseuille. The relationship is

$$1\ \frac{(lb)(sec)}{in.^2} = 68,948\ \frac{dyne\text{-}sec}{cm^2}\ (poise)$$

The proceeding discussion suggests that shear rate and shear stress could be measured by constructing an instrument which would extrude the polymer through a capillary while measuring the force and speed of the plunger. The viscosity would then be a simple calculation. This has been done and the instrument is called a capillary rheometer (Figure 3-10).

It is also possible to calculate viscosity using a rotational viscometer. The plastic is sheared between a rotating cylinder and a fixed wall at a constant speed. Upon equilibrium the torque and velocity of the rotating body are measured. A second type of rotational rheometer often used in the plastics industry is called a Brabender rheometer. It consists of two irregularly shaped rollers in a heated chamber. The data consist of values of the torque and plastic temperatures at constant rotational speed versus time. Analyses of this rheometer and conversion of its data to that obtained by capillary rheometry is found in Ref. 49 to 52.

The most widely used instrument is the capillary rheometer (53, 247) (Figure 3-10). It consists of a 14¼-in.-long barrel in which a 0.375 hole is bored. A plunger is closely fitted into the bored barrel. The bore is attached to a capillary whose diameter can be varied between 1/32 and 1/16 in. The length of the

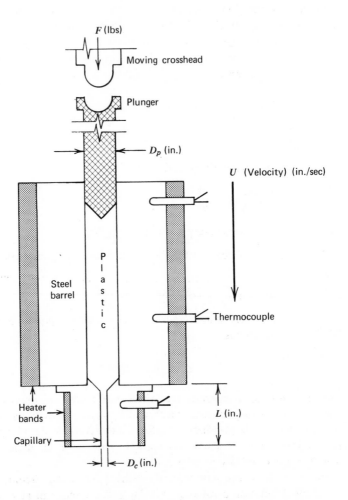

Figure 3-10 Calculations for a specific Instron® capillary rheometer.

Shear stress $(\tau) = \dfrac{FD_c}{\pi LD_p^2} = \dfrac{\text{lb}}{\text{in.}^2}$

Shear rate $(\gamma) \left(\dfrac{2}{15}\right)\left(\dfrac{UD_p^2}{D_c^3}\right) = \dfrac{1}{\text{sec}} = \text{sec}^{-1}$

Apparent viscosity $(\eta) = \dfrac{\tau}{\gamma} = \dfrac{(\text{lb}) \ (\text{sec})}{\text{in.}^2}$

Calculation $\eta = \left(\dfrac{15}{2\pi}\right) \left(\dfrac{FD_c^4}{LUD_p^4}\right)$

capillary is varied so that the L/D ratios can be selected from 3.5 to 188. So that the capillary geometry does not affect the results, minimum L/Ds of 60 are required for polystyrene and 30 for polyethylene.

The L/D of a melt indexer is 3.8. This is another reason why MI does not always correlate with comparative flow properties. The barrel and capillary sections are kept at a constant temperature and controlled by thermocouples. The head of the plunger is attached to an Instron® testing machine. Speeds start at 0.02 in./min and the maximum load is rarely more than 2000 lb.

The method of calculating shear stress, shear rate, and viscosity are show in Figure 3-10. The only two variables are the force and the velocity. All others are the physical constants of the rheometer.

Rheometer Corrections.   There are a number of corrections that have to be made to the raw data obtained in a rheometer. Since these corrections are only occasionally referred to in the literature on rheology and since, more importantly these corrections illustrate some of the properties of viscous elastic flow, a brief discussion on these corrections is warranted.

1. An important error which must be corrected is in the effect of forcing the polymer from the large diameter piston chamber into the much smaller capillary. The energy lost should only come from viscous flow, rather than any elastic effects of this compression. This error can be corrected by using capillaries of varying L/D dimensions and suitable graphic treatment. Also since polymers are sensitive to the shear rate, these determinations are made at different shear rate values. There are other elastic absorbences of energy in the system. These are usually negligible when large L/D capillaries are used.

2. Energy is required to accelerate the fluid velocity to obtain the average flow rate. This energy is lost after the plastic has been extruded. The effect of this is small enough to be ignored for most calculations.

3. When a plastic flows, the "friction" of the molecule is appreciable and can raise the temperature of the material. Because of the low thermal conductivity of the plastic the heat cannot be dissipated quickly enough. Consequently, the temperature of the material will increase and the viscosity will decrease. Under most circumstances this temperature rise is ignored.

4. The material is compressed significantly when it enters the capillary. As the pressure drops it expands and becomes a lower density material. This changes the volumetric flow rate which can be corrected from the equation of state of the polymer.

5. Since the apparent viscosity varies with the fourth power of the capillary diameter, accurate measurements are needed. The difficulties of accurately measuring a capillary tube have not been completely overcome. While this error can be of considerable magnitude, it does not appreciably effect the use of flow curves.

6. There is one other important correction. The shear rate equation has been derived for a Newtonian fluid but plastics are generally non-Newtonian. Calculated values obtained with data using this equation are called the apparent Newtonian wall shear rates. Rabinowitch derived a correction for converting the Newtonian wall shear rate to what is called the "true wall shear rate."

Plotting Rheological Results.   The results of rheological measurements-shear rate, shear stress, and viscosity—are usually shown graphically. The curves of each type material assume characteristic shapes which are used for their classification (Figure 3-11). The arithmetic plot of shear rate versus shear stress for a Newtonian material is a straight line whose slope is related to its viscosity.

Dilatant Material.   A dilatant material requires more than a unit increase in shear stress for a unit increase in shear rate. These materials are found in some plastisols and highly filled plastics. One explanation for this type flow assumes that at low flow rates the particles can glide over each other. At higher rates eddy currents causes one phase to ball up and require the other phase to jump over it. This causes the expansion or "dilatant" effect. As expected this is a reversible phenomena. Dilatant behavior does not normally occur in injection molding.

Bingham Materials.   A Bingham material is one that will not flow until a certain shear stress is applied. Once these materials flow their flow characteristics are usually between Newtonian and plastic. Examples of Bingham materials are concentrated polymer solutions which may form a gel structure, catsup, mustard, and toothpaste. Bingham materials are not normally met in injection molding.

Plastic Flow.   This flow (sometimes called pseudoplastic) is characteristic of molten polymers above their glass transition point. At low shear rate the flow is Newtonian (straight line). At shear rates in the processing range, a unit increase in shear stress will result in several unit increases of shear rate (curved line). At very high shear rates the flow is again Newtonian (straight line).

An arithmetic plot of the viscosity versus shear rate Figure 3-11 shows the Newtonian curve—a straight line parallel to the shear rate axis. This shows that the shear rate has no effect on the viscosity. The Dilatant curve shows that as the shear rate increases the viscosity increases. In a plastic material the viscosity drops exponentially with the shear rate, so that there is a different viscosity value for each shear rate.

Since viscosity is the measure of flowability, it is readily apparent that a plastic may be too viscous to fill a mold at low shear rates but perfectly suitable if the shear rates are increased. The flow can be approximated by a power law

$$\tau = K \ (\gamma)^{n} \tag{3-11}$$

which describes flow curves very well for large intervals of shear rate up to 4

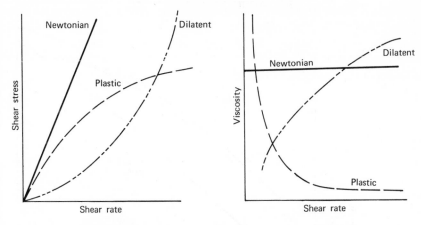

Figure 3-11 Arithmetic plots of shear stress/shear rate and viscosity/shear rate curves for (a) Newtonian liquids, (b) plastic fluids, and (c) dilatent materials.

magnitudes. The K and n are constants. When n is 1 the flow is Newtonian and K is the viscosity. The exponent n is called the flow index and its numerical value is a measure of the deviation from Newtonian flow. The greater the difference from 1 the less Newtonian is the system.

When n is less than 1 the material is plastic with the slope of the flow curves getting smaller with increasing shear rate. The viscosity decreases with increasing shear rate.

When n is greater than 1 the material is dilatant. The slope of the flow curve increases with increasing shear rate as does the viscosity.

It is evident that for reasons of accuracy and convenience these curves can best be plotted on log-log paper. In that instance the shear stress/shear rate curve shows a Newtonian material at an angle of $45°$ (slope of 1). A plastic will have a slope of less than 1 at every point on the curve and a dilatant material will have a slope more than 1 at every point of the curve.

Similarly log-log plots of viscosity/shear rate show that a Newtonian material will have a slope of 0 and be a straight line parallel to the shear rate axis. A plastic material will have a decreasing curve as the shear rate increases and a dilatant material will have an increasing curve as the shear rate increases.

Figure 3-12 shows a log-log plot for the shear stress/shear rate curve for four different plastic materials. The shear rate for compression molding ranges from 1 to 10 sec^{-1}, for calendering 10 to 100 sec^{-1}, extrusion 100 to 1000 sec^{-1} and injection molding 1000 to 10,000 sec^{-1}. It is impossible to characterize completely plastic flow over large shear rates by measuring it only at one shear rate point. The melt index is taken at very low shear rates. In some instances the

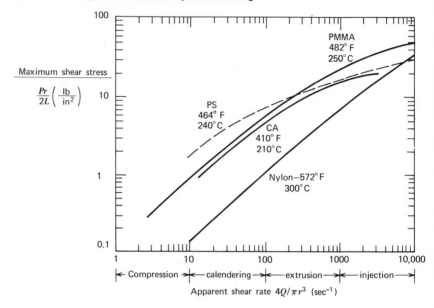

Figure 3-12 Log-log plots of shear stress/shear rate curves for four plastic materials.

flow curves cross at higher shear rates, giving erroneous results (the higher MI having poorer flow properties in the molding range). Therefore, in selecting a specific resin for a critical application, or in changing suppliers for the same material (particularly in the olefins), shear stress/shear rate curves of each material should be inspected.

Melt Fracture

It must be noted here that we are only discussing the viscous flow component of the plastic. There is an elastic component which is reversibly stored kinetic energy. When the shear stress exceeds the shear strength of the melt, melt fracture occurs. It is most likely to occur where there is a sudden increase in shear rate, such as at a restricted gate. Since it is reversible, it will usually disappear in the molded part. In some instances, particularly with a cold mold, it will result in surface blemishes which are overcome by lowering the shear rate. It is a significant problem in extrusion and there is considerable literature on melt elasticity and fracture (54–59).

The assumption was made that there is zero velocity at the wall of the capillary. Experience tells us this is not so, otherwise it would be impossible to change color. By using small irregularly shaped carborundum particles about

0.002 in. in size as tracers, velocity profiles for polyethylene melts were determined (60). Analysis showed only about 25% of the particles at the wall have 0 velocity. About 15% move with a moderate velocity and the balance at relatively slow velocities. The proposed mechanism suggests that the melt elasticity causes the slipping. The polymer just away from the wall deforms elastically and the polymer along the wall slips to reduce the elasticity. The flow in the center region of the tube had little shear and a uniform velocity. Slippage increases with molecular weight so that ultrahigh molecular weight polyethylenes become difficult to extrude (61). These considerations do not affect the practical use of rheometric data for injection molding.

Applications of Rheological Data

Before discussing experimental results and the applications of rheological data let us consider, conceptually, how polymer molecules flow. A plastic differs from a Newtonian material in that it contains long molecules. Its segments and chains are flexible. They intermingle and are held together by Van der Waal's type of forces.

All molecules have heat energy which result in vibration or movement within the bounds of molecular attraction. This Brownian movement, named after its postulator, tends to locate the polymer sections in a random position, this being the lowest energy level. This might be shown schematically in Figure 3-13. The plastic molecule is too large to move as a unit. Motion occurs in segmental units of the polymer.

If a force is applied in one direction to a polymer above its glass transition point, segments will begin to move in the direction away from the force, to relieve the stress. The net result is a movement of the viscous mass. As the polymer moves, it tends to untangle and orient itself parallel to the direction of flow. If the force is applied very slowly, so that the Brownian movement can cancel the orienting force caused by the flow, the mass of the polymer will move with a rate proportional to the applied stress. This is Newtonian flow (62).

As the plastic begins to move more rapidly the molecular segments and chains tend to orient in the direction of flow, as they did before. At this faster flow rate the Brownian movement is no longer able to return all of them to the random position. The orientation results in untangling the polymer and separating the molecules from each other, reducing the Van der Waal's forces (Figure 3-14). Obviously the untangled molecules will slide over each other more easily than the tangled ones, that is, lowering the viscosity. When the polymer is in the configuration of Figure 3-13, a unit increase of shear stress will give a unit increase of shear rate. Compare this with the situation in Figure 3-14. The polymer molecules are now further apart and the resistance to shearing is much less because of the decreased Van der Vaal's forces. Therefore he same unit

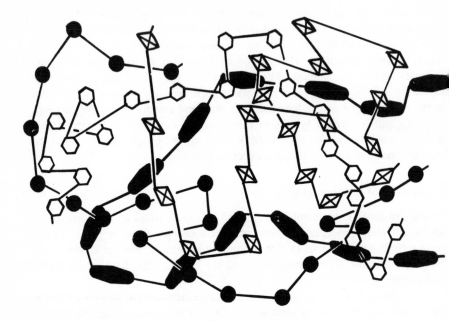

Figure 3-13 Schematic representation of polymer chains in a random pattern, the result of Brownian movement.

increase of shear stress will give a greater increase in the shear rate by at least several units. This is the plastic flow section of the curve in Figure 3-11. This process will continue until the polymer has reached its maximum state of orientation. At that time there is no further untangling and a unit increase in shear stress will again give a unit increase in shear rate, that is, Newtonian flow. This range is rarely met in injection molding. These shear rates are so high that they usually start to degrade the plastic.

Figure 3-15 shows an arithmetic plot of apparent viscosity and shear rate which illustrates plastic flow. At all temperatures, the viscosity decreases rapidly with increasing shear rate. Raising the temperature of the 0.7 MI-PE increases the separation of the molecules so that orientation (lower viscosity) starts more quickly. This effect of heating is shown by an almost identical curve displaced toward the left. As the shear rate increases the apparent viscosity of the materials at both temperatures tend to approach each other. This is saying, in effect, that the point has been reached where the overwhelming percentage of orientation has taken place and that additional shear rate will have relatively little effect on viscosity. This is true of most plastics in the processing range. It is interesting to note that most polymers have similar apparent viscosities,

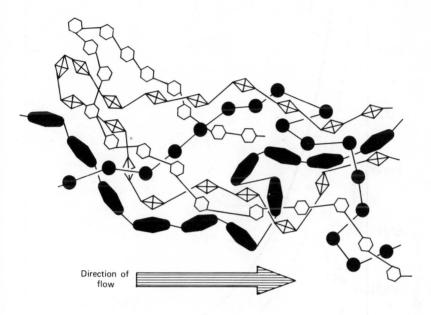

Figure 3-14 Schematic representation of Figure 3-13 when the polymer is flowing at a rapid speed and is oriented in the direction of flow.

700 - 7000 poise

10^{-1} to 10^{-2} lb sec/in.^{2}, at the shear rates (10^{-3} to 10^{-4} sec^{-1}) used in injection molding.

How Rheological Properties Relate to Molding Conditions

As the shear rate increases beyond the area of rapid change, the dependency of viscosity on shear rate decreases. For example, increasing the shear rate of the 0.7 MI-PE (428°F) from 200 to 400 sec^{-1}, decreases the viscosity by 4500 P. The same increase in shear rate from 900 to 1100 sec^{-1} decreases the viscosity by 350 P. Large variations in the viscosity caused by small changes in the shear rate will lead to molding difficulties. Surface imperfections, uneven filling, unequal densities, high stress levels, warpage, and differences in linear shrinkage may result. It is therefore important to mold at the part of the curve where the effect of shear rate is minimal.

When a mold does not fill it means that the viscosity is too low at the gate. This assumes the molding machine has enough injection speed to satisfy the flow

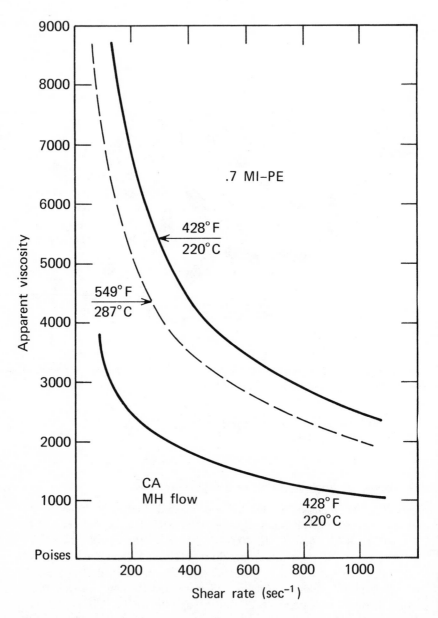

Figure 3-15 Arithmetic plot of viscosity/shear rate curves for (a) 0.7 MI polyethylene at two temperatures, and (b) MH cellulose acetate. (Ref. 63).

requirements (Q) and that the runner system is large enough to transmit this to the gate. One way of decreasing the viscosity is to raise the temperature (Figure 3-20). This is limited by the activation energy of viscous flow (which is discussed on p. 237) and the temperature at which the polymer degrades.

The second approach is to increase the shear rate. Figure 3-16 shows the effect of shear rate on viscosity for an acrylic. Assume a mold requires a viscosity of 0.07 lb-sec/in.^{2} to fill. When the shear rate is 100 sec^{-1} the material must be at 475° for this viscosity. If the shear rate is increased to 1000 sec^{-1}, the temperature is 400°F or 75°F lower. Consequently, the material will have to be cooled 75° less when molded at 1000 sec^{-1}. This is a major saving in molding time. Figure 3-17 shows the effects of shear rate on viscosity for some materials. Aside from the machine, the major limitation on increasing the shear rate is the possibility of polymer degradation. (As we have seen when discussing molding machines, high surface speeds rip the polymer apart.)

Shear rate can be increased by changing the geometry of the gating system and/or increasing the pressure (shear stress).

Both of these methods will lower the viscosity. If one substitutes for Q (Eq. 3-7) in the shear rate Eq. 3-9, one gets

$$\gamma = \left(\frac{4}{\pi R^3}\ \frac{\pi}{8}\right)\ \left(\frac{R^4\ \Delta P}{L\mu}\right)$$

$$\gamma = \left(\frac{R\ \Delta P}{2L}\right)\left(\frac{1}{\mu}\right)$$

The first group on the right hand side $R\Delta P/2L$ is, of course, the shear stress. Most materials are molded with short restricted gates to increase shear with diameters of about 1/16 in. diameter (0.003 in.^{2}) to 0.10 in. diameter (0.008 in.^{2}).

There are two other advantages to small gates. They break off easily requiring little if any cleaning, and they freeze off quickly so that the extra material packed into the cavity cannot flow back into the runner system. With large gates, however, injection forward pressure must be held for a considerably longer time to revent back flow, thus increasing cycle time. The shear stress also depends on the pressure drop between the runner and the cavity (ΔP). However, if the gate, cross-section R, is too large the machine system may not be able to deliver enough plastic to maintain a large pressure drop and in turn, shear stress would be significantly reduced. Beyond restricted gate sizes, gates must be

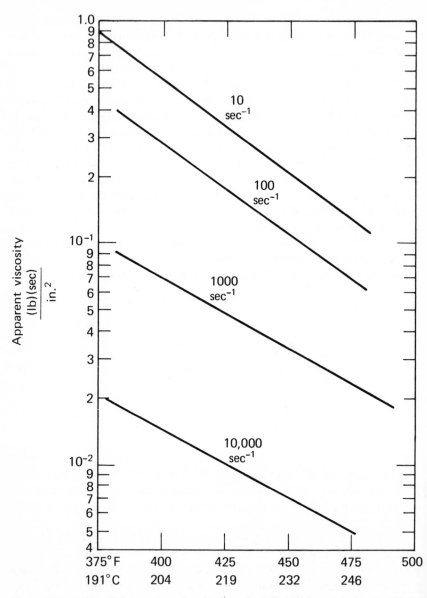

Figure 3-16 Effect of shear rate on viscosity for polymethyl methacrylate (Lucite 129 E. I duPont de Nemours & Company).

increased by many orders of magnitude to be moldable, so that R is increased enough to overcome the lowering of the shear stress caused by the lowered pressure drop (ΔP) of the larger gate.

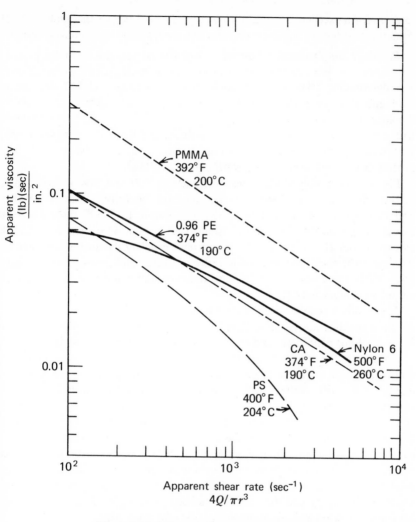

Figure 3-17   Effect of shear rate upon viscosity.

The effects of static pressure on viscosity have been investigated (64–66). The static pressure increase affects the flow much less than does ΔP, the pressure drop. Increasing the pressure drop from 100 to 1000 psi increases the flow rate from 0.01 to 1 in.^{3}/sec.

Increasing the static pressure from 10,000 to 25,000 psi raises the viscosity from 1000 to 4100 P for polystyrene, and from 1000 to 1220 P for polyethylene. Apparently one might expect a decrease in flow with very high pressures. An excellent anaylsis of these type data (67) shows that under normal conditions raising the injection pressure will always result in an increase in flow.

Rheology and Production Control.   Rheological data are useful for production control in the manufacture of resins, raw material control for processors, and determining relationships between polymer structure (MWD, branching, etc.) and flow properties (68, 69). To be of further value to the molder, rheological data obtained from capillary flow rheometers must be correlated with actual performances of molding machines and their extruder plasticizers. This would permit the molder to improve the quality of the molded part, processing rate, and selection of proper conditions (69a).

For example, polyethylene was extruded at 400 and 504°F four times, successively, with no change in the apparent viscosity. This means that degradation did not occur. Raising the temperature to 525°F showed a 29% reduction in viscosity after ten extrusions. Raising it to 646°F showed a 53.6% reduction after 10 extrusions (70). Data from a capillary rheometer were used to estimate operational performance in terms of output and processability of rigid PVC. Shear rates where melt fracture occurred and conditions for maximum output were predicted (71, 71a).

A simplified analysis of mold filling and some of the flow properties in injection molding are given in Refs. 72 and 73. Some of the effects of mold design, molding conditions, and materials on thermoplastic flow are discussed.

Looking at the schematic diagram of the rheometer, one is impressed by its similarity to an injection cylinder. The plunger corresponds to the injection plunger, the plastic material to the plastic in the "cylinder, nozzle, and runner system", and the capillary to the gate. There are some major differences however. The flow rate in a rheometer is steady, compared to the large variations in the mold. The material in the rheometer is at a constant temperature, compared to the nonisothermal state of the mold. The rheometer's material is heated by conduction while, in a screw machine, the material is heated primarily by shearing. Notwithstanding, there is good qualitative correlation between the results of rheometry and the injection molding process (74–77). Westover (78) has compiled extremely useful rheological data for many plastic materials.

The filling of a cold cavity with a hot non-Newtonian fluid was studied (79). Evidence was presented to show that the flow process depended on the rheological properties as measured in an isothermal, steady-state rheometer. Rheological data obtained for polystyrene in a capillary rheometer were correlated with the length of spiral flow over a broad range of temperature, but with otherwise constant conditions in a specific molding machine (80).

Spiral Flow Mold.   To evaluate better the moldability of material, a spiral mold was developed (81). These molds showed good correlation with rheological data (82), are an excellent method for comparing flow properties of different batches of the same material and of the same material from different sources.

Many types of molds using this principle have been developed for polymer evaluation. They have been used for studying the effect of gate size, cavity depths, flow rates, and evaluation of physical properties. A variation of this idea can be very useful in molding parts to tolerance. The shrinkage of a part is directly related to the amount of material molded in the cavity. A tab can be machined near the end of the gate. It should be long enough and thin enough so that it does not fill out even if the shot is flashed. Equally distant lines are machined and numbered successively. The length of the tab can be correlated with the critical dimensions. Control limits are set for the tab. If the moldings fall outside the specified limits an alert operator can call the foreman and prevent molding rejects.

When viscosities are taken at high shear rates approximating molding conditions (1500–2500 sec^{-1}) excellent correlation with spiral flow is obtained. Viscosities taken at low shear rates and constant shear stress (such as in the MI determination) did not show good correlation with spiral flow length.

Average Viscosity—The use of an expression for the "average" viscosity instead of the one point viscosities (MI) was developed. It correlated reasonably well with such practical molding tests as spiral mold length, minimum cycle temperatures, molding area diagram limits, and experiences in production runs on commercial parts (76).

Reference 83 states that for broad ranges of operating conditions in a wide variety of high pressure polyethylenes, the flow data obtained from a high shear rate viscometer and a commercial extruder were identical. Good correlation was obtained in using the data for predicting screw power consumption.

GLASS TRANSITION TEMPERATURE

Throughout this section we described plastic flow in terms of molecular properties. The differentiating characteristics of a polymer are the length of its molecules, the ability of its parts to rotate and be flexible, and the nature of the internal and external molecular forces. The ability of the C-C linkages to rotate freely permits the polymer molecule to move in segmental jumps. Mathematical models based on these assumptions are beginning to describe qualitatively these segmental motions.

There are two theoretical approaches to the flow of polymers. One approach treats viscous flow as a modified simple liquid. Here the viscosity of the system is caused by the frictional drag of molecular segments over each other when they move from position to position. This drag is due to exchanges of momentum between molecular segments.

The second approach, chosen by Eyring, examines the polymer as a disordered solid. Each polymer segment, containing approximately 20 to 40 monomer units, is assumed to move in a limited area bounded by the attractive and repulsive forces of its neighboring segments. Many defects or "holes" appear as a disorder in the disordered condition of the polymer. From time to time a segment will receive excessive kinetic energy and will be able to break the intermolecular forces and jump into a hole. Another segment will move into the hole just vacated. These are random movements. This approach is based on the assumption that there are unoccupied areas known as "free volume." The jump frequency of the segments depends on the energy level required to move them and the amount of free volume in the polymer. The free volume is distributed equally throughout. When an external stress is applied in one direction, the jumps will no longer be random but will move in such a manner as to relieve the stress. This causes the polymer to flow.

The Glass Transition Point

We are now in a better position to understand the glass transition point. If we take a molten polymer and cool it slowly, it gradually becomes elastic in nature, then it becomes leather-like, and, finally in a specific small temperature range, it abruptly changes into a solid. This point is called its glass transition point (T_g). At this point the polymer segments are no longer free to rotate. This applies to the longer molecular segments in the neighborhood of 20 to 40 monomer units long. Motions of shorter segments continue and other transition temperatures below T_g are found which indicate the cessation of a particular type of motion.

Figure 3-18 graphically shows the effect of free volume. Each small figure represents the rotational area of a polymer segment approximately 20 to 40 monomer units long. Below the glass transitional point there is not enough free volume for a significant number of segmental jumps. The amount of heat energy (temperature) is not enough to break the intermolecular forces. The specific volume (a measure of the free volume) increases relatively slowly with the increase in temperature.

At T_g the cohesive forces yield drastically and the polymer expands so that there is room for segmental rotation. Segment A moves into area A-1, B to B-1, C to C-1, and so on. Other segments move into the vacated spaces. The polymer now starts to flow, its characteristics being determined by the temperature increase above T_g and the crystalline structure, if any. Once this initial state (T_g) has been reached and the high energy barriers have been broken, a smaller amount of heat energy (T) will be required to expand a unit volume. This is in consonance with our understanding of the Van der Waal's forces which decrease as the sixth power of the distance.

There are a number of ways of determining T_g. The most common is to plot the specific volume against the temperature (Figure 3-19). At a particular point there is an abrupt change in the slope of the curve. This is the glass transition

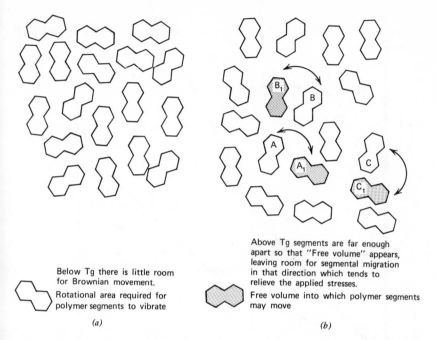

Below Tg there is little room for Brownian movement.
Rotational area required for polymer segments to vibrate

(a)

Above Tg segments are far enough apart so that "Free volume" appears, leaving room for segmental migration in that direction which tends to relieve the applied stresses.
Free volume into which polymer segments may move

(b)

Figure 3-18 Eyring "free volume" theory: (*a*) below T_g there is little room for Brownian movement, and (*b*) above T_g segments are far enough apart so that "free volume" appears, leaving room for segmental migration in that direction which tends to relieve the applied stresses.

temperature. As has been stated, below T_g the heat energy is not enough to break the intermolecular forces so that a unit raise in temperature gives a small unit increase in volume. Above T_g these barriers have been broken and the same unit increase in temperature will produce a much larger increase in volume. The second curve in Figure 3-19 shows a plot for a crystalline polymer. The amorphous section shows its own T_g. When the material reaches the melting point (T_m) the crystalline structure melts. Since crystals occupy the least volume, there is a very large increase in specific volume at that point.

Some of the other ways of determining T_g are dynamic mechanical methods, nuclear magnetic resonance, dielectric loss in polar polymers, the onset of brittleness, and refractive index (84, 85).

The most comprehensive listing of glass transition temperatures and melting points is found in Ref. 86.

Most commercial amorphous polymers have T_g from 115 to 300°F. Hence they act as solids at room temperatures. Most commercial crystalline polymers have glass T_g from -150 to 27°F. Therefore, they are usually ductile at room

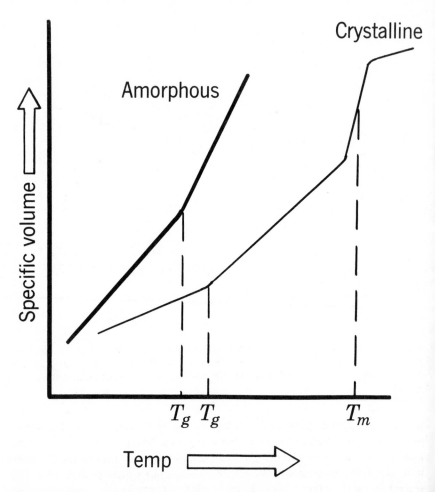

Figure 3-19 Plot of specific volume versus temperature for an amorphous polymer and a crystalline polymer, showing the glass transition temperature (T_g) and the melting point (T_m).

temperatures but maintain their rigidity because of the high modulus caused by crystallinity. Material such as nylon with low glass transition points and high crystalline melting points exhibit excellent mechanical properties.

From our conceptual model it would seem that as soon as the free volume is slightly larger than the moving segments, segmental displacement will occur (Figure 3-18). This in effect, is the glass transition temperature. From this we can infer how the T_g depends on polymer structure.

Polymer Structure and T_1

The following observations are made relating T_g to Polymer structure.

1. Polymers that have chains with stiff backbones or bulky side groups will have a high T_g. The moving segments will be large and a correspondingly large free volume would be required for T_g. This would require a correspondingly higher temperature than for a simpler polymer.

2. Polymers which have a high attractive force between the segments will require higher energy (T) to disassociate them, hence a higher T_g. The higher the molar cohesion energy, all other things being equal, the higher the T_g (87). Cross-linking the polymer would increase the size of the moving segments and raise T_g.

3. Polymer structures which include chains and side groups tend to spread the polymer allowing for more free volume. These tend to lower T_g. Similarly the addition of plasticizers will accomplish the same purpose.

Melting Point and T_g

There is a relation between T_g and the melting point (T_m). The ratio T_g/T_m generally is between 0.50 and 0.75. Symmetrical polymers such as polyethylene are in the lower range and nonsymmetrical polymers such as polystyrene are in the upper range. Table 3-5 shows the T_g and T_m of some thermoplastic polymers.

Relationship of Viscosity to Flow

One would predict from this model of polymer flow that increasing pressure would decrease the free volume and reduce the number of segmental jumps. This pressure increase should increase the viscosity. This has been found to be so (65), and might explain why increasing pressure when molding through restricted gates does not always produce as much increase in flow rate as expected. The increased pressure will increase the viscosity and tend to reduce the added flow effect of the pressure increase.

As one would expect the viscosity or resistance to flow is temperature dependant. The higher the temperature, the further apart are the molecular segments. This increases the free volume and decreases the electrostatic or Van der Waal's forces. The relationship is expressed by the Arrhenius equation

$$\eta = Ae^{(E/RT)}$$

(3-12)

where η = viscosity
A = constant depending upon the material
R = Gas Constant
T = °K
E = activation energy for viscous flow.

Table 3-5   Glass transition and melting points for some plastics as reported in the literature

Material	Glass Transition Temperature (T_g)		Melting Point (T_m)	
	°F	°C	°F	°C
Acrylonitrile-butadiene-	–	–	374	190
styrene	–	–	374	190
Acetal	-121	-85	347	175
Acrylic	158, 221	70, 105	320	160
Ethyl cellulose	109	43	–	–
Cellulose acetate proprionate	102	39	–	–
Cellulose tri-acetate	158	70	583	306
Cellulose acetate butyrate	122	50	–	–
PTFE (polytetrafluoro- ethylene)	-171, 68	-113, 20	626	330
Chlorinated polyether	–	–	358	181
CTFE (polychlorotri fluoroethylene)	95 to 113	35 to 45	428	220
FEP (fluorinated ethylene propylene)	52	11	–	–
PVF$_2$ (polyvinylidine fluoride)	-40	-39	340 - 410	171 - 210
Nylon 6/6	122	50	500	260
Nylon 6	122	50	437	225
Nylon 6/10	104	40	415 - 428	213 - 220
Nylon 11	115	46	360 - 381	182 - 194
Nylon 12	99	37	354	179
Polycarbonate	306	152	437	225
Polyethylene	-193, -4	-125, -20	230 - 286	110 - 141
Polypropylene	23, 113	-5, 45	342 - 349	172 - 176
Polystyrene GP	178 to 212	81 to 100	455	235
Polyvinyl chloride	158 to 176	70 to 80	392	200
4-Methyl pentene-1	104	40	464	240
Polyvinylidene chloride	-1	-17	410	210

From this equation one would expect that the plot of the log of the viscosity versus any linear function of the temperature would be a straight line over narrow temperature ranges and would describe flow fairly accurately. This holds true for a temperature range of approximately 100°F, Figure 3-20. In the plot of the log of viscosity against $1/T$ the slope of the line is the activation energy for viscous flow. It is related to the temperature coefficient of viscosity or the temperature dependency of the viscosity. The temperature coefficient of

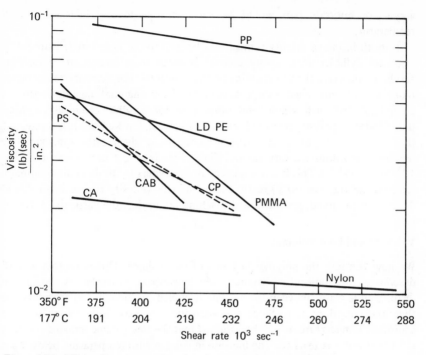

Figure 3-20   Effect of temperature on viscosity at a constant shear rate.

PMMA (Polymethyl Methacrylate)	CA (Cellulose Acetate)
CAB (Cellulose Acetate Butyrate)	CP (Cellulose Proprionate)
Nylon 6/6	PS (Polystyrene GP)
PE (Polyethylene, L.D.)	PP (Polypropylene)

viscosity per °F, for example, is 3.3×10^{-3} for low density polyethylene and 1.9×10^{-2} for cellulose acetate butyrate. The higher the temperature coefficient or the activation energy, the more dependant is the viscosity on the temperature.

It should be noted that activation energies or temperature coefficients must be calculated at a constant shear. The results will differ, depending on whether these energies or coefficients are determined at a constant shear stress or a constant shear rate.

A very useful plot is the log of the viscosity versus the temperature at a constant shear rate of $1000 \ sec^{-1}$, which approximates injection molding conditions (Figure 3-20). The slope of these lines is related to the temperature coefficient of viscosity. The greater the slope, the more sensitive is the viscosity to temperature. This has practical value in molding. For example, if a part in cellulose acetate or nylon (which both have small slopes) was not filling out, raising the temperature would have a minimal effect. Increasing the injection pressure or increasing the gate size would be more helpful. On the contrary an acrylic or butyrate part would be very sensitive to any increase in melt temperature.

In both instances attention must also be paid to the shear rate (Figure 3-17). Accurate cylinder temperature control is much more important in molding acrylic or butyrate than in molding nylon or acetate. Log viscosity/temperature plots at different shear rate are available from material manufacturers. A compilation of such graphs and other data for acrylics, cellulosics, nylons, polyethylenes, polystyrenes, and vinyls can be found in Ref. 78. It is interesting that the viscosity of most molding materials at shear rates approximating molding conditions are very similar. They range from 0.1 to 1 lb-sec/in.^{2} (7 $\times$ 10^3 to 7 $\times$ 10^4 P). This is not unexpected, as molders try to mold at a minimum temperature (i.e., viscosity) to fill the mold. This viscosity (flowability) should depend on the mold, gate size, and machine rather than the particular polymer.

Viscosity and Free Volume

We now consider the polymer in terms of free volume. This is another way of describing intermolecular distance, which controls viscosity. A qualitative distinction is made between the volume occupied by the polymer segments and that unoccupied or free. Since the viscosities are very high below the glass transition temperature, it can be assumed that the free volume remains constant until that point is reached. The increase in volume with temperature below T_g is caused by larger vibration of the segments. This increases the volume they occupy (thermal expansion) but does not increase the free volume.

Above T_g Doolittle (88) defined the free volume as the free volume at T_g plus the increase in temperature multiplied by the thermal coefficient of expansion.

$$f = f_g + (T - T_g)\,\Delta\alpha \qquad T > T_g$$

(3-13)

Where f = total free volume
      f_g = free volume at T_g
      T = temperature within 100°C of T_g (°K)
      T_g = glass transition temperature (°K)
      $\Delta\alpha$ = thermal coefficient of expansion.

This implies that above the T_g, free volume will occur proportional to the increase in temperature and the size of the coefficient of thermal expansion.

Williams, Landel, and Ferry (89) found that above T_g, the log of the viscosity varies inversely with the free volume. Therefore, the relationship of the viscosity and free volume at T_g and above can be written as

$$\ln\left(\frac{\eta}{\eta_g}\right) = \frac{1}{f} - \frac{1}{f_g} \qquad\qquad T > T_g \qquad\qquad (3\text{-}14)$$

This relation also holds within approximately $100°C$ above T_g.
Combining (3-13) and (3-14) gives the well-known WLF equation:

$$\log\left(\frac{\eta}{\eta_g}\right) = 2.303 \; \frac{1}{f_g + (T - T_g)\,\Delta\alpha} - \frac{1}{f_g}$$

$$\log\left(\frac{\eta}{\eta_g}\right) = \frac{a(T - T_g)}{b + (T - T_g)} \qquad\qquad T > T_g \qquad\qquad (3\text{-}15)$$

where

$$a = \frac{1}{2.303\,f_g}$$

$$b = \frac{f_g}{\Delta\alpha}$$

Empirically they found that when a is equal to 17.44 and b is equal to 51.6, not only the behavior of polymers was satisfied but also other organic and inorganic materials that exhibit a glassy transition point were satisfied. Equation 3-15 can now be written

$$\log\left(\frac{\eta}{\eta_g}\right) = \frac{17.44\,(T - T_g)}{51.6 + (T - T_g)} \qquad\qquad (3\text{-}16)$$

Knowing a we can now substitute

$$a = \frac{1}{2.303 f_g} = 17.44$$

$$f_g = 0.025 = 2\tfrac{1}{2}\%$$

This states that the glass transition temperature occurs when the free volume reaches $2\frac{1}{2}\%$ for all thermoplastics. If the percentage is the same for all plastics and is directly related to viscosity, one would expect that the viscosity of all thermoplastics would be the same at T_g. This is the case with the viscosity (in poises) being about 10^{13} P.

We can now calculate the coefficient of thermal expansion at T_g

$$b = \frac{f_g}{\Delta \alpha} = 51.6$$

$$\Delta \alpha = \frac{0.025}{51.6} = 4.8 \times 10^{-4}/°C$$

The observed coefficients of expansion are reasonably close and give support to the theory of free volume. Amorphous PE has a coefficient of 1.6×10^{-4}, and PVA 5×10^{-4}.

Substituting the viscosity of thermoplastics at T_g (10^{13} P) into (3-16) gives

$$\log \eta = 13 - \frac{17.44 \, (T - T_g)}{51.6 + (T - T_g)} \qquad T_g < T < (T_g + 100°C) \tag{3-17}$$

which will permit us to estimate the viscosity at any temperature within $100°$ above T_g. It can also be used to estimate molding temperatures if T_g and the viscosity at the shear rate required for molding is known.

Assume the viscosity of a particular polystyrene ($T_g = 100°C$) is 10^{-2} lb-sec/in.^{2} at the shear rate required for molding. What is the approximate material temperature required?

$$10^{-2} \frac{lb \, sec}{in^2} = 7 \times 10^3 \, P$$

Substituting (3-17)

$$3.85 = 13 - \frac{17.4 \, (T - 373)}{51.6 + T - 373}$$

$$T = 342°K = 319°F$$

It is therefore evident that knowing the T_g and the temperature of the material flowing into a mold it is possible to calculate its viscosity at the shear rate of injection. This is the theoretical justification for stating that at a given shear rate, the temperature is an accurate indication of the viscosity. It also forms the basis for feedback techniques in computer controlled equipment.

The WLF equation is an example of using experimental data determined at one temperature (T_g) for determining data at other temperatures. This is called a temperature superposition. A good example of viscosity – shear rate – temperature superposition is found in Ref. 90. Time is also a variable and time temperature superpositions are extremely useful in determining engineering properties of plastic material (91-93).

The WLF equation is remarkably accurate in describing amorphous polymers. From our concept of molecular structure we would expect that crystallinity would change the viscosity, because of the different, ordered, special relationship. This is the case, so that the WLF equation shows small deviations for crystalline polymers, depending on the amount of crystallinity.

CONTROL OF INJECTION MOLDING MACHINES—AUTOMATION

Injection machine control started with fixed logic systems based on limit switches and relays being tripped in a preset sequence. The next step can be termed variable logic where events were programmed to occur in sequence.

Most of the control systems today are based on the event type of control listed above. The latest systems, however, aim at controlling the process rather than the event. For example, rate of fill, pressure, and temperature can be controlled by some sort of analog or analog digital type system, reading either pressure, temperature, position and/or other variables. Cavity pressure is measured in many systems.

Computers may be used in either of these systems to store data or make calculations which alter the machine conditions which control the process and/or the events of one machine or many machines. (Process control must not be confused with production control which means control of downtime, counting parts, and relating production to accounting, inventory, and payroll).

Theoretical Basis for Automatic Control

The theoretical basis for the automatic control of injection molding machines has long been established. They have been discussed previously and are restated here for convenience. Hagan and Poiseuille independently developed the formula for the volumetric flow rate of a Newtonian liquid in a tube at a constant temperature.

$$Q = \left(\frac{\pi}{8}\right)\left(\frac{R^4}{L}\right)\left(\frac{\Delta P}{\mu}\right) \tag{3-7}$$

where Q = volumetric flow rate in tube
   μ = viscosity
   ΔP = pressure drop
   R = radius of tube
   L = length of tube.

For a given mold R and L represent the geometry of the runner, gate, and mold and are constant (A) so that 3-7 can be rewritten

$$Q = (A)\frac{\Delta P}{\mu}$$
(3-7a)

A characteristic of Newtonian fluids is that the viscosity does not change with the flow rate. Thermoplastics do not exhibit Newtonian flow. Viscosity is dependent on shear rate $(4Q/\pi R^3)$, which is a function of the volumetric flow rate.

For example, a graph of the viscosity versus temperature at different shear rates for an acrylic resin is shown in (3-16). at 425°F, increasing the shear rate from 10 sec^{-1} to 10,000 sec^{-1} decreases the viscosity from 0.34 to .01 lb-sec/in.^{2}. The viscosity at 375°F at 1000 sec^{-1} shear rate is the same as for 460°F at a-100 sec^{-1} shear rate.

Viscosity is a measure of the resistance to shearing stress. It depends on the composition of the polymer and the intermolecular distances. If a given volume of polymer is heated, its volume increases. Since the number of atoms and the internal atomic dimensions do not change, one must assume that the intermolecular spaces (distance between polymer segments) increase. The farther apart they are, the smaller are the Van der Waals' electrostatic forces. It is thus easier for the polymer segments to slide over each other.

This temperature dependence of viscosity is described by the Arrhenius equation

$$\eta = A e^{(E/RT)}$$
(3-12)

where η = viscosity
A = constant depending on material
R = gas constant
T = °K
E = activation energy of viscous flow.

The glass transition temperature, T_g, of a polymer, is that small temperature range which marks the change of its amorphous portion between the characteristics of a solid and viscous fluid. The viscosity of polymers at T_g is about 10^{13} P. Substituting this into the WLF equation

$$\log/\eta_t = 13 - \frac{17.44(T - T_g)}{51.6 + (T - T_g)}$$
(3-17)

which describes the viscosity for temperatures up to 100°C above T_g. It is interesting to note that the viscosities of most plastics at their molding temperatures fall into a narrow range of approximately 0.01 to 0.1 lb-sec/in.^{2}.

Equation 3-7a implies that if the volumetric flow rate and the pressure drop are constant at a given temperature the viscosity will be constant. Equations 3-12 and 3-17 imply that the temperature must be constant to maintain a constant viscosity. Therefore if the polymer is processed with the same Q, ΔP, and T from shot to shot, it will flow with the same characteristics and produce parts with identical properties. This assumes that the molecular weight and molecular weight distribution do not change substantially. This may not be a valid assumption (94). If molecular weight and molecular weight distribution do not remain constant, the viscosity would have to be checked before molding. Then, automatic molding requires the monitoring of Q, ΔP, and T by measuring the machine and mold conditions which affect them and automatically correcting any deviations from the preset standards.

It is only recently that instrumentation is available to monitor and record temperature, pressure, and speed during molding. Hydraulic servovalves and electrical control equipment are not new. The use of solid state controls has simplified computer control. It should be noted that a "computer" can also be a small relatively inexpensive unit developed just for controlling molding machines.

Instrumentation can now simultaneously record the nozzle pressure: two pressures in the mold; the screw position; the oil pressure behind the injection cylinder; and the oil pressure at the hydraulic motor turning the screw (95); and the screw position: plastic pressures in the cylinder, nozzle, runner, pregate, postgate, and the end of the cavity, and the plastic nozzle temperature (96); and the screw location: plastic cylinder pressure and nozzle pressure; two mold pressures; and the plastic temperature at the nozzle (97).

There is no suitable instrument to measure viscosity continually. The previous discussion shows that with Q and ΔP constant viscosity depends on temperature. Therefore the measurement and control of temperature is substituted for the measurement and control of viscosity. This is the procedure in automatic control.

Temperature Control

Conventional temperature control uses a thermocouple to measure the temperature of the steel barrel of the injection cylinder. This is not necessarily or usually the same as the molten plastic.

Good approximations of the plastic temperature are obtained by extruding the plastic into the air and plunging a needle thermocouple into the molten mass. In addition to its inaccuracy, the inherent disruption of the cycle prevents continual repetitive measurements under equilibrium conditions during molding. With increasing need for better data and processing control, thermocouples were inserted in the moving stream of plastic, primarily in the nozzle. This presented a number of technical problems which have been successfully solved (98 to 100, 246). Nozzles with indicating pyrometers and recording charts are now commercially available.

Plastic Pressure During Injection Molding

On almost all machines the only injection pressure measurement of the material during the molding cycle is of the hydraulic oil behind the injection cylinder. In single stage plunger machines it is an extremely poor indication of the pressure on the plastic in the nozzle and mold. In two stage machines and the reciprocating screws the correlation is much better, although far from accurate.

Recently pressure transducers have been developed that can withstand the temperatures and pressures in the nozzle and the mold (101 to 102c). Pressure transducers that fit behind knockout pins, and the appropriate amplification and recording charts, are commercially available. They measure the plastic pressure in the mold. In the event the transducer is not needed, a blank piece of steel can be rapidly interchanged with the sensing head. Thus the injection pressure on the material can be measured in the cylinder (nozzle) and the mold.

For a given machine under given conditions, the volume and the pressure of the oil delivered to the back of the injection ram controls the injection pressure in the cylinder and mold. A secondary factor is the plastic leakage past the injection plunger. This will be reasonably constant for a given material in a given temperature and pressure range. There are distinct pressure losses from the nozzle to the mold. For a given mold, the major parameters are (a) initial temperature of the material, (b) mold temperature, (c) resistance to flow, and (d) time.

A 4-oz plunger machine molded ABS in a 7/32 by 3/16 in. spiral mold. The material temperature was 445°F and the mold water temperature 150°F. The oil hydraulic pressure was 12,200 psi which corresponded to a pressure of 9000 psi on the plastic entering the mold (25% pressure loss). Pressure measurements were made after 20 sec at varying distances from the gate (103).

Distance from Gate		Pressure on the Plastic Material	
in.	mm	psi	kg/cm^2
0	0	9000	633
2¾	70	7300	513
7	178	5000	352

As this discussion is directed toward controls, it is assumed that the condition of the hydraulic system and injection cylinder are within the machine design.

Mold Filling

It is relatively easy to describe qualitatively mold filling. The initial fill is at very low pressure. The rate of fill depends on the machine settings and the hydraulic oil flow rate. When the mold is filled there is a rapid build-up of pressure. The constant A (Eq. 3-7a) represents the gate dimensions, $(\pi/8)\,(R^4/L)$. While the mold is filling the radius of the melt stream through the gate remains constant. Once the flow is reduced the material starts to cool effectively reducing the radius as the skin soliditfies It is safe to assume that if the mold temperature and the other molding conditions are constant this decrease will also be constant. Additional material flows into the mold (packing) until either the gate freezes or the injection pressure is stopped. This flow of material fills the volume generated by the shrinking of the plastic material as it cools. This amount is usually small compared with that added initially by the compression of the material due to the injection pressure. If the injection pressure is removed before the gate seals, the elasticity of the plastic causes it to discharge back through the gate into the runner.

It is evident that the five major factors affecting the amount of material in a molding are:

1. The injection fill time or the time to fill the cavities before the injection pressure builds up. The more rapidly it occurs the less time for cooling of the plastic. Cooling increases the viscosity and the pressure loss.

2. The injection pressure determines the initial amount of compression (added material) of the molding.

3. The injection (ram) forward time adds more material to compensate for shrinkage caused by cooling. Once the cavity is optimally packed, additional injection forward time serves only to add stresses at the gate. This is very dangerous as it cannot be detected immediately. Failures may develop a long time after molding.

4. Gate seal time. If the gate seals too rapidly incomplete packing will occur. If the gate does not seal quickly enough, the injection forward time has to be increased so that the material does not flow back into the runner. Ideally, the gate should seal when the optimum amount of material is in the mold.

5. Velocity of Fill.

In the previous experiment, the longer the ram forward time (measured after 40 sec) the higher the residual pressure for a specific molded part (maximum mold pressure 8000 psi) is (103):

Ram Forward Time (sec)	Residual Pressure	
	psi	kg/cm^2
18	0	0
21	200	14
23	700	49
24	1200	84
27	3000	211

The shrinkage (part dimensions), warpage, orientation, part removal, and stress are functions of the amount of material in the cavity. The tremendous advantage to the molder in being able to monitor and/or feedback and control this parameter is self-evident. The complexity of the problem is compounded because pressure increases plastic melt temperature (104). Increasing the pressure to 15,000 psi raised the melt temperature of a modified acrylic 10°F.

Equations 3-7, 3-12, and 3-17 are only valid for the initial flow into the mold which corresponds to a dynamic or flowing hydraulic system. Once the mold is filled the packing mechanism controls and is closely analogus to a static hydraulic system. It can be monitored by pressure transducers under knockout pins. It is controlled by varying the hydraulic pressure behind the injection ram. After the mold is initially filled, the control of the hydraulic pressure is switched from the nozzle sensors to the mold sensors. If the pressure/time profile is the same from shot to shot the parts will receive the same amount of packing.

Plastic Flow Rate

The third variable in (3-7a) is the volumetric flow rate (Q). This, in effect, is the speed of the injection plunger. It can be readily monitored by a wire wound potentiometer. It is controlled by the volume of oil supplied to the injection cylinder. Low pressures produce the required flow rate during the mold filling period. The speed of the plunger is controlled by varying a flow control valve or a variable delivery pump. Many machines use accumulators to deliver the large volume of low pressure oil needed for fast molding filling.

The interdependency of the volumetric flow rate and the pressure is shown by the amount of pressure (psi) required to fill a 2 by 9 by 0.075 in. cavity with polyethylene. The material temperature was 425°F and the mold temperature 125°F (73).

Volumetric Fill Rate		Pressure to Fill	
in.^{3}/sec	cm^3/sec	psi	kg/cm^2
4.5	74	1180	83
5.9	97	1300	91
9.6	157	1520	107
12.6	207	1745	123
18.5	303	2120	149

Approximately quadrupling the volumetric filling rate required less than doubling the minimum pressure requirement. It is the result of lowered viscosity caused by increasing shear rate.

The filling rate affects the molded part in a number of ways:

1. Gate freeze-off time. In normal molding the gate does not start to freeze until the main volume of a material has flowed into the cavity. This affects shrinkage.

2. Warping. Different rates of fill will pack different areas of the cavity in a different manner. This is particularly true of parts with varying wall sections.

3. The rate of filling is an important parameter in orientation. This affects the mechanical strength and the size of the part.

4. Flashing. If a mold is filled too rapidly the displaced air may not vent quickly enough. It spreads over the mold surface, effectively increasing the projected area. This force may then be large enough to open the mold, permitting the plastic to flow and flash. Sometimes the material does not flash, but the opening of the mold increases the thickness of the part. In that particular case increasing the injection speed increases the part dimension. Sometimes this is confused with shrinkage.

5. A high filling rate can cause cores to shift by unbalanced pressures. Usually this is caused by the slowness of venting on the side of the core away from the gate. In that instance a slower ram fill time will solve the problem.

6. In a properly balanced, multicavity mold the fill rate does not usually effect the rate of filling of the individual cavities. This is not so for very slow fill rates.

7. The location of the weld line can be varied by changing the filling rate.

8. Surface defects are occasionally caused by rapid filling. Slow filling may cause wrinkles, iridescent effects and flow marks.

The advantages of monitoring, comparing, and automatically changing machine conditions are:

1. Better mold parts.
2. More consistent parts.
3. Fewer rejects.
4. Faster production.
5. Less dependence on labor and man made judgments.
6. Closer tolerance parts.
7. Easier automation.

All systems thus far developed are essentially pragmatic. The parts are molded to certain specifications (dimension, physical properties, etc.) and the machine conditions are then used as the standard. This is what the molder has always been doing.

Conversion of existing equipment to fully automatic control would be difficult and costly. It would seem that adding nozzle temperature, nozzle pressure, and mold pressure indicators to existing equipment would considerably improve operations. Installations of this kind, which can be done by the plant personnel, costs approximately $3500. The next step would be to use the pressure of the plastic in the mold to control the injection pressure during packing. This would require an hydraulic servovalve to control the injection pressure and an interface device between the valve and the mold pressure transducer.

A number of systems have been developed using a specific dimension of a molded part to control the machine conditions. This has a number of disadvantages, most important of which are the time delay in measurement, the lack of correlation between the part variance and the condition causing it, and the difficulty of measuring a part at a consistent time during its time of maximum shrinkage rate. Other methods for controlling machines are found in Refs. 105 to 107c.

Theoretically, control of injection temperature, pressure, and speed will produce consistently molded parts. All other conditions on which they depend, such as mold temperature, screw speed, clamping pressure, and time delay between shots, must also be controlled. Present technology is available to monitor all these systems. Programming them for automatic control of the molding process is within the scope of present knowledge.

Computer Control

The theory of process and computer control are relatively simple. As an example, the plastic temperature is measured in the nozzle as it flows into the mold. The thermocouple reading is translated into computer language. The computer compares the actual plastic temperature with the programmed temperature. In a monitoring system, a discrepancy beyond the allowable programmed limit will either sound an alarm, shut off the machine, or take any action other than correcting the fault. In a feedback system, the error signal will cause a change in one or more of the machine settings depending on the previously programmed instruction.

The designing of these programs is the difficult part of computerizing the equipment. Let us examine the main factors controlling the temperature of the plastic in a reciprocating screw injection molding machine. A two-stage machine would have all these variables plus the temperature control of the shooting cylinder. Consideration of other types of machines would add additional machine design parameters.

1. The temperature setting of the temperature control units on the barrel at three or four positions.

2. The design accuracy of the temperature control system on the barrel.

3. The operating accuracy (as distinguished from the design accuracy) of the temperature control system of the barrel.

4. The effect of over and under voltage on the barrel control system. In New York City in the summer, for example, under voltages of 5% to 7% are common. The control system might not have been designed for such conditions.

5. The screw speed and related torque.

6. The back pressure on the material.

7. The length of time the material is in the barrel.

8. Differences in the molecular weight and molecular weight distribution of the plastic.

9. The shot size.
10. The screw running times as a function of the overall cycle.

The computer would have to be programmed to differentiate the causes so that appropriate corrections can be taken. In some instances this is not difficult. For example, assume that the temperature of the plastic has risen and is the result of the machine being left open too long between shots. If the computer program automatically compares the actual gate opening time with the programmed time, it could either stop the machine and indicate purge, or separate automatically those parts for further inspection. Similarly, if the temperature is too low and it coincides with a low voltage condition, corrections can be easily programmed. Changes in the composition of the material can be considerably more difficult to detect. The theoretical knowledge and technical equipment are available to build a computer controlled injection end that will provide plastic at a constant temperature, or in the event of a malfunction, stop the machine. Reference 108 describes the use of a computer in monitoring and controlling temperature using some of the parameters above.

IMPORTANCE OF ORIENTATION IN MOLDING

Orientation effects are very important. By orientation, we mean the alignement of the molecule in the direction of flow. The strength in that direction is that of the carbon–carbon linkage whose disassociation energy of 83 kcal/mole is much greater than the 2 to 5 kcal/mole Van der Waal type forces holding the polymer together perpendicular to the line of flow. Unfortunately oriented parts are often stressed in undesired patterns. Orientation can be used by the molder to produce parts with some very desirable properties by taking advantage of the increased strengths in the direction of flow (of course, plastics are never fully oriented, so that the above-mentioned strength ratio of 83/5 is not reached in practice).

Orientation in the Mold

Let us now consider what happens in the mold during molding (248). Gilmore and Spencer (109) constructed a glass wall injection mold for an edge gated disk 2 in. in diameter and 0.10 in. thick. They took motion pictures of the mold filling and packing. Intuitively one would expect a circular wave front from the gate. Instead, at the beginning of the flow, the plastic adhered to the cold wall and the wave front seemed to stretch the material around the circumference. As the mold fills, material seems to bounce off the wall causing the wave front to first straighten and then creep around the edges so that the circumferential parts seal just before the portion directly opposite the gate. After the cavity looks filled (plastic touching all the walls) additional material is forced in, primarily around the gate area and toward the warm center.

Photographs of molding through a 0.020 in. high by 0.030 in. wide by .040 in. land gate showed an initial jetting of a small diameter rod which spiraled and exhibited melt fracture. Then 0.18 sec after the jetting started, the characteristic wave front of material appeared. When the filling was slow, the jetted material cooled and was not completely melted and absorbed in the rest of the polymer. This produced characteristic surface imperfections.

Round and square rods were also put into the disk. The photographs showed characteristic weld lines developing. If the material was injected too rapidly air was entrapped as the material flowed around the insert causing burning. Other techniques, such as marking mold with red crayon and noting how the red marks are transfered by the plastic (110), have been used to establish flow patterns in molds and how they are effected by changing molding conditions.

Consider the molding of a rectangular plaque edge gated on the short side (Figure 3-21). Regardless of the orientation caused by flowing through the gate, the turbulence is enough to randomize the outside molecular layer of the polymer wave front. As this material hits the cold wall it freezes. This outside frozen layer will be completely unoriented. Consider the next layers of polymer. One end of the polymer molecule is frozen and anchored on the outside layers attached to the wall. The flow of the material is pulling the remainder of the molecule in the direction of flow. These layers will therefore be the most highly directional or oriented. For the same reason it would be the most highly shear stressed. As we approach the center of the part, the molecules will have less orientation caused by flow.

Since the outer layers thermally insulate the inner parts, they will remain warmer allowing more time for Brownian movement to disorient the parts. The shearing action stops when the shearing force provided by the injection cylinder stops. This will happen when either the injection pressure is removed, the gate freezes, or the part freezes. The shearing stress is greatest where the pressure is greatest during the dynamic or filling portion of the shot, which is at the gate. Birefringence which is directly proportional to this pressure, increases with increasing pressure (111). For this reason the gate area is usually the most oriented part of the piece, with the largest amount of residual stress. Here the material is flowing in a relatively cold state and because of the injection pressure has little opportunity to relax.

Measurement and Control of Orientation

Residual stresses were recognized early in the analysis of injection molding (112). They are studied by putting a clear injection molded part between two polarizing filters, using white light. When one of the filters is turned a

Skin layer
not oriented

Outside layer
most heavily
oriented

Inside layer
less heavily
oriented

Gate

Center
least
oriented

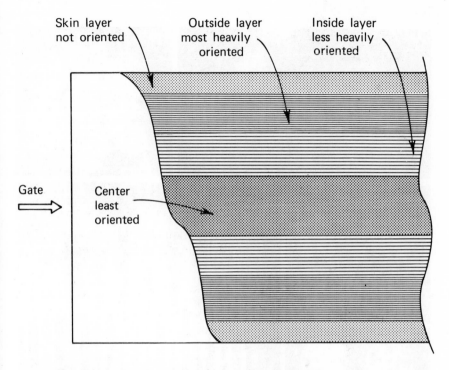

Figure 3-21   Schematic representation of orientation effects of flow of plastic into cavity.

characteristic series of repeating colored bands appear, each color of which indicates a certain amount of molecular orientation. This is called birefringence. Figure 3-22 shows some typical birefringence patterns of injection molded parts. It can be assumed that birefringence in molding is a measure of the molecular alignment or orientation.

Measurement of the orientation stress is then possible (113). Orientation stress is the stress given in pounds per square inch of cross section with which a heated sample tends to return to its preorientation dimensions. The higher degree of birefringence the more the orientation stress (112). Orientation stress has been plotted against birefringence and the resulting straight line shows a linear relationship. If parts are produced with a modest degree of orientation, internal stresses will also be reduced.

Materials that crystallize will also orient. The first layers of material solidfy and crystallize quickly on the cold mold wall. The small crystals act as nucleating agents. Since they are orientated, the crystals grown from them also will be oriented. A good description of this process for acetal homopolymers, which would apply to other crystalline material as well, is found in Ref. 114.

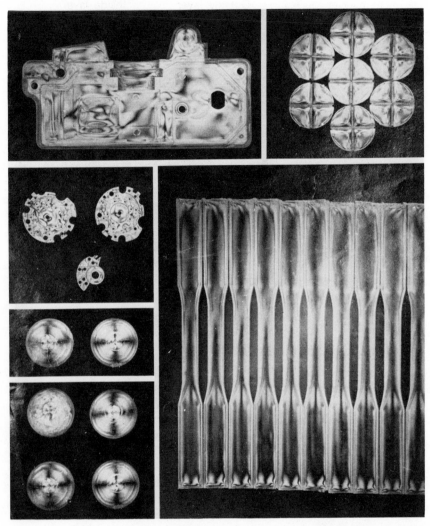

Figure 3-22   Birefringence patterns of injection molded parts (Picatinny Arsenal US Army)

If the explanation of orientation in the molded plaque of Figure 3-21 is correct a cross section of the slab should show the birefringence pattern of Figure 3-23. The walls should have no orientation as the cold mold freezes the random molecule. Orientation should rise abruptly near the walls and decline to zero at the center where there is no shear stress (dotted line). As the polymer relaxes and the Brownian movement takes over, the amount of birefringence

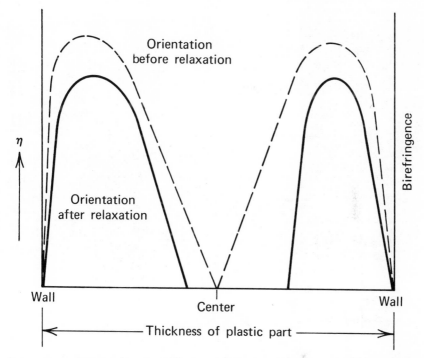

η

Birefringence

Orientation
before relaxation

Orientation
after relaxation

Wall

Center

Wall

Thickness of plastic part

Figure 3-23   Amount of birefringence before and after relaxation vs. distance from the wall for a molded flat section. Difference between solid and dotted line is the loss of orientation due to relaxation (Ref. 115).

should decrease (solid line). This has been found to be the case (115). The peak orientation was between 0.025 and 0.030 in. from each side of a 0.100-in. thick slab. The two peaks were not identical in each specimen, reflecting the difference in temperature of each side of the mold.

A simple way to prove what is shown in Figure 3-23 is to mill off approximately one-third of the surface on one side (Figure 3-24). This will produce a part which has randomly placed molecules on one side and highly oriented ones on the other. If the specimen is heated close to its T_g, the oriented molecule should assume a random location. The oriented position originally had the molecules lined up straight. After it has been heated and the molecules assume the random position (kinked, turned, and twisted in various directions), it should decrease its length and be smaller then the original nonoriented part. The specimen should now act like a bimetallic unit and bend in the direction of the oriented layer. This is what occurs (116).

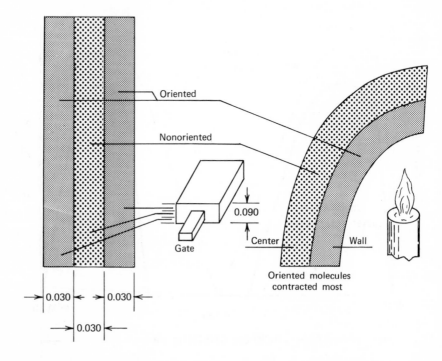

Figure 3-24 When one third of molded part is milled off, and the remainder neated, the part bends in the direction of the oriented side. This is caused by the larger shrinkage of the oriented portion (Ref. 116).

Factors Influencing Orientation

Orientation is the net result of the alignment of molecules in one direction caused by shearing action and the relaxation of this alignment caused by Brownian movement.

It follows that anything that will keep the plastic hot will allow increased Brownian relaxation and therefore decrease orientation. The following operating conditions are important:

1. High material temperatures and hot molds decrease orientation. A significant amount of relaxation can occur after the part is removed from the mold. Putting heavy section parts in hot water for slow cooling will decrease orientation.

2. Pressure also increases orientation. Higher pressures give higher shearing stresses which cause more molecular alignment.

3. Ram forward time has a strong effect on molecular orientation. If the gate

does not seal, the pressure on the ram produces continual flow and shearing. Longer ram forward times also maintain high pressure in the mold, inhibiting relaxation as the plastic cools.

4. The thickness of the part affects orientation. Because of the low thermal conductivity of plastics, thicker parts act as insulators keeping the center sections warmer for a longer time. This promotes relaxation, decreasing orientation. Often, with very large gates, there is a counter action caused by a longer maintenance of injection pressure. The relaxation effects seem to predominate.

5. Gate seal-off time. When the plastic flow stops, orientation stops and Brownian movement causes previously oriented material to relax. In turn, plastic flow stops when the injection pressure is removed either by retracting the injection plunger or by the gate sealing. Other things being equal, the time of gate sealing depends on its size particularly the minimum cross sectional dimension. As the material flows through the gate rapidly there is little or no reduction of the gate diameter caused by cooling. When the cavity is filled, the velocity of the material through the gate drops drastically, and cooling begins. While the mold and material temperature are influential, it is primarily the velocity which determines how quickly the gate will cool and seal off. Larger gates cool more slowly, and take longer to seal and thereby increase orientation.

6. Flow rates are critical. In molding a part with the same configuration but differing only in thickness, the volumetric flow rate through the gate will be the same since it is controlled by the machine conditions. The velocity in the mold will be different, the thicker the part (larger volume) the slower the velocity. The shear rates and shear stresses will be correspondingly lower for the thicker part, resulting in less orientation. The rate of mold fill affects orientation. A rapidly filled part will have its gate frozen more quickly. Once the flow stops, orientation stops and relaxation starts. (This effect decreases orientation compared to a slowly filled part.)

On the other hand, in a slowly filled part, there is more cooling during flow, freezing "in" the orientation, shearing action and consequently more orientation.

Table 3-6 summarizes the effects of mold and molding variables on orientation. Unfortunately the conditions producing the desirable results in one area may conflict with those in others. A good example of the use of these principles of orientation to solve a molding problem is illustrated in Figure 3-27 and the adjacent text.

The Effects of Orientation on the Physical Properties of Plastics

Polystyrene sheet which has a tensile strength of 6000 to 7000 psi is also quite brittle Heat the sheet slightly above its T_g and stretch it. Chill it while under

Table 3-6. Effect of mold and molding variables on orientation

	To Increase Orientation	To Decrease Orientation
Temperature		
Material temperature	Cold	Hot
Mold temperature	Cold	Hot
Part Cooling	Fast	Slow
Pressure		
Injection pressure	High	Low
Ram forward time	Long	Short
Mechanical		
Part thickness	Thin	Thick
Gate size	Large	Small
Other		
Fill rate	Slow	Fast

Table 3-7. Effect of orientation of the Izod impact strength for some styrenic materials (Ref 119).

| Material | Izod Impact Strength (ft-lb/in. of notch) | | |
	Base Property	In Direction of flow	Across Direction of Flow
SAN	0.4	0.41	0.24
Change (%)		+3	-40
General purpose PS	0.3	0.39	0.26
Change (%)		+30	-13
High impact PS	0.8	2.3	0.3
Change (%)		+188	-60
Very high impact PS	1.8	2.8	1.2
Change (%)		+55	-33
ABS	3.4	4.0	1.0
Change (%)		+18	-70

tension, to retain its orientation. The tensile strength now is 9000 to 12,000 psi, depending on the percentage elongation and processing temperature. The brittleness disappears. If the material were allowed to cool slowly its orientation would disappear and the properties would be similar to the starting sheet (113).

If 1/8 in. thick by 1 9/16 in. wide, high-density polyethylene parts are oriented by heating and stretching at 200°F, the ultimate tensile strength goes from 2500 to 11,000 psi. However, if the stretching temperature exceeded the crystalline melting temperature, the stretched sample, when cool, showed no increase in tensile strength. Obviously this is because the material flowed and was warm enough for the Brownian movement to disorient whatever orientation had taken place. If a sheet of this material, oriented in one direction, was placed in a diaphragm and hydrostatic pressure applied, it would burst at the normal stress of 3200 psi, with a split parallel to the direction of stretch. If the same unstretched sheet was heated to approximately 250°F and blown into a hemisphere (biaxial orientation) and then cooled, the bursting stress could reach as high as 21,000 psi (117).

In a molding example, general purpose clear polystyrene was molded in plaques 0.020 and 0.100 in. thick. Tensile strength and percentage elongation to failure was measured parallel and perpendicular to the flow and plotted against the amount of birefringence (Figure 3-25). The same was done for the notched Izod impact strength (Figure 3-26).

As expected the higher the degree of orientation (birefringence) the higher the tensile strength, percentage elongation, and notched Izod in the direction of flow. These represent the strength of the carbon–carbon linkages. Perpendicular to flow the higher birefringence indicates fewer carbon–carbon linkages and poorer the physical properties (118). The strength perpendicular to flow represents the Van der Waal's forces. Since there are other mechanisms involved in these tests besides orientation one expects a qualitative rather than a quantitative relationship between parallel and perpendicular flow.

References 118a and 118b show the effects of orientation on GP and HI polystyrenes.

Table 3-7 shows the effect of orientation on the Izod impact strength of a number of materials measured in the direction of and perpendicular to the flow. The "base" property is the impact strength as given in the manufacturers literature. Tensile strength and percentage elongation at rupture show similar anisotropic behavior.

How Orientation Can Be Used to Advantage

Some typical examples follow:

1. A dramatic example of the effect of orientation on properties was displayed when a protective cap was molded with a threaded metal insert (Figure

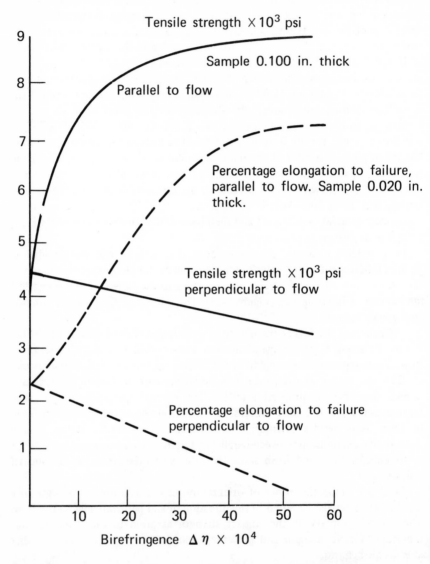

Figure 3-25 Effect of the direction of flow upon the tensile strength and percentage elongation to failure for GP polystyrene as measured by birefringence (Ref. 118).

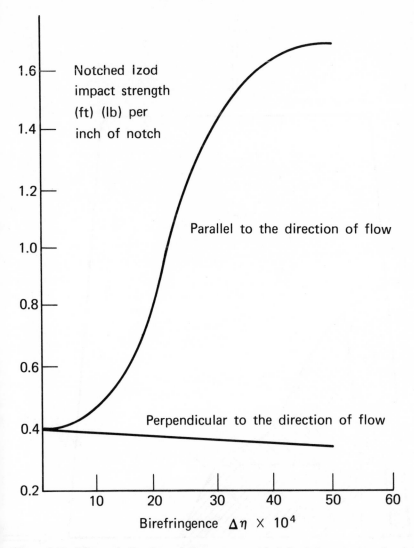

Figure 3-26 Effect of direction of flow upon notched Izod impact strength of GP polystyrene as measured by birefringence (Ref. 118).

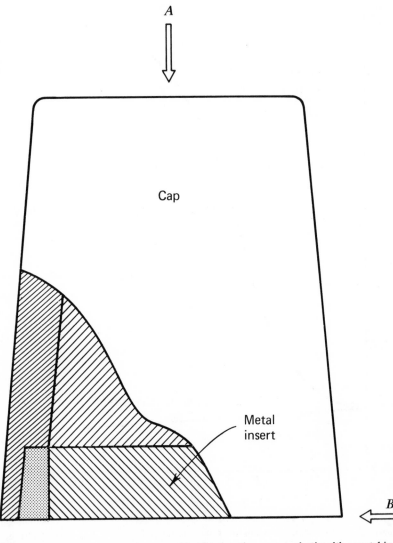

Figure 3-27 Effect of orientation caused by gate location upon a plastic with a metal insert molded inside.

3-27). Initially the gate was at point A. After several months in the field, a significant number cracked around the metal insert. The part could not permit thickening of the plastic at that point. The gate was changed to point B and the failures disappeared. When gating at point A, there was strong orientation in the direction of flow. This left the circumferential direction, perpendicular to the flow, as the weakest part of the molding. When the direction of flow was changed by gating at B, the orientation was around the circumference (or hoop direction) and the increase in strength in that direction made the difference. Venting did present a problem with the initial mold design but was overcome by building a cam type mold. While this is an extreme case, the effect of orientation is to be considered before designing parts and gating systems of molds.

    2. In another well-known example, a molded polyolefin hinge works because the polymer molecules are oriented in the direction of flow. If the hinge is made too thick, orientation will decrease to the point where it does not function correctly. If such a molded hinge is not flexed immediately, the Brownian movement will reduce orientation enough to lower the quality of the hinge (120, 123).

    3. Some outstanding nonmolded uses of orientation are in synthetic plastic fibers (nylon, acrylic and polyester) and shrink wrap film which is highly biaxially oriented. When the film is heated the orientation stress causes the sheet to shrink over the enclosed form.

    4. In molding tumblers conventionally, orientation in the direction of flow cannot be avoided. To overcome this, a mold was built with a rotating core to orient the material along the circumference mechanically. This resulted in parts with significantly greater craze resistance and greatly improved physical properties (124).

OTHER NON-ORIENTATION CONDITIONS WHICH CAUSE INTERNAL STRESS

Some other causes for internal stress follow:

    1. The geometry of the piece may cause differential shrinking. This is particularly true when there are sections of different thicknesses.

    2. Improper gate location is a possible problem. This causes the anisotropic effect of orientation.

    3. Molding conditions, such as packing at the gate, may induce stress. As the polymer cools, the volume available to receive the hot extra material forced in by the injection ram decreases rapidly—this material is required to minimize shrinkage. If too much material is forced in, it will be highly stressed and fail either immediately or at a later date.

4. Environmental stresses, such as heat, mechanical force, and UV, will accelerate failure.

Anything that will permit more material to enter, such as injection pressure and gate size, affects packing stresses. Ram speed, material temperature, gate size, and ram forward time all affect packing. Probably the main parameter is the ram forward time. In molding a center gated polyethylene tumbler with a ram forward time of 10 sec, there were no visible failures the first day and 7% 14 days later. Increasing the ram forward time to 25 sec gave a 1-day failure rate of 70% and a 14-day rate of 88%. While packing is concentrated at the gate it can extend throughout the part. Packing, which is a prime cause for sticking in the mold, is a result of better adhesion to the mold surface and probable deflection of the mold itself. Obviously the amount of feed and temperature will also affect the amount of plastic that flows into a cavity, hence the packing. Stresses induced by packing in a molded polycarbonate disc reduced the impact strength significantly. Some of the factors affecting packing in the disc were ram forward time, ram speed, material temperature, and gate size (125).

Relieve Stresses by Annealing

Build-in stresses can be relieved to some extent by annealing. Since annealing rarely changes the birefringent pattern, one can assume that it cannot affect orientation. It probably allows the polymer segments to move from their frozen position to a random or lower stress location. Aside from the annealing of nylons for dimensional stability, very little experiment work has been done on annealing (126, 127). In all instances annealing has improved the properties of the part. The extra handling is expensive so that it is not done unless there is a compelling technical reason.

Figure 3-28 shows the front and back of two similar general purpose polystyrene brush blocks 0.400 in. thick. They are molded four up with the runner system ending in a hook. The hook is used to suspend the shot in hot water at 160°F. The tank is large enough so that the shot remains in for 2 hrs. The block is inserted in a bristling machine which drills approximately 130 holes and forces nylon bristles into them. A crack in any hole caused by drilling or bristling is a reject. Without annealing, the reject rate is between 40 and 45%. With annealing it is less than 0.3%. Lowering the water temperature increases the reject rate. At 135° the reject rate is 30%. At 120° it is the same as no annealing.

Slowing down the cooling rate after molding is a much more efficient method of annealing than reheating the cold part. The part coming from the machine will almost always have its internal section at a higher temperature than can be obtained by reheating. If the whole molding was brought to that temperature it probably would distort in handling.

In a slow cooling experiment a rectangular plaque 1 in. wide by 5 in. long

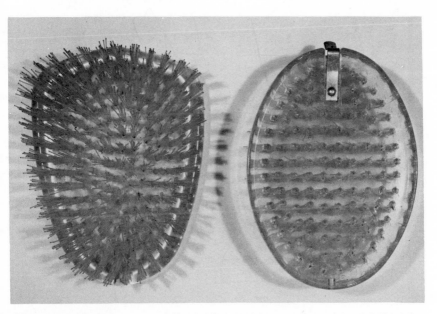

Figure 3-28 Polystyrene brush blocks 0.400 in. thick. Annealing reduces bristling reject rate from 45 to 0.3%.

by 0.150 in. thick is gated at the 1-in. end. The mold is kept above T_g until all the molecular segments have time to randomize their position. The mold is cooled to room temperature. The plastic will contract (shrink) as the material cools and the intermolecular distances decrease. After removal from the mold the linear shrinkage will be found to be the same both perpendicular and parallel to the direction of flow.

Molding the same plaque with the mold at room temperature produces a different effect. As the material flows it starts to cool and is oriented in the direction of flow. This shear stress stretches the C-C bonds in the direction of flow and freezes them. As the part cools, two events are simultaneously occurring. The intermolecular distances decrease as the plastic cools and the part contracts equally in all directions. The second mechanism is the return of the stretched C-C linkages to their normal position. Since there are more such linkages in the direction of flow there will be a greater shrinkage in that direction than perpendicular to it. The more orientation there is, the larger the effect. This has very significant results, particularly in molding the olefins and styrenes.

Consider for example molding a 4-in.-diameter center gated, polyolefin cover

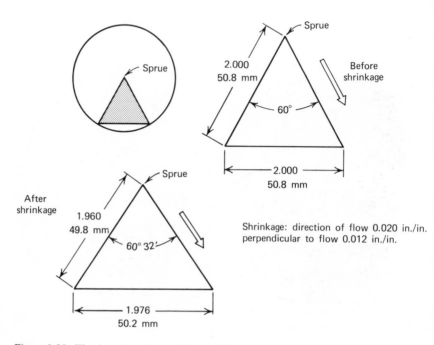

Figure 3-29 Warping of center-gated polyolefin part caused by different shrinkage parallel and perpendicular to flow.

for a container (Figure 3-29). Take a segment radiating from the center with a 60° included angle. When the material is hot it will form an equilateral triangle with a 2-in. dimension on each side. Polypropylene will typically shrink 0.020 in./in. in the direction of flow and 0.012 in./in. perpendicular to flow. When the material cools then, the radial dimension will be 1.960 and the chord 1.976. Simple trigonometry will show the angle is no longer 60°, but 60° 32′. For the whole 360° circle, the increase will be 3° 14′ or approximately 0.9%. Obviously the material has to go somewhere. Since it cannot lie in a flat plane, it will warp. If the thickness of the material and the ribbing provide enough strength, the part might not visibly warp, but would be highly stressed. The way to minimize such warp or stress is to mold under those conditions which would give the least orientation. Multiple gating is also effective, as in redesigning the cover (128).

The effect of orientation has a strong influence on gate selection. Consider a deep polyethylene box (Figure 3-30). If the part were center gated (Figure 3-30*a*) the box would be severely distorted, following the mechanism illustrated in Figure 3-29. It would be further complicated by the difference in flow lengths from the gate to point *x* and the gate to point *y*.

Using two gates diagonally opposed to each other (Figure 30*b*) would not

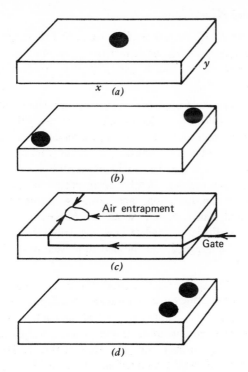

Figure 3-30 Effect of gate location on a deep polyethylene box. (a) Center gate: radial flow-severe distortion, (b) diagonal gate: radial flow-twisting, (c) edge gate: warp free air entrapment, and (d) end gates: linear flow-minimum warping.

solve the problem. An analysis of the orientation flow pattern shows that the box would twist. This is what happens.

It is evident that edge gating in the center of the short side (Figure 3-30c) would provide a straight flow with an even orientation pattern of shrinkage parallel and perpendicular to the flow. There would be no distortion. This is correct and gating there would result in a warp free box. However, a second problem appears. The material flows rapidly around the bottom perimeter of the box sealing the top off, before material can flow vertically and cover the top of the box. This causes air entrapment, resulting in incomplete filling, burning, surface blemishes, and/or poor welds.

The best solution is to have two gates on the top end of the box (Figure 3-30d). This gives maximum linear flow without air entrapment and produces a wrap free part. In most instances a satisfactory part could be molded with one gate located on the top end. For larger parts it is necessary to mulitple gate to assure relatively even orientation patterns and flow lengths.

Warpage of Molded Plastics Parts

Warpage is the result of unequal residual stress in the molded part, when the stress is strong enough to strain or distort the piece. Warping can be caused by poor part or mold design and/or poor molding conditions, and the basic material.

Part Design. The main cause for warping attributable to piece part design i uneven or too thin walls. The different thicknesses have different cooling rates densities, and orientation. Consequently large internal stresses can be built up a the part cools. (Different wall thicknesses may result from core shifting.)

In some instances, varying the mold temperature in different parts of the mold will reduce the warping slightly. Also, abrupt changes in cross-section develop high stress concentrations and resulting distortion. Therefore, radi should be large and planar sections should be blended as broadly as possible Where insufficient thickness causes warping, ribbing, or other forms o reinforcement can be designed into the part to prevent or minimize the warping (128).

Mold Design. The major fault in mold design concerning warping is due to incorrect gate location. This causes undesirable orientation (see section above) Inadequate cooling and the inability to control the cooling in separate section of the mold will also cause warping. Uneven cooling will cause a difference in orientation and density through the part. Uneven flow lengths from the gate may also cause trouble. As the material enters the mold it begins to cool as it contacts the mold surface. The further it flows the cooler it becomes. The part most distant from the sprue will solidify first and have the least amount of packing. They will be less dense than the part near the gate which is molten enough to receive more material. Mold cooling channel design can permit a higher mold temperature for the parts furthest away from the gate, when required. Uneven ejection and insufficient ejection area are also common causes for warpage.

It is best to gate into the heavy section (which may also be required to pack fully the thick section). Gating into the thinner section may cause warping because of the stresses resulting from the higher pressure needed for mold filling and packing.

Small gates tend to increase warpage in that these gates may not allow enough material to be packed into the cavity before sealing. Once gates have been opened to an optimum point, increasing the size will have no adverse effect on warping unless molding conditions permit packing at the gate.

Molding Conditions. In general, molding conditions that tend to reduce orientation will reduce warpage. Warpage, particularly in the olefins, is caused by differing densities in the part. For this reason rapid fill is important so that the material first entering the mold will not have much more time for heat transfer than the material last entering the mold. It is necessary to fill the mold with enough material to compensate for thermal shrinkage. Therefore higher injection pressures and longer injection forward times tend to minimize warping.

There is an optimum material temperature. If the material is too cold it will be impossible to pack completely the mold. Overheating the material causes more warpage. The harder the material is when it is removed from the mold the less the chance for the stresses to turn into strains. Therefore a longer cure time will decrease warpage. For the same reason lower mold temperatures decrease warpage.

Materials.   The nature of the material affects warpage. The narrower the molecular weight distribution the less warpage everything else being equal. Crystallinity will affect warpage. The higher the crystallinity the stronger the material and the less tendencey to warp. On occasions, different colors of the same material will warp differently. This is probably because the pigments act in varying degrees as nucleating agents.

Crystalline materials have higher shrinkage. For example, the mold shrinkage allowance for high density polyethylene is 2.0%, low density 1.6%, while polystyrene, which is amorphous, requires only 0.6%.

What Else To Do About Warpage

Annealing the parts will rarely help if the warping is caused by the piece part design or the mold design. If annealing helps the part, a much larger improvement can be made by correcting the molding conditions. Cooling and shrink fixtures are of little value in controlling warpage, although they are of some help with shrinkage.

Table 3-8 shows some of the conditions that affect warpage.

Cooling fixtures are used to control part distortion. A typical fixture can be a device which holds all or part of the molded piece rigidly but still allows it to have its normal shrinkage. An example would be a box with a rim. The box is put face down on the table and a weighted metal flange put over the rim. The part can shrink normally and any tendency to distort is reduced. Frequently, jigs of this kind or the parts placed in water will permit significantly shorter cycles.

A shrink fixture on the other hand, is a mechanical device which will limit the shrinkage of the molded part. Usually this can be done without unduly stressing the part, as the stresses are relaxed by cold flow. To obtain the full effect of a shrink fixture, the part must be left on for a long time. This requires a large number of shrink fixtures. Therefore, it is usually more economical to change the mold. A shrink fixture was successfully used when a 5/8-in.-thick low density polyethylene U-shaped luggage handle was molded. The two uprights of the handle bowed in on cooling. A simple shrink fixture was made by turning down the two ends of a piece of aluminum and inserting them in the handle

Table 3-8   Conditions that tend to reduce warpage

To Reduce Warpage	
Decrease	Orientation
Increase	Gate size
Low	Material temperature
Low	Mold temperature
Fast	Injection speed
Increase	Injection pressure
Increase	Injection forward time
Increase	Cure time
Minimal	Packing at gate
Narrow	Molecular weight distribution
Increase	Wall thickness
Increase	Ribs, fillets reinforcement

hole. The part and the fixture were left in cold water for an hour. Since it was a two cavity mold run slowly and the fixtures very inexpensive, turning the 200 pieces of aluminum was the most economical solution. Shrinkage of the plastic material is discussed in the next section.

SHRINKAGE

When a plastic material is heated it expands. Upon cooling to the same temperature it will contract to the same volume (neglecting the effects of crystallinity). *Note.* Mold shrinkage should not be confused with tolerance. Tolerance is the variation in mold shrinkage rather than the shrinkage itself. It is almost always the result of variations of molding conditions (assuming the material is the same).

The following discussion should also make the molding of heavy sections, 1/2 in. and over, more understandable. Shrinkage on cooling causes either sink marks on the outside or voids on the inside. The solution, obviously, is to pack in enough material to overcome the shrinkage. This is not always easy to do.

Sink marks are the result of insufficient material at a given spot. They are usually the result of poor part design. Since a thick part will shrink more than a thin part (Figure 3-39), there will either be a void on the inside or a depression on the outside in most thick parts.

PVT Relationship and Effector Shrinkage

When a plastic is compressed, its volume is reduced. It will return to its original size when the pressure is reduced to the original pressure. Gilmore and Spencer measured the thermal expansion and volume-pressure characteristic of polystyrene (129). Their results are graphed in Figure 3-31. The volume will change 0.00025 in.^{3}/in.^{3}°F. As might be expected the amount of compression depends on both the pressure and temperature. Gilmore and Spencer derived a modified Van der Waal equation of state for polystyrene relating the pressure, specific volume, and temperature.

$$(P + 27{,}000) \, (V - 1.422) = 11.18 \, T + 5134)$$

(3-18)

where P = pressure on material (psi)
       V = specific volume (in.^{3}/oz)
       T = temperature (°F).

Equation 3-18 states that temperature, pressure, and volume are dependent variables. To obtain a given volume (shrinkage) any number of combinations of pressure and temperature are possible. Any conditions that will affect the temperature or pressure (cylinder temperature, mold temperature, gate size, etc.)

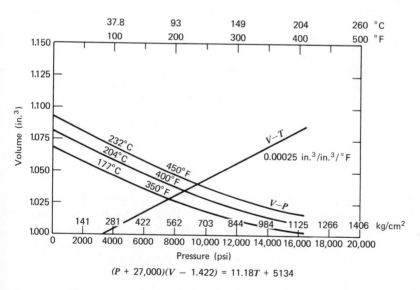

$$(P + 27{,}000)(V - 1.422) = 11.18T + 5134$$

Figure 3-31 Effect of temperature on volume, and of pressure on volume for general purpose polystyrene (Ref. 129).

will affect the shrinkage.

Some of the effects of temperature and pressure on polystyrene which were calculated from the equation of state (3-18) are shown in Table 3-9. Column A shows the conditions of polystyrene at room temperature (75°F) and no pressure.

Table 3-9   Effects of temperature and pressure on polystyrene calculated from equation of state (3-18)

	A	B	C	D	E	F
psi	0	0	0	13,000	13,000	0
T (°F)	75	400	75	400	400	75
Weight (oz)	0.613	0.613	0.567[b]	0.613	0.602[b]	0.602
Volume (in.³)	1.000	1.081[a]	0.925[b]	1.018[a]	1.000	0.982[l]
Density (oz/in.³)	0.613	0.567[b]	0.613	0.602[b]	0.602	0.613
Sp. Vol. (in.³/oz)	1.631	1.763	1.631	1.661	1.661	1.631
Side of 1 in. cube (in.)	1.000		0.974			0.994
Shrinkage (in./in.)			0.026			0.006

[a]From Figure (3-31).
[b]From density = weight/volume

A = polystyrene at room temperature and 0 pressure.
B = effect of raising temperature of 1 in.³ to 400°F.
C = taking 1 in.³ of B and cooling it to room temperature.
D = compressing B with 13,000 psi.
E = 1 in.³ of D.
F = cooling E to room temperature.

pressure. Column B shows the result of heating one in.³ to 400°F at no pressure. The weight will remain the same, 0.613 oz. From Figure 3-31 the volume will increase to 1.081 in.³. It could also be calculated by multiplying the temperature rise (325°F) by the coefficient of expansion (0.00025 in.³/(in.³)(°F). The density of the heated material is 0.613/1.081 or 0.567 oz/in.³ Column C shows the results of cooling 1 in.³ of this material to room temperature. From the density we know that 1 in.³ weights 0.567 oz. When the material cools to 75°F its specific volume is 1.631 (column A). Therefore its volume is 0.567 x 1.631 = 0.925 in.³. The cube root of the volume would give the side length, 0.974 in. This is a shrinkage of 0.026 in./in.

The conditions of column C applies to casting. Injection molding compresses

material in the hot stage. Column D shows the effect of 13,000 psi on the
erial of column B. From the graph (Figure 3-31) the volume has been
aced to 1.018 in.^{3}. It still weighs 0.613 oz so the new density equals
13/1.018 or 0.602 oz/in.^{3}. Column E shows the result of taking 1 in.^{3} of the
erial of column D. Column F shows the result of cooling the material in
amn E to room temperature. Multiplying the specific volume times the
ght (1.631 × 0.602) gives the volume, 0.982 in.^{3} which is equivalent to a
e with a side of 0.994 in. This is equivalent to shrinkage of 0.006 in./in.

It is easy to calculate the pressure-temperature combinations for molding
ystyrene with any shrinkage. For example, from Table 3-9 it is evident that
  time the specific volume is 1.661 at a given combination of molding
aperatures and pressures, the shrinkage will be 0.006 in./in. The result of
stituting this specific volume in (3-18) are shown in Figure 3-32.

Similarly the pressure-volume relationships can be determined at any
aperature. Figure 3-33 shows them at 400°F. Included is the shrinkage if the
terial under those pressure/volume conditions is brought to room temperature,
F.

The constants for the equations of state for several materials are shown in
le 3-10. Specific volume/temperature and specific volume/pressure curves are
ilable for thermoplastic urethanes (130), nylon 66 (131, 132), and
yleneproylene copolymer (133). Specific volume/pressure curves at molding
aperatures for acrylic, nylon 6/6, ABS, HDPE, PS, and PP are found in Ref.
). A more precise generalized equation of state is found in Ref. 134.

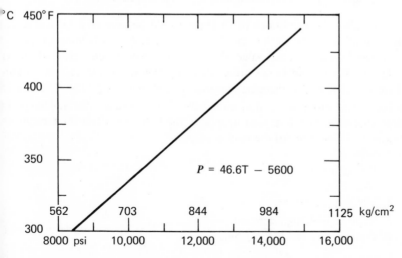

ıre 3-32 Approximate pressure-temperature relationships for molding polystyrene
ing with an 0.006 in./in. shrinkage at 75°F room temperature.

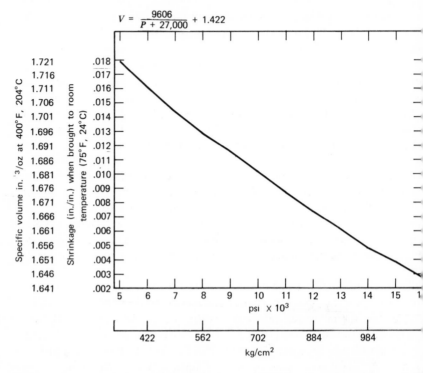

Figure 3-33 Approximate specific volume versus pressure for polystyrene at 400° (204°C) and resultant shrinkage when brought to room temperature (75°F, 24°C).

To realize the conditions of (3-18) it would be necessary to have a mol temperature the same as the material temperature. After the injection pressur has reached its determined value, the gate would have to be sealed off and th mold cooled to room temperature. Certainly this is not the case in moldin There is a continually changing temperature and pressure profile. The ultimat shrinkage (the difference in dimension between the mold and the molded part a room temperature) will depend upon how the following affect the pressure an temperature conditions of the molten polymer.

1. Material.
2. Part geometry.
3. Mold design.
4. Molding conditions.
5. Type of molding machine.
6. Condition of molding machine.

Table 3-10 Constants for equations of state for molding conditions

$$(P + a)(V - b) = \frac{R}{M}T$$

where P = pressure on material   atmosphere

V = specific volume cm^3/g

R = gas constant $(cm^3)(atm) / (g \text{ mole})(°K)$

T = temperature $°K$

M = molecular weight of polymer unit

	A	B	M	Ref.
Polystyrene	1840	0.822	104	135
Acrylic	2130	0.734	100	135
Ethyl cellulose	2370	0.720	60.5	135
Butyrate	2810	0.688	54.4	135
Polyethylene	3240	0.875	28.1	135
Polypropylene	1600	0.620	41	136

Injection Variables and Shrinkage

A picture of what happens during injection molding will make it easier to understand the affect of the variables on shrinkage. The hot material is injected into the cavity initially under low pressure. Cooling starts immediately, with the part in contact with the wall solidifying. Since the specific volume (the volume of a unit weight of plastic) decreases with the temperature, the solid will occupy less room than the molten polymer. If it were possible to have high enough rate of heat transfer through the plastic so that the part would solidify to room temperature throughout at the same time as the gate froze, there would be no shrinkage at all.

Pressure.   Actually, the material fills the cavity and pressure builds up rapidly. This compresses the material. During a production run the amount of pressure transmitted to the cavity is the deciding factor in controlling mold shrinkage. Normally this pressure will be maintained until the gate freezes, sealing off the cavity. At that moment the pressure begins to decay in the cavity as the plastic cools and shrinks. The effect of machine pressure can also be stopped either by retracting the injection ram or by the freezing of the runner system. In both those cases there is a possibility of material flowing back through the gate until the gate solidifies. At constant molding material temperatures, the most important variable affecting the dimension of the part is the pressure in the mold. The most important variable affecting the tolerance of the part is the variation of this pressure.

While the injection pressure is applied, the part continues to cool, increasing the thickness of the frozen layer. As the part solidifies its volume decreases and the injection pressure packs in more material. The thicker the frozen layer the lower the shrinkage. This is why higher mold temperatures give more shrinkage (Figure 3-34). One might expect that a higher mold temperature would permit higher mold pressures. While this is the case, the lowering of specific volume upon solidification is the overriding factor.

The part will continue to cool until the frozen skin is rigid enough for ejection. Increasing the cure time beyond this point will decrease shrinkage. It permits the formation of a thicker skin with greater rigidity. The inside will cool, reducing its volume and setting up a more highly stressed region, while the outside will maintain the dimension. Additionally, if the gate does not seal off more material can pack inside.

The part then cools outside the mold. About 90% of the shrinkage occurs within the first 6 hr, and almost all of the rest within the first 10 days. For

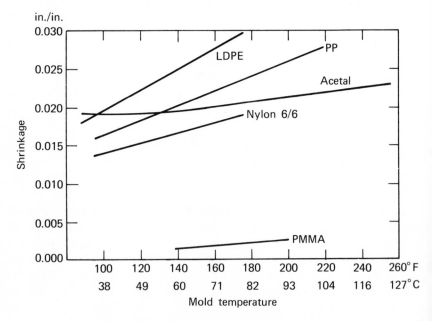

Figure 3-34   Effect of mold temperature on shrinkage.

crystalline materials lattice type shrinkage is caused by increasing crystallinity. In both crystalline and amorphous materials, stress relaxation at room temperature occurs for an extended period of time. For example, polyacetal showed an increase of shrinkage of 0.001 in./in. after 6 months (137) and polystyrene 0.0005 in./in. after 3 months (138). Annealing the part or having it in service above room temperature will increase the rapidity and extent of relaxation and crystallization. The acetal previously mentioned after 4 months at 180°F shrank an additional 0.002 in./in. and at 240° an additional 0.007 in./in. Hygroscopic materials will also change dimension depending on the amount of water absorbed.

Figure 3-35 shows the effect of the pressure on the injection plunger upon shrinkage. A more direct measure of the actual pressure in the mold at the time of gate sealing is shown in Figure 3-36. Both these graphs are qualitatively in consonance with (3-18). These pressures are the net result of many factors including the effect of temperature upon viscosity which in turn effect pressure transmission.

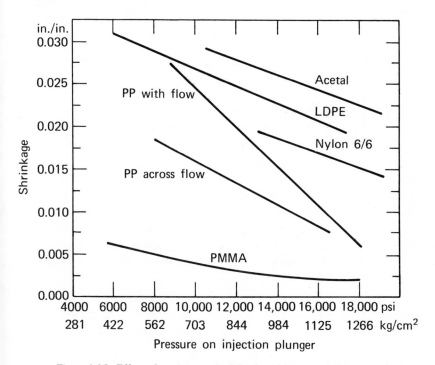

Figure 3-35 Effect of pressure on the injection plunger on shrinkage.

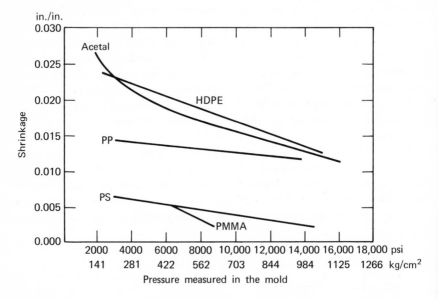

Figure 3-36   The effect of injection pressure as measured in the mold on shrinkage.

Assuming the gate does not seal, the longer the injection plunger forward time is maintained the more material can be packed in the mold and the lower the shrinkage. This is shown in Figure 3-37.

Temperature.   The higher the temperature of the material the more it has to shrink. Therefore one would expect the higher the material or cylinder temperature the more the shrinkage. In fact this is not found to be the case (Figure 3-38). Increasing melt temperature so decreases the viscosity that pressure transmission is significantly increased. The higher pressure overrides the effect of the higher temperature. In molds with short runner systems and large gates the shrinkage is reduced. Molds with small runners and gates show more pronounced shrinkage.

Part Thickness.   The part thickness has a significant effect upon shrinkage. The thicker the part the more shrinkage (Figure 3-39). The thicker materials have a lower proportion of frozen skin (solidified material). The hot interior has more hot plastic to shrink per unit volume.

It is interesting to note that in blow molding HDPE bottles it was virtually impossible to change the shrinkage by altering molding conditions (139). The manufacturing process only permitted a variation of 50 psi in the pressure and

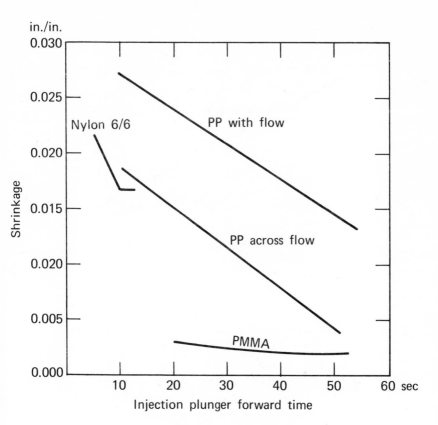

Figure 3-37 Effect of injection plunger forward time on shrinkage.

70°F in the temperature. The variables when inserted in the equation of state have such a small change on the specific volume as to make the difference in the volume of the blow molded bottle insignificant.

Gate Size.   The gate size is important in that it the material in the gate area must remain fluid to transmit the injection pressure for the required length of time. The mold temperature at the gate area has less importance as the gate becomes bigger. The size of the gate depends on the nature of the material and the thickness of the part. An excellent review of the effects of gate dimension on freeze-off times, shrinkage uniformity, impact strength, and minimum cycles for polypropylene is found in Ref. 140.

When determining gate size, be sure that the gate does not freeze up prematurely. This will give high shrinkage rates. Where high injection pressures (packing) would cause serious injection problems, small gates which freeze quickly are used.

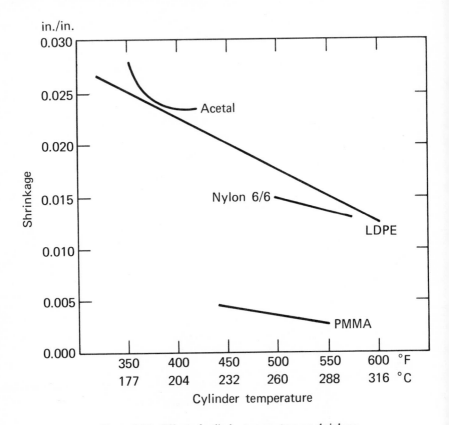

Figure 3-38   Effect of cylinder temperature on shrinkage.

Cavity Dimension Calculations

It is important to use the proper formula for computing cavity dimensions. where

$$D_c = D_p + D_p S + D_p S^2$$

where

D_c = dimension cavity (in.)

D_p = dimension molded part (in.)

S = shrinkage (in./in.).

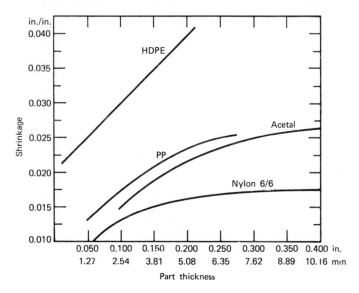

Figure 3-39   Effect of part thickness on shrinkage.*

he proper cavity size for a part 10.000 long with a 0.030 in./in. shrinkage is 0.309 not 10.300. If the mold is to be very hot the expansion of steel should e taken into account. Its coefficient of expansion is 6×10^{-6} in./°F.

Table 3-11 shows the shrinkage range used in mold building for thermoplastic naterials. A more accurate estimate can be made in conjunction with the naterial supplier. It should be emphasized again that the mold should be onstructed so that changing dimensions to correct for shrinkage can be done nost easily, that is, removing steel rather than adding steel.

Molding Heavy Sections.   The first prerequisite is a large gate-runner-nozzle ystem, so that injection pressure may be maintained for almost the full cycle ime. The outer skin of the part will be frozen by the relatively cool mold urface. Because of the poor thermal conductivity of the plastic, the center ortions cool much more slowly, and the gate must remain open until the center ection is relatively cool. This may require delicate mold temperature control nd possible heating of the runner system and the gate area. It is not always ossible to mold a thick section using conventional techniques.

Often intrusion molding will be successful (Chapter 1, p. 62). This technique lls the mold very slowly using very low pressure. Therefore, the cooling pattern

Figures 3-34 to 3-39 were based on data supplied by Avisun Corp., E.I. du Pont de Nemours & Co., Robinson Plastics Corp., Rohm & Haas Co., and from Ref. 96, 131, 137, 138, and 141. The information is most useful in a qualitative manner.

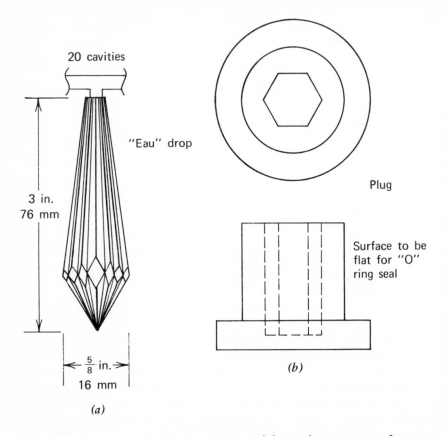

Figure 3-40 To prevent voids in heavy section parts, it is sometimes necessary to freeze one side only. The "eau drop" (*a*) was put half into water to freeze one side to prevent warping. The other side sank slightly but it was not seen because of the facets. The plug (*b*) had the outside placed into water to have a smooth surface for an "O" ring. The inside shrank which was acceptable because it only had to prevent an hexagonal nut from turning.

will be significantly different. Unfortunately surface blemishes caused by material folding will frequently appear. In normal molding these will be flattened out by the injection pressure. In intrusion molding a large part of the plastic may have solidified in those locations away from the gate. When the injection pressure is applied it will not be effective there. Mold temperature control is particularly effective in solving this problem.

Shrinkage Control-Examples

A plug molded in ABS (Figure 3-40*b*) had to have an absolutely smooth outside wall for an "O" ring. The inside hexagonal shape was used to hold a nut,

lerance not being important. The part had heavy sections. Since mechanical
rength was also important it was not possible to quench the part completely in
ld water because of the creation of internal stresses. The solution was to put
e plug in cold water, shoulder down, in a pan shallow enough so that the
xagonal part remained dry. This gave a smooth outside, sink marks on the
side, and a part with good mechanical properties.

A 20 cavity crystal polystyrene pendant (Figure 3-40a) required an excessive
cle to prevent voids. When the parts were put into water voids developed. The
imate solution was to put the shot in a very shallow pan of water where only
lf the part was immersed. This froze the pendant so that it did not curl out of
ape. The part exposed to air showed slight sinks which were successfully
mouflaged by the facets.

The effects of mold and molding variables on shrinkage and other properties
different materials are given in Ref. 130, 131, 137, 138, and 140 to 146. The
ect of molding conditions on shrinkage is summarized in Table 3-12.

Shrinkage is only one factor in the selection of conditions for molding.
hers would include cycle time, appearance, physical properties, clamping
ssure, degradation, and equipment. For example, increasing the pressure
creases shrinkage. However, increasing pressure also increases the residual
ssure in the mold. The increases the force required to open the mold which is
ectly related to the frictional forces between the plastic and the mold (Figure
1). It might be necessary to mold at lower pressures (increasing the shrinkage)
t to be able to eject the part from the mold. The molder might want to
ange the shrinkage to make a part stay on one or the other side of the mold
d change the mold to take care of the dimensional requirements. Knowledge
the mechanism of shrinkage will increase the molder's ability to produce the
uired part.

terature

There is a large amount of useful literature produced by raw material
ppliers. They fall into four broad categories:

(1) Material specification sheets. They show the properties of their different
aterials and the test methods used. Much of the information is published in the
nnual Processing Handbook of Plastic Technology. It is also available for the
neric materials in the Modern Plastics Encyclopedia. Pamphlets and booklets
e available for specific properties such as chemical resistance.

(2). Processing aids. This includes how to mold, fabricate, decorate, and
ch.

(3). Trouble shooting. These booklets suggest what corrective action may be
ken for the common molding faults.

Table 3-11 Approximate mold shrinkage (in./in.). Mold shrinkage is effected by part design, mold temperature, thickness, injection pressure, packing time, overall cycle time, orientation, gate size, gate design, gate location, glass content, and glass size (for further information consult the raw material manufacturer)[a]

		PART THICKNESS			No Thickness Specifications	
mm	3.2	6.4	12.7			
in.	1/8	¼	½	From	To	
ABS				0.005	0.007
ABS 30% glass	0.001	0.0015			
Acetal parallel to flow	0.021	0.025	0.026		
Acetal perpendicular to flow	0.018	0.019	0.020		
Acetal 30% glass	0.005	0.006			
Acrylic	0.004	0.007			
Cellulose acetate	0.006	0.006	0.008		
Cellulose acetate butyrate	0.005	0.006	0.006		
Cellulose acetate proprionate	0.005	0.006	0.006 ·		
Ethyl cellulose				0.005	0.009
Chlorinated polyether	0.004	0.008		0.004	0.008
Chlorinated polyether 30% glass				0.005	
CTFE				0.010	0.015
Ethylene vinyl acetate (EVA)				0.007	0.020
FEP	0.037	0.045	0.050		
FEP 30% glass				0.004	0.0045
Ionomer				0.005	0.020
Nylon 6				0.006	0.014
Nylon 6 30% glass				0.0035	0.0045
Nylon 66 parallel to flow	0.012	0.019	0.033		
Nylon 66 perpendicular to flow	0.020	0.020	0.028		
Nylon 66 30% glass				0.004	0.0055
Nylon 6/10	0.012 to 0.022	0.026 to 0.030	0.028 to 0.040		
Nylon 6/10 30% glass	0.0035	0.0045			
Nylon 6/12				0.007	0.012
Nylon 6/12 30% glass				0.003	
Polysulfone	0.007	0.009			
Polysulfone 30% glass	0.002	0.003			
Modified phenyline oxide (PPO)				0.005	0.006
Modified phenyline oxide 30% G1.	0.001	0.0025			
Polyaryl ether				0.007	
Polyaryl sulfone				0.007	0.009
Polycarbonate	0.006	0.007	0.008		
Polycarbonate 30% glass	0.0004	0.0025	0.003		
Polyethylene	0.015	0.022	0.028	0.015	0.050
Polyethylene 30% glass	0.004	0.0045			

| | PART THICKNESS | | | No thickness Specification | |
	mm 3.2 in. 1/8	6.4 1/4	12.7 1/2	From	To
Polypropylene 30% glass	0.004	0.0045			
Polyallomer	0.016	0.021	0.028		
Thermoplastic polyester 30% glass	0.001	0.0015			
Phenolics				0.007	0.012
Phenolics 30% glass				0.0005	0.004
4 Methyl pentene-1				0.015	0.030
Polystyrene				0.002	0.006
Polystyrene 30% glass	0.0005	0.001			
Styrene acrylonitrile (SAN)	0.002	0.007			
SAN 30%	0.0005	0.001			
Polyvinyl chloride (PVC)				0.003	0.008
PVC 30% glass	0.001	0.002			
Thermoplastic urethane	0.008	0.015	0.020		
Thermoplastic urethane 30% glass	0.004	0.005			

[a]Information courtesy of B. F. Goodrich Co., Celanese Plastic Company, Diamond Alkali Co., E. I. Dupont de Nemours & Co., Eastman Chemical Products Co., Hercules, Inc., Liquid Nitrogen Processing Corp., Mobay Chemical Co., Monsanto Co., Rohm & Haas Co., and Union Carbide Corp.

(4). Engineering and design properties. These range from single sheets to books. The author is partial to those companies who provide maximum information in these areas. It is this type of information that permits engineers to make maximum use of the properties and processing techniques of plastic.

Table 3-12 Conditions that affect shrinkage

To Decrease Shrinkage	
Lower	Wall thickness
Higher	Injection pressure
Longer	Injection forward time
Increase	Injection speed
Longer	Overall cycle
Higher	Material temperature
Lower	Mold temperature
Increase (see text)	Gate size
Narrow	Molecular weight distribution

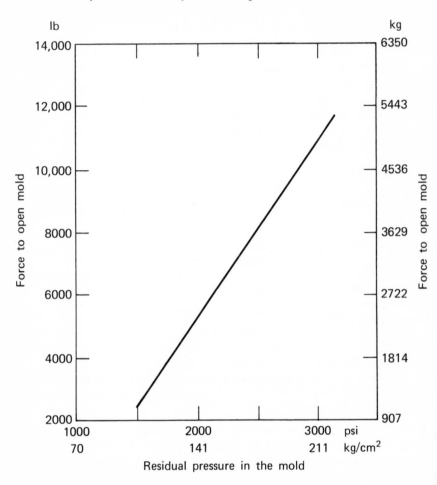

Figure 3-41 Relationship of force required to open mold and residual pressure in the mold (at the instant of mold opening) (Ref. 129).

It is strongly urged that the interested reader obtain such literature from the material supplier.

The introduction of a new material is accompanied by articles in the technical journals. They are useful in that they provide immediate access to the information. Typical articles on processing selected materials are urethane (147), ethylene-vinyl acetate copolymers (148), FEP (149, 149a), CTFE (150), rigid PVC (151), fibrous glass reinforced thermoplastics (FRTP) (152), polyaryl sulfone (152a), ethylene-propylene rubber (EPDM) (152b), Nylon 11 (152c), and Nylon 66 (152d).

VISCOELASTIC BEHAVIOR AND HOW IT RELATES TO PROPERTIES OF FINISHED PARTS

The engineer must understand that plastics parts will not behave as other materials do under stress. Thermoplastic polymers exhibit flow behavior which falls between the two extremes of ideal elasticity (Hookean solid behavior) and ideal flow (Newtonial liquid behavior). Therefore these materials are neither purely elastic nor purely viscous—they exhibit the characteristics of both. These rheological properties are called viscoelastic.

There are three components of viscoelasticity—elasticity, viscous flow, and retarded elasticity. (Retarded elasticity, a time dependent function, is almost completely recoverable—unlike viscous flow—the retardation resulting from the nonequality of stressed-biased and nonstressed movement of polymer segments. For more on retarded elasticity, see p. 290.

Stress-Strain Relationships

The stress-strain relationship of viscoelastic materials can be characterized by:

$$\text{stress} = \text{strain} \times f \text{ (time)} \tag{3-21}$$

This implies a linear relationship, and plastics which follow it are classed as linear viscoelastic materials (for low deformations, in the area of 10^{-3}, and small time intervals the stress-strain relationship is linear).

Some plastics are nonlinear over any deformations that are of practical value. A full discussion of the deformation properties of plastics and their engineering implications are not properly within the scope of this book. They are very briefly discussed later.

An obvious implication of (3-21) is that mechanical property measurements are dependent on the rate at which the measurements are carried out, as are the resultant deformations. Less evident, but equally true, is that measured mechanical properties depend on the previous history of the material (153, 156).

Viscoelastic Models

When a material exhibits linear viscoelasticity, its actions can be qualitatively explained by mechanical models. This is fully developed by Alfrey in his book (154) which is highly recommended. Viscoelasticity is discussed in varying degrees of depth in Refs. (155 to 160).

The models are a combination of springs, which represent ideal elastic solids (Hookean) and dashpots, which represent irreversible flow of viscous liquids (Newtonian). The elastic deformation component can be considered to consist of the stretching of the primary valence bonds, almost always C-C. Viscous flow represents the sliding of molecular segments past each other. The force used in viscous flow, overcomes the Van der Waal's forces and other electrostatic attractions and is not used to stretch any C-C linages.

The following notation is used in these models:

γ_T = total displacement (strain)

γ_E = elastic displacement

γ_F = flow displacement

G = shear modulus

S = shearing stress

μ = visoosity.

A Maxwell element consists of a spring and a dashpot in series shown in the top section of Figure 3-42. A material of this nature will exhibit an

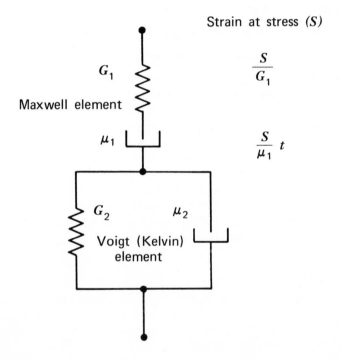

Figure 3-42   Maxwell-Voight (Kelvin) elements used to help describe visco-elastic flow.

istantaneous elastic deformation and a slower viscous flow. Its total displacement will be the sum of the two:

$$\gamma_T \;=\; \gamma_E \;+\; \gamma_F \tag{3-22}$$

The elastic displacement is the shearing stress divided by the shear
modulus S/G . The rate of change of the elastic deformation therefore is

$$\frac{d\gamma_E}{dt} \;=\; \left(\frac{1}{G}\right)\!\left(\frac{ds}{dt}\right) \tag{3-23}$$

The flow displacement depends on the stress and the amount of time the
stress is acting. The instantaneous rate of flow is the stress dividend by the
viscosity (S/μ).
The rate of change of the flow displacement is

$$\frac{d\gamma_F}{dt} \;=\; \frac{S}{\mu} \tag{3-24}$$

It is not possible to write directly an equation relating the stress to the
deformation, but the total rate of displacement of the plastic material is the sum
of the rate of displacement of the elastic deformation (Eq. 3-23) and the rate of
viscous displacement (Eq. 3-24).

$$\frac{d\gamma_T}{dt} \;=\; \left(\frac{1}{G}\right)\!\left(\frac{ds}{dt}\right)\!+\; \frac{S}{\mu} \tag{3-25}$$

This equation governs the mechanical response of the material at any sequence
of shear stresses or constraints upon the deformation itself.

Consider the situation where a part is rapidly deformed and maintained at a
given deformation (strain). (The amount of stress required to maintain this
deformation is measured). The rate of deformation, $d\,\gamma_T/dt$, is 0. Substituting in
3-25),

$$0 \;=\; \left(\frac{1}{G}\right)\!\left(\frac{ds}{dt}\right)+\; \frac{S}{\mu} \tag{3-26}$$

Integrating with the conditions that at the beginning (t = 0), the stress is 0
$S = S_0$), then

$$S \;=\; S_0\, e^{-\,(G/\mu)\,t} \tag{3-27}$$

If we take the specific time, where $t = \mu/G$ then (3-27) becomes

$$S = S_0\, e^{-1}$$
$$S = S_0\; 0.367 \tag{3-28}$$

The time μ/G is called the stress relaxation time and is given the symbol τ

$$\tau = \frac{\mu}{G} \tag{3-29}$$

It is the time equal to the decay of approximately two-thirds of the original stress. Equation 3-27 can be written.

$$S = S_0 e^{-(t/\tau)} \tag{3-30}$$

In summation, if a plastic part is elongated to a definite size by a stress (S), that stress required to maintain that elongation declines exponentially according to (3-30).

Retarded Elasticity

The Voigt (Kelvin) element represents the retarded (recoverable) elastic response (lower portion of Figure 3-42). The viscous element acts as a damping element, resisting the establishment of the elastic equilibrium. It resists the instantaneous elastic elongation upon the application of stress and the instantaneous contraction of the elastic element upon removal of stress. By contrast, the viscous part of a Maxwell element represents flow, which is part of the displacement, and is not recoverable. If a plastic part is rapidly deformed by a stress (S), its retarded elastic response is governed by

$$\mu \frac{d\gamma}{dt} + G\gamma = S \tag{3-31}$$

The stress (S) is removed so that $S = 0$. (The amount of strain is measured over the period of time of the experiment.) Substituting in (3-31),

$$\mu \frac{d\gamma}{dt} + G\gamma = 0 \tag{3-32}$$

Integrating with the condition that the time is measured from the application of the stress, then

$$\gamma = \frac{S}{G} \left(1 - e^{-(G/\mu)t}\right) \tag{3-33}$$

If μ/G is called the retardation time (τ), then

$$\tau = \frac{\mu}{G} \tag{3-34}$$

Then

$$\gamma = \frac{S}{G} \left(1 - e^{-(t/\tau)}\right) \tag{3-35}$$

When the stress is removed the part eventually returns to its original shape ($\gamma = 0$), following the exponential curve,

$$\gamma = \gamma_0 e^{-(t/\tau)} \tag{3-36}$$

with τ being the retardation time, or the time for the removal of approximately two-thirds of the strain.

Effects Of Time

Time is an important factor in viscoelastic behavior. Consider the effect of time on a Maxwell element with a stress (S) acting during a time (t). Then the elastic deformation is S/G and the viscous flow is $(S/\mu)(t)$

1. If t is very short, then the viscous flow will be very small, and the material will act essentially like an elastic body.

2. If t is very long, the viscous flow will be very large in comparison to the elastic deformation. The material will essentially behave like a liquid with the viscosity μ.

If t is in the order of magnitude of the relaxation time, μ/G both elasticity and viscous flow can be observed. This becomes very important in experimental and design work. For example, a given stress applied very rapidly (short t), might give rise to an elastic deformation which would not affect the usability of the part. Applying the same stress for a longer time might cause viscous flow which would permanently deform the part beyond the design requirements.

Similarly, for a Voigt element, if a stress (S) is applied for a time that is short in comparison to the retardation time, the plastic will act like a fluid with a viscosity of μ. If the time scale is long compared to the retardation time, the material will be able to return to its original position and appear to be elastic in nature.

Neither the Maxwell nor Voigt elements alone describe viscoelastic behavior. Combining them in series, (Figure3-42) gives a good qualitative description of the effects of stress, strain, and time. A graph of the strain versus time using Maxwell-Voigt elements will now be discussed (Figure 3-43).

At times before t_1, when no stress (S) has been applied, the elements are in configuration A, which is identical with Figure 3-42. At t_1, a stress (S) is applied. There is an instantaneous elastic deformation of the spring (G_1), S/G_1. This is shown as configuration B of (3-43). Continued application of S will cause the dashpot, μ_1, to move; its deformation depends on the length of time the stress is acting upon it (similarly for μ_2). However, the movement of μ_2 is in conjunction with the elastic element G_2 and it will ultimately be returned to its original position because of the elastic action of G_2. The actual viscosity of the Voigt elements μ_2 is much lower than that of the viscous element of the Maxwell unit μ_1. The Maxwell unit represents moving either entire molecules or large segments, while the Voigt μ_2 represents the moving of molecular segments. Figure C represents the maximum deformation just before the stress is released at t_2.

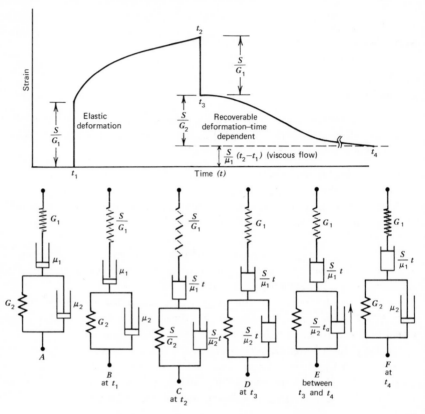

Figure 3-43   Maxwell-Voight model for viscoelastic flow.

Upon release of the stress, the elastic deformation of G_1 is recovered immediately and the model assumes the configuration of D at t_3. Between t_3 and t_4 spring G_2 contracts, its movement being retarded by μ_2. This is shown in E. Eventually, all of the strain in the Voigt element, S/G_2, will be recovered at t_4. The final total strain (deformation) will be only that of the viscous flow of μ_1.

Effects of Temperature

Temperature has a major effect on plastic flow through the viscosity functions. Some of its effects can be easily seen from inspecting the deformation-time equation of a Maxwell-Voigt element:

$$\gamma = \frac{S}{G_1} + \frac{S}{\mu_1} t + \frac{S}{G_2} \left(1 - e^{-\left(\frac{G_2}{\mu_2}\right)t}\right) \tag{3-37}$$

deformation   elastic   viscous   retarded flow
tion   response   flow

1. At low temperatures viscosities μ_1 and μ_2 are very high, so that the second and third terms are very small. The material behaves as an ideal elastic solid with shear modulus of G_1.

2. At low temperatures, if a continual stress is applied below the elastic limit, creep occurs. This is called "cold flow" if it happens at room temperature. It is primarily a viscous flow effect (the second term of Eq. 3-37) and is time dependent. The very last part of the flow is recoverable in that the retarded flow of the Voigt element will operate (the third term of Eq. 3-37). This is very small compared to the total flow.

3. If the temperature is very high, the viscosity is very low, and the material behaves like a liquid, whose strain follows the viscous flow component μ_1.

4. In a temperature range where the viscosity of the Voigt element μ_2 is low, the flow component will be very large. The plastic will flow steadily until the stress is changed, when it will exhibit a small elastic recovery.

The four element model just described is oversimplified. The viscous flow element is almost certainly non-Newtonian. The elastic response can be non-Hookean. The creep curves cannot be represented by creep curves with only one retardation time. Materials have many retardation times distributed over many decades of time. Nonetheless, they give an excellent qualitative understanding of the way plastic materials respond to stress, thus are most helpful in understanding their mechanical behavior.

It should be noted that G is the shear modulus, not the E of Young's modulus (tensile modulus). Assuming that polymers are not compressible (which is permissible for qualitative discussions), the Young's modulus is about three times the shear modulus. Conversions can be made by multiplying G by 3.

Creep and Deformation and Their Relationship to Part Design

The engineer is faced with the problem of creep and deformation in the design of parts. Quantitatively, stress-strain-time curves are used. It should be recognized that there is no accepted standard creep test. Since the ultimate properties of the part are in significant measure controlled by the method and nature of fabrication and the environment, a generous design safety factor must be included.

Some of the types of available curves are

1. Stress-strain.

2. Tensile creep modulus—time. These are particularly valuable because many engineering formulae require creep modulus.

3. Isometric stress—time. These graphs have the coordinates of tensile stress and time. The curves are for different strain percentages.

4. Isochronous stress-strain. These graphs have stress and strain as coordi-

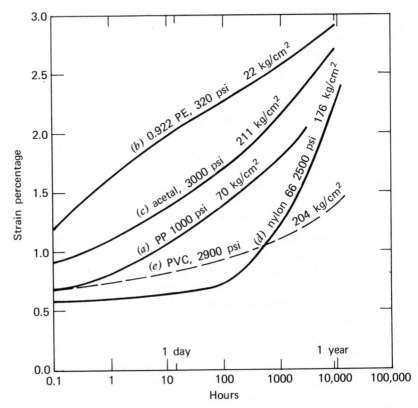

Figure 3-44 Percentage strain/time curves for 5 plastic materials at 20°C: (a) polyprc pylene homopolymer, 1000 psi, (b) 0.922 polyethylene 320 psi, (c) acetal copolymer, 300 psi, (d) nylon 66, 2500 psi, and (e) unplasticized polyvinyl chloride, 2900 psi (Ref. 2).

nates and the curves are for different times.

A superb collection of these graphs for polypropylene, LD polyethylene poly (4-methylpentene), acetal copolymer, nylon, acrylics, PVC, and polytetra-fluoroethylene (PTFE) is given by Ogorkiewicz (2). In addition there is a wealth of other curves such as tensile creep-time; creep curves-tensile loading; creep modulus-temperature; creep rupture stress in tension-time to failure; yield stress in tension-temperature; yield stress in tension-density; impact strength-temperature; impact strength-temperature showing the effects of different materials and fabricating conditions; loss angle-frequency temperature; apparent volume resistivity-temperature; apparent volume resistivity-time of electrification; linear thermal expansion-temperature; and specific information for certain materials. Figure 3-44 shows the percentage strain graphed against time for some polymers taken from Ref. 2.

Reference 161 has apparent creep modulus at 1, 10, 30, 100, 300, and 1000
for almost all materials with initial applied stresses usually in the 1 to 3000
i range.

Since injection molding engineers are often involved in design, either before
after the fact, references 153, 162 to 209, 91 to 93, 244, 245 and 249 to 251
late to creep and relaxation of thermoplastics.

OLDING OF THERMOSETTING MATERIALS

ae polymer molecules in thermoplastic materials are held together by
ectrostatic forces. When the material is heated, these forces diminish and the
astic softens and ultimately flows.

Thermosetting materials are held together by chemically cross-linking the
olymer molecules. The molded, cured parts are substantially infusible and
soluble. They cannot be resoftened by heat. Thermosetting molding materials
mmonly used are stable for long periods of time at room temperatures.
owever, as the temperatures increase the shelf life decreases. When they reach
e temperature range of 250 to 300°F, they "kick over" in a matter of seconds.
ais mechanism is very different than that of thermoplastics.

There are four methods of molding thermosets:

1. Compression molding is the oldest technique. The mold consists of a
vity into which is put cold material in powder form or cold preforms (a
mpressed pill of material). The force side of the mold is pushed into the cavity
 the hydraulic ram, compressing the material. The material receives its heat by
rculating steam in the mold. Pressures on the material normally range from
)00 to 5000 psi. Since there is relatively little flow, there is little or no
ientation or breakage of reinforcing materials. Holes cannot be molded
mpletely through. The parts have to be deflashed. Material loss is minimal.
aterial handling is easy, particularly with low bulk density fillers. This is the
owest method of thermoset molding. The process can be and is sometimes
ghly automated, running unattended 24 hr/day, 7 days/week. Automatic
mpression molding equipment is less expensive than transfer or inline injection
olding machines, since compression molding requires clamping forces approxi-
ately half those needed in transfer and injection for molding the same part.
ompression molding is least susceptible to contamination as there is no screw
 transfer system. This can be very helpful when trying to mold light colored
elamines.

2. The second method is identical with the first except that the material is
reheated by a radio frequency field or infrared heaters. This technique can
duce the molding time by as much as 30%. The RF machines are relatively

difficult to control and are sensitive to slight material changes. Preforms can be made and placed in the compression press.

3. The third method is transfer molding. Preheated material is put in an open pot. From there a hydraulic plunger forces the material through a runner system into the cavity (209a). In plunger molding, a plunger in a cylinder is used as the transfer medium. A variation is to use a screw to preheat the material. The plunger system is then like a two stage screw machine where the mold is similar to an injection mold. The process is used for molding inserts, intricate shapes, parts with varying wall thicknesses, and parts where dimensional tolerances are important. The material is more compacted and therefore has higher compressive and flexural strength. If preforms are used and transferred there are only four process variables—mold temperature, charge size, preheat time, and transfer pressure. The cost of making preforms is approximately 1¢/lb, plus a cost of about 50¢/hr for operating the RF equipment. Molding cycles are generally shorter than methods 1 and 2, presumably because of the frictional heat generated by the high velocity of the material as it is transferred.

4. The logical extension of the screw type transfer press is to injection mold thermosetting materials with an inline screw, which is the fastest method by far for processing thermosetting materials.

Injection Molding of Thermosets

The mechanics of thermoplastic and thermoset molding on a reciprocating screw are very similar. The molding conditions differ (209b to 211). The material in powder form is gravity fed into the hopper, which eliminates the need for preforming and preheating equipment. The material is heated by conversion of mechanical energy into heat energy through the screw. The many controls make this the most sensitive method. The material can be heated closer to its reaction temperature, and there is no wasted time in transferring the material. As soon as enough material for the shot has reached the premolding temperature, it is injected into the mold. This, too, allows for a higher premolding temperature. High injection pressures are available resulting in good mechanical properties. The injection speed and pressure can be controlled in at least two stages. The material is injected into a hot mold where the cross-linking starts. It is usually molded in a horizontal machine so that it can be ejected with gravity removal of the molded parts. The machine and operation can be fully automated.

The Screw Injection Machine   The main differences in the machine are in the heating cylinder and screw. The nozzle and the injection head are designed for quick removal in the event that the thermosetting material sets up in the screw or nozzle. This used to be a problem but has been overcome by specially designed materials and improved technique.

The heating cylinder is made of the same material as the thermoplastic cylinder. Instead of electrical heating bands, the cylinder is jacketed for circulating water. There are three separately controlled zones on the barrel and one for the nozzle. The maximum temperature, therefore, would be 212°F. Typical temperatures for molding general purpose phenolic are 160°F rear zone, 160°F middle zone, 200°F front zone, and 200°F nozzle temperature. The temperature of the material, however, depends not only on the barrel temperature but also on the screw speed and back pressure. For example, a phenolic molded with a barrel temperature of 160° had a stock temperature of 240° (212).

Most screws for thermoset molding have a compression ratio in the range of 1 to minus 0.8 in. The higher compression ratios used for molding thermoplastics would heat the material enough to kick it over and cure in the cylinder. The screw L/D ratios are considerably smaller because the three different screw functions for thermoplastic molding are not required. The screw tip is of the smearhead type (Figure 1-21), which bottoms out at the end of the injection motion. This is very important to prevent gradual reaction of the material in the nozzle, and erratic press action. Other than provisions for reversing the screw rotation, the balance of the press is identical in both types of molding.

Molds.   The molds for thermosets are similar to those for thermoplastics. The basic difference is that thermosetting materials require hot molds to heat the material to initiate the reaction while thermoplastic molds require cooling channels to remove the heat.

Mold temperature control should be held within ±5°F. Typical mold temperatures are shown in Table 3-13. The mold should be insulated from the platen to prevent heat transfer. Asbestos filled fibre board ¼ to 3/8 in. thick can be used. The molds are best heated electrically. The amount of heat required is

$$\text{watts} = \frac{(\text{weight})\,(\text{sp. heat})\,(\Delta T)}{(\text{time})\,(3.412)} + 2A$$

Heat is dissipated from the exposed surfaces of the mold. An approximation of 2 W/in.^{2} of exposed surface is used ($2A$).

The approximate number of watts to raise a steel mold ΔT degrees in 1 hour is

$$\text{watts} = 0.01\,(l\,w\,h\,\Delta T) + 4\,(l\,w + h + w\,h) \tag{3-38}$$

where l = length mold (in.)
w = width mold (in.)
h = height mold (in.)
ΔT = temperature increase (°F)

A mold 10 by 10 by 10 in. to be raised from 70 to 370°F in 1 hour would

Table 3-13   Approximate mold temperatures for reciprocating screw molding of thermosetting materials

Material	Temperature °F	°C
Urea	295-310	146-154
Melamine	310-340	154-171
Diallyl pthalate	330-350	166-177
Filled alkyd	350-365	177-185
Epoxy	350-370	177-188
Phenolic	350-390	177-199

require 4200 W in its electrical heating system.

If the mold temperature is too low, the material may not react completely, hold its shape, or fill correctly. If the temperature is too high the molding may not fill out, the surface of the parts may be dull, parts may be porous, flow lines may show, parts may stick, and parts may flash.

The mold should be adequately vented. Vents 0.003 × 3/16 in. are typical The mold should be hardened to approximately R_c 50. Since thermoset materials are usually more corrosive and errosive than thermoplastics, flash chroming is advised after the mold is in good operating condition.

Gates and Runners.   Gates and runners are similar to those in use for injection molding. The same type of gating can be used, including restricted and submarine gates. Full round runners are much more important with thermosets. Runners and gates should be polished. A rough approximation for initial runner and gate sizes are

$$A_R = 0.0004W + 0.03 \qquad (3\text{-}39)$$

where A_R = cross-sectional area runner (in.²)

   W = weight of parts fed by runner (g)

$$A_g = 0.0002W + 0.005 \qquad (3\text{-}40)$$

where A_g = cross-sectional area gate (in.²)

   W = weight of part (g)

For example, a runner feeding two 100-g parts would be approximately 3/8 in. in diameter. A gate for a 150-g piece would be about 7/32 in. in diameter.

When the thermoset material goes through restricted and small orifices, the temperature increases. It is a function of the hole size and injection speed. Increasing the speed from 5 to 35 cc/sec through a 1/8 in. hole raises the temperature of a phenolic 18°F (213).

The runner system cannot be reground and reused. Many parts are better center gated than edge gated. To overcome these difficulties, a "cold" runner

old has been developed (214, 215). This is a manifold which is kept cooler
an the mold, at a temperature which prevents the premature kicking over of
e material.

Precuring.   Precuring of the material inside of the nozzle and in the screw
e no longer a serious problem. One of the methods used to prevent buildup in
e nozzle is to reverse the rotation of the screw. This sucks the uncured
aterial back from the nozzle into the screw (U.S. Patent 3,427,639).
metimes either because of poor operating, power failure, or machine trouble,
e thermoset will cure on the screw. The cylinder head is removed, the barrel
ated to 200°F to expand it so that the screw can turn, the cold material is fed
to the hopper. This will purge out the cured material. The temperature is then
opped, the head replaced, and the molding restarted.

Flow Rating.   The flow rating for thermosets is based on the ASTM D 731
st. This records the minimum force in pounds required to close a specifically
signed cup mold, compression molded under specific conditions. From this,
bitrary numbers from two to 32 have been assigned, the lower numbers
dicating stiffer flow. Reciprocating screws use materials with flow values close
12. Another flow test is described in Ref. 216.

The Thermoset Molding Process. There are three phases to thermoset
olding; melting, flowing, and gelling. As the material is heated, the viscosity
ops. As the heating continues, cross-linking occurs and becomes the overriding
ctor in the viscosity. The more the cross-linking, the higher the viscosity. A
ot of viscosity versus time shows a U shaped plot (217). The rate of gellation
ross-linking) is predominantly temperature controlled. In molding, the
timum temperature is one that gives the lowest viscosity. This is the
mperature just below the point where cross-linking becomes the predominant
echanism (211 - 213, 217 - 219). This is quite different from thermoplastic
olding where raising the temperature decreases the viscosity. In thermoset
olding two factors are balanced. First, the molder wants to get the material hot
ough to lower its viscosity, so that it flows in the mold rapidly, within the
essure limitations of the equipment. The higher the material temperature the
ticker it will reach the reaction temperature in the mold. Second, the molder
ust be careful not to reach that portion of the temperature-viscosity curve
here the rate of gellation increases the viscosity so much that the injection
essure capacity of the machine is too low to force the material into the mold.

Characteristic signs of overheating are increasing injection time after each
cle, irregular forward movement of the screw during injection, irregular
ovement of the screw during screw rotation, and unusual decomposition when
rging. If these occur, the cylinder should be purged and the screw speed-back
essure-barrel temperature systems should be adjusted downward.

The main factors affecting the temperature of the melt in the cylinder are the barrel temperature, screw speed, and back pressure. If the barrel temperature is too low the plastic will remain granular and not start to flux. If it is too high it will permit the material to overheat and precure. Screw speeds for thermoset molding are in the area of 50 rpm. It is an important consideration. If too fast the screw will retract too quickly and not give sufficient time for adequate heat transfer from the barrel. Back pressure is very important in temperature control. Raising the hydraulic back pressure from 0 to 75 psi for a two-step wood flour filled phenolic compound increased the stock temperature from 260 to 280°F (212). In thermoset molding it is very important that the material be injected immediately upon screw return. The longer it remains in the barrel the more cure and the higher the viscosity. This is in direct contrast with thermoplastic molding where a longer resident time would increase the temperature and lower the viscosity.

Cycle Time.   Overall cycles are similar for thermoplastics and thermosets. A two cavity food blender base was molded in different materials. The shot was 12.4 in.^{3}, with a 1/8 in. normal cross-section and almost 1/4 in. rib section. The overall cycle in polycarbonate was 57 sec; ABS 55 sec; polyester premix 49 sec; polystyrene, alkyd, and glass filled phenolic 45 sec; polysulfone 40 sec; GE phenolic 28 sec; and melamine 27 sec.

Shrinkage and Finished Part Problems

Because of the different mechanism, molding conditions have different effects on shrinkage. For a general purpose phenolic (220) injection pressure, injection holding pressure, and injection hold time have little effect on mold shrinkage. Increasing the curing time increased the amount of cross-linking. This made the part more rigid and better able to resist the internal contractions caused by cooling. After-baking increased mold shrinkage. The longer the time and the higher the temperature the more the shrinkage. This effect varied with the type of material as the mechanism is the removal of volatiles and increasing polymerization.

Increasing the injection time decreased the shrinkage. The mechanism is probably similar to that of thermoplastics. Increasing mold temperature could increase or decrease shrinkage. There are two mechanisms, one which increases shrinkage is primarily related to the volume change during contraction. Higher mold temperatures also increased the evolution of gas and moistures. The opposing parameter is that higher mold temperatures increase the cure which gives a more rigid material.

The phenolic materials showed strong anisotropic characteristics. As with many thermoplastics, mold and piece part design must take orientation effects into account.

Further information is available for screw injection molding of phenolics

(212, 213, 220a, 220b, 220c) DAP (218, 220c) Urea and Melamines (214, 220d). Alkyd and polyester (220e), and reinforced thermosets (220f).

HANDLING OF FINISHED PARTS

An intrinsic advantage of injection molding is its ability to produce economically large numbers of parts. More often than not, quantity determines whether the part will be molded or produced by another process. Handling, sorting, and protecting the part must be considered in the beginning of the project. Handling of the part encompasses more than removing it from the machine or the machine area. Jigging, fixturing, and secondary operations are also part of handling procedure. This is particularly true of parts to be used on automatic assembly equipment.

Figure 3-45 shows an automatic assembly machine for inserting brass bushings into finials (lamp top). A metallic thread is used because a plastic thread would not be strong enough at the high temperatures above the bulb. The bushings are fed by a vibrator on the top right, go down a track, are heated by an induction heater, brought forward on a slide, and pressed into the plastic by a ram. The plastic parts are selected by the vibratory feeder on the left, picked up by a shuttle, and dropped into the retaining pocket of the rotary table. Two of the parts bushed in this machine are shown at the right side of Figure 8-11. The top right finial was designed with a large diameter ring above the center of gravity, so that regardless of the direction of feed the part would fall by gravity and could be supported by a simple open track leading down from the vibratory feed. The part on the lower right was designed before the assembly machine was conceived. It would not feed simply along a track and required expensive experimentation before a method of feed was ultimately found.

Not all applications are as critical. However, when a part has to be fabricated, decorated or metallized the method of holding and handling should be determined before the mold is built. Packing is a part of handling. The cost of shipping, for example, can quadruple depending on whether a large part is stackable.

If the part is removed manually by the operator, it is usually placed on a table. At that time it can be degated manually, with the use of hand tools, with degating fixtures or ultrasonically (221). From there the part will be sorted, and either packed or removed by a conveying system. In designing the layout of a multicavity mold, ease of sorting is considered. Often a cycle is slowed because the operator does not have enough time to handle the parts.

In automatic molding parts either fall directly into a container or are removed from under the mold by specialized belt type conveyors. These conveyors are also useful for an operator-tended machine. They can lift the parts on to a table

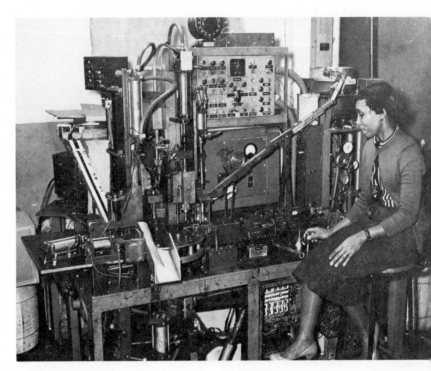

Figure 3-45 Bushing Machine for Automatically Heating and Inserting Metal Bushings into Plastic Parts (Robinson Plastics Corp.)

for inspection instead of having the operator continually bend over to retrieve them (222). From the conveyor they can be routed to sorting, inspection, packing, or other operation. Automatic handling of small parts is not unique to the molding industry. Extensive systems have been developed and are readily available commercially.

The introduction of larger machines and parts that weigh over 30 lb require special handling techniques. An operator cannot successfully and consistently cope with the heavier parts. Various shuttle systems have been developed. The most promising method of handling these parts seems to be the use of industrial robots (223). They are mechanisms with movable arms terminating in some tool which can grasp the plastic. Suction devices and claws are commonly used. They are usually hydraulically powered and electronically controlled. They can be used on more than one machine depending on their proximity and the cycling. It is estimated that replacing one operator on a two shift operation will return the cost of the machine in two years.

PART DESIGN

Part design is properly the subject of another book. It is the function of the plastics engineer to interpret the plastic part in terms of the end user's need. To do so a minimal amount of information is required. The following list is helpful in that respect.

General Information

1. When is the production of the part required.
2. What information is available about the part.
   a. Part drawing.
   b. Assembly drawing.
   c. Sample, with history of its use.
3. What is the function of the part.
4. What is the physical system in which the part will operate.
5. What will happen if the part fails.
6. What is the required service life of the part and system.
7. Can the part or the system be altered.
8. Are specific standards, tests, or certifications required.
9. What are the quality control requirements.
10. What are the nonmechanical tolerance requirements.
11. What secondary operations of any kind will be performed.
12. Will the part be used immediately or stored.
13. How will the product be packed.
14. How will the product be shipped.

Mechanical Information

1. Is the load static or dynamic.
2. Is the load constant or changing.
3. Is the load continuous or intermittent.
4. Is the load concentrated or distributed.
5. What is the magnitude of the load.
6. What is the direction of the load.
7. What is the rate of application of the load.
8. Are there size, weight, or shape limitations.
9. What are the possibilities of outside (nondesigned) forces.
10. What is the maximum deformation allowable.
11. What is the maximum deformation for part failure.
12. What are the effects of friction and wear on moving parts.
13. What mechanical tolerances are permitted.
14. What mechanical tolerances are attainable.

Environmental Information

1. Temperature range of the operation of the system.
2. Variations of temperature.
3. Durations of varying temperature; if it affects the ultimate use of the part.
4. Humidity conditions.
5. Exposure to weathering.
6. Exposure to sunlight.
7. Exposure to radiation.
8. Chemical environment: concentration and time.
9. Permeability requirements.

Electrical Information

1. AC or DC.
2. Voltage used.
3. Frequency used.
4. Tracking requirements.
5. Conductivity requirements.
6. Surface and volume insulation requirements.

Appearance and Esthetic Information

1. Is the shape "pleasing" to those responsible for the product.
2. Is the shape functional for human use (i.e., will the plastic contact lenses stay in place).
3. Is it "human engineered" (i.e., does a plastic chair have the right contour and height).
4. Color requirements.
5. Surface appearance (polished, matte, decorated, etc.).
6. Finish requirements for all surfaces (i.e., the surface finish of the inside of a float is not esthetically important).
7. Effect of surface finish on mechanical, environmental and, safety properties.
8. Optical requirements, including clarity and light transmission.

Economic Information

1. Cost of part.
2. Cost of part made by different processes.
3. Cost of part made in different materials (nonplastic as well).
4. Quality standards as related to economics.

5. Reliability (replacement cost of part) as related to economics.
6. Expected working life.

afety Information

1. What will happen if the part fails. This depends on its use. For example, a battery case failure in the space program is not the same as the identical failure in a portable drill.
2. What will happen if the part burns. Will it spread the fire or produce toxic gases.
3. Has color been used to promote safety.
4. Has design been used to promote safety.

ltimate Questions

1. What is the best process.
2. What is the best material.
3. What is the most economical process.
4. What is the most economical material.

A designer, of course, would always prefer to have such things as uniform all thickness, adequate coring, generous radii, and elimination of entrapped aterial. These and many other things are not all possible all the time. There is a ealth of information available from material manufacturers and a significant mount of literature on the subject, of which Refs. 224 to 236C are typical. ome of the things to consider when designing a part follow:

1. Wall thickness.
2. Coring.
3. Taper.
4. Radii.
5. Reinforcements (ribs, fillets).
6. Entrapped material.
7. Gating.
8. Weld lines.
9. Orientation.
10. Warping.
11. Venting.
12. Entrapped gases.
13. Parting line.
14. Holes.
15. Undercuts.
16. Threads.
17. Inserts.

18. Surface finish (including lettering).
19. Mechanical postmolding operations.
20. Finishing.
21. Decorating.
22. Degating.
23. Tolerances.

MOLDING RECORDS

Few activities are more fruitful for the molder than keeping complete an accurate molding requirement forms. They not only are useful as a record o past performance but also are needed for planning. Often special parts, such a polyethylene bags, bushings, decals, printed instructions, and special carton have to be procured in advance. It is rather dismaying to schedule a part fo production without having all the requirements. A typical form is shown i Figure 3-46. They are inexpensive to print and should be tailored to th requirements of each plant. It is advisable to have two copies, one for the plar and one for permanent record. The plant copy should be printed on heavy stoc (30 lb) while the file copies are better printed on ordinary paper for easie reproduction. Some plants have the factory copies returned after the moldin run is complete and redistributed with the molding order.

ECONOMICS

Quotations for the same item by custom molders may vary considerably. For th most part this is inherent in the process and the method of quotation. The pric is determined by the material cost, the number of parts produced per hour, an certain intangible factors.

The basic material cost is the weight times the cost of material. To this added a "waste" factor. Material is wasted in handling, regrinding sprues an runners, and regrinding rejects at the machine. Additional material loss come from changing colors. To this is added the cost of cleaning the material fee system, hopper, and grinder. There is additional loss in rejects from the custome and overruns not shipped. This waste factor normally varies between 3 and 10% In instances where rejects cannot be reground because of inserts or decoration this factor may be higher. In instances where the ratio of the cost of material the molding cost is very high an additional factor may be added. For example, a part used $1.50 worth of material and 30¢ worth of molding time, it would t unrealistic to quote $1.80 for the part because of the cost of financing th material and the possible material loss caused by wastage and rejects.

MOLDING REQUIREMENTS

Mold _____ No. of Cavities _____

Date _____ Location _____

Production standards
   pieces/hr   shots/hr   allowable rejects/hr
_____
_____

Weight of parts   g = lb/thousand parts
Weight runner   g = lb/thousand parts
Weight shot   g = lb/thousand parts

Plastic material
Packing
Supplies required
Power tools required
Hand tools required
Jigs and fixtures required
Testing equipment required
Mold lubrication
Method of gate removal
Drilling, machining, or other operations required
Finishing operations
Sprue bushing radius
Nozzle length
Ko Bar length
Hydraulic cylinders required
Special clamps required
Special equipment for mold
Electrical interlocks
Safety requirements
Quality control and other standards applicable
Foremans check list
Operator's check list
Routing after molding

Figure 3-46(a)   Information on "Molding Requirements" form. (Robinson Plastics Corp.)

Date								
Machine No.								
Front (Temp.)								
Middle (Temp.)								
Rear (Temp.)								
Nozzle (Temp.)								
Mold (inj.) (Temp.)								
Mold (cl.) (Temp.)								
Screw speed (rpm)								
Screw back press (psi)								
Clamp pressure (psi)								
Inj. pressure (1)								
Inj. pressure (2)								
Plunger speed (in./sec)								
Inj. forw. time(sec)								
Booster time (sec)								
Clamp time (sec)								
Overall time (sec)								

(b) Machine Control Settings Reverse Side of Molding Requirements Form

The second cost element is the hourly rate charged for a machine divided by the number of parts manufactured hourly. The hourly rate usually includes profit. To this is added the extra costs that are allocated to the particular job. This might include special packing, decorating, machining, and such.

The third component is intangibles which are essentially business judgments. Some of them follow:

1. Difficulty of the job.
2. Length of the job.
3. Customer attitude. A customer who will overlook certain specifications if a part is usable or will make extra effort to salvage parts or solve problems is less costly to do business with than one who is difficult and unreasonable, even though technically correct.
4. Evaluation of future business from the customer.
5. Will it use unfilled plant capacity.
6. Credit standing of the customer.
7. Promptness of payment.
8. Competitive picture.
9. Economic conditions.
10. Special requirements of the customer. Occasionally parts will be run at little or no profit to help a customer maintain a price structure.

The two main variables in costing are the estimates of machine production and the method of allocating expenses. Very often quotations differ because of the type and condition of the equipment available for the job. The experience of the molding plant in that type of item or operation can work in either direction. They might have been able to improve production beyond what normally would be expected. Contrary wise they might quote based on low production rates which they know are required by the job, but not readily obvious before actually running the part.

There are certain costs which ary easily allocated to a machine. These include direct labor, utilities, and depreciation. It is not advisable to use the actual depreciation on the machine. This may give a distorted cost. A fully depreciated large machine might be priced below its true cost and receive a lot of work at too low a price, while a smaller machine with maximum depreciation might be overpriced and have its hourly rate raised to the point where it is no longer competitive. A better figure to use would be its estimated replacement rate.

Assume a machine costs $40,000 and will be replaced in 5 years at an estimated cost of $60,000. It would be wise to charge that machine $12,000 a year for depreciation.

Usually the other direct and indirect manufacturing costs, selling, and general and administrative costs are allocated by one of four methods.

Horsepower rating of the machine.

Productive hours.

Productive capacity (maximum ounces per shot)

Floor space.

These are all value judgments reflecting history and management philosophy. It is important they be constantly evaluated, if a plant is to remain competitive. References 237 to 242d contain valuable and detailed information for those with further interest.

The relationship between the custom molder and his customer is discussed in references 242e to 242g.

The injection molding industry has a very low percentage of profit (243). It would seem that the large capital investment required, the high risks caused by the lack of fundamental knowledge, and the type of operation would entitle the industry to larger profits. A move in that direction is indicated to provide finance for the engineering advances and improved operations tha are within the capabilities of the injection molding process.

REFERENCES

1. "First Order Glass Transition," E. W. Merril, and D. A. Gibbs, *Chem. Eng. News*, September 31, 1963, p. 41.

2. *Engineering Properties of Thermoplastics*, R. M. Ogorkiewcz, Wiley-Interscience, New York 1970.

3. "Effects of Basic Polymer Properties on Injection Molding Behavior," R. B. Staub, *SPE-J*, April 1961, p. 345.

4. "Relationship Between Structure and Mechanical Properties of Polyolefins," S. Matsuoka, *PES*, July 1965, p. 142.

4a. "Melting Rates of Crystalline Polymers Under Shear Conditions," D. W. Sundstrom and Chi-Chang Young, *PES*, January 1972, p. 59.

5. "Meaning and Measurement of Crystallinity in Polymers – A Review," S. Kavesh and J.M. Schultz, *PES*, November 1969, p. 452.

5a. "Processing Conditions and Properties of Semicrystalline Polymers. I. Effect of Mechanical Properties of a Chlorinated Polyether," J. R. Collier, A. R. Cruz-Saenz, W. Gentile, and N. Dinos, *PES*, November 1971, p. 452.

6. "Polymer Morphology," P. H. Geil, *Chem. Eng. News*, August 16, 1965, p. 72.

7. "Structure & Properties of Crystalline Polymers," G. C. Oppenlander, *Science*, March 22, 1968, p. 1311 (Vol. 159, No. 3821); copyright 1968, American Association for the Advancement of Science.

8. "Theoretical Aspects of Polymer Crystallization with Chain Folds; Bulk Polymers," J.D. Hoffman, *SPE Trans.*, October 1964, p. 315 - 362.

9. "Thermodynamic and Morphological Properties of Bulk Crystallized Polymers," L. Mandelkern, *PES*, October 1967, p. 232.

10. *Polymer Single Crystals*, P. H. Geil, Interscience Publishing Co., New York, 1963.

11. *Textbook of Polymer Science,* F. W. Billmeyer, Jr., Interscience Publishers, New York, 1962.

12. *The Structure of Polymers,* M.L. Miller, Reinhold Publishing Co., 1966, Chapter 1; "Molecular Weight;" Chapter 2; "Molecular Weight Distribution,"

13. Structural Interpretations of Mechanical Properties of Polycrystalline Polymers," R. S. Stein, *SPE Trans.,* July 1964, p. 179.

14. "Crystallinity in High Polymers," G. Natta, *SPE Trans.,* April 1963, p. 99.

15. "Meaning and Measurement of Crystallinity in Polymers," S. Kavesh, and J. M. Schultz, *PES,* September 1969, p. 331.

16. "You Can Control Brittleness of High-Density Polyethylene," T. S. Brazier, and B. Maxwell, *MP,* October 1961, p. 125.

17. *J. Appl. Polymer Sci.,* **1**, S. Mohlberg, J. Roth, and R. A. V. Raff, 1959, p. 114.

18. "Crystallization Modified Polypropylene," C. J. Kuhre, M. Wales, and M. E. Doyle, *SPE-J,* October 1964, p. 1113.

19. "Nucleated Nylon" L. L. Scheiner, April 1967, p. 37.

20. "Using Polyethylene Morphology to Select Cooling Cycles, T. W. Huseby, and S. Matsuoka, *MP,* November 1964, p. 117.

21. "Effect of Processing Conditions Upon Specific Gravity of High Density Polyethylene," J. P. Fogerty, and E. Poindexter, *SPE-J,* April 1958, p. 41.

22. "How Molding Conditions Affect Polypropylene," H. Robb, *MP,* November 1960, p. 116.

23. "How the Crystallization of Nylon Affects Processing and Properties." R. T. Steinbuch, *MP,* September 1964, p. 137.

24. "Relations Between Physical Properties and Molecular Structure of Polypropylenes," N. H. Shearer, Jr., J. E. Guillet, and H. W. Coover, Jr., *SPE-J,* January 1961, p. 83.

25. "The Transparency of Crystalline Polymers," R. Pritchard, *SPE-Trans.,* January 1964, p. 66.

26. "Influence of Annealing on the Density and Tensile Properties of Polypropylene," R. S. Schotland, *PES,* July 1966, p. 244.

27. "Effects of Thermal History on Some Properties of Polyethylene," J. B. Howard, and W. M. Martin, *SPE-J,* April 1960, p. 407.

28. "Time-Dependence of Crystal Orientation in Crystalline Polymers," R. S. Stein, *PES,* October 1968, p. 259.

29. "Heat Transfer Calculations in Crystalline Plastics," W. E. Gloor, *SPE Trans.,* October 1963, p. 270.

29a. "Thermal Conductivity of Polyethylene: The Effects of Crystal Size, Density, and Orientation on the Thermal Conductivity," D. Hansen and G. A. Bernier, *PES,* May 1972, p. 204.

30. "Calculation of Melting and Freezing Rates of Acetal Resins," P. N. Richardson, Technical Papers, *SPE,* Volume 6, 1960, p. 1-1.

31. "Heat Transfer Effects During Processing of Nylon Resins," E. Heyman, *SPE-J,* October 1967, p. 37.

32. "Cooling Molded Parts-A Rigorous Analysis," S. Kenig, and M. R. Kamal, *SPE-J,* July 1970, p. 50.

32a. "Heat Transfer to Molten Polymers," O. M. Griffin,, *PES,* March 1972, p. 140.

32b. "Thermal Transport in the Contact Melting of Solids," O. M. Griffin, *PES*, July 1972, p. 265.

32c. "How Radiation Affects Plastics," G. A. Bohm, W. F. Oliver, and D. M. Pearson, *SPE-J*, July 1971, p. 21.

32d. "Improving Load Bearing Properties of Reinforced PE by Gamma Radiation," R. A. V. Raff, D. J. Schecter, and R. V. Subramanian, *MP*, September 1971, p. 74.

33. "Chemically Cross-Linked Polyethylene," B. C. Carlson, *SPE-J*, March 1961, p. 265.

34. "A Study of Crosslinked Compounded Polyolefin Systems," R. J. Boot, and T. J. Jordan, *PDP*, September 1963, p. 11.

35. "Injection-Molding Crosslinkable Polyolefins," K. S. Tenney, *SPE-J*, March 1970, p. 68.

36. "The Promotion and Retardation of Cross-Linking in Thermally Processed Polyvinyl Chloride Systems," C. H. Fuchsman, *SPE-J*, June 1961, p. 590.

37. "Effect of Molecular Weight on Properties of HDPE," A. F. Margolies, *SPE-J*, June 1971, p. 44.

37a. "Ultra High Molecular Weight Polyethylene Offers Ultra High Resistance to Abrasion," *PDP*, June 1972, p. 22.

38. "How Additives Affect High Molecular Weight PE," D. W. Larsen, *SPE-J*, June 1971, p. 40.

39. *Polymers and Resins*, B. Golding, D. Van Nostrand Co., Inc., Princeton, N. J., 1959.

39a. "The Influence of Molecular Weight on the Melt Rheology of Polypropylene," D. P. Thomas, *PES*, July 1971, p. 305.

40. "Melt Flow Indexing of Polypropylene," R. V. Charley, *Br. Pl.*, September 1961, p. 476.

41. "Reliability of the Melt-Flow Index for Production Control," M. Narkis, and A. Ram, *PES*, January 1967, p. 24.

42. "Guide to Moldability Tests," C. W. Deeley, and J. F. Terenzi, *MP*, August 1965, p. 111.

43. "Effect of Molecular Weight Distribution upon Melt Rheology of High Density Polyethylene," H. L. Wagner, and K. F. Wissbrun, *SPE-Trans.*, July 1962, p. 222.

44. "Effects of Molecular Weight Distribution on High Density Polyethylene Container Molding," M. G. Leegwater, *SPE-J*, November 1969, p. 47.

45. "Melt Index Equivalent – A New Flow Parameter," R. J. Martinovich, P. J. Boeke, and R. A. McCord, *SPE-J*, December 1960, p. 1335.

46. "The Effect of Molecular Weight Distribution on the Flow Properties of Poly-ethylene," D. R. Mills, G. E. Moore, and D. W. Pugh, *SPE Trans.* January 1961, p. 40.

47. "The Influence of Molecular Weight Distribution on Melt Viscosity, Melt Elasticity, Processing Behavior and Properties of Polystyrene," D. F. Thomas, and R. S. Hagan, *PES*, May 1969, p. 164.

48. "Properties of High Density Polyethylene with Bimodal Molecular Weight Distribution," H. H. Zabusky, and R. F. Heitmiller, *SPE Trans.*, January 1964, p. 17.

49. "Predicting the Processability of Plastics," W. T. Blake, *PDP*, September 1966, p. 22.

50. "A Rheological Interpretation of Torque-Rheometer Data," J. E. Goodrich, and R. S. Porter, *PES*, January 1967, p. 45.

50a. "New Rheometer is Put to the Test," C. Macosko and J. M. Starita, *SPE-J*, November 1971, p. 38.

51. "An Analysis of Brabender Torque Rheometer Data," L. L. Blyler, and J. H. Daane, *PES*, July 1967, p. 178.

52. "Conversion of Brabender Curves to Instron Flow Curves," G. C. N. Lee, and J. R. Purdon, *PES*, September 1969, p. 360.

53. "A High Shear Rate Capillary Rheometer for Polymer Melts," E. H. Merz, and R. E. Colwell, *ASTM Bull.*, September 1958, p. TP 211.

54. "Melt Fracture-Extrudate Roughness in Plastic Extrusion," J. P. Tordella, *SPE Tech Papers*, 1956, p. 285.

55. "Flow Behavior and Turbulence in Polyethylene," R. F. Westover, and B. Maxwell, *SPE-J*, August 1957, p. 27.

56. "Causes of Melt Fracture," R. M. Schulken, Jr., and R. E. Boy, Jr., *SPE-J*, April 1960, p. 423.

57. "Polymeric Melts-Steady-State Flow, Extrudate Irregularities and Normal Stresses," A. B. Metzner, E. L. Carley, and I. K. Park, *MP*, July 1960, p. 133.

58. "New Light on Melt Elasticity," B. Maxwell, and R. A. McCord, *MP*, September 1961, p. 116.

59. "New Aspects of Melt Fracture," J. J. Benbow, and P. Lamb, *SPE-Trans.*, January 1963, p. 7.

60. "Velocity Profiles for Polyethylene Melts," B. Maxwell, and J. Galt, *MP*, December 1964, p. 115.

61. "The Significance of Slip in Polymer Melt Flow," R. F. Westover, *PES*, January 1966, p. 83.

62. "Lower Newtonian Viscosities of Polystyrene," K. K. Chee, and A. Rudin, *PES*, January 1971, p. 35.

63. "Significant Flow Properties of Thermoplastics," L. B. Ryder, *PT*, February 1960, p. 35.

64. "Hydrostatic Pressure Effect on Polymer Melt Viscosity," B. Maxwell and A. Jung, *MP*, November 1957, p. 174.

65. "Effect of Hydrostatic Pressure on Polyethylene Melt Rheology," R. F. Westover, *SPE Trans.*, January 1961, p. 14.

66. "Properties of Polymers at High Pressures," W. I. Vroom, and R. F. Westover, *SPE-J*, August 1969, p. 58.

67. "Effect of Static Pressure on Polymer Melt Viscosities," J. F. Carley, *MP*, December 1961, p. 123.

68. "How Resin Structure Affects Extrudability," R. F. Heitmiller, *MP*, January 1963, p. 140.

69. "Flow Properties of Polyethylene Melts," R. A. Mendelson, *PES*, September 1969, p. 350.

69a. "Prediction of High Density Polyethylene Processing Behavior from Rheological Measurements," M. Shida and L. V. Cancio, *PES*, March 1971, p. 124.

70. "Thermal and Shear Degradation in Polyethylene Extrusion," H. Schott, and W. S. Kaghan, *MP*, March 1960, p. 116.

71. "Rheology of Rigid PVC Defined in Terms of Commercial Production Conditions," R. A. Paradis, *SPE-J*, July 1967, p. 54.

71a. "Putting the Heat on Vinyl Resins," C. L. Sieglaff and R. L. Harris, *SPE-J*, June 1972, p. 48.

72. "Flow Properties of Thermoplastics in Injection Molding," R. B. Staub, *SPE-J*, February 1962, p. 181.

73. "An Analysis of Injection Mold Filling of Polyethylene," R. B. Staub, *SPE-J*, April 1960, p. 429.

74. "Application of the Rheology of Monodisperse and Polydisperse Polystyrenes to the Analysis of Injection Molding Behavior," R. S. Hagan, D. P. Thomas, and W. R. Schlich, *PES*, October 1966, p. 373.

75. "Using Rheology to Predict Moldability, D. P. Thomas, *PDP*, September 1970, p. 21.

76. "Moldability of Plastics Based on Melt Rheology," *SPE-Trans.*, January 1963: Part 1; "Theoretical Development" F. E. Weir, p. 32; Part 2; "Practical Applications," F. E. Weir, M. E. Doyle, and D. G. Norton, p. 37.

77. "The Injection Molding of Thermoplastics," M. R. Kamal and S. Kenig, *PES*, July 1972; I – Theoretical Model, p. 294; II – Experimental Test of the Model, p. 302.

78. *Processing of Thermoplastic Materials*, E. C. Bernhardt, (Ed.), Reinhold Publishing Co., New York, 1959.

79. "Injection Molding, A Rheological Interpretation," Part 1; R. L. Ballman, T. Shushman, and J. L. Toor, *MP*, September 1959, p. 105; Part II; October 1959, p. 115.

80. "A Rheological Study of the Injection Molding of Polystyrene Polymers," G. Pezzin, *SPE Trans.*, October 1963, p. 260.

81. "Spiral Flow Molding," L. Griffiths, *MP*, August 1957, p. 111.

82. "Comparison of Spiral-Cavity Mold Flow with Laboratory Scale Flow Tests on Thermoplastics," J. J. Gouza, and G. G. Freygang, *SPE-J*, November 1961, p. 1211.

83. "Polyethylene Flow Data from Melt Viscometer and Commercial Extruder Measurements," H. P. Schreiber, *SPE Trans.*, April 1961, p. 86.

84. "Glass Transitions in Polypropylene," D. L. Beck, A. A. Hiltz, and J. R. Knox, *SPE Trans.*, October 1963, p. 279.

85. "Controlling Polycarbonate Molding with D.T.A.," S. Keefe, *PDP*, May 1969, p. 31.

86. *Polymer Handbook*, J. Brandrup, and E. H. Immergut, John Wiley & Sons, New York, 1966; "Melting Points III-32," "Glass Transition Temperatures-III-61."

87. "Correlation of Physical and Polymer Chain Properties," W. E. Wolstenholme, *PES*, April 1968, p. 142.

88. "Studies in Newtonian Flow II. The Dependence of the Viscosity of Liquids on Free Space," A. K. Doolittle, *J. Appl. Phys.* 22, 1951, p. 1471.

89. "The Temperature Dependence of Relaxation Mechanisms in Amorphous Polymers and Other Glass-Forming Liquids," M. L. Williams, R. F. Landel, and J. D. Ferry, *J. Am. Chem. Soc.*, 1955, p. 77 and 3701.

90. "Prediction of Melt Viscosity Flow Curves at Various Temperatures for Some Olefine Polymers and Copolymers," R. A. Mendelson, *PES*, July 1968, p. 235.

91. "Creep Behavior of Chlorinated Polyether," M. G. Sharma, and L. Gesinski, *MP*, January 1963, p. 164.

92. "Alpha nylon-6 Mechanical and Processing Characteristics," G. A. Toelcke, M. N. Riddell, G. N. Bellucci, R. J. Welgos, and J. L. O'Toole, *MP*, December 1968, p. 117.

93.   "The Nature of Time Effects in Solid Polymeric Systems," R. D. Andrews, *PES*, July 1965, p. 191.

94.   "Material Parameters for Injection Molding," W. R. Schlich and R. S. Hagan, *SPE-J*, July 1966, p. 45.

95.   "Pressure Losses in the Injection Mold," D. C. Paulson, *MP*, October 1967, p. 119.

96.   "Plastic Pressure Recording Analysis of Injection Molding Machine Pressure Variables," D. C. Paulson, and W. M. Viilo, *SPE Tech. Pap.*, 1968, p. 291.

97.   "Process Control in Injection Molding," D. C. Paulson, *SPE Tech. Pap.*, 1969, p. 381.

98.   "Polymer Temperature Measurement in Injection Molding," T. S. Huxham, *SPE-J*, August 1962, p. 899.

99.   "A Method for Measuring Plastic Melt Temperatures During Injection Molding," W. Tychesen, and W. Georgi, *SPE-J*, December 1962, p. 1509.

99a.   "Meet The Ribbon Thermocouple," J. Nanigian, *SPE-J*, February 1971, p. 51.

99b.   "New Melt Temperature Probe," A. G. Forton, *Br. Pl.*, March 1972, p. 74.

100.   "Temperature Profile of Molten Plastic Flowing in a Cylindrical Duct," H. Schott, and W. S. Kaghan, *SPE-J*, February 1964, p. 139.

101.   "A New Concept in Melt Pressure Measurements," M. R. Harris, *SPE-J* February 1966, p. 69.

102.   "Accurate Measurement of Nozzle Pressures During Injection Molding," H. W. Ashton, and R. T. Cassidy, *SPE-J*, March 1963, p. 295.

102a. "Calibration of Melt Pressure Transducers," G. D. Eggleston, *PDP*, July 1971, p. 23.

102b. "Pressure Transducers—Which, Why, and Specmanship," D. B. Hoffman and N. Sarasohn, *PDP*, July 1972, p. 20.

102c. "Monitoring Mold Cavity Pressure for Precision Molding," W. Masters, T. Livermore, and K. Arthur, *MP*, December 1971, p. 58.

103.   "Pressure Effects in the Spiral Mold," T. W. Huseby, *MP*, December 1962, p. 112.

104.   "The Effect of Pressure on Plastic Melt Temperature," R. Burkhalter, W. Tychesen, and W. Georgi, *SPE-J*, December 1962, p. 1515.

105.   "Computer-Controlled Injection Machine," R. Currie, *MP*, October 1968, p. 122.

106.   "Programmed Injection," V. Gardner, *Plastics*, March 1967, p. 285.

107.   "Injection Machine Control," R. Nightingale, *Br. Pl.*, October 1969, p. 131.

107a. "Improvements in Control Technique, Design, and Operation of Injection Moulding Machines," P. Wippenbeck, *Kunstoffe*, April 1970, p. 7 (English translation).

107b. "Adaptive Control System for Injection Moulding Machines," A. C. Haynes and L. W. Turner, *Br. Pl.*, February 1972, p. 59.

107c. "Computers in Injection Molding—the Key is the Interface," S. C. Stinson, *PT*, April 1972, p. 44.

108.   "One Approach to Automatic Plastics Machines," H. E. Harris, *SPE-J*, May 1970, p. 27.

109.   "Photographic Study of the Polymer Cycle in Injection Molding," G. D. Gilmore and R. S. Spencer, *MP*, April 1951, p. 117.

110.   "Plastic Flow in Injection Molds," F. J. Rielly, and W. L. Price, *SPE-J*, October 1961, p. 1097.

111.   "Strain-free Injection Molding," L. I. Johnson, *MP*, June 1963, p. 111.

112. "Residual Strains in Injection Molded Polystyrene," R. S. Spencer and G. D. Gilmore, *MP*, December 1950, p. 97.

113. "Orientation Characterization in Plastic Film and Sheet," C. T. Hathaway, *SPE-J*, June 1961, p. 567.

114. "Molecular Orientation in Injection Molding Acetal Homopolymer," E. S. Clark, *SPE-J*, July 1967, p. 46.

115. "Orientation in Injection Molding," R. L. Ballman, and H. L. Toor, *MP*, October 1960, p. 113.

116. "Effects of Processing Techniques on Structure of Molded Parts," W. Woebcken, *MP*, December 1962, p. 146.

117. "Why Biaxially Oriented Pipe?" W. E. Gloor, *MP*, November 1960, p. 111.

118. "The Effect of Orientation on the Physical Properties of Injection Moldings," G. B. Jackson, and R. L. Ballman, *SPE-J*, October 1960, p. 1147.

118a. "Gaging the Benefits of Controlled Biaxial Orientation," L. S. Thomas and K. J. Cleereman, *SPE-J*, Part 1 – General Purpose Polystyrene, April 1972, p. 61; Part 2 – High Impact Polystyrene, June 1972, p. 39.

118b. "Influence of Molecular Orientation on the Impact Behaviour of Non-modified Polystyrenes," B. Carlowitz and N. Kallus, *Kunstoffe*, May 1970, p. 19 (English translation).

119. "ASTM Data vs. Parts Data," R. E. Eshenaur, *SPE-J*, May 1965, p. 466.

120. "How to Make the Polypropylene Hinge," J. L. Vermillon, *MP*, July 1962, p. 121.

121. "How to Form Integral Hinges in Polyethylene," S. A. Woods, *MP*, September 1964, p. 149.

122. "Coined Hinges in Molded or Extruded Parts of Acetal and Nylon," J. Mengason, *SPE-J*, August 1969, p. 72.

123. "Hinges Give More Design Scope," W. G. Miller, *Br. Pl.*, November 1969, p. 120.

124. "Injection Molding of Shapes of Rotational Symmetry with Multiaxial Orientation," K. J. Cleereman, *SPE-J*, Part 1, October 1967, p. 43; Part II, January 1969, p. 55.

125. "Effects of Packing on Impact Resistance of Molded Polycarbonate," R. Rager, *MP*, November 1967, p. 131.

126. "Annealing Injection Molded Styrene," R. I. Dunlap, F. J. Pokigo, and S. E. Glick, *MP*, August 1950, p. 83.

127. "The Temperature Behaviour of Polycarbonate," G. Peilstocker, *Br. Pl.*, July 1962, p. 365.

128. Molding Very Thin Sections from Polypropylene and Polyallomer," J. K. Harkleroad, *SPE-J*, November 1966, p. 29.

129. "Role of Pressure, Temperature, and Time in the Injection Molding Process," G. D. Gilmore, and R. S. Spencer, *MP*, April 1950, p. 143.

130. "Mold Shrinkage of Thermoplastic Urethanes," E. J. Lui, *MP*, November 1967, p. 115.

131. "Molding and Annealing Shrinkage of 66 Nylon," E. Heyman, *SPE Tech. Pap.*, 1969, p. 373.

132. "Thermodynamic Properties of Nylon 6/6," R. G. Griskey, and J. K. P. Shou, *MP*, June 1968, p. 138.

133. "Thermodynamic Properties of Ethylene-propylene Copolymer," G. N. Foster III, N. Waldman, and R. G. Griskey, *MP*, May 1966, p. 245.

134. "A Generalized Equation of State for Polymers," H. L. Whitaker, and R. G. Griskey, *J. Appl. Polymer Sci.*, ii, 1967, p. 1001.

135. R. S. Spencer, and G. D. Gilmore, *J. Appl. Phys*, **21**, 1950, p. 523.

136. "Pressure-Volume-Temperature Behavior of Polypropylene," G. N. Foster III, N. Waldman, and R. G. Griskey, *PES*, April 1966, p. 131.

137. "Dimensional Behavior of Acetal Copolymer Moldings," A. G. Serle, *MP*, June 1965, p. 115.

138. "Mold Shrinkage of Thermoplastic Materials," S. E. Glick, *SPE-J*, May 1951, p. 9.

139. "Post-Molding Shrinkage of HDPE Bottles," J. D. Frankland, J. R. Weeks, and A. J. Fox, *SPE-J*, July 1970, p. 38.

140. "Gating for Polypropylene," C. W. Williamson, *SPE Tech. Pap.*, 1963, p. VI-4.

141. Dimensional Stability of Acrylic Resins," A. J. Ragolia, *MP*, July 1966, p. 107.

142. "Injection Molding Polypropylene," P. R. Jardine, *MP*, June 1962, p. 137.

143. "Effect of Process Variables on Mold Shrinkage," R. F. Williams, Jr., and L. H. Pancoast, Jr., MP, September 1967, p. 185.

144. "Effect of Molding Conditions on Shrinkage of Modified Polystyrene," R. G. Hochschild, *SPE-J*, April 1961, p. 358.

145. "New Inexpensive Mold for Polymer Evaluation," P. L. Barrick, R. H. Crawford, B. L. Espy, W. F. Robb, and E. E. Sawin, *SPE-J*, January 1964, p. 69.

146. "A Guide to Molding High-density Polyethylene," R. B. Staub, *MP*, October 1961, p. 110.

147. "Processing Urethane Elastomers," K. A. Pigott, C. L. Gable, W. Archer, Jr., E. C. Haag, and S. Steingiser, *SPE-J*, December 1963, p. 1281.

148. "Injection Molding Ethylene-Vinyl Acetate Copolymers," J. Fischer, and K. S. Tenney, *SPE-J*, April 1967, p. 83.

149. "Injection Molding FEP Flurocarbon Resin," H. A. Larsen, G. R. DeHoff, and N. W. Todd, *MP*, August 1959, p. 89.

149a. "Study on Teflon FEP Injection Molding Method," Tsutomu Machida, *Jap. Pl.*, April 1970, p. 25.

150. "Processing of CTFE Plastics," R. P. Bringer, and G. A. Morneau, *PDP*, August 1968, p. 17.

151. "Molding Rigid PVC," J. M. Sherlock, *SPE-J* October 1970, p. 46.

152. "How to Mold FRTP Resins," T. P. Murphy, *MP*, June 1965, p. 127.

152a. "How to Injection Mold Polyarylsulfone," R. L. Burns, and B. H. Spoo, *PDP*, April 1972, p. 26.

152b. "Rapid Molding of Large, Paintable EPDM (ethylene-propylene rubber) Rubber Parts," R. E. Knox, *MP*, February 1972, p. 56.

152c. "Processing Nylon 11," Pierre de Sigy, *Jap. Pl.*, January 1971, p. 33.

152d. "Designing Nylon 66 for Optimum Injection Molding Performance," W. C. Filbert, Jr., *PT*, June 1971, p. 35; September 1971, p. 48; November 1971, p. 36.

153. "Behavior of Polymers under Complex Load Histories; Linear Rule of Cumulative Damage," D. C. Prevorsek, G. E. R. Lamb, and M. L. Brooks, *PES*, October 1967, p. 269.

154. *Mechanical Behavior of High Polymers*, T. Alfrey, Jr., Interscience Publishers, New York, 1948.

155. *Viscoelastic Properties of Polymers,* J. D. Ferry, John Wiley & Sons, Inc., New York, 1961.

156. *Mechanical Properties of Polymers,* L. E. Nielsen, Reinhold Publishing Corp., New York, 1962.

157. *Physical Properties of Polymers,* F. Bueche, Interscience Publishers, New York, 1962.

158. *Physics of Plastics,* P. D. Ritchie, (Ed.), D. Van Nostrand Co., Inc., Princeton, N. J., 1965.

159. *Polymer Processing,* J. M. McKelvey, John Wiley & Sons, Inc., New York, 1962.

160. *Plastics Rheology,* R. S. Lenk, McLaren & Sons, Ltd., London, 1968.

161. "Creep Data for Plastics," *Modern Plastics Encyclopedia,* 1969/1970, p. 939.

162. "Mechanism and Mechanics of Creep of Plastics," W. N. Findley, *SPE-J,* January 1960, p. 57.

163. "Stress Relaxation and Combined Stress Creep of Plastics," W. N. Findley, *SPE-J,* February 1960, p. 192.

164. "Creep of High Polymers," M. L. Williams, W. H. Howard, *SPE-Trans.,* January 1962, p. 74.

165. "General Formula for Creep and Rupture Stresses in Plastics," S. I. Goldfein, *MP,* April 1960, p. 127.

166. "Hydrostatic Creep of Solid Plastics," W. N. Findley, A. M. Reed, and P. Stern, *MP,* April 1968, p. 141.

167. "A New Creep Law for Plastics," J. R. McLoughlin, *MP,* February 1968, p. 97.

168. "Creep Behavior of Transparent Plastics at Elevated Temperatures," F. L. McCrackin, and C. F. Bersch, *SPE-J,* September 1959, p. 791.

169. "Creep of Thermoplastics in the Glassy Region: Stress as a Reduced Variable," R. L. Bergen, Jr., *SPE-J,* October 1967, p. 57.

170. "On the Characterization of Nonlinear Viscoelastic Materials," R. A. Schapery, *PES,* July 1969, p. 295.

171. "Device for Measuring Stress Relaxation of Plastics," R. J. Curran, R. D. Andrews, Jr., and F. J. McGarry, *MF,* November 1960, p. 142.

172. "Some Information on Stress Relaxation of Polyethylene Melts with Respect to Processing Properties," Y. Kosaka, and T. Fujiki, *Japan Pl.,* January 1970, p. 38.

173. "Short Term Stress Relaxation Studies at Low Initial Strain Rates," G. C. Karas, *Br. Pl.* September 1961, p. 485.

174. "Tensile Characteristics of Thermoplastics," R. M. Ogorkiewicz, and M. P. Bowyer, *Br. Pl.* September 1969, p. 125.

175. "Deformation of Thermoplastics Under Different Types of Tensile Loading," R. M. Ogorkiewicz, L. E. Culver, and M. P. Bowyer, *SPE-J,* March 1969, p. 43.

176. "Dynamic Mechanical Properties of High Polymers," L. E. Nielsen,May 1960, p. 525.

176a. "Cyclic Creep Test Data for G. R. Polypropylene," L. C. Cessna, Jr., *SPE-J,* February 1972, p. 28.

177. "How Plastics React Under Rapid Loading," G. R. Rugger, E. McAbee, and M. Chmura, *SPE-J,* December 1958, p. 31.

178. "High Speed Testing—A New Dimension in Polymer Evaluation," M. Silberberg, and R. H. Supnik, *SPE Trans.,* April 1962, p. 140.

179. "High Rate Tensile Properties of Plastics," E. McAbee, and M. Chmura, *SPE-J,* January 1963, p. 83.

180. "Computing High-Rate Tensile Strength from Static Strength Data," S. Goldfein, *MP,* August 1964, p. 149.

181. "Deformation and Failure of Plastics and Elastomers," T. L. Smith, *PES,* October 1965, p. 270.

182. "Long-Term Hydrostatic Strength Characteristics of Thermoplastic Pipe," F. W. Reinhart, *PES,* October 1966, p. 285.

183. "Repeated Impact Tests," J. R. Heater, and E. M. Lacey, *MP,* May 1964, p. 123.

184. Estimation of Long-Time Performance of Extruded Plastic Pipe from Short-Time Burst Strength," S. Goldfein, *MP,* May 1960, p. 139.

185. "Prediction of Impact Resistance from Tensile Data," R. M. Evans, H. R. Nara, and E. G. Bobalek, *SPE-J,* January 1960, p. 76.

186. "The Strain Response of Plastics to Complex Stress Histories," S. Turner, *PES,* October 1966, p. 306.

187. "Designing Thermoplastic Structural Components," W. D. Harris, W. W. Burlew, and F. McGarry, *SPE-J,* November 1960, p. 1231.

188. "Failure by Buckling," G. Menges, and E. Gaube, *MP,* July 1969, p. 96.

189. "Straining Behavior of Cellulose Acetate Plastics," K. Ito, *MP,* July 1961, p. 129.

190. "Nonlinear Viscoelastic Behavior of Cellulose Acetate Butyrate," M. G. Sharma, and P. R. Wen, *SPE Trans.,* October 1964, p. 282.

191. ,"Combined Tension-Torsion Creep Experiments on Polycarbonate in the Nonlinear Range," J. S. Y. Lai, and W. N. Findley, *PES,* September 1969, p. 378.

192. "The Effect of Biaxial Stress on the Creep Properties of Polymethyl Methacrylate," M. Zaslawsky, *PES,* March 1969, p. 105.

193. "Interpretation of the Tensile Creep Response of an ABS Polymer," R. S. Moore, and C. Gieniewski, *PES,* May 1969, p. 190.

193a. "Analyzing Stress in ABS," R. P. Tison, *SPE-J,* April 1972, p. 70.

194. , "Tensile Creep Data for Polyolefins," A. C. Morris, and R. A. Gill, *Plastics,* October 1967, p. 1250.

195. "Polypropylene Creep and Impact Data," A. C. Morris, and A. Richardson, *Br. Pl.,* March 1968, p. 92.

196. "Tensile Creep and Recovery of Compression Molded Polypropylene Samples," A. A. Hiltz, *SPE Trans.,* January 1963, p. 72.

197. "Influence of Normal Stress on Creep in Tension and Compression of Polyethylene and Rigid Polyvinyl Chloride Copolymer," D. G. O'Connor, and W. N. Findley, *SPE Trans.,* October 1962, p. 273.

198. "The Ductile Failure of Polyethylene Pipe," S. A. Mruk, *SPE-J,* January 1963, p. 91.

199. "Creep in Polyethylene at Elevated Temperatures," R. R. Dixon, *SPE-J,* April 1958, p. 23.

200. "Creep Characterizations of Modified Styrene Materials," G. B. Jackson, and J. L. McMillan, *SPE-J,* February 1963, p. 203.

201. Stress-strain Behaviro of Polystyrene-Glass Fiber Composites," R. V. Viventi, H. T. Plant, and R. T. Maher, *MP,* January 1968, p. 129.

202. "Fatigue Endurance and Creep of Glass Fiber-Fortified Thermoplastics," J. E. Theberge, *MP,* June 1968, p. 155.

203. "Engineering Properties of Glass-Reinforced Thermoplastics," M. N. Riddell, and J. L. O'Toole, *MP,* May 1968, p. 150.

204.   "Deformation Properties of RTP; K. Hattori, and M. A. Todd *MP*, November 1969, p. 122.

204a. "Thermoplastic Stiffness Evaluation" (polycarbonate and polyphenylene oxide), G. W. Weidmann and R. M. Ogorkiewicz, *Br. Pl.*, December 1971, P. 53.

205.   Time-Dependent Compressive Properties of PTFE," E. D. Jones, G. P. Koo, and J. L. O'Toole, *MP*, November 1967, p. 137.

206.   "Creep Resistant PTFE," M. N. Riddell, G. A. Toelcke, and J. L. O'Toole, *MP*, October 1970, p. 140.

207.   "Stress-Strain-Time-Temperature Relationships for Polymers," T. L. Smith, *ASTM Special Tech.*, No. 325, 1962, p. 60.

208.   "How to Use Time-Dependent Property Data, J. H. Faupel, *MP*, August 1964, p. 119.

209.   "Time-Modified Moduli for Polysulfones," T. E. Bugel, *MP*, June 1966, p. 99.

209a. "How Transfer Molds Fill," R. J. Hishaw and T. F. Stegmaier, *SPE-J*, March 1971, p. 44.

209b. "A Practical Approach to Thermoset Rheology," W. G. Frizelle and J. F. Norfleet, *SPE-J*, November 1971, p. 71.

209c. "High Speed Thermoset Molding," R. Paci and S. S. Helliwell, *SPE-J*, March 1971, p. 49.

209d. "Processing Technique and Economics for Thermoset Injection," Shoji Matsuda and Teruo Tsujita, *Jap. Pl.*, April 1968, p. 45.

209e. "Effect of Moisture on Flow of Thermoset Molding Compounds," J. E. Hess, *MP*, November 1971, p. 60.

210.   "How Process Conditions Affect Thermoset Injection Molding," R. H. Beck, Jr., *SPE-J*, May 1971, p. 43.

211.   "Selecting Molding Conditions for Thermosets," B. L. Talwar, and L. T. Ashlock, *SPE-J*, September 1970, p. 42.

212.   "New Data on Injection Molding Phenolics," L. D. Fishbert, and D. C. Longstreet, *MP*, June 1969, p. 92.

213.   "Screw Injection Molding of Phenolics," Y. Morita, *SPE-J* August 1966, p. 57.

214.   Injection Molding the Ureas and Melamines," R. L. Hughes, *SPE-J* March 1970, p. 26.

215.   "Injection Molding Thermosets Through the Cold Runner Method," L. N. Davis, *PDP*, March 1971, p. 24.

216.   "Predicting the Molding Behaviour of Thermosets," D. W. Sundstrom, and L. A. Walters, *SPE-J*, April 1971, p. 58.

217.   "Understanding Thermoset Flow," P. J. Heinle, and M. A. Rodgers, *SPE-J*, June 1969, p. 56.

218.   "How Time and Temperature Affect the Cure of DAP," S. Y. Choi, *SPE-J*, June 1970, p. 51.

219.   "The Reciprocating Screw Injection Machine for Thermoset Processing, J. M. Grigor, Jr., *SPE Tech. Papers*, 1967, p. 763.

220.   "Determining Shrinkage in Injection-Molding Phenolics, L. D. Fishberg, D. C. Longstreet, and A. R. Olivo, *SPE-J*, March 1970, p. 34.

220a. "Reducing Cycle Times for Injection Molding Phenolics," L. D. Fishberg and D. C. Longstreet, *MP*, October 1971, p. 132.

220b. "Practical Guide to Engineering Parts in Phenolics," D. L. Messinger, *PDP*, June 1971, p. 19.

220c. "Injection of Phenolics and DAP Materials," W. G. Miller, *Br. Pl.*, March 1971, p. 67.

220d. "Injection Moulding Urea and Melamine Plastics," F. J. Parker, N. A. Milburn, R. L. Hughes, and K. Shelley, *Br. Pl.*, March 1971, p. 70.

220e. "You Can Injection-Mold Alkyd and Polyester Compounds at High Speeds," G. B. Rheinfrank, *SPE-J*, September 1971, p. 30.

220f. "Practical Guidelines to Designing Parts in Reinforced Thermosets," *PDP*, January 1971, p. 16.

221. Ultrasonic Degating of Molded Parts," J. R. Sherry, *MP*, April 1970, p. 136.

222. "Handling Finished Parts," R. E. Slyh, *SPE-J*, March 1970, P. 43.

223. "The Role of Robots in Molding Plastics," D. G. Harrer, *SPE-J* March 1970, p. 61

224. "Designing for Nylon 6.6," L. Horvath, *Plastics*, April 1968, p. 415.

225. "Designing for Acetal Homopolymers," L. Horvath, *Plastics*; Part 1, May 1968, p. 532; Part 11, June 1968, p. 679.

226. "Designing for HD Polyethylene," O. P. Phillips, *Plastics,* July 1968, p. 786.

227. "Designing for Polypropylene," G. J. Nichols, *Plastics,* September 1968, p. 1018.

228. "Designing for Glass Reenforced Nylon 6," K. Fischer, *Plastics*; Part I, November 1968, p. 1290; Part II, December 1968, p. 1396.

229. "Designing for Polycarbonates," H. Streib, K. Protoschill, and D. Schauf, *Plastics*; Part I, February 1969, p. 178; Part II, March 1969, p. 304.

230. "Designing for ABS," C. Waxman, *Plastics*, April 1969, p. 418.

231. "Designing for Nylon 11," J. M. Adsett, *Plastics*; May 1969, p. 537.

232. "Designing for Modified PPO," W. H. T. Adriaens, *Plastics*, June 1969, p. 731.

233. "Designing for Toughened Polystyrene," R. W. Stone, *Plastics*, July/Aug. 1969, p. 878.

234. "Fundamental Design for Injection Moulding," J. D. Robinson, *Plastics*, April 1967, p. 446.

234a. "Thermoplastic to Replace Thermosets," *Br. Pl.*, October 1971, p. 98.

235. "Design Tips for Polyolefin Parts," J. N. Scott, J. V. Smith, and D. L. Alexander, *MP*, June 1960, p. 130.

236. "General Principles for Designing with Plastics," P. Grafton, *MP* Encyclopedia, 1970/1971, p. 54.

236a. "Engineering with Plastics," M. M. Hall, *Br. Pl.*, October 1971, p. 85.

236b. "Optimizing Performance with Engineering Thermoplastics," *Br. Pl.*, February 1972, p. 41.

236c. "How to Computerize Your Materials Selection Program," H. L. Miller and S. V. Gogela, *SPE-J*, January 1972, p. 26.

237. "How to Determine Material Cost," G. R. Smoluk, *MP* May 1960, p. 119.

238. "Determining True Machine-Hour Cost," R. L. Daniel, and M. J. Cooper, *SPE-J* April 1967, p. 75.

239. "Estimating Costs of Molded Parts," C. L. Ward, *SPE-J*, December 1969, p. 48.

240. "Controlling Custom Molding Costs," M. Steed, *PT* April 1970, p. 49.

241. "Cost Factors of Injection Molding," R. E. Bell, *PT*, April 1970, p. 53.

242. "The Role of Machine-Hour Rates in the Make-or-Buy Decision," E. G. Slatcher, *SPE-J*, December 1969, p. 63.

242a. "Keep Tabs on Your Molding Costs," W. R. Reinbacher, *SPE-J*, April 1972, p. 49.

242b. "One Approach to Keeping Costs Down and Profits Up," J. L. Witt, *SPE-J*, February 1971, p. 38.

242c. "Increasing Efficiency in the Molding Plant," W. R. Reinbacher, *PDP*, August 1971, p. 16.

242d. "Moulding for Profit," *Br. Pl.*, April 1972, p. 59.

242e. "How to Deal With Custom Molders," A. A. Schoengood, *SPE-J*, June 1971, p. 22.

242f. "Custom Processors Rate Their Customers," *MP*, May 1971, p. 56.

242g. "Customers Rate Their Custom Processors," *MP* December 1971, p. 48.

243. Tarnell Reports, Tarnell Co., New York.

244. "Application of Time-Temperature Superposition Principle to Long Term Engineering Properties of Plastic Materials," J. T. Seitz, and C. F. Balazs, *PES*, April 1968, p. 151.

245. "Stress Relaxation and Creep Measurements of Some Thermoplastic Materials," R. L. Bergen, Jr. and W. E. Wolstenholme, *SPE-J*, November 1960, p. 1235.

246. "Measuring True Melt Temperature in Extruders," H. T. Kim, and J. P. Darby, *SPE-J*, August 1970, p. 31.

247. "Capillary Rheometry," R. L. Ballman, and J. J. Brown, *Bull SA-2* Instron Eng. Corp., Canton, Mass., p. 20.

248. "How an Injection Mold Fills," I. T. Barrie and P. Lamb, *SPE-J*, August 1971, p. 64.

249. "Stress-Strain Behavior of Styrene-Acrylonitirile/Glass Bead Composites in the Glassy Region," L. Nicolais and N. Narkis, *PES*, May 1971, p. 194.

250. "Stress-Time Superposition of Creep Data for Polypropylene and Coupled Glass-Reinforced Polypropylene," L. C. Cessna, Jr., *PES*, May 1971, p. 211.

251. "Why Cold Rolling Improves the Tensile Strength of Polycarbonate," A. D. Murray and K. C. Rusch, *SPE-J*, August 1971, p. 42.

CHAPTER 4

Materials and Their Properties

PLASTIC MATERIALS – STATISTICS

The fourth largest volume of material used in the United States, when measured by cubical content, is plastics. Concrete is first, followed by steel and wood. It is estimated that plastic will overtake wood by 1975 and steel by 1980. Sales increased from 5 -½ billion lb in 1960 to 10 billion in 1965 and 24 billions in 1972. It is estimated that in 1980 the sales will be 45 billion lb. In 1970 the production of six plastic materials exceeded 1 billion lb each. They were low density polyethylene, PVC, polystyrene, high density polyethylene, phenolic, and polypropylene. These figures are even more impressive when one considers that polypropylene was commercially introduced in 1957, high density polyethylene in 1954, low density polyethylene in 1942, polystyrene in 1937, and PVC in 1933. Table 4-1 shows the approximate sales of thermoplastic and thermosetting materials for 1970, plus the estimated poundage of thermoplastics used for injection molding.

Table 4-2 shows the major markets for plastics in 1970, and includes about two thirds of the annual consumption. No accurate estimates of consumption are available for the military which is a large users of plastics. As we see from Table 4-2, packaging consumes the largest amount of plastics. The major uses for molded materials were in molded containers and lids which used about 250 million lb of styrenic materials, and 140 million lb of olefins. Thermoplastic closures accounted for approximately 70 million lb.

The building and construction industry is the second largest user of plastics. The main usages are in resin bonding, insulation, flooring, and extrusions. The extrusions are used for piping and sheet material. The main usage for molding is for pipe fittings, lighting, and bathroom fixtures.

Table 4-1   Approximate sales of thermoplastic and thermoset materials
for 1970, plus the estimated poundage of thermoplastics used
for injection molding (1)

Thermoplastics	1970 Sales (million)		Used for Injection Molding (millions)	
	lb	kg	lb	kg
Low density polyethylene	4,180	1,896	620	281
PVC	3,050	1,383	90	41
Polystyrene and copolymers	2,819	1,279	1,960	890
High density polyethylene	1,625	737	375	170
Polypropylene	1,010	458	500	227
Rigid PVC	678	308	8	4
ABS	504	229	300	136
Acrylic	390	179	85	39
Cellulosics	175	79	70	32
Nylons	103	47	70	32
RTP	64	29	64	29
Acetal	56	25	54	24
Polycarbonate	40	18	37	17
Fluroplastics	23	10	10	5
Subtotal	14,717	6,675	4,253	1,927
Thermosets				
Reinforced plastics	911	414		
Urethane foam	905	411		
Phenolic	888	402		
Urea-melamine	638	290		
Polyester	613	278		
Epoxy	155	70		
Subtotal	4,110	1,865		
Total	18,827	8,540		

The electrical and electronic industries are also large users of plastics. Approximately 900 million lb were used for extruded insulation on wire and cable. About 150 million lb were injection molded. The telephone system is a major user of molded parts and wire. Connectors, plugs, sockets, switch plates, and housings are some typical examples of injection molded parts.

Each year sees an increase in the amount of plastics used in automobiles. Approximately 1 billion lb were used in 1970. This included 160 million lb of

Table 4-2   Major markets for plastics 1970 (1)

	Million lb		Millions kg	
Packaging				
Film and sheeting	1,595		723	
Foamed	897		407	
Bottles and tubes	707		321	
Coatings	600		272	
Closures	114		52	
Adhesives	51	3,987	23	1,808
Building				
Piping, fittings	697		316	
Resin bonding	655		297	
Insulation	407		185	
Flooring	395		179	
Panels and sidings	170		77	
Decorative laminates	148		67	
Lighting fixtures	108		49	
Vapor barriers	100		45	
Wall coverings	90		41	
Plumbing and bathroom	55		25	
Profile extrusion	57		26	
Glazing and skylight	37	2,919	17	1,324
Electrical/Electronic		1,263		573
Transportation				
Passenger cars	924		419	
Others	214	1,138	97	516
Housewares		967		439
Furniture		735		333
Toys		576		261
Appliances				
Major	356		161	
Small	190	546	86	248
Agriculture		197		89
Total		12,328		5,591

polyurethane foam, 120 million lb of RP, and 50 million lb of phenolics. The more than half of the 670 million lb balance, was injection molded, although a significant amount was extruded for use as fabrics. The use of plastics in the automotive field should increase rapidly as automatic techniques and stronger materials become available.

Housewares is another billion pound market for plastics, with polyethylene used for about half the applications. About a quarter of this billion pounds is polystyrene. It is hard to imagine the modern kitchen without plastics, most of which are molded. The use of plastic for the bathroom is beginning to generate significant volume.

The most rapid recent growth in the plastic industry has been in furniture. The shortage and expense of wood and the difficulty in replacing skilled craftsmen has finally led to the large scale use of plastics. The initial successful attempt was to imitate wood. Plastic is now being developed as a material in its own right and a much more rapid expansion is foreseen. About 300 million lb are used in foam, 255 million in upholstery, and about 110 million lb are injected molded. The major use for molded components is high impact polystyrene panels which are finished to look like wood.

About 90% of the material used in toys is plastic, most of which are injection molded. Used in toys are 255 million lbs. of polystyrene and 235 million lb of oletins. Since plastic has penetrated most of the market, its expansion will follow the advances of the toy industry.

The use of plastics, particularly in small appliances, is rapidly expanding. Radios, TV's, vacuum cleaners, blenders, mixers, and hand tools are major users of injection molded parts. The advent of glass reinforcements and large tonnage molding machines are making major advances in replacing metals. This will be particularly true in the larger appliances such as dishwashers, refrigerators, and air conditioners.

There are relatively few injection molded parts used in agriculture, other than for containers. The primary tonnage is used for extruded pipe and film. The pipe is used for irrigation and the film is used for mulching, fumigation, and bagging.

THE PROPERTIES OF PLASTIC MATERIALS

The balance of this chapter consists of two parts. The first part deals with the standards by which plastic materials are described. The second part, starting on p. 372, surveys the materials commonly used in injection molding, referring to their chemistry, preparation, properties, advantages and disadvantages, major applications, and economic data.

STANDARDS

To communicate effectively, regardless of the method, the same sign, sound, or symbol must be understood the same way by both parties. This, in effect, is a standard. The standard for words is the dictionary. Industry standards develop through the government and private organizations whose results are voluntarily accepted by industry.

The viscoelastic nature of plastic required the development of a new set of standards. Industry has been fortunate in that the groups involved with plastic standards have acted promptly and with exceptional ability. The major groups involved in plastic standards follow:

1. American Society for Testing & Materials (ASTM), 1916 Race St., Philadelphia, Pa. 19103. ASTM's issuing of plastic standards began in the 1930s with ASTM's Committee D-9 on electrical insulating materials. In 1937; the present Committee D-20 was formed and is responsible for some 200 standards. And these are the most widely used in the industry. Volume 26 contains standards for specification of thermoplastic and thermosetting materials, plastic pipe and fittings, reinforced plastics, filament sheeting, cellular plastics, compounding ingredients, chemicals, and reinforcements. Volume 27 contains general methods for testing of plastics. Included are mechanical properties, permanence properties, analytical methods, specimen preparation, definitions, statistical techniques, conditioning, electrical insulating materials, rubber and rubber-like materials, color tests, and general methods of testing. Purchase of these books is strongly recommended. Very brief descriptions of some of these tests are available in Refs. 2 and 3.

2. U.S. Government. There are three main sources for government standards.

a. U. S. Dept. of Commerce
   National Bureau of Standards
   Washington, D. C. 20234
b. General Services Administration
   Federal Supply Service
   Washington, D. C. 20245
c. Military specifications

Government specifications can be incredibly difficult to find. For example, there are about 150 specifications for nylon scattered through tens of thousands of items in the index. To simplify this the Plastics Technical Evaluation Center at Picatinny Arsenal has published "Government Specifications and Standards for Plastics, Covering Defense Engineering Materials and Applications (revised)," which can be bought at CFSTI, 5285 Port Royal Road, Springfield, Vt., 22151.

3. American Standards Association (U.S.A. Standards Institute), 10 East 40th Street, New York, N. Y. 10016. The ASA was founded in 1918 and is a

coordinating body for all United States standard groups. It represents us at the International Organization for Standardization (ISO), which is the world's clearing house for the development of international standards, headquarters in Geneva, Switzerland.

4. The Society of the Plastics Industry (SPI), 250 Park Avenue, New York, N. Y. 10017. Voluntary commercial standards are developed in cooperation with SPI. Their purpose is to establish quality criteria, standard methods of tests, rating, certification, and labeling of commodities and to provide uniform basis for fair competition.

5. Underwriters Laboratory (UL), 1285 Walt Whitman Road, Melville, Long Island, N. Y. 11746. The Underwriters Laboratory is a nonprofit corporation supported by fire insurance companies in the United States. It "maintains and operates laboratories for the examination and testing of devices, systems, and materials, as to their relation to life, fire, and casualty hazards and crime prevention." While it is a private organization and can act unilaterally, it takes action in consultation with the industry. The UL label is important for commercial acceptance of electrical products.

6. Society of Plastics Engineers, Inc., 656 W. Putnam Ave., Greenwich, Conn. 06830. Through its plastic engineering property group, The Society of Plastic Engineers periodically publishes *Plastics Standards Digest* updating the various standards accepted and proposed since its previous publication (4, 4a). It also publishes a guide for uniform testing and reporting of identification properties of plastics (5).

7. Other Standards

> National Electrical Manufacturing Association (NEMA)
> 155 East 44th Street
> New York, N. Y. 10017

> Society of Automotive Engineers, Inc. (SAE)
> 485 Lexington Avenue
> New York, N. Y. 10017

> Plastic Pipe Institute
> 250 Park Avenue
> New York, N. Y. 10017

> Building Codes

Importance and Limitations of Tests and Standards

Before discussing standards relating to plastic properties the following caution is advised.

Standards reflect values obtained under precisely controlled conditions on specially prepared samples. For practical applications they must be interpreted by qualified plastics engineers.

The purpose of testing is to see that raw materials and finished products meet specifications to maintain quality and uniformity of the product. The purpose of the standard is to measure a single property under strict control. The plastic engineer who is faced with such questions as whether the part will do the job, or what is the minimal usable wall thickness has a different set of problems. He is concerned with the interaction of many physical properties. These are controlled by the inherent properties of the material plus the result of processing conditions. The test specimen, on the other hand, is either compression molded with minimal orientation or is molded with orientation in the direction of the testing.

The speed of testing is rigidly controlled for standards. Plastics exhibit a time dependency. Almost all of them exhibit a critical rate of loading above which they behave as brittle plastic solids and below which they behave as ductile tough materials. The Izod impact test has a hammer speed about 11 ft/sec at the point of impact. A foul ball on a nylon face mask might have an impact velocity of 500 ft/sec. While the speed of impact and the rate at which the object is loaded are not fully equivalent, tests must be fairly close to actual conditions to be useful for designing. Figures from standard tests have to be based on the knowledge of the test, previous experience, and, in some instances, intuition.

It is understandable, therfore, that a plastics engineer cannot always say that a particular design will work. His suggestion to make a sample or a limited run for tests is sensible and practical. After the parts are run tests will have to be developed which will correlate with actual usage. For example, looking at the physical properties of polypropylene one would expect it to be a material of choice for shades around incandescent bulbs. Its heat distortion point is in the range of 220°F, while the measured temperature is about 150°F. It is not a material of choice because the radiations from the bulb will cause trace metallic impurities in the plastic to catalytically oxidize it, causing it to crumble within a short time.

Generalized tests of emperical nature have been developed. One of the most popular is the falling ball impact. A steel ball of appropriate weight (2.3 lb, 2½ in. in diameter) is allowed to drop from a given height on the plastic part. If it was dropped from 1 ft it would have an impact energy of 2.3 ft/lb. As the height is doubled, the velocity is 1½, but the energy, $(Mv^2/2g)$ is doubled. Doubling the weight does not change the velocity but does double the energy. Since the test is emperical, it does not make much difference. Failure of the part in usage can be correlated to the amount of impact it can absorb. This is a useful, nondestructive, rapidly applied test which can be used to control manufacturing quality. References 6 to 13 discuss problems of testing and the evolution of specific tests for specific products.

The major tests in each category are discussed and evaluated. They are primarily ASTM tests which have been accepted as the industry's standards.

Some of the properties most often used are presented in tabulated form. All values are taken from the literature or the published data of the material manufacturer. Because of the many different grades of each material there is always a range of values reported. A median figure is selected. References for particular poperties and the evaluation of test methods are given. Again, the information in this chapter, while correct, is generalized. If more specific and current information is required source materials should be consulted.

GENERAL PROPERTIES

The general properties of plastics can arbitrarily be classified as

Density
Specific Volume
Cost
Refractive index
Haze transmittance
Luminous transmittance
Clarity
Color

Density

Density is the mass per unit volume. The specific gravity is the ratio of the weight of a given volume of a substance to an equal volume of water at 23°C. For practical purposes they are identical. They are usually determined by weighing the plastic in air and in water (ASTM D-792), or by dropping a piece of the unknown plastic into a liquid column with a density gradient (ASTM D 1505). The column contains samples of known density. It contains the most dense liquid at the bottom and has glass floats of known density floating at various levels in the column. Density is usually reported as grams per cubic centimeter or ounces per cubic inch. The specific volume is the reciprocal of the density and is usually reported as cubic centimeters per gram or cubic inches per ounce.

Because the density is indicative of the character of the material (e.g., in polyethylene) it is widely used in process control. From the molder's point of view it is directly related to cost. For example, changing the material from acrylic to polystyrene not only lowers the price because of the difference in material cost per pound, but also because polystyrene is about 9% less dense. Thus if the part weighed 100 g in acrylic it would weigh approximately 91 g in polystyrene.

The cost per cubic inch is sometimes a better comparator for part cost from different materials than the cost per pound. The figure used in Table 4-3 are based on posted truck load prices of the natural material as of June 1971. These are inaccurate for many reasons. In most materials the price varies with demand. Competition, economic conditions, patent status, and improved manufacturing facilities all have strong influence on prices. There are also many different grades of the same material, such as UV stablized, heat stablized, and antistatic which also affect the price.

OPTICAL PROPERTIES

Refractive Index.   The index of refraction of a material is the ratio of the velocity of the light in a vacuum to that of the specimen. It is expressed as a ratio of the sine of the angle of incidence to the sine of the angle of refraction. It is used as a measure of purity, identification and for optical design (ASTM D-542).

Haze and Luminous Transmittance.   These tests measure the light transmitting properties of clear plastic. They form the basis for comparing the transparency of different types and grades of plastic. Haze is defined in the ASTM D-1003 test as the percentage of transmitted light passing through the specimen which deviates more than $2.5°$ from the incident beam by forward scattering. Luminous transmittance is defined as the ratio of transmitted to incident light.

There are test methods for measuring the yellowness of clear plastic (ASTM D-1925), gloss (ASTM D-523), reflectance, and surface irregularities (ASTM D-637).

Clarity and Color.   Clarity can also be less rigorously defined as transparent, translucent, and opaque. Table 4-4 shows the main classification of the natural material. For example, acrylic is always transparent, while high impact polystyrene ranges from translucent to opaque, depending on the material and its thickness. All materials can be made opaque by the addition of suitable colorants.

Color is a basic specification for plastic parts. One of its major properties is its ability to be integrally colored, almost without restriction. The science of coloring, color, and the effects of color are beyond the scope of this chapter.

In viewing a colored object one sees the result of three different factors.

1.   The quality of the light illuminating the object.
2.   The ability of the object to absorb certain portions of the light and to reflect the others
3.   The sensitivity of the eye to the reflected illumination.

Normal daylight is a mixture of many colors. It is not constant and varies according to the time of day, atmospheric and cloud conditions, and direction of

Table 4-3 General properties of plastic materials typical values; (prices are of June 1971)

	Density[a] (g/cc) D-792	Specific Volume (cc/g)	Specific Volume (in.³/lb)	Water Absorp- tion[b] D-570	$/lb	¢/in.³ [c]
ABS	1.08	0.926	25.7	0.25	0.36	1.40
ABS 20% GR	1.2	0.833	23.1	0.12	0.65	2.82
Acetal cepolymer	1.41	0.709	19.6	0.22	0.65	3.32
Acetal 20% GR	1.55	0.645	17.9	0.45	0.74	4.13
Acetal homopolymer	1.42	0.704	19.5	0.25	0.65	3.32
Acetal 20% GR	1.55	0.645	17.9	0.25	0.74	4.13
Acrylic	1.2	0.833	23.9	0.2	0.46	2.00
Modified acrylic MMA	1.16	0.862	24.0	0.3	0.21	0.88
Cellulose acetate	1.22	0.820	22.7	2.6	0.56	2.47
Cell. acet. butyrate	1.15	0.870	24.1	0.9-2.2	0.62	2.57
Cellulose proprionate	1.17	0.855	23.7	1.2-3	0.62	2.62
Ethyl cellulose	1.13	0.885	24.5	1.3	0.66	2.7
Chlorinated polyether	1.4	0.714	19.8	0.01	4.50	22.7
CTFE	2.1	0.476	13.2	0.00	4.90	37.1
Polyvinylidene fluoride	1.75	0.571	15.8	0.04	3.36	21.3
FEP	2.12	0.472	13.1	0.01	4.65	35.5
Ionomer	0.93	1.07	29.8	0.1-1.2	0.45	1.51
Nylon 6	1.16	0.862	23.9	1.6	0.80	3.35
Nylon 6 30% GR	1.35	0.741	20.5	1.3	0.995	4.85
Nylon 6/6	1.14	0.877	24.3	1.5	0.80	3.29
Nylon 6/6 30% GR	1.35	0.741	20.5	0.7	0.995	4.85
Nylon 11	1.4	0.962	19.8	0.4		

Table 4-3 (Continued)

	Density (g/cc) D-792	Specific Volume (cc/g)	Specific Volume (in.³/lb)	Water Absorption D-570	$/lb	¢/in.³
Nylon 6/10	1.09	0.917	25.4	0.4	1.20	4.72
Nylon 6/10 30% GR	1.35	0.741	20.5	0.2	1.61	7.85
Polysulfone	1.24	0.806	22.3	0.22	1.49	6.68
Modified phenylene oxide PPO	1.08	0.926	25.6	0.066	0.70	2.74
PPO 30% GR	1.29	0.775	21.5	0.06	0.90	4.20
Polyaryl ether	1.14	0.877	24.3	0.25	0.85	3.37
Polyaryl ether 30% GR	1.37	0.730	20.2		1.30	6.44
Polyarylsulfone	1.36	0.735	20.4	1.8	25.08	122.94
Polycarbonate	1.2	0.833	23.1	0.16	0.80	3.46
Polycarbonate 30% GR	1.43	0.699	19.4	0.07	1.30	6.70
Polyethylene LD	0.910-0.925	1.10-1.08	30.4-30.0	<0.015	0.14	0.46
Polyethylene MD	0.926-.940	1.08-1.06	30.0-29.4	<0.01	0.16	0.54
Polyethylene HD	0.941-0.97	1.06-1.03	29.4-28.6	<0.01	0.18	0.62
Polyethylene 30% GR	1.18	0.847	23.5		0.40	1.70
Polyethylene High MW	0.94	1.06	29.4	0.04		
Ethylene vinyl acetate (EVA)	0.935	1.07	29.6	0.09	0.28	0.94
Polypropylene	0.9	1.11	30.8	<0.01	0.21	0.68
Polypropylene 30% GR	1.12	0.893	24.7	<0.01	0.41	1.66
Polycopolymer	0.9	1.11	30.8	<0.01	0.26	0.84
Polyallomer	0.9	1.11	30.8	<0.01	0.28	0.91
4-Methyl pentene-1	0.83	1.20	33.3	0.01		
Polystyrene GP	1.07	0.935	25.9	0.03-0.1	0.155	0.60

Table 4-3 (Continued)

	Density (g/cc) D-792	Specific Volume (cc/g)	Specific Volume (in.³/lb)	Water Absorption D-570	$/lb	¢/in.³
Polystyrene GP 30% GR	1.33	0.752	20.8	0.05-0.1	0.39	1.87
Polystyrene HI	1.07	0.935	25.9	0.05-0.6	0.205	0.79
PTMT	1.31	0.763	21.1	0.08	0.68	3.22
Styrene-acrylonitrile (SAN)	1.07	0.935	25.9	0.25	0.23	0.89
Styrene-acrylonitrile 20% GR	1.22	0.820	22.7	0.08	0.39	1.72
Styrene-butadiene	0.93-1.1	1.07-.909	29.6-25.2	0.2-0.4	0.37	1.64
Rigid polyvinyl chloride (PVC)	1.4	0.714	19.8	0.07-0.4	0.25	1.26
Chlorinated PVC	1.5	0.666	18.4	0.02-0.15	0.50	2.71
PVC/PP	1.4	0.714	19.8	0.07-0.4		
Vinylidene Chloride	1.7	0.588	16.3	0.1		
Urethane elastomer	1.2	0.833	23.1		1.21	5.25
Phenol-formaldehyde (no filler)	1.3	0.769	21.3	1.6	0.24	1.13

[a]To obtain density as ounces per cubic inch, multiply density (g/cc) 0.5781; and as pounds per cubic inch, multiply density (g/cc) by 0.03613.

[b]Water absorption; 1/8 in. thick, percentage increase in weight — 24 hr.

[c]To obtain ¢/in.³, multiply density (g/cc) by 3.613 and cost of material ($/lb).

334

Table 4-4   Light transmitting properties of plastics
(the transparent materials can be made translucent
or opaque by the addition of suitable pigments)

TRANSPARENT	TRANSLUCENT	OPAQUE
Acrylic (PMMA)	Acrylonitrile-butadiene-styrene (ABS)	ABS/PC
Cellulose acetate (CA)		Acetal
Cellulose acetate butyrate (CAB)	Casien	Acrylate-styrene-acrylonitrile (ASA)
Cellulose proprionate (CP)	Chlorinated polyether	Alkyd
Cellulose nitrate	Chlorinated polyethylene	Diallyl pthalate (DAP)
Epoxy	FEP	Melamine
Ethyl cellulose (EC)	Nylon (PA)	Phenolic
Ethylene-vinyl acetate (EVA)	Polypropylene (PP)	Phenylene oxide (PPO)
Ionomer	Polyethylene (PE)	Polyaryl ether
Methacrylate-butadiene-styrene (MBS)	Polystyrene (HI)	Polyimide
4-Methyl pentene-1	Urethane	Urea
Polyarylsulfone	Vinylidene fluoride	
Polycarbonate (PC)		
Polyester		
Polysulfone		
Polyvinyl acetate (PVAc)		
Polyvinyl chloride (PVC)		
Polystyrene (GP)		
Styrene-acrylonitrile (SAN)		
Styrene-butadiene		

the light. Daylight contains all the different spectrum colors. Incandescent lamps are seriously lacking in green, blue, and violet. The fluorescent lights normally found in factories do not have the full spectrum of colors. It is therefore necessary to have a daylight balanced light source or use daylight itself. If the colors match in daylight they are most likely to match in other lights. To most people, a blue object looks blue because all the other colors are absorbed by the part. Partial color blindness is comparatively common. It is only a great handicap if the person does not realize he is incapable of judging certain colors.

There are three factors to color — purity, brightness, and hue. If we start with a basic color, such as red, and put no other color in it we say it has full purity. If we add white to it, we lower its purity. If we add grey to it, we lower its brightness. If we add a different color, such as orange, we change its hue.

To measure color, a spherical concept is used. The vertical axis starts with white on the top, goes through all colors of grey to black on the bottom. This is a measure of the brightness. The equator has a circle of red, orange, yellow green, blue, violet, purple, and back to red again, all blending in. This indicates the hue. Any point on any radius would indicate the purity. Therefore by specifying the physical point in this sphere by three coordinates (x, y, and z) one can establish the color. There are two basic systems; one is ICI system, which measures the energy distribution, and the other is the Munsell system which is one of color specification rather than measurment.

The two instruments for measuring color are colorimeters, which measure transmitted or reflected light through various colored filters with a photo cell, and spectrophotometers which use a prism to break the light into small wavelength bands which are then measured with a photo cell.

References 14 and 15 discuss the theory and measurements of plastic coloring. References 16 to 25d cover the coloring of thermoplastics. Reference 24 gives pigments and dry color formulae for polystyrene. A series of charts listing colorants and pigments used in plastic, plastics in which they are usable and the manufacturer's source is found in Ref. 25.

Coloring Plastics

There are four methods of obtaining colored material — milled in color, dry colored, color concentrates, and liquid color.

Milled in color has the advantage of the best dispersion and the ability to match any color. It is the most costly method of coloring. It involves a reextrusion of the material with the colorant which, under some conditions, may lower its properties, and requires a large inventory, compared with the other methods, and a significantly longer lead time for production.

Dry coloring consists of tumbling the basic resin with a colorant that may be dispersed in a liquid. Sometimes wetting agents are used. Certain automatic material handling systems meter in the colorant during feeding. This is the most economical way to color. It requires a minimum inventory of the uncolored material. Colorants are usually packed in bags preweighed for mixing with 50, 100, or 200 lb of the resin. If bulk colorants are available, the color can be easily changed or matched to a customer's specification.

Dry coloring has a number of disadvantages. The wrong colorant might be used or the colorant might not be completely emptied into the tumbling equipment. Further more, there is a lot of material handling and chance for loss; it is basically a messy procedure so that dusting may cause contamination problems. Then it is more difficult to clean the feed system than with milled in material. Colors which require several pigments, particularly in small amounts, are very difficult if not impossible to color evenly, and dispersion can be significantly poorer than in milled in material and in many instances unacceptable. To improve dispersion restricted gates should be used. If this does not help, or is not possible, many different types of dispersion nozzles are available. These nozzles work on the principle of turbulent flow caused by increased material velocity through a small orifice. In plunger machines they are usually necessary, but are rarely required in screws which have superior mixing capability.

A compromise between milled in material and dry coloring are color concentrates or master batches. Here, high concentrations of colorant are extruded into the same type of resin from which the part will be molded. The concentrate is usually arranged so that about 5 lb of concentrate are required to color 100 lb of base resin. While this is a more expensive way than dry coloring it overcomes a number of its disadvantages. Complex colors which do not work in dry coloring will usually work with color concentrates. The contamination, dusting, and cleaning problem are minimized as they are for milled in material. Compared to milled in material, concentrates are usually less expensive, require lower material inventory, and occupy less space.

The fourth method pumps liquid colorant directly into the screw at the bottom of the hopper. Very accurate pumps, such as the diaphragm type, are required. It is advantageous because colors can be changed without removing the basic resin from the hopper, and the hopper does not have to be cleaned if the same plastic is used for the next color.

Material is shipped in 50 lb bags, 300 lb drums, 1000 lb gaylords, tank trucks, and tank cars. It can also be conveyed pneumatically through pipes and tubes. Modern material handling systems provide considerable savings (25e, 25f). The technology is changing rapidly and those interested in system installations should consult manufacturers and material handling specialists.

Gates, runners, and rejects are reground in grinders (granulators). They come in many forms, including augur types. Reground material can be blended back into the hopper, either automatically or manually, or collected for further use

(25g). Grinders are dangerous and noisy and should be a focal point in the safety procedures of the plant. Regrinding does not normally affect the material's properties when reused (25h).

MECHANICAL PROPERTIES

The mechanical properties of plastics, which are summarized in Table 4-5, can be arbitrarily classified as

> Tensile strength.
> Tensile modulus.
> Tensile strength at yield.
> Percent elongation at failure.
> Flexular strength.
> Flexular modulus.
> Compressive strength.
> Compressive modulus.
> Impact strength.
> Deformation under load.
> Hardness.
> Scratch and abrasion resistance.
> Friction.

Tensile Strength, Tensile Modulus, Tensile Strength at Yield, Percent Elongation, and Percent Elongation at Yield

A typical test specimen for determining tensile strength (ASTM D-638) consists of a bar 1/8 in. thick, ¾ in. wide, and 8½ in. long. It is reduced to a ½ in. width for about one third of its length. Measurements are taken in this area.

The specimen is firmly clamped in a machine which moves at a controlled rate -0.1, 0.2, 2, or 20 in./min. The stress is automatically plotted against the strain (elongation). From this one can calculate tensile strength, percentage elongation, and modulus of elasticity.

The determination of these values is simple. What they mean and how to use them is not so clear. The tensile properties, or the strength in tension, are the most important single indication of the strength of a material. They represent the force necessary to pull apart the material and how the material moves or stretches before breaking. The mechanical properties of a particular polymer depend on three factors: (a) the structure of the polymer, (b) the physical state of the polymer which includes its relation to its T_g and its degree of crystallinity, molecular weight, molecular weight distribution, and extent of chain branching, and (c) the experimental conditions under which the properties are measured. The mechanical tests used to describe a property utilize simple shapes and, where possible, test one property at a time. Actual parts, for which the design data are used, are exceedingly complex in their shape and are the result of the combination of many forces. The usual procedure is to consider each load as if it existed alone and add up or impose one upon the other. With a lot of experience, this is not as difficult as it sounds. In many cases though, the only practical method is to build a prototype using the initial calculations.

To understand better the significance of mechanical properties one must remember that plastic deformation consists of four phenomena:

1. Ideal Hookean elasticity (all deformation is fully and instantaneously recoverable when the force is removed).
2. Delayed recovery, where the stretched polymer bonds return to their original position.
3. Viscous flow, which represents permanent deformation.
4. Creep and stress relaxation caused by the effect of the load over a long period of time.

This time dependency, plus the effects of temperature and orientation, make designing and predicting long term properties of plastics more difficult than that of conventional material.

Ductility Versus Brittleness.   Polymers may be classed as ductile or brittle at a given temperature. A ductile material fails by the molecules sliding over each other and flowing. This is typified by high elongation before failure. A brittle material fails by shearing action and the fracturing of the molecular bonds. This is typified by low elongation. This property is directly related to T_g and the temperature under consideration. At higher temperatures the molecule is more likely to uncurl and flow than to fracture. It therefore lowers the tensile strength and increases the percentage elongation. Obviously time will be a factor. If the impact is applied very rapidly there may not be time for ductile yielding. For example, the tensile properties of high density polyethylenes are extremely sensitive to the testing speed. In fact, it is possible to obtain almost any desired value by manipulation of the conditions of the test (26). As will shortly be discussed the geometry of the part will also affect the type of failure.

Stress-strain Data.   Stress-strain data are usually reported as a graph (Figure 4-1). As the material is stressed it moves (strain). In the early part of the curve the stress and strain are linear; that is, they increase in proportion to each other. It is in this area that the material acts as an elastic solid. This is the stress area in which plastic products are usually designed. Here, also, the tensile modulus (or modulus, modulus of elasticity, elastic modulus, and Young's modulus) is measured. It is the ratio of the nominal stress to the corresponding strain-below the proportional limits of a material. It is expressed in units of force (usually

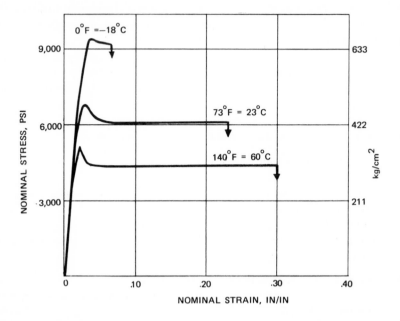

Figure 4-1 Stress-strain properties in tension of ABS. Specimens cut from extruded sheet transverse to principal flow direction, low orientation (Lustran ABS-261, Courtesy Monsanto Chemical Co.).

psi). It is measured at a speed of 2 to 10 in./min. The higher the value the stiffer the material. The modulus decreases with the temperature. For example, the material in Figure 4-1 has modulus of 410,000 psi at 0°, 390,000 psi at 73°, and 380,000 psi at 140°.

As the material continues to stretch, a maximum stress will be recorded based on the original cross-sectional area. This maximum tensile stress during a tension test is called the tensile strength and is expressed as pounds per square inch. At 73° the ABS material in Figure 4-1 has a tensile yield strength of 6600 psi. It is calculated by dividing the maximum load in pounds by the original minimum cross-sectional area in square inches. Some materials such as acrylics will break during this portion of the curve, while still exhibiting elastic behavior. Other materials, such as polyethylene, nylon, or ABS, exhibit a ductile characteristic. They have a yield point at which the material continues to elongate with no increase in the stress. The stress decreases to a relatively constant value (tensile strength at failure). The material moves until it breaks, indicated by the arrows, Figure 4-1. At 73° the material had a tensile strength at failure of 6000 psi, while its tensile strength at the yield point was 6600 psi. The percentage elongation is the distance between two marks on the specimen bar at rupture, divided by the original distance between the marks × 100. The percent elongation of yield at 73° for the ABS material illustrated is 2.2% and at failure, 20%.

Flexural Properties

One method of measuring the flexural properties of plastics is to place a test bar, usually 5 in. long by ½ in. wide by 1/8 in. thick, on two supports. A force is applied vertically at the center of the bar (ASTM D-790). The flexural strength is defined as the strength of the material in bending, expressed as the tensile stress of the outermost fibers (bottom of the strip) at the time of failure. For materials which do not break, the flexural yield strength is reported when the specimen exhibits 5% strain. This is true for most thermoplastics. From these data the modulus of elasticity in flexure is calculated.

Comparing the tensile, compressive, and flexural strength at fracture and the corresponding moduli of elasticity shows that the flexural data are the highest. This makes the data suspect. In this test, when the head bends the plastic, the layer touching the head will have a compressive stress. The bottommost part will be stretched and have a tensile stress. It can readily be seen that one level somewhere between top and bottom is theoretically unstressed, and the amount of stress in any other section would be proportional to the distance from this line of zero stress. This would only hold true in the proportional limits of the material. Since the specimen usually fails by splitting on the bottom section (which is under tension) the maximum strength should be no higher than the ultimate tensile strength. This should be the limiting value of the flexural strength. The figures do not show this. The formula used to calculate these flexural properties assumes that the modulus of elasticity in tension and compression are the same and that the material does not exceed the proportional limit. The first part is true but the latter part is not. Once past that point, the stress is no longer proportional to the strain and the formulae used for calculating these figures are no longer applicable. These figures are only of value if the application is at a stress level below the proportional limit, that is, when the strain varies with the stress. The rate of stress is important. At low rates, the criterion for flexural failure is related to the compressive strength of the material. As the rates increase tensile forces come into play and at the highest rates the tensile strength of the material is involved (27).

A second method of measuring the stiffness and flexure uses a test strip supported in a vise which is attached to a motor that rotates 60°/min (ASTM D

747). The unsupported end contacts a weighted, movable arm. As the bar is rotated the force is transmitted through it to the movable arm, which turns a scale and pointer. This is a measure of the magnitude of the force transmitted. The stiffer the material, the more force transmitted. This test measures the elastic and viscous elements of the material, but does not distinguish between them. It is most useful in comparing the relative stiffness of various materials. Typical values in psi are low density polyethylene 18,000, medium density polyethylene 35,000, high density polyethylene 120,000, general purpose polypropylene 190,000, and rigid PVC 250,000.

Fatigue Properties.   The fatigue property of plastics is a general term describing their behavior under repeated cycles of stress or strain which cause a deterioration of the material which may result in a failure. Vibrating loads (e.g., opening and closing a refrigerator door) may cause failure even though the stress is well under the static strength of the material. A relatively small number of cycles of stress will cause failure if the stress is near the ultimate static strength.

The "fatigue endurance limit" is the maximum stress at which the material will sustain vibratory loads indefinitely. This value may be greater or less than the elastic limit and often lies between one third and one half of the tensile strength at break.

Fatigue failure depends on

1. Magnitude of stress.
2. Frequency of stress.
3. Type of stress (tension, compression, bending).
4. Temperature.
5. Environment.
6. Geometry of the parts.

Parts are tested under constant deflection or constant load (ASTM D-671). Machines for measuring fatigue apply their stress directly, flexurally, or torsionally. The results are reported graphically by plotting the stress versus the number of cycles to failure.

Creep and stress relaxation properties are reported graphically (ASTM D 674). A discussion of them is found on p. 287.

Compressive Properties

The compressive properties of plastics describe their behavior when they are subjected to forces that squeeze or crush. The usual test is to take a 1/2-in. square bar 1 in. long, or a 1/2-in. diameter rod 1 in. long, and compress it at a constant rate, recording the force and the change of length (ASTM D-695). The compressive strength is equal to the maxium load divided by the original cross-sectional area. A useful value is the compressive stress at 1% deformation.

This is obtained by dividing the force by the original corss sectional area at the time when the specimen has been compressed 1%. The compressive modulus can be calculated from this test. As a practical matter the tensile modulus can be used in its place. The compressive strength is not too important a specification because plastics rarely tail under compressive loads alone. They are useful in comparing materials.

Shear Strengths. The shear strength of plastic is measured by punching a hole through a sheet of plastic whose thickness may vary from 0.005 to 0.5 in. (ASTM D-732). A hole is drilled in the center of the specimen through which the punch is bolted. The rest of the specimen is held solidly. The punch is moved at the constant speed of 0.05 in./min and the maximum load required to shear the plastic recorded. The shear strength is the maximum load divided by the area sheared (circumference of the punch times the sample thickness). As there may be some flow as well as shearing, the shear strength for design purposes is considered to be one half the tensile strength. Some typical shear strengths (psi) are high density polyethylene 3500, general purpose ploystyrene 7000, acrylic 8000, nylon and acetal 10,000, and cellulose filled melamine 11,000.

Impact Strength And Toughness

Impact strength and toughness are difficult to define and even more difficult to measure. Impact resistance, in general, means the resistance of the plastic to breaking when subjected to stresses at high rates at loading. They are usually measured by Izod and Charpy tests (ASTM D-256), drop tests, falling dart tests and tensile impact tests (ASTM D-1822). Toughness is a more general term and not related directly to any particular speed. To describe it one would need a series of stress-strain curves over a wide range of velocities. Figure 4-2a shows such a graph for a hard tough material. The shaded area under the curve is considered a measure of the toughness of the material. As the stress is increased the material strains in an elastic manner. It then elongates for a considerable period before it fails. In contrast Figure 4-2b is a soft weak material which strains under low stress and yields very quickly. Figure 4-2c is a hard brittle material which strains similarly to a, but to a lesser extent, and breaks very quickly without much elongation. Figure 4-2d is a hard strong material similar to the hard brittle material, except that it accepts much more stress and has a much larger area under the curve. Figure 4-2e is a soft, tough material which yields under low stress and elongates considerably. There is a relatively large area under the curve.

Izod Tests. The Izod impact tester consists of a pendulum with a cylindrical striking edge, 1/32 in. in radius, and a vise to hold the specimen bar, usually 2 1/2 in. long by 1/2 in. wide by 1/8 in. thick. The pendulum is so designed that it will have a velocity of 11 ft/sec when it hits the specimen. After

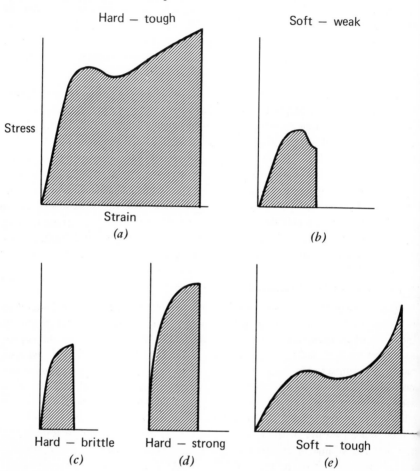

Figure 4-2 Types of stress-strain curves. (*a*) Hard-tough; (*b*) soft-weak; (*c*) hard-brittle; (*d*) hard-strong; (*e*) soft-tough.

breaking the specimen the pendulum continues to travel. From the physical constants of the machine and the distance of travel, the energy loss of the pendulum in breaking the specimen is calculated. The specimen is notched on the 1/8 in. section with a 45° angle and 0.010 radius at the tip. The impact values are calculated on the basis of a notch 1 in. long, regardless of the fact that the specimen's notch may only be 1/8 in. long. The Izod tester is also used with unnotched specimens, although this is not a part of the ASTM procedure.

Some plastics, such as low and medium density polyethylene, plasticized vinyls, and FEP, do not completely break and therefore have no Izod reportable impact values. The test is sensitive to the radius of the notch, the alignment of the part in the vise, the pressure holding the part in the vise (28), and the preparation of the sample. A variation of the Izod impact test is the Charpy method (ASTM D-256 B). The specimen, which is usually notched, is supported horizontally at both ends as a simple beam, and is broken by a single blow midway between the supports on the side opposite to the notch.

The Falling Dart Test.   In this test, a steel ball is held a certain height above a specimen and releases magnetically. By adjusting the height and weight of the ball the feet per pound impact and the velocity or rate at loading can be varied. The specimens are inspected for failure.

The Drop Test,   as the name implies, is a pragmatic test. The sample is loaded under typical usage conditions and dropped under typical service conditions. Examples of this test are filling a food container with the food and dropping it 7 ft, which is the highest it would be stored in a food market. Another example would be to pack a molded luggage case and drop it on its corner on concrete from the height of a baggage compartment of an airplane.

Tensile Impact Test (ASTM D-1822).   This test consists of mounting one end of the specimen on a pendulum and the other end on a cross head attached to the pendulum. The pendulum swings; the cross head hits a stationary stop; at the lowest point of its arc the pendulum continues moving rupturing the specimen. The pendulum continues its swing and indicates by the height of its swing the balance of its initial energy that is left. The results are reported as feet per pound per square inch of surface. This test removes the notch sensitivity factor of the Izod and is usable on materials too thin or flexible for that test.

Results of Tests.   All these impact tests record the amount of energy absorbed by the sample. This is not necessarily the amount of energy causing failure. Using pressure transducers, oscilloscopes, and high speed cameras, a clearer insight of the impact behavior of thermoplastics was obtained (29). Force-time impulse curves showed three characteristic types of breaking. In the first type, the specimen deforms uniformly until the maximum pressure is reached and the part breaks catastrophically. This is characteristic of brittle materials. The second type shows elastic deformation, flow and finally tearing. The third type shows elastic deformation, very little if any flow, and failure by tearing. Those materials with high impact strength absorbed approximately 70% of the total impulse after yielding. This may be desirable depending on the application. The results show that impact strength is not a good single measure of the mechanical behavior of the impact properties of thermoplastics. Detailed analysis of the Izod test (30-32) points out some of its limitations. A study of thermoplastic behavior during impact is found in Ref. 33. A comparison of the Charpy and Izod impact tests is found in Ref. 33a.

Obviously the speed of testing is an important consideration. When a material is tested in a standard machine the stress spreads essentially uniformly throughout the specimen. When the specimen load is applied at high speeds, one end of the test bar may reach maximum stress before the remainder is stressed at all. This has long been recognized (34) and the results from high speed testing equipment are helpful in selecting material for those applications.

Some of the limitations of impact testing, particularly when related to toughness, follow (35, 36).

The Izod impact test does not aid in product design when specific abuse resistance is needed. It does not supply basic design data. It cannot measure stress and strain and has a fixed strain rate much higher than encountered in normal abuse. It compares the relative shock resistance of materials.

The falling dart test is difficult to correlate with the results of other impact tests. It is difficult to reproduce, the criteria for determining failure are not precise, and the material behavior can only be quantitatively studied at the point of failure. It is also a slow and costly testing procedure.

The drop test cannot be related to other design problems. It is a go/no-go system which only tells if the part failed or not, and is a difficult test to reproduce and does not define the type of failure. There are many uncontrolled variables. It is costly in time and in the number of samples required for a test.

Impact tests that can predict the service of a particular part under a particular condition do not exist now nor are they foreseeable in the immediate future. It will come about from a fuller understanding of the relationship of molecular properties to mechanical behavior (37). Impact testing has stimulated research into the physical nature of polymers and their reaction under specific test conditions. This, coupled with the rapidly expanding amount of information from existing applications, enables engineers to make good qualitative recommendations for most applications.

Deformation Under Loads

The deformation of plastic under load is measured by inserting the specimen between two anvils which are loaded at 1000 psi. The gauge is read 10 sec and 24 hr after loading. The percentage deflection is measured. The test is run at 73.4°F, 122°F or 158°F (ASTM D-621.) The test measures the ability of the plastic to withstand continuous short term compression. This is of value when plastics are used in fastening devices. It does not take into account creep or the long term stability of the plastic. It is a measure of the rigidity at these specific temperatures. The test can also be run at different temperatures and longer times which yield practical design information.

Hardness

Hardness represents various properties more or less related to each other. It represents a combination of resistance to indentation, scratching, abrasion (38), and marring. The two principle kinds of hardness are resistance to indentation and resistance to rubbing over the surface (abrasion).

Rockwell Test for Hardness.   The standard method for testing for hardness is to use a Rockwell tester. A standardized procedure is described in ASTM D 785. The sample is placed on a steel anvil and a steel ball with a 10-kg load is forced into it. The dial at this low load is set at zero. A high load is released over the ball indenting the sample for 15 sec. The high load is removed and partial recovery takes place. After another 15 sec the hardness is read on the dial of the instrument with the low load still being applied. Readings on the R scale use a 60-kg high load with a 1/2-in. ball and on the M scale 100-kg high load with a 1/4-in. ball. The higher the number, the harder the material. This in effect is measuring the tensile strength, tensile yield, and modulus (39). There are other methods for measuring indentation types of hardness. They cannot be readily converted by a table and values must be determined experimentally.

Durometer Test for Hardness.   For softer material a durometer is used (ASTM D-2240). This in effect measures the indentation under a given load. It is primarily used for PVC and softer ployethylenes.

Abrasion Resistance.   Abrasion resistance is the ability of a plastic to withstand mechanical action by friction which tends to remove material from its surface. This can be caused by rubbing, wearing, scraping, eroding, or grinding. There are essentially two types of friction, sliding friction and rolling friction. Because of the many different ways abrasion can be applied and the many materials with which plastic comes into contact, the testing and interpretation of abrasion resistance is very complex.

The three most common tests are (*a*) impinging of abrasives by dropping them through a tube (ASTM D-673) (the loss in gloss is measured), (*b*) by rotating the specimen on a turntable (Taber Machine) against a pair of loaded abrasive wheels which has a unique rotary sliding action (ASTM D 1044), and by measuring the loss in light transmission by diffusion. These last two tests are primarily used for clear plastic. The Taber machine can be used for checking mar resistance of opaque materials (40). (*c*) The use of loose abrasives or abrasive wheels by an Olsen Wearometer is used for ASTM D-1242 test. The test measures the amount of material lost by abrasion.

Friction

Many applications of plastic depend on their frictional properties. The understanding of friction in plastics is considerably more difficult, primarily because of the difficulty of estimating Young's modulus plus the complicating effect of temperature on the properties of the material. The rate of heat removal becomes important, and cleanliness, surface condition, load, and speed also effect the results. It is therefore very difficult to get a reproducible coefficient of friction.

A frictionometer for plastics is described in Ref. 41. The frictional properties of plastic are described in Refs. 42 to 44. The frictional properties of specific plastics are shown in Refs. 45 to 49. A study of plastic bearings is found in Refs. 50 and 51, and of low friction materials in Ref. 52.

THERMAL PROPERTIES

The thermal properties of plastics can be classified as

Thermal expansion.
Specific heat.
Thermal conductivity.
Thermal diffusivity
Melting and softening points.
Deflection temperature of plastics under load.
Brittleness temperature.
Flammability.
Polymeric decomposition.

Some of the properties for different plastic materials will be found in 4-6 and in the unnumbered tables.

Thermal Expansion

As materials are heated there is increased molecular vibration and larger distances between the atoms. This results in an increased volume. The amount of this expansion depends on the material. Glasses, particularly quartz, have a very low coefficient of expansion. The coefficient of expansion for copper is 9.2×10^{-6} in./in./°F, while that for steel is 6.3×10^{-6}. Plastics expand about 10 to 15 times more than metals, the coefficient for general purpose polystyrene being 4.3×10^{-5}.

The coefficients listed above are linear coefficients. If the material expands equally in all directions (isotropic) the coefficient of cubical expansion would be the cube of the linear expansion. These coefficients are constant over the temperature ranges used for plastic except when they pass through a transition point. In that instance, calculations have to be adjusted for the different coefficients.

The linear coefficient is of extreme importance when the plastic is attached

Table 4-5   Mechanical properties of plastic material (these are typical values)

	Tensile Strength D-638 (1000 psi)	(kg/cm²)	Tensile Modulus (10⁵ psi) D-638	Percentage Elongation at Yield D-638	Flexural Strength D-790 (1000 psi)	(kg/cm²)	Flexural Modulus (10⁵ psi) D-790	Izod[a] D-256	Rockwell Hardness (L, R, M Scales) (Shore-D Scale)
ABS	6-8	422-562	3-4.2	5-60	6-13	422-914	2-4	3-8	R 75-115
ABS 20% GR	8-19	562-1336	5.9-10	3	16-27	1125-1898	9-13	1-2.4	M 65-100
Acetal copolymer	10	703	4.1	70	13	914	3.8	1.5	R 120
Acetal homopolymer	8.8	619	5.2	25.	14	984	4.1	1.4	M 80
Acetal 20% GR	11.	773	10.	3	15	1055	8.8	0.8	M 80
Acrylic	7-11	492-773	3.5-4.5	2-7	13-14	914-984	4.3	0.4	M 100
Modified acrylic MMA	10	703	4.6	3	12	844	5	0.3	M 105
Cellulose acetate	2-9	141-633	0.7-4	6-10	2-16	141-1125		0.4-5	R 35-125
Cellulose acetate butyrate	2.5-7	176-492	0.5-2	40-90	2-9	141-633		0.5-12	R 30-116
Cellulose proprionate	2-8	146-562	0.6-2	30-100	3-11	211-773	1.5-3.4	1-6	R 10-122
Ethyl cellulose	2-8	141-562	1-3	5-40	4-12	281-844		2-8	R 50-116
Chlorinated polyether	6	422	1.6	60-160	5	352	1.3	0.4	R 100
CTFE	4.5-6	316-422	1.5-3	80-230	9.3	654	2.2	2.6	R 85
Polyvinylidene fluoride	5.5-7.5	387-527	1.2	100-300	7	492	2	4	R 80
FEP	3	211	0.5	250-330			1	N B	R 25
Ionomer	2-4	141-281		150-450				6-11	R 55
Nylon 6	7-12	492-844	1.1-3	100-400	7-15	492-1055	1	2	R 103-120
Nylon 6 30% GR	21	1476	10	2.2	26	1828	10	3	
Nylon 6/6	9-12	633-844	1.8-4	60-300	N B		1.8	1-2	R 108-120
Nylon 6/6 30% GR	20-28	1406-1968	10-12.8	5-10	30	2109	10-12	2	R 110-120
Nylon 11	8	562	1.9	300	N B		1.4	1.8	R 110-120

Table 4-5 (Continued)

	Tensile Strength D-638 (1000 psi)	(kg/cm²)	Tensile Modulus (10⁵ psi) D-638	Percentage Elongation Yield D-638	Flexural Strength D-790 (1000 psi)	(kg/cm²)	Flexural Modulus (10⁵ psi) D-790	Izod D-256	Rockwell Hardness (L, R, M Scales) Shore-D Scale
Nylon 6/10	8.5	597	3.8	85-300	N B		1.6	1.2	R 110-120
Nylon 6/10 30% GR	18	1125	9	2.5	23	1617	8	1.6	R 120
Polysulfone	10.2	717	3.6	50-100	15.4	1083	8-18	1.3	M 100
Modified phenylene oxide PPO	8-10	562-703	3.7	20-30	13	914	16	1.8	R 118
PPO 30 % GR	17	1195	9-12	6	20	1406	20	1.5	L 108
Polyaryl ether	7.5	527	3.2	25-90	11	773		8	R 117
Polyarylsulfone	13.	914	3.7	13	17	1195		5	M 110
Polycarbonate	8-9.5	562-668	3.3	100-130	12.5	879	3.4	12-18	M 70
Polycarbonate 20% GR	9	633	8.	110	13.5	949	3.4	16	M 91
Polyethylene LD	0.6-2.3	42-162	0.14-0.38	90-800			0.08-0.6	N B	D 45
Polyethylene MD	1.2-3.5	84-246	0.25-0.55	50-600	4.8-7	337-492	0.6-1.2	0.5-15	D 55
Polyethylene HD	3.1-5.5	218-387	0.6-1.8	30-1000			1-2.6	0.5-20	D 65
Polyethylene 30% GR	8	562	8.0	3	9.5	668	7	2	R 70
Polyethylene high MW	2.5-3.5	176-246	0.2-1.1	300-500			1.3	N B	
Ethylene vinyl acetate (EVA)	1.5-2.8	105-197	0.02-1	750-900			0.01-0.2	N B	D 20
Polypropylene	4.3-5.5	302-387	1.6-2.3	200-700	6-8	422-562		0.5-2	R 85-110
PTMT	8.2	576		250	12	844	3.1	N B	R 117
Polypropylene 30% GR	7.3	513	0.8	3	10	703	7.6	2	R 110

Table 4-5 (Continued)

	Tensile Strength D-638 (1000 psi)	(kg/cm²)	Tensile Modulus (10⁵ psi) D-638	Percentage Elongation at Yield D-638	Flexural Strength (1000 psi) D-790	(kg/cm²)	Flexural Modulus (10⁵ psi) D-790	Izod D256	Rockwell Hardness (L, R, M Scales) Shore-D Scale
Polypropylene copolym									
copolymer	2.5-4.5	176-316	1-1.7	200-700	5-7	352-492		1-20	R 50-95
Polyallomer	3-4	211-281		400-500					R 50-85
4-Methyl pentene-1	4	281	2.1	15				0.8	L 70
Polystyrene GP	5-12	352-844	4-6	1-2.5	9-14	633-984	4.5	0.3	M 72
Polystyrene GP 30% GR	9-15	633-1055	12	1.3	11-20	773-1406	8-10	4.	M 80
Polystyrene HI	1.5-7	105-492	1.5-5	2-80	5-12	352-844	3.8	0.5-11	R 50-100
Styrene-acrylo-nitrile (SAN)	9-12	633-844	4-6	3	14-19	984-1336	6	0.4	M 85
SAN 20% GR	9-20	633-1406	4-14	3	22-26	1547-1969	8-18	4	M 100
Styrene-butadiene	0.6-3	42-211	0.01-0.5	300-1000	N B		0.4-1.5 N B		
Rigid polyvinyl chloride (PVC)	5-9	352-633	3.5-6	100-1000	10-16	703-1125		0.4-20	D 70
Chlorinated PVC	7.5-9	527-633	3.6-5	5-65	15-17	1055-1195	3.4-6	1-6	R 120
PVC/PP	5-8	352-562	4	100-140	11-15	773-1055	3.5-5	0.4-32	R 110
Vinylidene chloride	3-5	211-352	0.6	10-250	4-6	281-422	1	1	M 60
Urethane elastomer	6.5-8	457-562		600					D 52
Phenol-formaldehyde (no filler)	7.	492	7-10	1.5	9	633	10	0.25	M 125

[a]Izod impact— ft lb/in. of notch (½ X ½ bar).

351

to or confined in another material, particularly metal. Some examples are molded in metal inserts, plastic handles attached to metal or held by metal screws, plastic cases enclosing metal objects, and plastic sandwich structures.

The stress can be calculated by

$$S = E \left(a_p - a_M\right) \Delta T \tag{4-1}$$

where

S = stress (psi)

E = modulus of elasticity (psi)

a_P = linear coefficient of thermal expansion of the plastic (in./in./°F)

a_M = linear coefficient of thermal expansion of the attached material (in./in./°F)

ΔT = Change in temperature (°F.)

Plastic failure caused by these conditions is a function not only of the thermal expansion but also of the tensile strength, modulus of elasticity, thermal conductivity, and heat capacity. If the part has a large residual stress, the application of heat may cause severe strains. Annealing the part may be of value. In designing parts one must consider long term creep properties. For example, if a hollow handle is held by a screw through its length and maintained at a high temperature for a long period of time, the material may creep to relieve the stresses. When the handle is brought to a lower temperature, it will become loose.

Linear expansion is usually measured in a quartz-tube dilatometer (ASTM D 696). A test specimen is placed in a quartz-tube and heated. A dial indicator measures its linear expansion. The cubical expansion is measured by immersing the specimen in a long tube of mercury and measuring the change in volume of the mercury in a capillary section of the tube (ASTM D-864).

Specific Heat

The specific heat or heat capacity of a material is defined as the amount of heat required to raise a unit mass of the material 1°. It is expressed as Btu/(lb)(°F) or cal/(g)(°C). They are based on the heat capacity of water which is 1.000 in either system. Numerically, therefore, the specific heat is the same in each system. Most materials have a lower specific heat than water. The specific heat at room temperature for ABS is 0.25, polystyrene 0.3, and polyethylene 0.5. The specific heat is important in that it determines the amount of heat required to raise the plastic to molding temperature and the amount of heat to be removed so that it can be removed from the mold.

The specific heat varies with the temperature. In crystalline materials it peaks

c

at the melting point. In both amorphous and crystalline materials a plot of the specific heat versus temperature changes slope at a transition point. A more useful value is the enthalpy or heat content, which is an integration of the specific heat and temperature rise. For example, if the enthalpy of polystyrene at 360° is 217 Btu/lb and at 80° is 22 Btu/lb, the amount of energy required to raise the temperature from 80 to 360°F would be 195 Btu/lb. Therefore, raising 1 lb of polystyrene from 80 to 360°F in 1 hr would require 57 watts.

Tables of enthalpy for high density polyethylene, polypropylene, ethylene-propylene copolymer, acrylic, polyvinyl chloride, GP polystyrene, nylon 6/6, nylon 6/10, polytetrafluorethylene, and polychlorotrifluorethylene are found in Ref. 53. Enthalpy-temperature graphs for low and high density polyethylene and general purpose and high impact polystyrene are found in Ref. 77 of Chapter 3.

Thermal Conductivity

Thermal conductivity is the rate at which heat is transferred by conduction through a unit area of unit thickness, when a temperature gradient exists perpendicular to the area. It can be measured in a number of ways (54). The guarded hot plate method (ASTM C-177) uses guard heaters to control the flow of heat so that a zero temperature gradient exists in all directions except the one of desired heat flow. The coefficient of thermal conductivity (k) is the amount of heat transferred (Btu/hr) for a square foot of surface material, 1 in. thick, per °F temperature difference between the two sides. The coefficient of thermal conductivity (k) in these units (Btu) (in.)/(hr)(°F)(ft^2) for some materials are:

	(Btu)(in.)/(hr)(°F)(ft^2)	(cal)(cm)/(sec)(°C)(cm^2) x 10^{-3}
Aluminum alloy	840	289
Beryllium copper	800	276
Brass	720	248
Steel	310	107
Window glass	6.0	2.07
Nylon	1.7	0.59
Polycarbonate	1.4	0.48
Acrylic	1.3	0.45
Wood	1.2	0.41
Asbestos sheet	1.2	0.41

Metals are good conductors of heat because of the availability of valence electrons to transfer energy. Because plastics do not have this mechanism to any large degree, they are excellent insulators. It is obvious, too, that aluminum, beryllium, and brass molds are much more efficient than steel in removing heat.

The rate of heat transfer is given by Fourier's law.

$$q = (k)(A) \ \frac{(\Delta T)}{x} \tag{4-2}$$

where

q = rate of heat transfer (Btu/lb)
k = coefficient of thermal conductivity (Btu)(in.)/(hr)($^{\circ}$F)(ft^2)
A = area (ft^2)
ΔT = temperature differential ($^{\circ}$F)
x = thickness (in.).

Fourier's equation applied for steady-state conditions which is usually the case when plastics are used as insulating material. The rate of heat transfer of the plastic is the determining factor in processing time. Here steady-state conditions no longer exist (55). The thermal conductivity changes at transition points. It is temperature dependent, particularly for crystalline materials. Amorphous polymers show relatively little change. One would expect that the thermal conductivity increases with pressure as the molecules are closer together. At 300°F the thermal conductivity of polystyrene at zero pressure is 1.75, while at 16,000 psi it is 1.9.

The thermal diffusivity measures the rate at which a temperature disturbance moves from one point to another in a body. It is the thermal conductivity divided by the product of the density and specific heat at constant pressure.

$$a = \frac{k}{\rho \, C_p} \tag{4-3}$$

where

a = thermal diffusivity
k = coefficient of thermal conductivity
ρ = density
C_p = specific heat at constant pressure.

The variations of thermal conductivity and specific heat with temperature for many materials is given in Ref. 56.

Melting and Softening Points

The melting points has been discussed previously (p. 180). For semicrystalline materials it is determined by the disappearance of birefringence when viewed under a microscope (ASTM D-2117). In amorphous materials there is no melting point but a glass transition temperature (p. 233). Both are rough indications of the ultimate use temperature of the plastic (196).

The Vicat softening point test was developed for the polyethylenes and has been used for other materials. The temperature is that at which a flat ended cylindrical needle (1-mm^2 cross section) under a 1000-g load (1422 psi) penetrates 1 mm. The temperature is raised 120°C/hr (ASTM D-1525). When used on harder material they correlate with the deformation temperature under load (ASTM D-648). The Vicat temperatures are in effect the maximum short term temperatures to which a part made of plastic should be subjected under no load. They do not relate at all to long term heat stability. They do indicate relative temperature limits for different materials or grades of a particular material. Typical Vicat temperatures are

	°F	°C
Nylon 6/6	485	252
Acetal	338	170
Polypropylene	300	149
Acrylic	240	116
Polyethylene HD	260	127
Polyethylene MD	230	110
Polyethylene LD	185	85
Polystyrene GP	220	104
Cellulose acetate	220	104

Deflection Temperature of Plastics Under Load

One of the most thoroughly misunderstood tests is the deflection temperature under load (DTL) (ASTM D-648) formerly known as the heat distortion temperature. A plastic sample is placed on supports 4 in. apart. A load to produce either 66 or 264 psi outer fibre stress is placed on the center. The temperature is raised 2°C/min and the DTL is reported at that temperature where the bar has deflected 0.010 in. While this is reported as a thermal property it is in effect a mechanical one. It is the temperature at which the flexular modulus is 35,000 psi when the applied stress is 66 psi and 140,000 psi when the applied stress is 264 psi. Just like the melt index, it is a measure of a property at a particular point. It does not differentiate between the combined effects of creep, thermal history, and change of the modulus with the temperature. To be useful these factors must be known separately. For example, the olefins have good DTL under low loads, but they drop very rapidly as the load increases. It is useful if one is comparing materials under the specific test conditions. It is primarily used for research and control purposes.

Brittleness Temperature

The specimen size for brittleness tests is 1 1/4 in. long by 1/4 in. wide by 0.075 in. thick. It is held in a vise and struck by an arm under specified conditions (ASTM D-746). The temperature at which 50% failure occurs is recorded as the brittleness temperature. The test is valuable in comparing the relative merits of different plastics at low temperature. It does not signify, at all, the lowest temperature limits for the use of the plastic.

FLAMMABILITY

The rapid expansion of the use of plastics in construction, clothing, home furnishings, appliances, transportation, and packaging has led to a proper emphasis on their flammability properties. Both the Federal government and consumer agencies have become directly involved. The relative newness of the material, its potential threat to other materials, and its economical methods of fabrication have added emphasis to this movement. The safety of people and property has always been overwhelming concern of the American plastic industry.

The flammability of a material is related to two distinct properties—ignition and burning. Ignition describes how easily and under what conditions the material ignites. Burning describes the rate of surface flame spread, amount of heat released, amount of smoke released, and toxicity or nontoxicity of the by-products (57, 58). All these are important in determining the total hazard of a material. Until recently the primary concern has been the burning rate as it relates to construction materials.

Ignition Temperatures

The basic work on the ignition temperatures of plastic was done in a hot air furnace developed by Setchkin (59). He differentiated between the two types of ignition temperatures. The *flash ignition temperature* is the lowest initial temperature of air passing around the specimen at which sufficient combustible gas is evolved to be ignited by a small external pilot flame or heat source. The *self-ignition temperature* is the lowest initial temperature of air passing around the specimen at which, in the absence of an ignition source, ignition occurs of itself, as indicated by an explosion, flame, or sustained glow. Ignition temperatures of plastics and other materials from Ref. 60 are shown below.

	Flash Ignition Temp.		Self-Ignition Temp.	
	°F	°C	°F	°C
Polystyrene	680	360	925	496
Polyethylene	645	341	660	349
Ethyl cellulose	555	291	565	296
Nylon	790	421	795	424
Polyester-glass fiber	750	399	905	485
Styrene acrylonitrile	690	366	850	454
Styrene methyl methacrylate	625	330	905	485
Polyvinyl chloride	735	391	850	454
Polyvinylidene chloride	>990	>532	>990	>532
Urethane foam (rigid polyether)	590	310	780	416
Polystyrene beads	565	296	915	491
Polystyrene beadboard	655	346	915	491
White pine (shavings)	500	260	500	260
Paper news print (cuts)	445	230	445	230
Cotton batting	490	254	490	254

Flammability Tests

Nonelectrical flammability tests use an open flame and a test strip. In one test (ASTM D-635) the specimen is 5 in. long, 1/2 in. wide, and about 1/4 in. thick. In this test, the strip is held horizontally with the 1/2 by 1/4 section making a 45° angle with the horizontal. The specimens are inscribed 1 and 4 in. from the unsupported end. A 3/8 in. Bunsen burner with a 1 in. flame is placed to contact the end of the specimen and held for 30 sec. The burner is removed. If the specimen does not ignite it is classified as nonburning. If the specimen burns partially and extinguishes itself it is called self-extinguishing and the length of the burned portion reported. If the specimen burns it is classified as burning and the 4 in. mark is used to calculate the rate of burning.

UL uses a more severe test in that the strips are extended vertically. A material is classified self-extinguishing as follows: if the material extinguishes itself within 5 sec and does not drip flaming particles it is classified SE-O. If it extinguishes itself within 25 sec and does not drip flaming particles it is classified SE-1. If it extinguishes itself within 25 sec but releases flaming particles or drip during that time, it is classified SE-II. UL has ratings for ignition due to electrical sources. They are for hot-wire ignition, high-current-arc ignition, high-voltage-arc ignition, and high-voltage-arc tracking rate (61).

UL has developed a temperature index for specific materials which establish a general working temperature limit for materials in a large number of electrical applications. These temperatures are used as a basis for evaluating a particular application and are not, therefore, specific for all applications (62). Reference 63 has a UL temperature index chart for materials listed by their generic name and supplier. Reference 64 has an excellent listing of thermoplastic and thermosetting materials, by generic material, trade name, designation, and description. The results of the UL flammability tests, ASTM flammability tests, and arc tracking tests are given. Other flammability tests have been developed which are more usable for experimental work and determining the mechanisms of flammability (65). A summary of smoke and flame testing activity is given in Ref. 65a.

Flame retardancy in plastics is either a function of the basic polymer, where exposure to heat and oxygen will not support combustion, or is obtained by altering a polymer to increase its flammability resistance or adding a flame retarding compound. They are primarily compounds of phosphate, chlorine, bromine, and antimony and are used many times in combination. The phosphates are used as plasticizers and flame retardants for the vinyls and cellulosics. They are sometimes used in acrylics, nylons, styrenes, polyesters, epoxies, phenolics, alkyds, and urethanes. Antimony oxide combined with chlorinated paraffins are used for olefins, styrenes, epoxies, and urethanes. These are some typical examples. The high cost of flame retardant additives has accelerated research (66-71). An excellent chart of flame retardants and the materials in which they are used is found in Ref. 72.

Polymer Decomposition

The study of polymer decomposition is important in understanding flammability, degradation mechanisms, long term aging, and the development of new material (73, 74). It is complicated because the plastic is never identical and often very difficult to purify. Molecular weight and other determinations are painstaking and difficult. References 75 to 79 are typical. A brief survey of the developments of more than 16 heat resistant organic polymers and literature references are found in Ref. 80.

A common question is "how high a temperature will this part stand?" It is not possible to answer this theoretically. The mechanical and physical environments, method of manufacture, and original specification of the material all have a bearing. As important is the definition of what properties are important. The complexity of the problem is illustrated in Ref. 81 where the tensile strength in two thicknesses, tensile impact strength, notched Izod impact strength, dielectric strength in two thicknesses, arc resistance, flammability, and dimensional stability of polysulfone was studied at high temperatures to 410°F. One year data were extrapolated by Arrhenius type plots. Even with this

Table 4-6 Thermal properties of plastic materials (these are typical values and vary with the different grades of material

	Deflection Temp (°F) 264 psi 66 psi D 648		Deflection Temp ((°C) 18.6 kg/cm² 4.64 kg/cm²		Thermal Conductivity[a] C-177	Thermal Expansion (in.)(in.) (°C) x 10⁻⁵ D-696
ABS	200	210	93	99	7	8
ABS 20% GR	230	240	110	116		3
Acetal copolymer	230	316	110	158	5.5	8.5
Acetal 20% GR	320	325	160	163	6	
Acetal homopolymer	255	330	124	115	5.5	8.1
Acrylic	180	190	82	88	7	7
Modified acrylic MMA	221	239	105	116	5	5.4
Cellulose acetate	140	150	60	66	6	13
Cellulose acetate butyrate	160	170	71	77	6	14
Cellulose proprionate	170	200	77	93	6	14
Ethyl cellulose	140	150	60	66	5.5	13
Chlorinated polyether	210	285	99	141	3.1	8
CTFE		258		126	5	6
Polyvinylidene fluoride	195	300	91	149	3	8.5
FEP		160		71	6	9
Ionomer	100	110	38	43	5.8	12
Nylon 6	150	330	66	166	6	8.3
Nylon 6 30% GR	390	420	199	216	6	4
Nylon 6/6	185	395	85	202	6	8
Nylon 6/6 30% GR	480	490	249	254	5	2
Nylon 11	180	300	82	149	7	15
Nylon 6/10	180	300	82	149	5	9
Modified phenylene oxide PPO	240	260	116	127	5	5.2

359

Table 4-6 (Cont.)

	Deflection Temp (°F) 264 psi / 66 psi D-648		Deflection Temp (°C) 18.6 kg/cm² / 4.64 kg/cm²		Thermal Conductivity[a] C-177	Thermal Expansion (in.)(in.)(°C) x 10^{-5} D-696
Modified phenylene oxide 30% GR	300	310	149	154	4	2.2
Polyaryl ether	300	320	149	160	7	6.5
Polyarylsulfone	525		274		4.6	4.7
Polycarbonate	275	280	135	138	4.6	6.6
Polyethylene LD	100	112	38	44	8	17
Polyethylene MD	115	142	46	61	9	15
Polyethylene HD	120	165	49	74	11.5	12
Ethylene vinyl acetate (EVA)	93	146	34	63		18
Polypropylene	132	215	57	102	2.8	8
Polypropylene 30% GR	280	310	138	154		
Polypropylene copolymer	127	206	53	96	3	8.5
Polyallomer	130	175	54	80	3	9
4-Methyl pentene-1		239		115	4	12
Polystyrene GP	175	195	80	91	3	3
Polytetramethylene terephthalate (PTMT)	122	302	50	150		9
Polystyrene GP 30% GR	223	231	106	111		4.5
Polystyrene HI	200	207	93	97	2	4-20
Styrene-acrylo-nitrile (SAN)	205	215	96	102	2.9	3.7
SAN 20% GR	230	240	110	116		2.7

Table 4-6 (Cont.)

	Deflection Temp (°F) D-648		Deflection Temp (°C)		Thermal Conductivity[a] C-177	Thermal Expansion (in.)(in.)(°C) x 10^{-5} D-696
	264 psi	66 psi	18.6 kg/cm²	4.64 kg/cm²		
Styrene-butadiene					3.6	13.5
Rigid polyvinyl chloride (PVC)	160	165	71	74	4.	7
Chlorinated PVC	218	232	102	111	3.3	7.2
PVC/PP	160		71			
Vinylidene Choride	140		60		3.	19
Phenol formaldehyde (no filler)	250	260	121	127	4.5	5

[a]Thermal conductivity [(cal) (cm)/(sec) (cm²) (°C)] x 10^{-4}

information, the design engineer would essentially be taking an educated guess on predicting long term heat aging properties of a particular part for a particular application. This illustrates some of the problems the industry must solve to utilize fully the engineering potential of plastics.

FLOW PROPERTIES

The flow properties of plastics are measured with extrusion rheometers (Figure 3-10) and instruments similar to the melt indexer used for the olefins (Figure 3-6). The limitations of this test are essentially that it measure the flow properties at very low shear rates which may not be typical of its properties under processing conditions (p.207). Flow properties can also be measured by dissolving the plastic in a suitable solvent and determining the viscosity of the solution.

ELECTRICAL PROPERTIES

The fine insulating properties of plastics have resulted in plastic's application in electrical and electronic devices. Cables, capacitors, plugs, sockets, switches, electrical appliance housings, TV, radio, radomes, and microwave applications are some of the major usages. Because of this sparcity of free electrons, plastics are effective insulators. (Some semi-conducting polymers have been developed however.) As with all tests, they represent results under special conditions, and have to be interpreted for a special application. Electrical tests are particularly sensitive to temperature, humidity, mechanical stress, frequency, geometry of the electrodes, geometry of the parts, and contamination.

The electrical properties of plastic can be classified as:
Dielectric strength.
Dielectric constant.
Dissipation factor.
DC resistance
Arc resistance.

Some of these properties for different plastic materials will be found in Table 4-7.

Dielectric Strength

One of the most important electrical properties of plastics is its ability to withstand voltage impingement. It is defined as the voltage difference per unit of thickness which will cause catastrophic failure of the dielectric material. It is measured by placing a conditioned specimen between two heavy cylindrical brass electrodes (ASTM D-149). In the short time test the voltage is increased at a uniform rate (0.5 - 1.0 kV/sec). In the step-by-step test the initial voltage is

50% of the breakdown voltage shown in the short time test. In both instances the rate of increase of the voltage is specified according to the material. Breakdown occurs when there is a sudden flow of current through the specimen. The breakdown is caused by a combination of electrical discharges, thermal effects, and the intrinsic degradation of the plastic. Dielectric strength is very dependent upon thickness. Generally speaking values drop approximately as the reciprocal of the square root of the thickness. The dielectric strength decreases with the time of application of the voltage. This is probably due to the gradual breaking down of the plastic. Humidity tends to reduce dielectric strength with some plastics such as glass filled melamine and is insignificant with others, such as polyethylene. The dielectric strength generally decreases rapidly with increasing AC frequency. The 60-Hz figures are not usable at higher frequencies.

Dielectric Constant

The dielectric constant is a measure of the material's ability to store an electrical charge. It is the ratio of the capacity of the condenser made with the plastic over the capacity of an identical condenser using air (ASTM D-150). Dielectric constant varies according to frequency, temperature, and humidity, depending on the particular plastic. For insulating purposes a low dielectric constant is desirable. This reduces the power loss through the insulator. In condenser applications a high value is desired to reduce the physical size of the capacitor.

Dissipation Factor

The dissipation factor also measured by the test above is the ratio of the in-phase power (real power) to the 90° out-of-phase power (reactive power). Dissipation factor is a measure of the conversion of the reactive power to real power which shows up as heat. In all applications it is desirable to keep the dissipation factor which is strongly affected by temperature, humidity, impurities, and frequency as low as possible. The loss factor is a product of the dissipation factor and dielectric constant. It is proportional to the energy loss in the dielectric.

DC Resistance

Electrical resistance is reported as the volume resistivity, surface resistivity, and a combination of both - insulation resistance (ASTM D-257).

The volume resistivity is defined as the resistance in ohms between two opposing faces of a 1 - cm cube. The surface resistivity is the resistance between two electrodes on the surface of an insulating material. It is reported in ohms per square centimeter. The insulation resistance is the ratio of direct current voltage applied to the electrodes to the total current between them. All three are temperature, frequency, and voltage dependent.

Table 4-7 Electrical properties of plastic materials (these are typical values and vary with the different grades of materials.

	Volume Resistivity ohm/cm 50% RH 23 °C	Dielectric Strength Volts/mil	Dielectric Constant 60 hz	Power Factor 60 hz
ASTM	D 257	D 149	D 150	D 150
ABS	3×10^{16}	400	2.5-5	0.006
ABS 20% GR	1×10^{16}	500	3.2	0.006
Acetal copolymer	10^{14}	500	3.7	0.006
Acetal copolymer 20% GR	10^{15}	500	3.9	0.004
Acetal homopolymer	10^{15}	380	3.7	0.005
Acetal homopolymer 20% GR	10^{14}	400	3.7	0.005
Acrylic	10^{14}	450	3.6	0.04
Modified acrylic MMA	10^{18}	500	3.0	0.03
Cellulose acetate	10^{13}	500	6	0.03
Cellulose acetate butyrate	10^{13}	300	5	0.02
Cellulose proprionate	10^{14}	350	4	0.03
Ethyl cellulose	10^{13}	450	3.6	0.01
Chlorinated polyether	10^{13}	400	3.1	0.01
CTFE	10^{18}	550	2.6	0.001
Polyvinylidene fluoride	2×10^{14}	260	8.4	0.05
FEP	2×10^{18}	550	2.1	0.0003
Nylon 6	10^{15}	480	4.2	0.05
Nylon 6 30% GR	10^{15}	510	4.4	0.09
Nylon 6/6	10^{15}	480	4.4	0.03
Nylon 6/6 30% GR	10^{14}	485	4.2	0.006
Nylon 11	4×10^{13}	425	3.3	0.03

Table 4-7 (Cont.)

ASTM	Volume Resistivity ohm/cm 50% RH 23 °C D 257	Dielectric Strength Volts/mil D 149	Dielectric Strength 60 hz D 150	Power Factor 60 hz D 150
Nylon 6/10	10^{14}	500	3.9	0.04
Nylon 6/10 30% GR	10^{17}	500	4.2	0.02
Polysulfone	5×10^{16}	425	3.1	8×14^{-4}
Modified phenylene oxide PPO	10^{16}	475	2.6	0.0004
Modified phenylene oxide PPO 30% GR	10^{17}	510	2.9	0.0009
Polyaryl ether	1.5×10^{16}	400	3.1	0.006
Polyarylsulfone	3×10^{16}	350	3.9	0.003
Polycarbonate	2×10^{16}	400	3	0.0009
Polycarbonate 30% GR	6×10^{15}	480	3.5	0.001
Polyethylene LD	10^{16}	800	2.3	5×10^{-4}
Polyethylene MD	10^{16}	900	2.3	5×10^{-4}
Polyethylene HD	10^{16}	500	2.3	5×10^{-4}
Polyethylene high MW	10^{16}	700	2.3	5×10^{-4}
Ethylene vinyl acetate (EVA)	10^{15}	700	3	0.01
Polypropylene	10^{16}	560	2.4	5×10^{-4}
Polypropylene copolymer	10^{17}	560	2.3	2×10^{-4}
Polyallomer	10^{15}	825	2.3	5×10^{-4}
4-Methyl pentene -1	10^{16}	800	2.1	7×10^{-5}
Polystyrene GP	10^{16}	600	2.5	1×10^{-4}
Polystyrene GP 30% GR	10^{16}	425		0.001

Table 4-7 (Cont.)

	ASTM	Volume Resistivity ohm/cm 50% RH 23 °C D 257	Dielectric Strength Volts/mil D 149	Dielectric Constant 60 hz D 150	Power Factor 60 hz D 150
Polystyrene GP HI		10^{16}	450	3.7	0.006
Styrene-acrylo-nitrile (SAN)		10^{16}	450	3.	0.007
Styrene-acrylo-nitrile (SAN) 20% GR		10^{17}	510	3.6	0.006
Styrene-butadiene		10^{14}	470	3	0.002
Rigid polyvinyl chloride (PVC)		10^{16}	600	3.4	0.01
Chlorinated PVC		10^{15}	1400	3.	0.02
PVC/PP		10^{15}	350	3.4	0.009
Vinylidene chloride		10^{15}	500	5	0.04
Urethane		10^{5}	400	8.2	0.08
Phenol-formaldehyde (no filler)		10^{12}	350	6	0.05

Arc Resistance

The arc resistance test (ASTM D-495) gives values of little use for design purposes. They are satisfactory for broadly comparing the arc resistance rankings of different materials. It is a high voltage, low current, arc type test, attempting to approximate service conditions. The general types of failure are incandescence (which carries the current until it cools again), flammability, tracking, and carbonization until sufficient carbon is present to carry the current. Corona insulation failure is due to the continuous erosion of the insulator caused by ionized air in voids existing in the insulation, which are induced by the current in the conductor (81a).

ENVIRONMENTAL PROPERTIES

The environmental properties of plastics consist of

Environmental stress cracking.
Water absorption.
Water and gas permeability.
Weathering.
Ultraviolet (UV) resistance.
Resistance to bacteria.
Chemical resistance.

These properties are related to the chemical structure of the polymers (82).

Environmental Stress Cracking

When a plastic reaches a given stress level it will crack or fail. This level will vary for a given material depending on the temperature, time, and environment. (Environmental effects are not peculiar to plastics, but appear in glasses, metals and ceramics). A stress cracking environment is one which has no effect on the surface or properties of the part when there is no stress. When a stress is applied the environment will cause cracking and/or affect other mechanical properties at a reduced stress. While internal stresses caused by excessive orientation, weld lines, laminations, and contamination will cause failure, they are not included in this category. It is restricted only to the effect of the environment. The samples are tested either under constant strain or constant stress. Many different methods have been devised to expose the test specimen, and mechanical properties of the plastic are usually measured. Further discussion of this subject is beyond the scope of this section. This test has particular significance in the olefin family of plastics. More information is available in Ref. 83 to 89.

Water Absorption

The moisture content of a plastic is related to its electrical and mechanical properties and its dimensions. Most plastics are not hygroscopic. The moisture content is measured by drying a sample, weighing it, immersing it in water at a given temperature, reweighing it, and reporting the percentage increase in weight (ASTM D-570). Nylon, in particular, changes its properties and dimensions with its water content (90).

Permeability

Permeability is important in packaging applications. The amount of gas permeability depends on the nature of the polymer molecule, its crystallinity (the amorphous material is more permeable), the nature of the gas molecule, the composition and size of the barrier, the temperature of the system and the pressure difference across the barrier. The transmission rate varies exponentially with the reciprocal of the absolute temperature. The most common way of measuring this property is through the PVT relationship which is the basis of ASTM D-1434. One unit of measurement for describing permeability (gas transmission rate - GTR) is (cm^3) $(mil)/(100 in.^2)$ $(24 hr)$ (atm) or the volume of gas penetrating through 100 in.^{2} of a given film thickness in 24 hr at a given pressure differential.

Water vapor permeability is measured by putting the material to be tested over the mouth of a dish, which contains either water or dessicant. The assembly is put in a controlled humidity and temperature cabinet and the change in weight is used to calculate the water vapor movement through the material (ASTM E-96). Results are expressed as grams of water transmitted in 24 hr through 100 in.^{2} of exposed area $g/(24 hr)(100 in.^2)$.

Weathering

Some of the factors included in weathering are the ultraviolet and infrared radiations of the sun, oxygen, ozone, atmospheric pollutants, erosion caused by dust, by rain and by hail, and temperature cycling. Obviously this conglomeration of properties will be difficult to measure and to describe. The standard test procedure puts the specimens at a 45° angle facing south. The report describes the weather, exposure area, period of time, and any other pertinent information. After a preselected time the properties of interest are measured and compared against those of the specimen before it was exposed (ASTM D-1435). Synthetic laboratory accelerated weathering tests (ASTM E-42) subjects the specimen to carbon arc or xenon arc light and have cyclic changes in conditions of temperature, relative humidity, and UV energy, with or without direct water spray. There is a general agreement between data obtained from this test and outdoor weathering tests, which permit qualitative conclusions about existing or new materials. Another test is the Atlas Type Fadometer ® which is primarily used to check color stability. Typical studies on the weathering properties of materials and weather testing equipment are given in Ref. 91 to 99e and 192.

Ultraviolet degradation mainly comes from sunlight and fluorescent bulbs. The radiation causes free radical formation and catalyzes the decomposition. Evidence of failure is pitting, crumbling, surface cracking, crazing, brittleness and breakage.

Resistance to UV is increased by adding chemical stabilizers (100-103, 193-195). Generally, ultraviolet absorbers (or stabilizers) absorb energy in the UV range and dissipate it harmlessly, and retard or eliminate the initiating action of the radiation. Stabilizers such as carbon black can act as a screen by making the material opaque and hence limiting the depth of UV penetration. This limits the degradation to a thin layer on the surface. An additive which will block the harmful results of the free radical formation or other chemical mechanism can also be used.

There are other additives which will stabilize plastic against the effects of heat, oxygen, and ozone (104). Antistats, materials added to reduce the collection of static dust, are successful for some materials, particularly the olefins. Brighteners, internal lubricants, plasticizers and, of course, colorants can also be added to the material.

Biochemical Environments

Most plastics are resistant to bacteria and fungi. Some, particularly the thermosetting, are not. Additionally, plasticizers, stabilizers, and other additives may not be resistant. This is the case for polyvinyl chloride (105). Because of the use of plastics in outdoor environments, such as for pipe, building materials, transportation, and the military, microbiologic resistance is important (106, 107). The biocide is added to the polymer in processing (108). Biocides are used, for example, in molding dishes and in the cosmetic industry. Such additives that are used in food or cosmetic products may require FDA approval.

Chemical Resistance

The chemical resistance of plastics are measured by immersing the specimens in the chemical to be tested for a given time (usually 7 days), and then reporting changes in weight, appearance, dimensions, and strength properties (ASTM D 543). This is a test under no stress. The manufacturer of each material has data sheets or booklets describing its resistance to standard agents or other products. The major chemical resistance characteristics are given in Table 4-8 and will be described for each material in the following section. The chemical and thermal environmental resistance of many glass reinforced thermoplastics is given in Ref. 108a.

Additives

Additives are added to plastics to improve processability and to improve properties of the plastics (108b, 108c). A good review of additives is found in Ref. 108d. These additives are also discussed in different parts of this book as they relate to the particular subject being discussed, for example, UV stability, reinforcement, and colorants.

Lubricants which are not discussed elsewhere are generally added to improve the flow characteristics of the polymer during processing. In that instance, the lubricants act internally. It is also possible to lubricate so that it will affect the external lubricity of the plastic. In general, long carbon chains will not be readily soluble in the plastic and act as external lubricants. Short carbon chains are more readily dispersed and act as internal lubricants. Lubricants can be very helpful when molding problems are caused by marginal filling and flow properties (51, 109, 191). Lubricants also aid mold release.

AGING

Aging of plastics material may be defined as the slow, irreversible changes with time only in a direction that is usually detrimental to its usefulness. This seems to be true of all organic matter, including living organisms. The irreversibility of the properties caused by aging is permanent and is not recovered when the cause is removed. For this reason water absorption by nylon is not called aging, although it may occur over a long period of time. The nylon can recover its original properties if the water is removed at the end of the "aging" period. A discussion of aging is not within the scope of this book. References 110 to 113 discuss some of its aspects.

NON-DESTRUCTIVE TESTING

The tests described previously, which are directed toward assessing the relative merits of different plastics in comparative terms, are essentially destructive tests, with no direct link between the test results and serviceability of the part. Furthermore, different samples which are not identical have be be used for different tests.

Consequently, various nondestructive mechanical tests have been developed, some of which give information both on the molecular structure and physical properties of the material. The dynamic properties of polymers depend on the movements of the segments of the molecules when subjected to a stress. If the stress is applied sinusoidally, the resultant strain will be sinusoidal but may lag should be consulted; (see Appendix for abbreviations)

Normally unaffected by

Water	Alcoholics Ketones	Gasoline	Weak Acids	Weak Alkalis	Strong Acids	Strong Alkalis	Detergents Grease Oil
ABS	C1-polyether	Acetal	ABS[a]	ABS	ABS[a]	ABS	ABS
Acetal	CTFE	C1-polyether	C1-polyether	C1-polyether	CTFE	C1-polyether	Acetal
C1-polyether	FEP	CA	CTFE	CTFE	FEP	CTFE	CA
CTFE	Ionomer	CAB	FEP	EC	4-Methyl-pentene[a]	FEP	CAB
EC	4-Methylpentene	CTFE	4-Methylpentene	FEP	PE	4-Methylpentene	C1-polyether
FEP	PA	FEP	PA	Ionomer	Polysul-fone		EC
Ionomer	PC	PA	PC	4-Methyl-pentene	PP	PE	Ionomer
4-Methyl-pentene	PE	PC	PE	PA	PS[a]	Polysulfone	PA
PC	PP	PMMA	PMMA	PC	SAN[a]	PP	PC
PE	PS	PVC	Polysulfone	PE		PPO	PE
PMMA	PVC		PP	PMMA		PS	PMMA
Polysulfone			PPO	Polysulfone		PVC	Polysulfone
PP			PS	PP		SAN	PP
PPO			PVC	PPO			PVC
PS			SAN	PS			PVC
PVC				PVC			SAN
SAN				SAN			

[a]Except oxidizing acids.

behind the stress. From careful study of the spectra, information on th mechanical properties and their variation with molecular structure can b determined (114). The techniques fall into three types.

1. The free vibration test displaces a sample of the material slightly from it equilibrium position and releases it. The gradual decays of the oscillations ar measured.

2. Forced vibration with resonance applies a cyclic impulse at frequencie which embrace the resonant frequency.

3. Forced vibration but at the resonant frequency. Similar testing has bee done using sonic frequencies (115).

Nondestructive testing can also be done with microwaves which propagat through plastic and reflect and scatter from internal discontinuities and surface and interact with the molecular segments (116). Thickness gaging, flaw detection, chemical composition, moisture, orientation, and specific gravity ar some of the variables that have been measured. Another nondestructive tes method is available only for transparent materials. Birefringence has bee discussed previously (p. 253). It is relatively simple to build a light box with polarized material for testing and control use (117). Other methods includ ultraviolet and infrared spectrophotometry, differential thermal analysis, an thermogravinmotric analysis. The latter two are destructive tests but use ver small amounts of material.

MATERIALS DESCRIPTION

This section of the chapter surveys the materials commonly used in injectior molding. Their chemistry, preparation, properties, advantages and disadvantages major applications, and economic data is discussed only briefly since many good books have been published on each of this subject. Furthermore, any discussior is limited by the many variations of the many commercial materials. The reade is referred to the existing books, extensive literature, and the informatior available from the suppliers of raw material. The references at the ends o chapters 3 and 4 are useful for more information about those materials in which the reader is interested.

Two annual publications, the *Annual Processing Handbook of Plasti Technology*, and the *Modern Plastics Encyclopedia*, provide extensive curren information about materials. For more information, use the bibliography at th back of this chapter (p. 421). This literature is readily understood and does no require more than elementary chemistry.

The word polymer means many parts (poly = many; meros = parts). Polymer are characterized by high molecular weights and repeating chemical units. Th average molecular weight of commercial polymers range from the tens of thousands to a million. A monomer (mono = single) is the original reactant which is chemically transformed into a polymer. If a single type monomer is used the resulting product is called a homopolymer (polyethylene, polystyrene). If more than one monomer is used the resultant product is called a copolymer (acrylonitrile-butadiene-styrene).

High polymers occur naturally. Cellulose is a primary example. It provides the strength and structural properties for trees and the entire vegetable kingdom. Other natural high molecular weight materials include starch, wool, and proteins. Some of the latter differ from commercial high polymers in that they are of unique composition and have biological activity which permits them to be reproduced identically.

A material such as a metal or salt will have a molecular weight and properties (neglecting the effect of crystallization) which are identical regardless of the methods of manufacture. Such materials are called monodisperse. Polymers are examples of polydispersed materials. As will be seen when monomers are polymerized they do not react to form identical high molecular weight polymers. The number of monomer units that react together to form the polymer vary over a broad spectrum.

Many plastics form side chains. These vary in their location, length, and chemical configuration. Those polymers that cross link do so in an irregular manner. The spatial (stereo) configurations of some polymers differ widely. One might consider the material as a mixture of homologous polymeric constituents. In other words, a typical polymer is a mixture of different molecular weight materials with similar chemical configurations. Because of this, the properties of a polymer vary with the method and conditions of polymerization and processing.

Plastics

The word plastic is used to indicate the polymerized material with plasticizers, colorants, stabilizers, fillers, and additives which is ready for processing on production equipment (molding machines, extruders, formers, etc.) and the products that are derived therefrom. A good definition of a plastic material is "a material that in its finished state contains as an essential ingredient, a synthetic polymer of high molecular weight, is a flexible or rigid solid but not an elastomer in its finished state, and at some state in its manufacture or in its processing into finished articles can be shaped by flow or in-situ polymerization or curing" (See *Whittington's Dictionary of Plastics* in Bibliography, p. 421).

The characteristics of a plastic derive from the very large size of the molecule and the ability of its segments to rotate around their bonds. This permits entanglements which in a certain sense are physical or mechanical in nature. Secondary or Van der Waals' type bonding forces are also able to develop. The large size permits partial crystallization of certain polymers. Because of the many entanglements, orientation effects develop when the material flows. Finally, because many of these variables are controllable to some degree, plastics can be developed with specific properties.

Polymer Classification

Polymers can be classified as addition or condensation polymers (118). A small number of polymers are formed by a combination of condensation and addition. An addition reaction is one where the polymerization of the monomer does not yield any by-product. A condensation reaction yields a by-product of low molecular weight such as water or methanol.

Addition Polymers

Most plastics used today are addition polymers. They include polyethylene, polypropylene, polyvinyl chloride, polyvinyl acetate, polymethyl methacrylate, and polystyrene. It will be noted that these are all derivatives of ethylene. These are called vinyl monomers and are characterized by the general formula

$$
\begin{array}{ccc}
\text{H} \quad \text{R}_1 & & \text{H} \quad \text{R}_2 \\
\text{C}=\text{C} & \text{and} & \text{C}=\text{C} \\
\text{H} \quad \text{H} & & \text{H} \quad \text{R}_3
\end{array}
$$

where R_1 R_2 R_3 represent groups replacing H, such as Cl, Fl, ⬡⟩

All chain reactions have three stages – initiation, propagation and termination. The initiation of vinyl monomers is caused by free radicals. They usually come from the thermal or photochemical decomposition of certain organic compounds. They have an odd number of electrons so that one is not paired. When the appropriate free radical is introduced to vinyl monomers, it adds to the double bond forming a free radical on the other side of the double bond. This regeneration of an unpaired electron causes the chain reaction. Thus one free radical starts a chain which propagates a complete polymer molecule. Initiator efficiencies can reach 100%. One might expect that the chain reaction would continue until all the monomer was used up. This is not the case. One of two things will happen. In one instance the free radical of one chain will react with the free radical of another chain sharing their electrons and stopping the chain reaction (where the free radical attaches itself to the double bond of a vinyl monomer). This is called combination or or coupling. Since there is a very small number of free radicals compared to the available monomer the chances of the free radical attaching to a monomer are much greater than that of attaching to the other end of a free radical. This is why large polymers can be formed. The other method of terminating chain reactions is called disproportionation. This happens by the transfer of a hydrogen atom from one growing chain to another with the termination of one chain and the formation of an unsaturated group on the chain losing the hydrogen atom. There are other mechanisms as well, such as chain transfer, which lead to the formation of branched configurations.

There are other types of chain polymerization without the use of free radicals. The three major methods are cationic, anionic, and stereospecific (stereoregular) polymerization. Cationic polymerization uses Friedel-Craft catalysts ($AlCl_3$ $AlBr_3$ $SnCl_4$), which are strong electron acceptors. High polymerization rates and polymers of high molecular weights are characteristic of ionic polymerization. Anionic polymerization uses Grignard reagents, alkali metals, and alkali metal alkyls as catalysts. This process is primarily used for rubber and butadiene.

The third type use is what is commonly called Ziegler catalysts. They are primarily reduced transition metals supported on an inert base such as molybdenum oxide on an alumina support. The work of Ziegler and Natta resulted in the production of a large number of stereospecific polymers, including polypropylene. Polyethylene made by free radical methods required pressures of 1 to 3,000 atm and temperatures approaching to 250°C. These resulted in polymers with considerable branching. Using a Ziegler type catalyst it could be polymerized at normal pressures and temperatures and yield an almost linear polymer. Its composition could be altered by the choice of catalyst (119). For example, polypropylene can be made in three forms (Figure 4-3). The isotactic form has the CH_3 groups all on one plane which is essentially the composition of most commercial polymers. If the CH_3 units alternate regularly, one in the plane above and one in the plane below, it is called syndiotactic. Both these forms of polypropylene can crystallize. If the CH_3 groups are randomly located the polypropylene is called atactic and does not crystallize.

Condensation Polymers

Most of the condensation polymers are primarily the cross linked resins such as phenol-formaldehyde polymer.

Urea-formaldehyde plastics are formed by the reaction of one mol of urea and two mols of formaldehyde. They react together to form dimethyl urea. Under the appropriate conditions the dimethyl urea condenses with itself, eliminating water, resulting in a typical cross-linked network. The OH groups and the H groups attached to the nitrogen will continue to react until the whole mass is fused together.

Melamine and formaldehyde react to form dimethylolmelamine. Its reactant groups condense eliminating water and formaldehyde. The formaldehyde is used

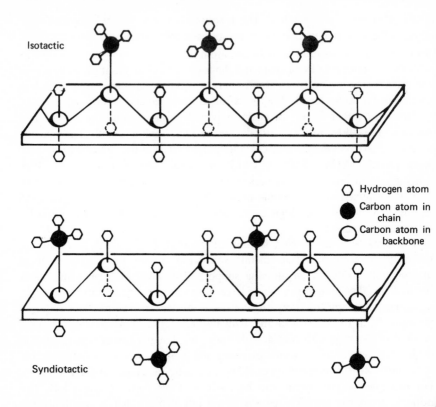

Figure 4-3 Isotactic and syndiotactic polypropylene. Isotactic form has the CH_3 units all on one plane. The syndiotactic form has the CH_3 units alternating from one plane to the other. The atactic form (not illustrated) has the CH_3 units randomly located.

in the reaction. The final polymer is cross-linked. For more examples of thermosetting condensation polymers, see "Thermosetting Materials" see p. 411.

Cellulose is a natural occuring polymer and the basis for many commercial plastics. Figure 4-4 shows the major commercial plastics derived from it. Figure 4-5 shows some of the formula for different nylons. The number and name is related to the number of carbon molecules in the repeating part of the chain. For example, nylon 6 has six carbon atoms. Nylon 6/10 has six carbons to the left of the central nitrogen and 10 carbons to its right.

The chemical formula for the repeating chain unit for some polymers with oxygen in their chain are shown in Figure 4-6. As has been noted most of the plastics used are derivatives of ethylene. Figure 4-7 shows carbon chain polymers based on ethylene with a single substitution for one hydrogen atom. Figure 4-8

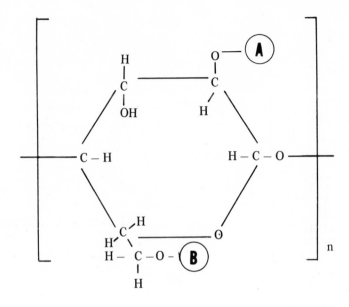

When Ⓐ is:	And Ⓑ is:	Plastic
H	H	Cellulose
$-\underset{O}{\overset{\parallel}{C}}-CH_3$	$-\underset{O}{\overset{\parallel}{C}}-CH_3$	Cellulose acetate
$-\underset{O}{\overset{\parallel}{C}}-CH_3$	$-\underset{O}{\overset{\parallel}{C}}-CH_2-CH_3$	Cellulose acetate proprionate
$-\underset{O}{\overset{\parallel}{C}}-CH_3$	$-\underset{O}{\overset{\parallel}{C}}-CH_2-CH_2-CH_3$	Cellulose acetate butyrate
$-\underset{O}{\overset{\parallel}{C}}-CH_2-CH_3$	$-\underset{O}{\overset{\parallel}{C}}-CH_2-CH_3$	Cellulose proprionate
$-CH_2-CH_3$	$-CH_2-CH_3$	Ethyl cellulose
$-NO_2$	$-NO_2$	Cellulose nitrate

Figure 4-4   Plastics derived from cellulose. *Note*: Third hydroxyl unit (OH) is partially ester-ied in final plastic.

377

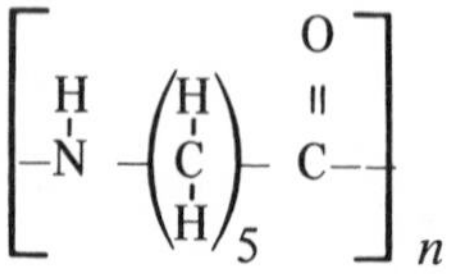

$$\left[\begin{array}{c} \text{H} \\ -\text{N} \end{array} - \left(\begin{array}{c} \text{H} \\ \text{C} \\ \text{H} \end{array}\right)_5 - \begin{array}{c} \text{O} \\ \| \\ \text{C} \end{array}-\right]_n$$

Nylon 6   Polycaprolactam (poly-6-amino caproic acid

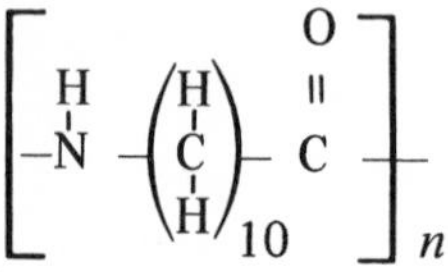

$$\left[\begin{array}{c} \text{H} \\ -\text{N} \end{array} - \left(\begin{array}{c} \text{H} \\ \text{C} \\ \text{H} \end{array}\right)_{10} - \begin{array}{c} \text{O} \\ \| \\ \text{C} \end{array}\right]_n$$

Nylon 11 Poly-11-aminoudecanoic acid

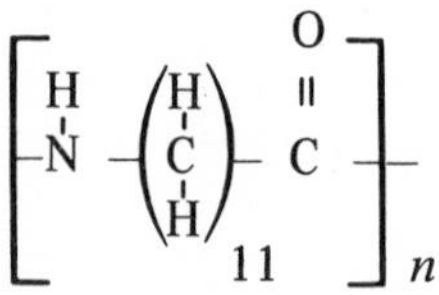

$$\left[\begin{array}{c} \text{H} \\ -\text{N} \end{array} - \left(\begin{array}{c} \text{H} \\ \text{C} \\ \text{H} \end{array}\right)_{11} - \begin{array}{c} \text{O} \\ \| \\ \text{C} \end{array}\right]_n$$

Nylon 12 Poly-12-aminolauric acid

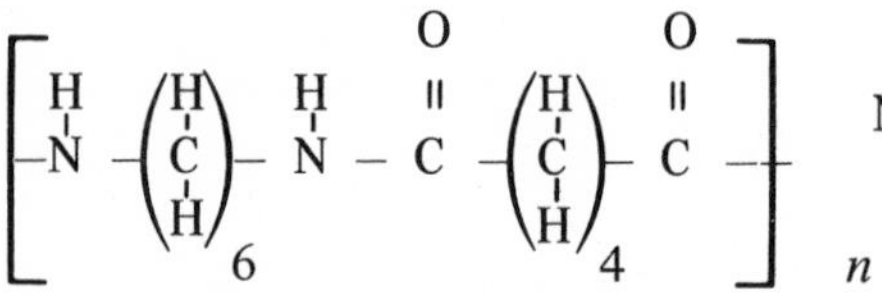

$$\left[\begin{array}{c} \text{H} \\ -\text{N} \end{array} - \left(\begin{array}{c} \text{H} \\ \text{C} \\ \text{H} \end{array}\right)_6 - \begin{array}{c} \text{H} \\ \text{N} \end{array} - \begin{array}{c} \text{O} \\ \| \\ \text{C} \end{array} - \left(\begin{array}{c} \text{H} \\ \text{C} \\ \text{H} \end{array}\right)_4 - \begin{array}{c} \text{O} \\ \| \\ \text{C} \end{array}\right]_n$$

Nylon 6/6   Polyhexamethylene adipamide

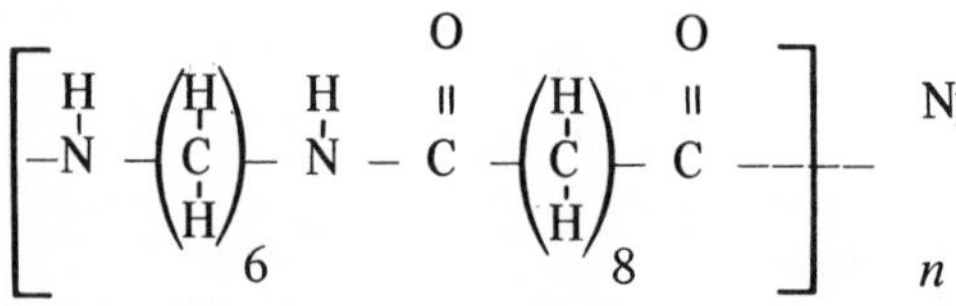

$$\left[\begin{array}{c} \text{H} \\ -\text{N} \end{array} - \left(\begin{array}{c} \text{H} \\ \text{C} \\ \text{H} \end{array}\right)_6 - \begin{array}{c} \text{H} \\ \text{N} \end{array} - \begin{array}{c} \text{O} \\ \| \\ \text{C} \end{array} - \left(\begin{array}{c} \text{H} \\ \text{C} \\ \text{H} \end{array}\right)_8 - \begin{array}{c} \text{O} \\ \| \\ \text{C} \end{array}\right]_n$$

Nylon 6/10   Polyhexamethylene sebacimide

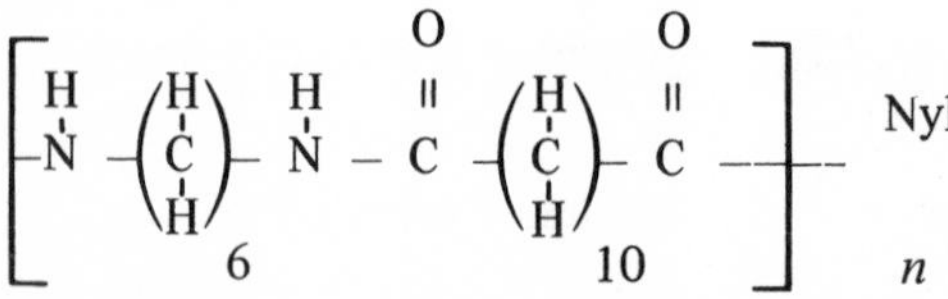

$$\left[\begin{array}{c} \text{H} \\ -\text{N} \end{array} - \left(\begin{array}{c} \text{H} \\ \text{C} \\ \text{H} \end{array}\right)_6 - \begin{array}{c} \text{H} \\ \text{N} \end{array} - \begin{array}{c} \text{O} \\ \| \\ \text{C} \end{array} - \left(\begin{array}{c} \text{H} \\ \text{C} \\ \text{H} \end{array}\right)_{10} - \begin{array}{c} \text{O} \\ \| \\ \text{C} \end{array}\right]_n$$

Nylon 6/12   Polyhexamethylene laurinamide

Figure 4-5   Formula for some nylon polymers. The number of the nylon corresponds to the number of carbon atoms in the chain of the repeating unit. Nylon 6 has six carbon atoms in the repeating unit. Nylon 6/12 has six carbon atoms on the "left" "side" and twelve on the "right" side of the "N" in the repeating unit.

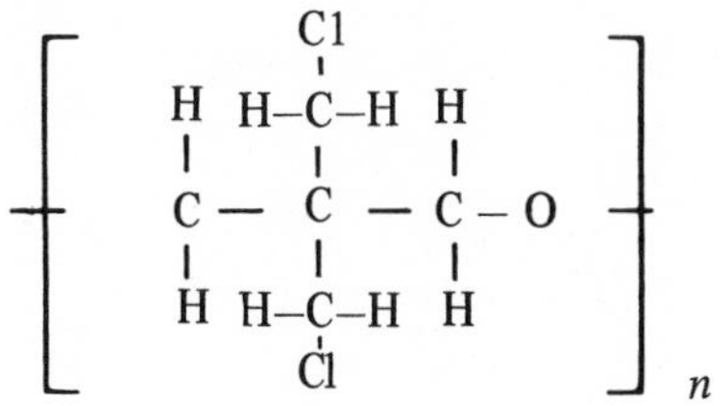

Chlorinated polyether

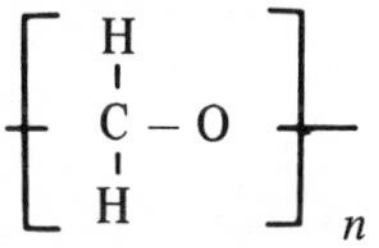

Polyacetal (homopolymer)

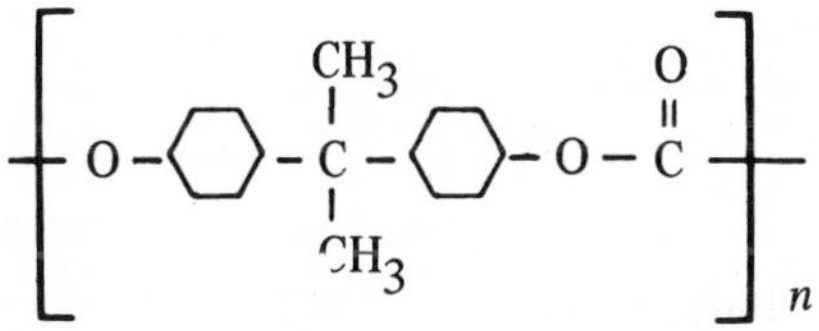

Polycarbonate

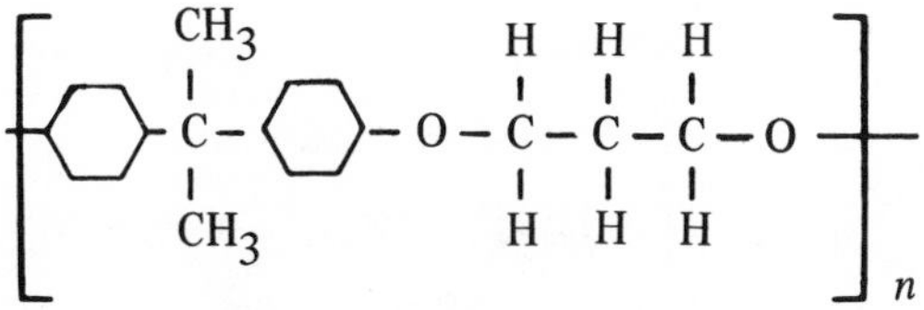

Polyhydroxyether (phenoxy)

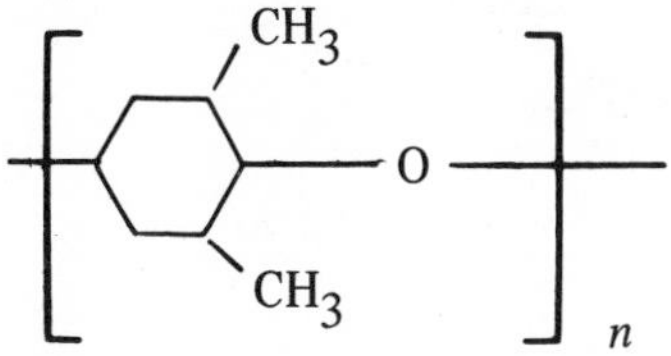

Polyphenylene oxide

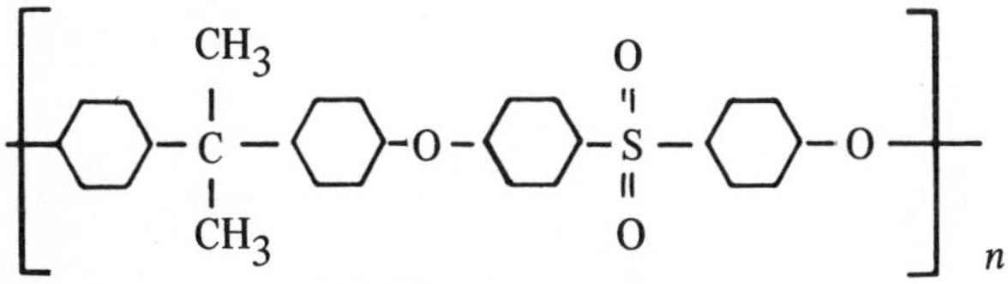

Polysulfone

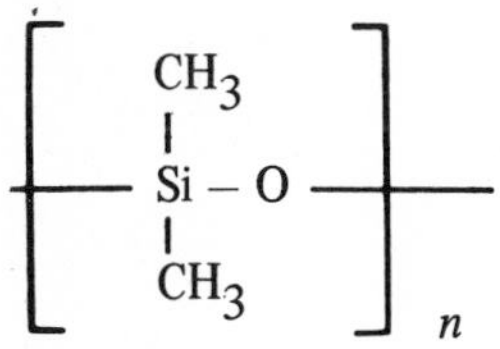

Silicone

Figure 4-6 Polymers with oxygen in their chain.

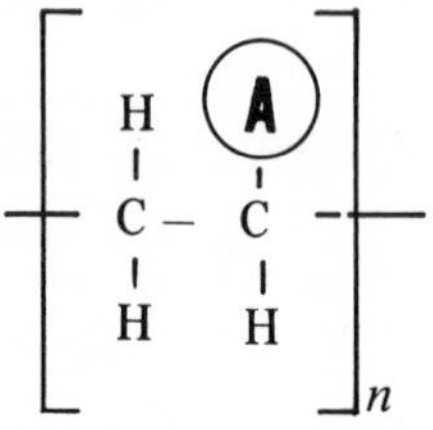

When Ⓐ is:	Plastic
−H	Polyethylene
−CH$_3$	Polypropylene
−Cl	Polyvinyl chloride
−F	Polyvinyl fluoride
−OH	Polyvinyl alchol
−CN	Polyacrylonitrile
−O−C(=O)−CH$_3$	Polyvinyl acetate
−O−CH$_3$	Polyvinyl ether
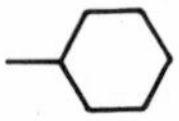 (cyclohexyl ring)	Polystyrene
−C(=O)−O−CH$_3$	Polymethyl acrylate
−CH$_2$−CH(CH$_3$)−CH$_3$	4-Methyl pentene-1

Figure 4-7 Polymer chains based on ethylene with a substitution for a single hydrogen atom.

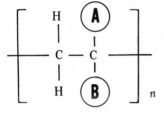

When Ⓐ is:	And Ⓑ is:	Plastic
–H	–H	Polyethylene
–F	–F	Polyvinylidene fluoride
–Cl	–Cl	Polyvinyl dichloride
–CH₃	–C–O–CH₃ O	Polymethyl methacrylate
–CH₃	⬡	Alpha methyl styrene

Figure 4-8 Polymer chains based on ethylene with a substitution for two hydrogen atoms.

shows carbon chain polymers based on ethylene with substitution for two hydrogen atoms. Figure 4-9 shows carbon chain polymers based on ethylene with substitution for all four of the hydrogen atoms.

COPOLYMERS

Homopolymers contain only one repeating unit. Even so their properties differ considerably because of different molecular configurations, molecular weights, and molecular weight distribution. Copolymers have two or more monomer units incorporated into the single polymer. The addition of even a small amount of a second polymer can have a profound effect on its properties. There are a number of possible ways for the monomer units to react with each other (Figure 4-10). If the two monomers, A, B, are on a 1 – 1 basis, each attached only to the other, the resultant polymer is called an alternating linear copolymer. If they attach to each other in a random manner, which is usually the case, we get a random linear copolymer.

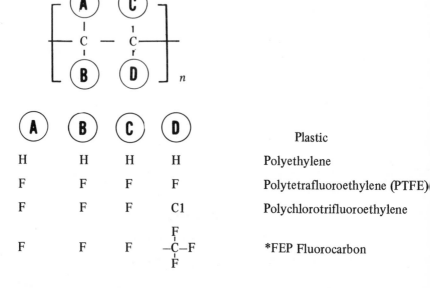

A	B	C	D	Plastic
H	H	H	H	Polyethylene
F	F	F	F	Polytetrafluoroethylene (PTFE)
F	F	F	C1	Polychlorotrifluoroethylene
F	F	F	$-\underset{\text{F}}{\overset{\text{F}}{\text{C}}}-\text{F}$	*FEP Fluorocarbon

Figure 4-9   Polymer chains based on ethylene with a substitution for four hydrogen atoms.

*Appears on every fourth carbon in place of F.

Block Copolymer

A block copolymer has one long segment or block of the same monomer unit attached alternately to blocks of other monomer units. These blocks are no necessarily of the same length. This in effect is the alternate joining of the end of two different homopolymers of comparatively short length. If a polymer i grafted unto the backbone of another polymer, it is called a graft copolymer.

Many copolymers cannot be readily classified into any one classification. For example, graft copolymers can have block copolymers on their chains. Block copolymers can consist of random homopolymers in part of its unit instead of straight homopolymers. For example, ABS (acrylonitrile-butadiene-styrene) i made by adding styrene and acrylonitrile monomer to an already polymerized butadiene. The styrene and acrylonitrile form a copolymer (random or block) which form a graft polymer by attaching itself as chains to the polybutadiene backbone. This, and an amount of the styrene-acrylonitrile copolymer, form the ABS material.

Stereoblock polymers have blocks of polymers of different stereostructures alternating with each other. Stereospecific polymers are those whose chains are arranged in a specific stereo pattern. The five types are cis, trans, isotactic, syndiotactic, and tritactic.

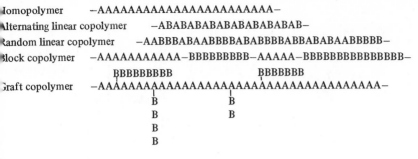

Figure 4-10 Different types of polymer arrangements.

Random Copolymers

An example of random copolymerization is that of ethylene and vinyl acetate (EVA). The ratio of the monomers in the final copolymer can be made to vary widely, from 400 parts ethylene and one part vinyl acetate to one part ethylene and 28 parts vinyl acetate. The copolymer will differ both in the length of the individual chain and its composition. This is due to the changing ratios of the monomers and polymers during polymerization. The monomers are used up at different rates. Therefore the ratio of the unreacted monomer is continually changing. Since the structure and composition of the chain depends upon this ratio, polymer chains formed at one time in the reaction will differ from those formed at another time. In many instances the composition of the final copolymer can be controlled (197).

Properties of Copolymers

Obviously the properties of copolymers depend on the percentage of each monomer and the way they form the polymer. Random polymers tend to show properties that are intermediate between those of the two polymerized monomers. Block or graft polymers with long groupings of individual polymers show properties of each of the monomers. As the size of the blocks gets smaller the properties tend to approach that of the random polymer. The properties of the block and graft polymers are therefore superior to those of random polymers (120).

This can be explained in terms of molecular structure. One would expect that the random addition of another monomer unit would keep the first monomer unit from packing as closely together as it would if it were polymerized alone. In block or graft copolymers which have long units, the units can compact closely as they would if they were polymerized alone. Therefore the random copolymer will occupy more space than the block or graft copolymer. Because of this extra volume and lower molecular attractive forces the random copolymer will usually have a lower density, T_g, T_m, hardness, tensile strength, tensile modulus, and elastic properties.

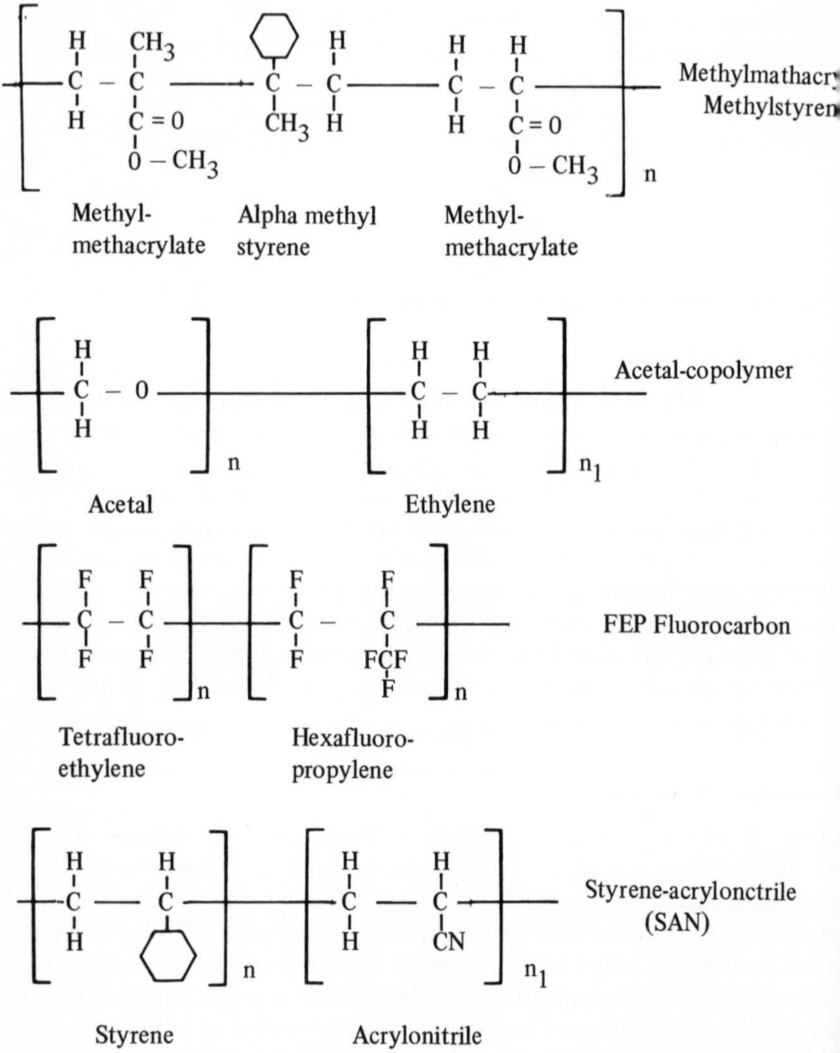

Figure 4-11 Monumer units of copolymers formed by the reaction of two or more polymers. The type of polymerization (graft, random, and block) is not shown.

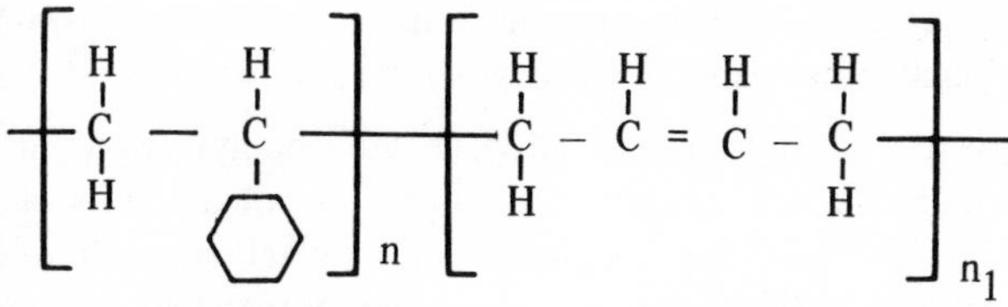

$$\left[\begin{array}{c} \overset{H}{\underset{H}{C}} - \overset{H}{\underset{\bigcirc}{C}} \end{array}\right]_n \left[\begin{array}{c} \overset{H}{\underset{H}{C}} - \overset{H}{C} = \overset{H}{C} - \overset{H}{\underset{H}{C}} \end{array}\right]_{n_1}$$ Styrene-butadiene

Styrene   Butadiene

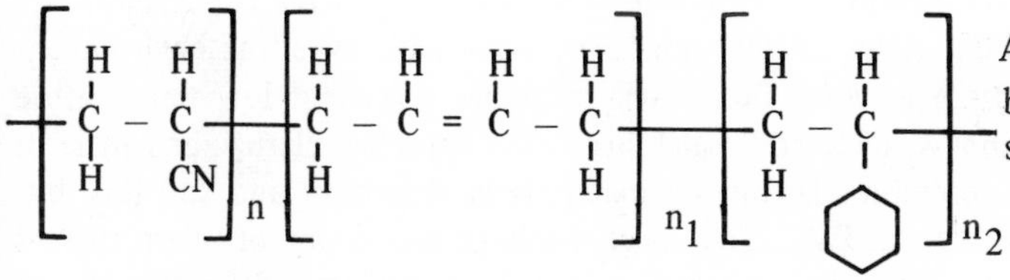

$$\left[\begin{array}{c} \overset{H}{\underset{H}{C}} - \overset{H}{\underset{CN}{C}} \end{array}\right]_n \left[\begin{array}{c} \overset{H}{\underset{H}{C}} - \overset{H}{C} = \overset{H}{C} - \overset{H}{\underset{H}{C}} \end{array}\right]_{n_1} \left[\begin{array}{c} \overset{H}{\underset{H}{C}} - \overset{H}{\underset{\bigcirc}{C}} \end{array}\right]_{n_2}$$ Acrylonitrile-butadiene-styrene (ABS)

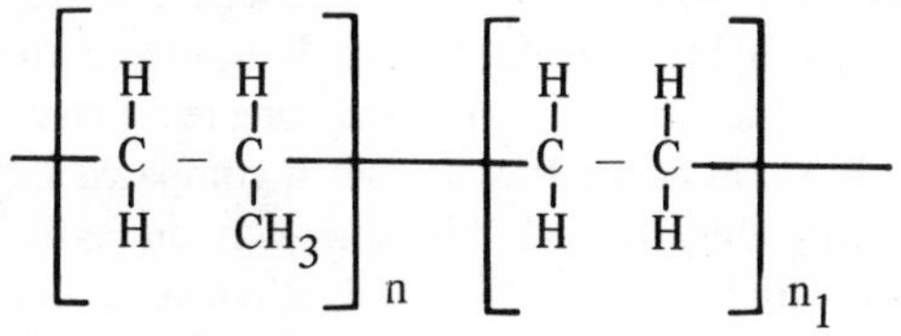

Acrylonitrile   Butadiene   Styrene

$$\left[\begin{array}{c} \overset{H}{\underset{H}{C}} - \overset{H}{\underset{CH_3}{C}} \end{array}\right]_n \left[\begin{array}{c} \overset{H}{\underset{H}{C}} - \overset{H}{\underset{H}{C}} \end{array}\right]_{n_1}$$ Propylene-ethylene

Propylene   Ethylene

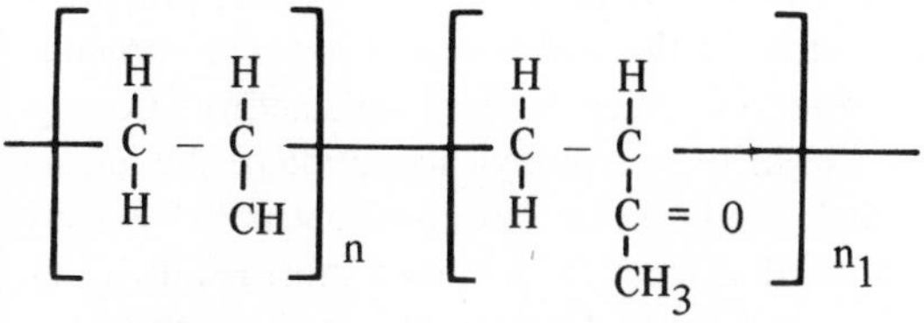

$$\left[\begin{array}{c} \overset{H}{\underset{H}{C}} - \overset{H}{\underset{CH}{C}} \end{array}\right]_n \left[\begin{array}{c} \overset{H}{\underset{H}{C}} - \overset{H}{\underset{\underset{CH_3}{C}=0}{C}} \end{array}\right]_{n_1}$$ Ethylene-vinyl acetate (EVA)

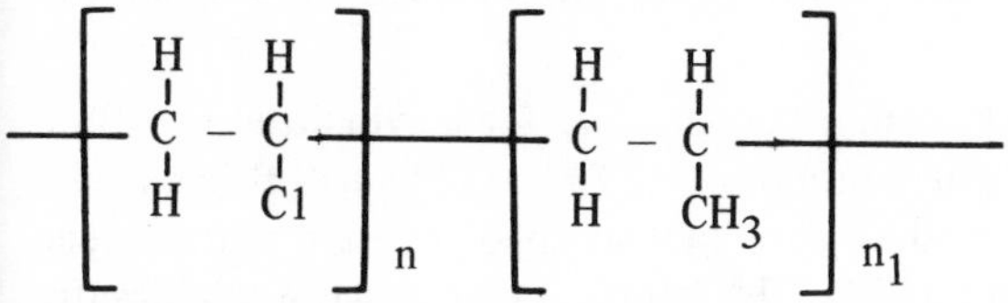

Ethylene   Vinyl acetate

$$\left[\begin{array}{c} \overset{H}{\underset{H}{C}} - \overset{H}{\underset{Cl}{C}} \end{array}\right]_n \left[\begin{array}{c} \overset{H}{\underset{H}{C}} - \overset{H}{\underset{CH_3}{C}} \end{array}\right]_{n_1}$$ Vinyl chloride-propylene (VCP)

Vinyl chloride   Propylene

The chemical formulas of the monomer units of some copolymers are shown in Figure 4-11. No attempt has been made to indicate the type of polymerization — graft, random or block. Some examples of copolymers follow:

Propylene and Ethylene.   Propylene and ethylene are copolymerized in varying amounts to form block and random copolymers. Small amounts of polyethylene are used to increase the toughness and impact strength of homopolymerized polypropylene. They also improve other properties such as stress cracking and lower brittleness temperature (121, 122).

Ethylene-Vinyl Acetate.   Ethylene-vinyl acetate (EVA) copolymers are made with varying ratios of the monomer. They have excellent environment stress crack resistance, good flex crack resistance, excellent low temperature properties, toughness, dielectric sealability, FDA approval, clarity, and little or no odor. They resemble elastomeric materials in flexibility and softness, but require no plasticizers as PVC. They accept a large proportion of fillers, such as calcium carbonate, talc, clay, asbestos, wood flour, and glass fiber. The use of EVA reflects its advantages when compared with molded properties of PVC and rubber products. Some typical injection molded uses are closures, gaskets, bumpers, shoe soles, and handle grips. Ethylene can also be copolymerized with acrylate esters (123). It has improved impact strength, stress cracking resistance, thermal stability, low temperature, flexibility, and flex life when compared to polyethylene. The copolymer of polyethylene and 1-butene was an early commercial application of copolymerization. Compared to polyethylene it had high stiffness and toughness, excellent stress cracking resistance, and superior performance under long term loads (124).

Vinyl Chlorides.   Unplasticized polyvinyl chloride has excellent properties but it is difficult to process. To overcome this additives and lowered molecular weight techniques are used, neither of which is fully satisfactory. Copolymerizing it with up to 10% by weight of polypropylene maintains the impact strength, dimensional stability, and toughness of PVC, but lowers its viscosity and increases its thermal stability so that it can be processed more readily (125, 126). ABS will also copolymerize with vinyl chloride-propylene copolymer to improve process ability (127). Copolymers of vinyl chloride and long chain alkyl vinyl ether when blended with PVC gives the mixture excellent processability without degrading any PVC characteristics (128). Acrylic modifiers also improve PVC processing (128a).

Acrylics.   By substituting methyl methacrylate for acrylonitrile in ABS a transparent material with a light transmission of 89% is obtained. It has similar properties to ABS. Methyl methacrylate can be copolymerized with styrene (129) and alpha-methylstyrene (130). The latter copolymer has a significantly higher deflection temperature under 264-psi load of 248°F. This compares with 190°F for PS and 203° for PMMA.

ACS is a polymer in which acrylonitrile and styrene are graft polymerized on a chlorinated polyethylene. It is similar to ABS except that the butadiene is replaced by the chlorinated polyethylene. The polymer has excellent weathering resistance and superior impact and heat resistance when compared to ABS. This is caused by the lack of the double bond of butadiene in the ACS resin. It processes as well as ABS with a much lower chlorine liberation than PVC (131).

The homopolymer of poly(4-methylpentene-1) is of little commercial application, but when copolymerizing it with other monomers it became a commercial plastic (TPX).

ACRYLICS POLYMERS

The first American cast acrylic sheets were produced in 1931. Molding powder was commercially introduced several years later.

Properties

Polymethyl methacrylate (PMMA) is a hard, rigid, transparent engineering material with a number of outstanding properties.

1. It has the best transparency and optical properties of commercially available thermoplastics. It is colorless with a 92% light transmission, 1% haze, and an index of refraction of 1.49. It comes in a full range of transparent, translucent, and opaque colors.

2. It has the best weather ability of all transparent plastics and probably of all plastics. It is exceptionally resistant to sunlight and outdoor exposure with a loss of 1% in transmission over 5 years. It is highly resistant to ultraviolet radiation, particularly from fluorescent lighting.

3. It has good dimensional stability.

4. It has good mechanical properties, but on the low side for engineering materials.

5. It is classified as slow-burning, releasing little, or no smoke.

6. It has outstanding decorative possibilities. It is exceptionally easy to fabricate (cut, polish, cement), and has a long history of successful applications.

7. It is uneffected by human tissues (dentures and temporary fillings) and is usable with food.

8. It has excellent gloss and a good "feel." Because of the nature of the material and the fine publicity by its two major suppliers, it has public acceptance as a quality material.

Acrylics have been modified by copolymerization and alloying. Copolymerizing with polystyrene is done for cost reduction. These polymers are superior to styrene but do not have the properties of acrylic. Other copolymers are made to increase the impact strength by as much as a factor of 5. They have a harder surface, excellent indoor aging properties, though inferior to acrylic in outdoor weather ability. PMMA has been alloyed with PVC to give an extraordinarily tough material which is self-extinguishing. Its heat distortion point is lowered and its outdoor weather ability superior to PVC, but not equal to the unmodified acrylic. PMMA is resistant to dilute acids and alkalis, petroleum oils, aliphatic petroleum oils, aliphatic hydrocarbons, dilute alcohols, and detergents. It is not resistant to concentrated alkalis and oxidizing acids, the lower ketones, esters, aromatic and halogenated hydrocarbons, and lacquer thinners.

Tables of the mechanical properties of PMMA are in Ref. 132 and the methods of polymer manufacture in Ref. 133.

Applications

Some of its major applications are lenses, outdoor lighting, indoor lighting, beverage dispensers, housings for vending machines, housings for gas and electric meters, and decorative items. In the automotive industry its applications include tail-lights, stop light lenses, instrument panels, nameplates, dials, and signal lights. In the appliance and houseware it includes knobs, control panels, and nameplates.

ACETALS

Acetal homopolymers were first available in commercial quantities in 1960. In 1963 copolymers of acetal and ethylene, and glass filled material became available. The homo- and copolymers have similar properties. The homopolymers are more rigid, with higher tensile and flexural strength, higher resistance to fatigue, but lower elongation. The copolymers have better long term temperature characteristics and resistance to hot water. The chemical resistances are similar, except that the copolymer is unaffected by strong bases. Compared to nylons, acetals have better fatigue resistance, creep resistance, stiffness and water resistance, but lower impact strength and abrasion resistance. Acetals have the physical properties typical of a crystalline high molecular weight polymer which is the class of materials to which they belong.

Properties

1. Homopolymers have the highest fatigue endurance of any unfilled thermoplastic. They have excellent resiliance and resistance to repeated impact. These properties are probably due to the crystalline structure. They make excellent springs, gears, and levers.

2. Homopolymers have high rigidity, a large part of which is maintained at elevated temperatures and high tensile strength.

3. The copolymer is one of the most creep resistant plastics, therefore showing good dimensional stability.

4. The impact strength of both is essentially constant from -40 to 212°F. Combining this with its resiliancy, snap fit assemblies are practical.

5. Aside from nylon 6/6, acetal has one of the lowest coefficients of friction. Under light loads, unlubricated, it is 0.15 against metal and 0.35 against itself. The homopolymer can be filled with TFE fibers for an even lower level of friction. The self-lubricating properties of acetals are particularly valuable where cleanliness is important as in foodstuff conveyors, pharmaceuticals, and textiles. Its abrasion resistance is second to nylon 6/6. It has one of the hardest surfaces of all thermoplastics which can be molded with a high gloss. Acetals natural color is white translucent and it can be purchased in opaque colors.

6. The copolymer has unusually good thermal properties, with excellent resistance to high intermittent temperatures. Maximum suggested continuous use temperature in air is 220°F and in water, 180°F. These properties have resulted in the use of acetals in hot water service and plumbing fixtures.

7. Acetals have excellent chemical resistance, particularly to most organic reagents. It is unaffected by washing in common solvents such as acetone and ethyl alcohol at room temperatures. It is affected by strong mineral acids. The copolymer is not affected by strong alkalis which attack the homopolymer.

8. It has normal electrical but poor outdoor weathering properties in the natural color.

9. The stiffness, impact resistance, and mechanical strength of the acetals are enhanced by adding glass. A 20% glass filled material is ideal for metal replacement. At 180°F it has better creep resistance than die cast zinc. Its mechanical, electrical, and insulating properties, corrosion resistance, and lower cost of fabrication have resulted in replacing a significant number of die cast applications.

Applications

Some of the major applications of acetals are for industrial parts including gears, springs, cams, levers, clips, housings, and bearings. The automotive industry uses it for seatbelt buckles, dashboards, clips, and large body components. The glass reinforced material is used under the hood in carburetors and pump housings. Water and plumbing applications include pumps, filter housings, sprinklers, potable water tanks and components, shower heads, valves, ballcocks, and fittings. Aerosol containers, toys, pen and pencil housings, and food contact applications are other wide uses of acetals. Reference 134 contains charts of the engineering properties of the copolymer, and Ref. 225 of Chapter 3 has information about the homopolymer.

ACRYLONITRILE–BUTADIENE–STYRENE (ABS)

The first ABS appeared commercially in 1948 with the graft polymer being developed in 1954. It has an excellent balance of high impact resistance, tensile strength, creep resistance, low temperature toughness, heat resistance, chemical and environmental stress resistance, electrical characteristics, and good surface gloss; it is also easily fabricated and decorated and processes very well. It is not surprising that it is one of the most widely used engineering plastics.

ABS is a group of plastics rather than an individual one. Broadly speaking, the acrylonitrile contributes the chemical resistance and rigidity, the butadiene contributes the impact strength and toughness, and the styrene adds to the rigidity and makes it easy to process.

Properties

The mechanical properties of ABS are excellent. Toughness, which is a combination of impact resistance and ductility, is relatively hard to define. Notwithstanding ABS is one of the toughest thermoplastic materials. A typical high impact ABS has a notched Izod of 7.5 at room temperature. The material has excellent ductility and low temperature characteristics and with an Izod of 3.0 at $-40°$F. It has good tensile strength at about 7000 psi. It has excellent low creep properties being able to support tensile loads of 1000 psi without any significant dimensional changes at room temperature.

ABS materials have good surface characteristics. They are highly resistant to abrasion and scuffing as evidenced by their use in luggage, telephone handsets, and tote boxes. ABS can be molded with a high surface gloss. It is easy to color, and it comes in translucent and opaque colors, and is easy material to fabricate – drill, tap, cut, polish, and so on. ABS can be easily decorated by all methods. Of particular interest is its ability to be electroplated. A beautiful metal surface is obtained which is not affected by corrosion because there is no bimetallic-electrolyte system. Since injection molding produces a part requiring little or no finishing, its combination with metal plating leads to wide economies compared to starting with a metal die casting. Plumbing fixtures, boating accessories, and automotive grills are some of the outstanding applications of this process using an ABS base.

It has good electrical properties which, because of its low water absorption, are little affected by humidity. They also do not vary much with temperature.

ABS weathers fairly well. Prolonged exposure to sunlight will affect a thin layer of the material causing color changes and brittleness. Substituting acrylic elastomers for the butadiene gives a polymer (ASA) with properties similar to ABS but much better weatherability (134a). The mechanical properties are less affected. When pigmented black, the material maintains approximately 75 to 80% of its impact strength after 3 years of exposure. In lighter colors, it is only 50 to 60%. The tensile strength of either the black or colored is only slightly affected under the same conditions.

ABS shows very good resistance to chemical environments, both with and without external stress conditions. They are highly resistant to staining and are unaffected by coffee, catsup, blood, ink, lipstick, motor oil, greases, rubber heels, and crayons. They are resistant to alkalis, acids (except concentrated oxidizing acids), many food products, and pharmaceuticals. They are attacked by low molecular weight organic materials such as carbon tetrachloride, ethylene dichloride, ethyl acetate, and toluene.

Applications

The automobile industry is the major user of ABS. Typical uses are defroster and heater housings, instrument clusters, grills, and body panels. ABS is used widely in luggage, housings, houseware housings, TV cabinets, and tote boxes. Other usages are as football helmets, hardware fittings, shoe heels, and slide rules. A major use just starting is furniture. Complete end tables and chairs are being molded and have achieved good consumer acceptance. The manufacture of ABS is described in Ref. 135, and some of its engineering design properties in Ref. 230 of Chapter 3.

CELLULOSICS

The cellulosics are the only commercial polymers not made synthetically. They are modifications of the natural polymer, alpha-cellulose. The four cellulosics molded are cellulose acetate (CA), introduced in the late 1920s, cellulose acetate butyrate (CAB) introduced about 1938, cellulose (acetate) proprionate (CP) introduced in 1950, and ethyl cellulose (EC).

All the cellulosics are made by completely esterifying the three hydroxyl (OH) groups of the cellulose. They are subsequently hydrolyzed to produce a moldable resin. CA, for example falls between the di- and tri- acetate in composition, with 50 to 55% acetyl by weight. CAB is a mixed ester of acetate and butyrate. CP is also made as a mixed ester with the acetate. Higher acetylization gives better moisture resistance but poorer flow. Lower acetylization gives better impact resistance and better flow.

Properties

The cellulosics are usually selected because of a combination of useful properties.

1. They are among the toughest of plastic and rarely break even under rough handling.

2. Their transparency and freedom from color permits almost unlimited color formulation and special color effects including metallics. This combination of toughness and transparency at moderate cost is the outstanding characteristic of these materials.

3. Basically they are hard stiff materials which can be compounded with plasticizers to increase flow for processing and increase flexibility and toughness in the molded product.

4. They have high surface gloss which is not particularly abrasive resistant.

5. They are characterized as slow burning, being somewhat less than would be presented by common newsprint paper in the same form and quantity. CA is available in self-extinguishing formulas.

6. CAB is the only cellulosic used for outdoor applications. It is specially formulated and used in outdoor lighting and signs.

7. CA absorbs about 2.5% moisture with the corresponding dimensional change. The other cellulosics absorb approximately half as much.

8. They are slightly affected by weak acids and alkalis and not resistant to strong acids and alkalis. They are affected by low molecular weight polar compounds such as ketones, alcohols, and esters of low molecular weight alcohols. They have excellent resistance to aliphatic hydrocarbons, detergents, greases, and oils.

9. They have good dielectric strength, high dielectric constants, good volume resistivity, and high dissipation factor.

10. They mold very easily in plunger and screw equipment. They also process well by all the other plastic fabricating techniques. They are easy to cut, cement, drill, tap, and polish.

Flow characteristics are reported according to ASTM D-569. The material is put in an injection cylinder of given dimensions. The temperature at which 1 in. of the material extrudes in 2 min under specified conditions is called the flow temperature of the material. For example, an "M" symbol means a material with a flow temperature of $145 \pm 5°C$, "H" a flow of $155 \pm 5°C$. (The flow steps are separated by $5°C$.) This flow characteristic is primarily used for CA and CAB. CP does not correlate too well, but the material is classified so an MH flow molds similar to an MH flow in CA. Generally the hard flow materials have higher hardness, stiffness, and tensile strength. As the flow becomes softer, these properties decrease, but the impact strength increases. CP is not available in the softer flows, and CA has the highest density 1.23 to 1.34, with CAB and CP 6% less and EC 10% lower.

The continuous useful temperature range for the cellulosics are: CA, −25 to 120°F; CAB, −40 to 130°F; CP, −40 to 130°F; and EC, −40 to 150°F. They all can stand higher short term exposures, some formulations reaching service temperatures of 2000°F.

Applications

Acetates, being the least expensive of the cellulose esters, are chosen when the properties of the others are not required. It is also chosen when a self-extinguishing formulation is needed. It has a very slight odor. CP and EC have no odor. CAB has a strong rancid odor which can be objectionable at high temperatures.

CAB is selected when outdoor weatherability and dimensional stability is needed. It is also indicated when a soft flow is wanted, as CP does not have that characteristic. CAB has excellent low temperature properties.

CP is chosen for its greater hardness, stiffness, and tensile strength. EC is different than the other cellulosics in that it is an ether rather than an ester. Its outstanding characteristic is its low temperature properties and its ability to maintain its toughness up to 200°F. It does not come in the colorless form of the other cellulosics, being amber and slightly hazy. It is not as resistant to acids, but resistant to weak and strong bases.

Some of the use for CA are sunglass frames, pen barrels, tool handles, brush backs, ornaments, toys, and dolls. The flame resistant grades are used in small appliance housings and other electrical applications. CAB is used for keys on adding machines, tool handles, steering wheels, knobs, and electrical applications. The weather resistant formulas are used for outdoor lighting, signs, and letters. CP is used for toothbrush handles, brush handles, safety goggles, indoor lighting applications, knobs, pen barrels, steering wheels, and toys. EC is used for helmets, gears, and many military applications.

NYLONS (POLYAMIDES)

Nylon is a generic term for polyamides. They consist of linear polymers with the amide group; $\left(\begin{smallmatrix} H & O \\ | & \| \\ -N-C- \end{smallmatrix} \right)$ in between methylene ($-CH_2-$) groups of varying lengths (Figure 4-5). Carrothers received the patents on polyamides in 1937, filament production was started in 1939, and molding material became available in 1941. It is interesting to note that even today many people do not associate nylon stockings with plastics.

Nylon consists of a family of materials, the most common of which is 6/6, 6, 6/10 and 6/12. Nylon 11 and 12 are commercially available. Nylon 13, 6/13, 13/13, 7, 8, and 9 are available in experimental or limited commercial quantities.

Properties

Nylons are characterized by good toughness, tensile strength, and resistance to creep, particularly in the high temperature range. They have excellent wear properties, low coefficient of friction, and exceptional chemical resistance, particularly to aromatic hydrocarbons, greases, and oils.

Nylon is an hygroscopic material. This has significant effects on its physical properties. To be molded satisfactorily nylon must have less than 0.3% moisture content. At 100% humidity, equilibrium is 8 1/2%. For 6/6 nylon after molding there are two distinct processes happening at the same time. Aging or annealing causes the polymer to shrink, while moisture absorption causes it to expand. The interaction determines the final dimensions. The rate of water absorption is relatively slow and changes are more visible from winter to summer than from week to week. The time to reach equilibrium conditions depends on the geometry of the piece and processing conditions. With proper design, nylon can be molded to close tolerances. Moisture acts as a plasticizer, reducing the tensile strength, stiffness, and increasing elongation. It increases the general energy absorbing characteristics of the material including its impact strength. For example, the Izod notched impact strength of 6/6 nylon increased from 1 to 5 ft lb/in. of notch as the moisture content increases from 1 to 4%.

Some of the more important properties of nylon follow:

1. Probably nylon's outstanding advantage is its toughness and ability to withstand repeated impacts. This property maintains itself over a wide temperature range. It is essentially a ductile material failing by yielding. Nylons are materials of high rigidity whose properties are increased by glass reinforcement. The tensile strength unfilled is about 12,000 psi, rising to 30,000 psi for a 40% glass reinforced material. These properties are characteristic of unplasticized, highly crystalline materials. Densities of unfilled nylons range from 1.02 to 1.14.

2. The combination of excellent abrasion resistance and exceptionally low coefficient of friction makes the nylons very useful for gears, cams, and bearing surfaces. The resiliency of the nylons permits the load to be spread over large surface areas. It can be used without lubrication which is valuable in food textile, and home appliance applications. When lubricated, it does not require continuous lubrication. Molybdenum disulphide is commonly added to lower the coefficient of friction.

3. One of the most important properties of nylon is chemical resistance. Nylons are particularly resistant to detergents, fuel, greases, and oils, including animal and vegetable oils. They are extremely resistant to aromatic and chlorinated hydrocarbons, ketone, and ester solvents as well as alcohols. They are not affected by alkalis or dilute acids and are not resistant to strong acids, bleaching agents, and hydrogen peroxide. They are soluble in phenol, which is used as a cement.

4. Because of its high crystallinity, nylons have excellent high temperature resistance. When stabilized against thermal degradation with a combination of cupric salts and alkali metal halides they perform well in temperature applications from 200 to 250°F. Nylons are self-extinguishing and classified in group II by UL.

5. Electrical properties are not outstanding and are affected by water. Because of its other physical properties nylon is often used in electromechanical applications such as coil forms, wire jackets, cable clamps, and switches. It is generally used below 500 V and 400 Hz/sec.

6. Nylons can be machined and fabricated readily. With one exception (an amorphous condensation product of adipic acid and an alkyl substitute hexamethyl diamine) they are translucent. Within that limitation they can be molded in a full range of colors. Outdoor weathering characteristics can be improved by carbon black and certain stabilizers. These properties are not outstanding.

Some of the comparative properties of the different nylons are described below. As one would expect, they all exhibit many similar properties. The choice is often dictated by economic and processing conditions.

Nylon 6/6 is the most widely used (152d, 224 of Chapter 3). It is the strongest of the three most popular types having the highest tensile strength, flexural modulus, and heat resistance. It absorbs less moisture than nylon 6 and is about 25% stiffer. It has good weatherability when properly pigmented with carbon black.

Nylon 6 is similar to nylon 6/6 in many respects though it has lower physical properties. Its lower melting point permits it to be processed at about 80°F lower than nylon 6/6. Nylon 6 can be self-plasticized with its monomer, yielding a tougher, softer, and more ductile material than nylon 6/6. However, it will absorb more water. Nylon 6 and 6/6 are in the same price range and compete directly with each other (228 of Chapter 3).

Nylon 6/10, absorbs about one half as much water as nylon 6/6 and 1/3 as much as nylon 6. This greatly improves its electrical properties and stability. It is less crystalline and more flexible than the other two nylons. It is used primarily for its low moisture absorption and corresponding improvement in physical properties.

Nylon 6/12 which is gradually replacing 6/10 retains its physical and electrical properties over a wide humidity range. At 50% RH it has 8% lower absorption, at 100% RH, 14% lower. It is about 12% stiffer than 6/10.

Nylon 11 has very low water absorption for nylon, low density (1.03), and relatively high impact strength. When fully plasticized it is especially shock resistant and for most practical purposes, unbreakable. It has inferior mechanical properties compared to other nylons but maintains its hardness and abrasion resistance (136, and 152d, 231 of Chapter 3).

Nylon 12 is similar to nylon 11 and has the typical properties of nylon. It has the lowest density (1.02), water absorption (0.95% at room temperature and 65% RH), and melting point (352°F) of commercially available nylons. It has properties similar to the olefins, which include excellent low temperature flexibility. There is no embrittlement at -110°F. Because of its low moisture absorption it has good dimensional stability (137, 138).

Nylon 13/13 and 6/13 have recently been introduced and are based on erucic acid derived from a vegetable seed oil. They are characterized by low water absorption, low molding temperatures, and good moldability (139).

All nylons are partially crystalline therefore cloudy or white and not transparent. An amorphous polyamide made by condensation of terepthalic acid and an alkly-substituted hexamethylene diamine is permanently transparent in thick sections. It transmits 85 to 90% of visible light with an index of refraction of 1.566. It has high impact resistance and rigidity, and is dimensionally stable under heat.

Applications

Some of the typical applications of nylons are gears, bearings, cams, coil forms, housings, combs, battery cases, gaskets, football face guards, and soldering iron handles. A general discussion of polyamides is found in Ref. 140, of their engineering properties in Ref. 141, and their manufacture in Ref. 142.

POLYPHENYLENE OXIDES (PPO)

Phenylene oxides were introduced commercially in 1964. The modified form was introduced in 1966. The latter material has the properties of the unmodified only to a slightly lesser extent. Since it is considerably less expensive, it has replaced the unmodified material except where the most stringent operating conditions are encountered. These would include hot water fittings, medical and surgical parts, washing machine components, and special components for the electrical and electronic industries. The information that follows applies to the modified material.

Properties

Modified PPO is a low density (1.06), and tough, rigid material which maintains its properties to approximately 265°F. It has outstanding resistance to aqueous environments and excellent dielectric properties. Some of its characteristics follow.

1. It has excellent mechanical properties. Its tensile modulus is one of the highest of engineering materials, 380,000 psi. Its tensile strength is about 10,000 psi, with good creep resistance. The glass filled material, for example, has a 1/4% strain at room temperature under 2000 psi stress after 1000 hr.

2. Its thermal characteristics are one of its most important properties. It has good low temperature characteristics to -40°F and is usable at temperatures as high as 265°F. It exhibits good load bearing properties at the higher temperatures. It has one of the lowest coefficients of thermal expansion, 3.3×10^{-5} in/in/°F. It is rated self-extinguishing and nondripping, group I by UL.

3. PPO is outstanding in its resistance to aqueous environments. It absorbs 0.07% water in 24 hr at room temperature and 0.14% in boiling water at equilibrium. Parts can be steam sterilized without any significant change in properties.

4. PPO is not attacked by dilute acids and alkalies or by detergents. It is normally resistant to alcohol and organic liquids including natural oils and fats. It is attacked by halogenated and aromatic hydrocarbons which can be used for solvent cementing.

5. Modified PPO has excellent dielectric properties which are enhanced by its low water absorption and good thermal resistance. Special formulations have low dielectric loss behavior at frequencies as high as 10^9 Hz.

Applications

Some of the major applications are in water handling equipment such as pumps, valve handles, shower heads, filters, automotive internal grills, air conditioning housings, instrument support brackets and bulb holders, machinery housings for copiers, readout equipment, dictating equipment, terminal block connectors, switch housings, appliance housings, and deflection yokes and other internal TV parts (232 of Chapter 3).

POLYCARBONATE (PC)

Polycarbonate, a condensation polymer of bisphenol A and phosgene, was discovered by H. Schnell in 1956 and became commercially available several years later. Polycarbonate is one of the toughest, most dimensionally stable thermoplastics over a wide temperature range. Combined with optical clarity and good electrical properties, it is a very useful engineering material. Some of its outstanding properties follow.

Properties

1. It has exceptionally high impact strength from -200 to 280°F. It has a higher impact resistance at -65°F than most thermoplastics have at room temperature. The notched Izod at room temperature is from 12 to 16 ft-lb/in. of notch. It has a critical thickness between 0.140 and 0.160 in. Below that the material elongates tremendously before failure with the aforementioned high impact strength. Above the critical thickness it tends to break cleanly with a typical Izod value of 2 to 3 ft-lbs/in. of notch. PC is notch sensitive and susceptible to crazing under strain.

2. The PC is one of the most dimensionally stable thermoplastics. The unfilled material can sustain a steady load of 2000 psi at room temperature. The glass reinforced material can at least double that. This dimensional stability, as with its other properties, is maintained over a wide temperature range. The mold shrinkage is a uniform 0.006 in./in. and the same in both directions. It therefore can be molded to close tolerances. The glass filled PC has the lowest post molding shrinkage of any commercial polymer-less than 0.05%.

3. The material has a water white transparency with a 90% light transmission. Its index of refraction is 1.586. It is available in all colors. Some of its properties can be diminished by overloading pigment, particularly in the whites and ivory.

4. PC is self-extinguishing group II in its regular grades and self-extinguishing group I in its flame retardant grades.

5. PC is unaffected by water below 140°F and can be used in boiling water on a limited basis. Polycarbonates are unaffected by greases, oils, detergents, aliphatic hydrocarbons, most mineral acids, and the higher alcohols. It is soluable or attacked by chlorinated hydrocarbons, most aromatic solvents, esters, and ketones. Properly stabilized, it has good outdoor weatherability. It is physiologically inert.

6. PC are excellent electrical insulators over a wide range of humidity and temperature conditions. Because of their low moisture absorption high mechanical properties, good heat distortion, low long term creep, and self-extinguishing properties, they are extensively used in electrical and electronic applications.

Applications

Some of the major uses of polycarbonates are in outdoor lighting, glazing, lenses, tool and appliance housings, aircraft interiors, and trim, bezels, and lamp housings for automobiles (143, 229 of Chapter 3).

POLYETHYLENE (PE)

Polyethylene was discovered in March 1933 by Imperial Chemical Industries of England. Pilot plant operation started in 1937 and commercial production in September 1939. Because of the military requirements of World War II, the knowhow was given to American companies. Polyethylene was first made under high temperature and very high pressure conditions, yielding a highly branched, low density material (LDPE). In 1954, Ziegler discovered catalysts which could produce polyethylene at low temperatures and pressures. This type material was linear in nature and is known as linear or high density polyethylene (HDPE).

Almost eight billion pounds of polyethylenes were produced in the United States in 1972, which makes it by far the largest volume plastic produced in this country. Seventy percent was LDPE and the remainder HDPE. The outstanding property of polyethylene is its low cost per cubic measure relative to other plastics. This combined with good toughness, excellent electrical properties, and chemical resistance, ease of fabrication and copolymerization ability account for the high usage.

Properties

Polyethylene is a linear polymer of methyl groups whose different properties ultimately depend on the method of manufacture. These properties, many of which have been discussed in Chapter 3, depend on the crystallinity, which is a function of the amount of branching, and which is measured in effect by the density (Table 3-1), the molecular weight (Table 3-3), and the molecular weight distribution (Table 3-4). For example, increasing crystallinity will increase the density, rigidity, tensile strength, chemical resistance, and opacity, and reduce the permeability and impact strength. Additionally, polyethylene forms copolymers readily, for example, with vinyl acetate, polypropylene, ethyl acrylate, and acrylic acid. For these reasons polyethylene and its copolymers can be tailored to fit specific applications.

Because of the large number of polyethylene variations it is difficult to be very specific about the properties of polyethylene at as they encompass a broad range. Within that reference the following distinguishing properties of polyethylene are listed.

1. Polyethylenes are among the lowest density plastics used, ranging from 0.91 to 0.97 (lighter than water). The lower the density the less material per unit volume and therefore the lighter the molded part. This added to the low cost of the resin makes it very competitive with other low cost materials such as paper and glass.

2. The impact behavior of polyethylene depends primarily on the crystallinity (density) and the molecular weight (melt flow). The low density poly-

ethylenes are too flexible at room temperature to be broken by standard impact tests. High density polyethylene ranges from 0.7 to 20 ft-lb/in. of notch. The ASTM brittleness temperature may be as low as -80°F. From this it is apparent that impact resistance is not normally a problem with polyethylenes. The tensile strength is relatively low ranging from 600 psi for the low density to 5000 psi for high density. Under loads polyethylenes are subject to considerable creep and stress relaxation. These properties too depend on crystallinity and molecular weight. For example, a 0.3 MI LDPE under a stress of 400 psi moves from a 3 1/2% at a 1000 hr to a 5 1/4% elongation after 7 years.

3. Polyethylenes have very low water absorption.

4. Polyethylenes exhibit poor weatherability unless they contain 2 to 2 1/2% of properly dispersed carbon black. Under those conditions their weatherability is excellent.

5. Polyethylenes have exceptional electrical properties with very low power factors which are virtually unaffected by frequencies and by temperature changes up to 140°F. This combined with their excellent stability in aqueous media and good weatherability contribute to their large use for extrusion as wire and cable insulators including submarine cable.

6. Under no load conditions polyethylene has good heat stability. Even small loads, however, cause distortion and flow even at moderate temperatures.

7. Polyethylenes are essentially inert, unaffected by strong and weak acids and alkalies, detergents, alcohols, and ketones. They swell with chlorinated and aromatic hydrocarbons including gasoline and oils. They are susceptible to environmental stress cracking which is the result of adverse environmental conditions and is limited, by definition to those chemicals which do not attack polyethylene by swelling. These are usually long term phenomena and are dependent on the stress cracking agent, temperature and stress.

Applications.

Polyethylene finds its way into many applications. Some of the typical injection molding applications are containers, covers, pails, housewares, toys, furniture disposable medical and laboratory ware, and closures. References 144 and 220 of Chapter 3 relate to the manufacture and processing of high density polyethylene; Ref. 198 has engineering information on low density-poly ethylene; Ref. 199 discusses the manufacture of polyethylenes.

POLYPROPYLENE (PP)

Before 1954, attempts to polymerize propylene to a high molecular weight commercial plastic were unsuccessful. Natta, using Ziegler type catalysts achieved a commercial polypropylene. Commercial production started 3 years later in 1957 reaching the billion pound level some 13 years later.

The reason for the rapid growth of PP can be found in its combination of properties rather than any special one. It has a good combination of rigidity and toughness, high rigidity at elevated temperatures up to 250°F, can be steam sterilized, good abrasion resistance and low coefficient of friction, nonhydroscopic, excellent electrical properties, unique flex properties, good chemical resistance without stress cracking, one of the lowest densities (0.90), high surface gloss, and ease of fabrication.

Properties.

As with the polyethylenes, PP can be made with different properties depending on polymerization techniques and purification of the polymer. The properties in large measure depend on the nature and amount of crystallinity which in turn relates to the amount of isotacticity, syndiotacticity, and atacticity. The atactic form, which is amorphous and will not crystallize at all, has to be removed during the manufacturing process. A second way to control its properties is to copolymerize it with other material particularly polyethylene (122). There are two major concentrations of polyethylene, about 7 and 15% by weight. The way in which the polyethylene molecule is introduced will also affect the properties. A description of some of the characteristics of the homopolymer and copolymers follows.

1. Polypropylene has the highest tensile strength, over 5000 psi, of the olefins. It has a high tensile modulus which can be increased from 160,000 to 550,000 psi by reinforcements such as asbestos and talc. Its notched impact test is relatively low for most grades though special impact materials are as high as 15. This does not correlate with service conditions where impact failure at room temperature is exceedingly rare. Unnotched impact tests give a much better correlation with the very high impact material reaching 35 ft-lb in. of width. PP is very notch sensitive and care should be taken in part design to eliminate sharp corners. Copolymerizing with small amounts of polyethylene greatly improves the impact resistance. It has good low temperature impact strength with a 4 ft-lb/in. unnotched rating at -100°F.

2. Polypropylene has unusually good fatigue resistance. Hinges are made approximately 0.015 thick that will flex almost indefinitely with very high tear strength (Refs. 120, 121 and 123 of Chapter 3). Because of its flexibility snap fits are practical. Under cuts up to 0.040 readily strip from the mold.

3. It has excellent abrasion resistance and a very low coefficient of friction, comparable to that of nylon. However, lubricating nylon surfaces reduces the friction much more than for polypropylene. It is used for low load bearing applications.

4. Because of its high melting point, about 330°F, and its high crystallinity it has excellent high temperature properties. It is usable up to 250°F, and can be readily steam sterilized.

5. It has poor weatherability which can be improved by UV stabilizers or carbon black.

6. Polypropylene has the typical excellent chemical resistance of the olefins. It is resistant to dilute acids, concentrated acids (except oxidizing acids), alkalis, alcohols, detergents, and water. It resists aromatic hydrocarbons, chlorinated hydrocarbons, greases, and oils at room temperature but is attacked at about 140°F. It does not exhibit the environmental stress cracking of polyethylene. Its water absorption is very low with an equilibrium immersed water content of 0.03%.

7. The homopolymer and copolymer can be compounded to be rated slow burning, self-extinguishing, and nondripping. It is recognized by UL for continuous electrical service at 105°C for 0.055 thickness and higher.

8. PP is an excellent electrical insulator and has extremely low dielectric losses. Because of the self-extinguishing properties of certain grades it has found wide use as support for electrical components.

9. Aside from 4-methyl pentene-1 (density 0.83) it is the lightest of all commercial polymers.

10. With proper treatment it can be electroplated or vacuum metallized. It can be hot stamped, painted, and printed as well.

Applications

Some typical applications are containers, housewares, hinged applications, hospital and institutional ware, closures, battery cases, tote boxes, seats, disposables, washing machine agitators, pump parts, auto fender liners, splash guards, door panels, underhood applications, fan shrouds, and accelerator pedals. Information on polypropylene and its copolymers will be found in Refs. 145, 146, 227 of Chapter 3 engineering tables in Ref. 147, and its manufacture in Ref. 148.

POLYSTYRENE (PS)

General purpose polystyrene (GP-PS) was introduced in the United States in 1937. The manufacture of PS is discussed in Ref. 149. Rubber modified high impact styrene became available in 1948. In 1972 over 3 billion lb were used. This rapid growth is the result of a combination of properties. It is low in cost, rigid and strong enough for household use, and has excellent clarity and unlimited color range. The material can be decorated by all processes, has good electrical properties, and is easy to fabricate.

PS is a linear polymer commercially available in an atactic state, therefore morphous and uncrystallizable. Isotactic polystyrene, which is crystalline, is not commercially available.

Properties

The properties of PS can be altered by the addition and dispersion of rubber for high impact materials, methylstyrene for heat resistance, methyl methacrylate for light stability, acrylonitrile (SAN) for chemical resistance, and acrylonitrile and butyldiene (ABS) for very high strength. Some of its characteristics follow:

1. Polystyrene is clear and colorless with 90% light transmission and a 1.59 refractive index. It is available in all colors with many specialized effects. The general purpose (GP) styrene molds with an exceptionally high gloss surface. Rubber modified or high impact (HI) styrene is a translucent, off-white color which molds with a semi-gloss. Polystyrene is one of the most rigid plastics with tensile modulus of 450,000 psi and a high (7000 psi) tensile strength. Polystyrenes have low elongation at break and must be designed for low strains. Under these conditions they have excellent dimensional stability. GP styrene, however, is very brittle with notched Izod impact strength of 0.3. HI polystyrenes are much tougher with impact strengths of about 4 and super high impact can reach as much as 8 ft-lb/in. of notch. Styrenes have low heat distortion temperatures. GP styrene has a long term temperature rating of 145°F while the high heat materials reach 160°F. This is one of its limiting design factors. Polystyrene burns with a yellow sooty flame. Self-extinguishing grades are available, but other properties of the material are sacrificed.

2. Polystyrene has excellent electrical properties, with the power factor, volume, and surface conductivity near zero. These properties are only slightly affected by humidity and temperature.

3. PS is tasteless and odorless and can be used with most foods. It is unaffected by water with a 0.05% water absorption for GP-PS. GP-PS is unaffected by alkalis, acids (except oxidizing acids), low molecular weight alcohols, and detergents. It is attacked by ketones and aromatic and chlorinated hydrocarbons. Other agents such as acetone and heptane will produce severe stresses which cause cracking under very low loads.

4. Polystyrene has poor outdoor weathering properties. It has excellent indoor environmental resistance. When light stabilized for ultraviolet radiation it an excellent material for lighting panels and louvres. Other additives are saturated aliphatic hydrocarbons for easy flow, saturated fatty acids for mold release and alkyl amines and amides for antistatic properties.

5. It is a light weight material with a density of about 1.05.

6. Glass filled styrene has dramatically improved impact properties, rigidity, strength, and heat distortion. It competes with many engineering materials.

7. Polystyrene is exceptionally easy and rapid to process. It exhibits stron
orientation effects which can be controlled. It is very easy to fabricate an
decorate.

Applications

Some of its many uses include containers, packaging, toys, housewares, combs
closures, disposable dishes, disposable cups, lighting applications, tape cassettes
and furniture.

STYRENE-ACRYLONITRILE (SAN)

Copolymer of styrene and acrylonitrile exhibits most of the properties o
polystyrene. Its main advantage is significantly increased chemical resistance
particularly to food, and somewhat higher mechanical and thermal properties.

SAN has excellent resistance to acids, alkalis, salts, and dilute alcohol. It ha
good resistance (better than styrene) to many foods and oils, gasoline, cleanin
agents, detergents, cosmetic creams, and lotions.

It has significantly higher tensile strength, 11,000 psi in the direction of flov
and 9000 psi across it. It has better stress corrosion resistance, higher resistanc
to crazing around metal inserts, harder surface, and higher heat resistance tha
unmodified polystyrene.

Its major uses relate to its resistance to foods and other chemicals. It is use
for cups, tumblers, dishes, trays, picnic ware, cosmetics, and packaging items.

POLYVINYL CHLORIDE (PVC)

PVC was first observed in 1835, and became commercially available in German
in 1931 and in the United States in 1933. It ranks second in the United State
following polyethylene with sales of 4.3 billion lb in 1972. General informatio
on PVC is found in Ref. 150, its properties in Ref. 151, and its manufacture i
Ref. 152. Because of thermal stability, PVC can not be molded in a plunge
machine and requires a screw machine. It thermally decomposes readily yieldin
HC1. While this effect can be masked to some extent by additives, the HC1 i
still a corrosive material which will corrode the screw, cylinder and mold. Fc
these reasons comparatively little PVC is injection molded as the unmodifie
polymer.

It can be readily copolymerized and alloyed. The two systems used fc
injection molding are plasticized ones which yield elastomeric moldings and
nonplasticized material which is stabilized and yields a rigid molding. PVC is als

supplied as a dispersed resin, such as a plastisol or organisol, water dispersed resin or latexes, low molecular weight resins dissolved in organic solvents, and polyblends which are physical mixtures of several materials.

Properties

PVC is a clear linear polymer with a good combination of properties, which can be varied to suit many applications. It can be formulated to have high impact strength and good stiffness, good weatherability, excellent chemical resistance, good electrical properties, and can be self-extinguishing and non-dripping. It is lower in cost than some engineering materials but this is often balanced by the problems of processing PVC and its high specific gravity. A discussion of plasticizers and their effects on PVC is beyond the scope of this section. They radically alter the properties of the resin. Both plasticized and rigid PVC use additives such as stabilizers, lubricants and pigments to improve or change processing properties.

The rigid PVC compounds have good tensile yield strength, 7500 psi, are rigid, tensile modulus 400,000 psi, and can have Izod impact strengths as high as 8. The materials, however, are notch sensitive. They have good abrasion resistance and wear well. It is one of the most dense thermoplastics, 1.40, although this density is varied by the plasticizer and filler content.

Rigid PVC is essentially inert to concentrated and dilute mineral acids, alkalis, detergents, greases and oils, and alcohol solvents. It is attacked by ketones, and aromatic hydrocarbons. These properties have made it very useful for chemical containers.

Properly stabilized, it has excellent resistance to weathering. Some cable and wire applications have been in use for a decade. It has low water absorption characteristics but some of its plasticizers are not resistant to fungi and rodents. Special formulations are required for such applications exposed to those environments. Vinyls can be formulated from excellent non-conductors to conducting materials. Its electrical and elastomeric properties have led to very wide use as a covering for wire and cable.

Applications

Plasticized PVC is used in electric cord sets, handle grips, vacuum cleaner parts, and beach shoes. Unplasticized PVC is used for pipe fittings, phonograph records, housings, and chemical containments.

OTHER MATERIALS

The preceding section discussed some of the properties of the most widely used thermoplastics. Many other commercially available polymers are used for specific and often limited applications. A very limited discussion of some of them follows.

Chlorinated Polyether (CPE)

This is a linear highly crystalline polymer containing about 46% chlorine by weight, introduced in 1959. Its outstanding property is its chemical resistance. No solvent has been found which will dissolve it at room temperature. A few highly polar polar solvents will dissolve it at elevated temperatures. It is resistant to attacks by alkalis, inorganic acids (except fuming sulphuric and nitric), alcohols, ketones, aromatic hydrocarbons, detergents, grease, and oils. Its corrosion resistance extends to elevated temperatures, 250°F, and it has exceptional creep resistance under load at high temperatures. For example, at 1000 psi load and 212°F, it will creep from 2 ½ to 3% in 1 ½ years. It also has considerably lower thermal conductivity than other corrosion resistant thermo plastics. This property often eliminates the necessity for thermal insulation. CPE has excellent electrical properties and is self-extinguishing. It has good abrasion resistance, but relatively poor mechanical properties. Its main uses are valves flanges, pumps and meters where corrosion resistance is important. Its manufacture and properties are discussed in Ref. 153.

Ionomers

Commercially introduced in 1965, ionomers describe a class of polymers which combine ionized carboxyl groups attached to the olefin polymer chain with metallic ions. This produces a cross-linkage which is normally associated with thermosetting materials. Since the ionic bonds are thermally reversible, these materials are processed as thermoplastics. They combine transparency with flexibility toughness and resistance to oils and greases.

The commercial material has no plasticizer, and has high impact strength high elongation, good flexibility, and high resilience. They are excellent at low temperatures with a brittleness temperature measured as low as -220°F. Their upper usable range is approximately 150°F, and they are not especially good for structural applications. Ionomers are low in density (0.935 - 0.96), have good abrasion resistance, and, when properly stabilized for UV or with carbon black display good weatherability. Most ionomers have good electrical characteristic over a broad frequency range. They are highly resistant to chemical stress cracking agents, resist alcohols, ketones, and most organic solvents, and are excellent for greases and oils.

Typical molded applications are shoe heels, football shoe soles, golf ball covers, tool handles, and hammer heads.

Poly(4-Methyl Pentene-1)

The homopolymer has little commercial application, but when copolymerized with other monomers has become a commercial plastic (TPX). It is manufactured with a Ziegler type catalyst and was introduced commercially in 1965.

It has the lowest density (0.83) of any commercial polymer. It has excellent transparency, with light transmission of 90%, comparable to that of polystyrene and acrylic. Because it is essentially an olefin it has an excellent chemical resistance and not affected by alkalis, most acids and many organic chemicals. It has outstanding resistance to waters, greases, and oils and it has excellent electrical properties. One of its outstanding properties is its high crystalline melting point of 464°F. It retains its properties to a maximum service temperature of about 260°F. It can go considerably higher on a short term basis.

Because of these properties, one of its major applications is in food processing and serving. The TPX parts can be used as food containers in hot air and microwave ovens, with excellent resistance to hot fats, sight glasses, food equipment such as coffee making machines, electrical appliances, microwave equipment, coaxial cable connectors, hospital ware, and syringes (154 − 156).

Polyaryl Ether

Introduced commercially in 1968 polyaryl ether has a combination of high heat, high impact, and excellent flow properties. Its heat deflection temperature at 264 psi is 300°F, exceeded only by polysulfone (345°F), unmodified PPO (375°F), and polyarylsulfone (525°F). In thin sections it has one of the highest impact strengths of all thermoplastic engineering materials. At room temperature, it is 8 ft-lb/in. notch and at -250°F, 2.5. The material has good processing flow properties and can easily fill thin sections. It is shear rate sensitive, its viscosity dropping considerably with increasing shear rate.

Some of its applications are in automotive components, radio and TV parts, instrument housings, impellers, and safety helmets.

Polyaryl Sulfone

Polyaryl sulfone introduced in 1967, consists of aryl groups linked by oxygen and sulfone groups. The amorphous polymer, with a T_g of 550°F is different from the polysulfones in that it contains no aliphatic carbon-hydrogen bonds. The aromatic structure gives excellent resistance to oxidation and the other linkages permit chain flexibility for processing.

The outstanding property of polyaryl sulfone is high heat resistance. At 500°F it has a tensile strength over 4000 psi and a flexural modulus of 250,000 psi. This is not the result of a crystalline structure, but rather an inherent property of the polymer below its T_g. It is resistant to oxidative degradation.

Testing at 500°F for 2000 hr showed a complete retention of its tensile strength. Polyaryl sulfone also has significant mechanical properties as low as -320°F. At room temperature it is a stiff tough material with a notched Izod of 5, and 40 in unnotched tests.

The material is resistant to hydrolysis. Treating it with steam at 300°F for 1000 hr did not change its tensile properties. Furthermore, it has excellent electrical properties and it is resistant to hydrocarbons, acids, bases, refrigerants and oils. Applications are limited, for the moment, by high cost. It can be used to replace metals and ceramics where fabrication costs are high, and other plastics which cannot withstand high temperatures (157 and 152a of Chapter 3).

Polysulfone

Introduced in 1965, with commercial production starting in 1966, the polymer consists of benzene groups alternately linked by sulfone, ether, and isopropylidene linkages. As in polyaryl sulfone, its properties result from its chemical structure. It is a light amber transparent with a wide range of coloring possibilities. It is a crystalline polymer with a T_m of 364°F.

Its outstanding property is its high tensile strength at high temperatures. At room temperatures its tensile strength is about 10,000 psi, which drops to 6000 psi at 300°F. Its flexural modulus at room temperature is 400,000 psi which drops to 300,000 psi at 300°F.

It has excellent high temperature creep resistance. At 3000 psi stress and 200°F the creep is less than 2% after 1 year. It has a relatively low notched Izod, 1.3, but unnotched will not break with most applications. It has good electrical characteristics. It has excellent resistance to dilute and concentrated acids, alkalis, alcohols, detergents, greases, and oils. It is soluble in ketones and aromatic and chlorinated hydrocarbons. Some of its molded applications are double insulated power tool housings, underhood automotive parts, computer parts, household appliances, coffee makers, high intensity lamps, shower heads, and hot chocolate dispensers. A general discussion of the material and its properties are found in Ref. 158 and 159.

Polybutadiene - Styrene

This block polymer of styrene and butadiene, introduced in 1966, is a thermoplastic rubber which does not require vulcanization. It derives its rigid properties from the aggregates of polystyrene, and its elastomeric properties from butadiene. Another elastomeric thermoplastic is obtained by graft polymerizing butyl rubber to high and low density polyethylene (160) and to ethylene and propylene (Ref. 152b of Chapter 3).

Polybutadiene-styrene is a true elastomer in that when it is stretched to twice its length and released, it returns to its original size. This contrasts with PVC or EVA which partially flow for a permanent set. The plastic has excellent low temperature flexibility being usable at -120°F, but poor property retention at high temperature. As with rubber it has a high coefficient of friction. It is chemically resistant to water, alcohols, dilute acids, and dilute alkalis, and soluble in esters, ketones, and hydrocarbons, and its properties are essentially those of a rubber. Molded parts find uses as bottle nipples, face masks, shoe soles, and general purpose rubber goods replacements (161).

Elastomeric Polyurethanes

Thermoplastic urethanes which can be injection molded are based on polyesters and polyethers. They have high impact resistance, high elasticity, resilience, and vibration deadening properties. This is combined with high load bearing capacity, high tear strength, good tensile strength, resistance to abrasion, and good low temperature properties. When properly stabilized they have excellent resistance to weathering, oxygen, and ozone. They are fully resistant to oils and fuels.

Some of its applications include bumpers, horseshoes, shock absorbers, gears, sprockets, and tires. (A general discussion of thermoplastic urethanes is found in Ref. 162. A comparison of the flexibility of polyurethane, plasticized PVC, acrylate copolymer, and chlorinated polyethylene is found in Ref. 163.) A general discussion of the properties and manufacture of thermoplastic, thermosetting, and foamed polyurethane is found in Ref. 164 and 165.

Polymethylene Terephthalate (PTMT)

Thermoplastic polyester of terepthalic acid and 1.4-butanediol called poly-tetramethylene terephthalate (PTMT) has become commercially available (165c). PTMT has excellent heat distortion under load, (302°F at 66 psi), good impact resistance (no break in an unnotched Izod test), excellent toughness, good fatigue resistance and electrical properties, and is classified as self-extinguishing by UL. It has low surface friction and wear with excellent dimensional stability under load.

PTMT is chemically resistant to alcohols, ethers, high molecular weight esters, aliphatic hydrocarbons, aqueous solutions of salts, acids and bases, oils, and fats, high molecular weight ketones, and detergents. It is attacked by strong acids and bases, low molecular weight ketones, substituted aromatic compounds, and aromatic hydrocarbons.

Some of the suggested applications are gears, bearings, tool housings, safety helmets, seats, bumpers, and electrical connectors.

Fluorine containing Plastics

Fluorinated ethyl propylene (FEP) is a copolymer of tetrafluorethylene and hexafluoropropylene which can be injection molded. (Tetrafluorethylene cannot be injection molded. It is sintered.) Density of FEP ranges from 2.14 to 2.17. Its outstanding properties are chemical inertness, resisting almost all chemical attack except molten alkali metals and various fluorine compounds. It has the lowest coefficient of friction of any known solid and it maintains its useful properties from -300 to 400°F. With a maximum electrical service temperature of 400°F the material shows no heat aging at this temperature, and outstanding electrical properties over a wide range of temperature and humidity conditions are exhibited (Ref. 149a of Chapter 3).

FEP is used in the electrical and electronic field, for insulators, valve fittings and pumps for corrosive fluids, bearing seals and rings.

Chlorotrifluoroethylene (CTFE)

This is the only fluorine containing polymer which also contains chlorine. It is chemically inert, has good electrical properties, is thermally stable and usable from -400 to 400°F, has very low coefficient of friction, and for all practical purposes absorbs no moisture. Compared to FEP, CTFE resins are harder and more resistant to creep, and have lower melting points, less permeable, higher coefficient of friction and less resistant to swelling by solvents.

The CTFE is optically clear in thicknesses up to 1/8 in.; UV absorption is very low, which aids its weatherability. It has the lowest permeability to water vapor of any plastic, and is impermeable to many liquids and gases, particularly in thin sections. Some of its applications are liquid sight glasses, seals, valve seats and packing for liquid oxygen equipment, circuit breaker components, tube sockets, self-lubricated gears and bearings, and packaging for hard to hold liquids. When copolymerized with ethylene its processing characteristics improve (165a).

Polyvinylidene Fluoride (PVDF)

Crystalline high molecular weight polymer with a density of 1.76 melts at 340°F. PVDF has an excellent balance of properties. Compared to the other fluorplastics it has greater mechanical strength, abrasion resistance, and reduced creep. These properties are maintained from -80 to 300°F. It has less chemical inertness than the fully fluorinated polymers and good electrical properties and excellent weathering properties. Its main uses are in corrosion resistant environments, and its molded products include pumps, impellers, seals, gaskets and electrical connectors.

THERMOSETTING MATERIALS

Thermosetting materials are relatively new for injection molding equipment. When compression or transfer molded, their properties were largely determined by the nature of the resin and the type and quantity of the fillers. Even a brief discussion of the properties of all the thermosetting materials is beyond the space limitations of this section. Specially formulated phenolic materials for injection molding are available and used in some quantities and is now discussed.

Thermosetting materials essentially differ from thermoplastics in their better properties at elevated temperatures. They do not have the color range, flexibility, and impact strength characteristic of most thermoplastics.

Most thermosets are condensation polymers formed by the splitting out of one or more low molecular weight products, usually water. These are essentially organic chemical reactions such as esterification or amide formation. If there are bifunctional monomers the polymer will have a linear chain. If the monomers are polyfunctional there will be a network type polymer resulting from cross linking of the functional groups. Phenol-formaldehyde, urea-formaldehyde, and melamine formaldahyde are characteristic of the latter group. (See "Condensation Polymers" page 375).

Phenolics, the first commercial plastic, developed by Baekland in 1909, are produced by reacting phenol and formaldehyde. With suitable compounding and fillers phenolics have excellent electrical characteristics, heat resistance, impact strength, dimensional stability, and resistance to water, acids, and alkalis. Unfortunately all these properties cannot be put into one phenolic at one time.

Tensile strength is about 9000 psi, with a low Izod notched of 0.4. The heat deflection temperature at 264 psi can be as high as 390°F, and temperature has relatively little effect on the rigidity. The water absorption properties of wood filled phenolics are poor, but are improved by changing to mineral filled materials. Outdoor weatherability is good depending on the material composition. The mechanical properties are little affected by UV. Resistance to mineral acids is fair; resistance to alcohols, aromatic and chlorinated hydrocarbons, grease and oils is good.

Phenolics are used in appliances, handles, knobs, blower wheels, toaster end panels, circuit breakers, switches, relays, light sockets, automobile ignition components, and electrical connectors.

Epoxy resins are polyethers with epoxide groups in the polymer before they are cross-linked. The most widely used is a condensation product of an epoxide and bisphenol A (Figure 4-12) which results in the liberation of HCl. Other compounds containing hydroxyl (OH) can be used in place of bisphenol A. The epoxy resin can be cross-linked with many materials including phenol-formaldehyde, urea-formaldehyde, polyamides, and acids or acid anhydrides. The most common curing agents are amines which react with the epoxide by opening the

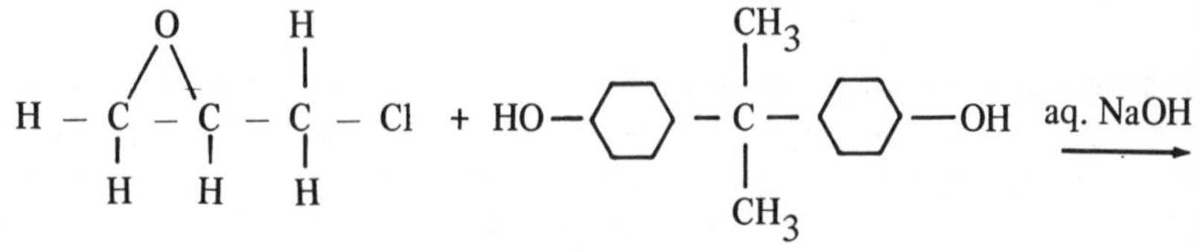

Epichlorohydrin   (Epoxide)   Bisphenol A

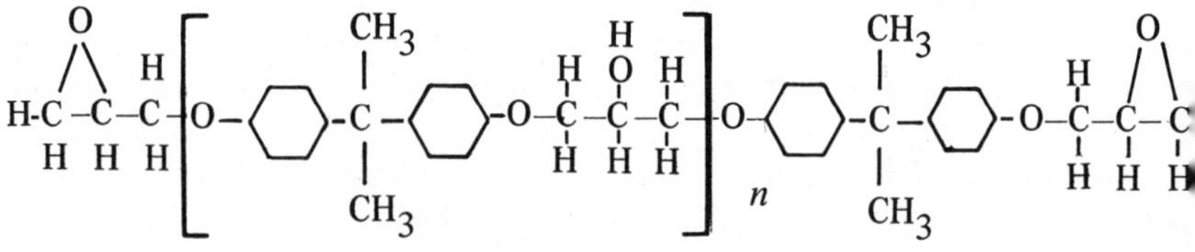

Epoxide polymer

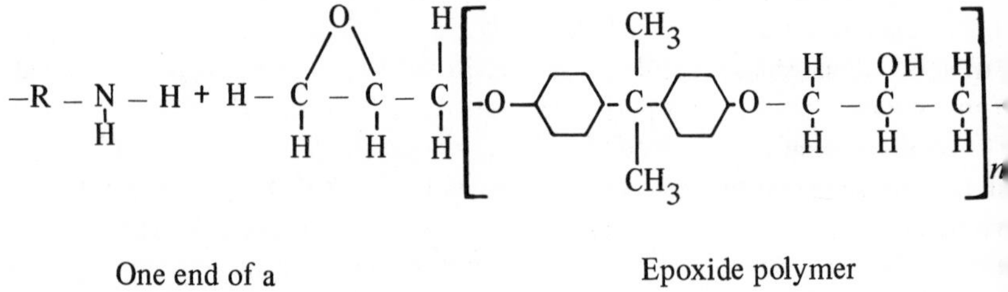

One end of a   Epoxide polymer
polyfunctional
amine
(diethylenetrianine)

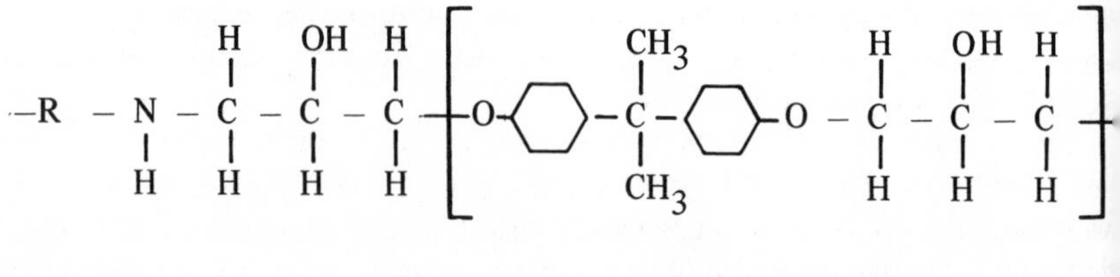

Cross-linkage occurs by the reaction of different amine groups with the epoxide polymer.

Figure 4-12 Epoxy Resins Formation and cross-linking with a polyfunctional amine of an epoxide polymer.

ring. An epoxy-acrylate resin is described in Ref. 165b.

Polyester is a general term for plastics that are formed by the esterification condensation of polyfunctional alcohols and acids. These cover a wide range of polymers. Saturated polyesters are high molecular weight linear thermoplastics, such as polyethylene terepthalate (Figure 4-13). Another thermoplastic poly-ester is polycarbonate which is a condensation product of bisphenol A and phosgene. It could be regarded as a polyester of carbonic acid.

Three-dimensional or cross-linked polyester resins are made with polyfunc-tional monomers. They are characterized by a vinyl unsaturation in the polyester background which allows subsequent curing or cross linking by copolymeriza-tion with another monomer. For example, ethylene glycol is condensed with maleic acid to form polydiglycol maleate. To this is added styrene monomer which cross-links. Alkyd resins are thermosetting and are based on unsaturated polyesters. They are formed from dibasic acids and polyhydric alcohols. The most common is called glyptal and is based on phthalic anhydride and glycerol.

Polyurethanes are formed by the reaction of polyhydroxy compounds, such as glycol and diisocyanates (Figure 4-14). A large family of plastics are derived including thermosetting resins, elastomers, casting resins, and foams.

Additional information on thermosetting, injection moldable materials is found in Ref. 209a to 220f of Chapter 3.

FIBROUS-GLASS REINFORCED THERMOPLASTICS (FRTP)

Thermosetting materials have long been reinforced with glass fibers or glass matting. In 1959, a U.S. Patent (166) was issued for a continuous glass roving which was sized to increase glass-plastic adhesion. The glass was impregnated into plastic which was then pelletized. This material which could be injection molded, was the commercial start of fiberglass reinforced thermoplastics. The large improvement of properties, which approached that of many metals, opened up new areas of application. Typical uses are housings for electrical tools, ducts, automotive hardware, dashboards, pump bodies, battery cases, bearings, valve parts, hammer handles, electrical switches, degreasing racks, and such.

Components of FRTP Systems

The system requires three components—plastic, coupling agents, and glass. All thermoplastics can be used. The coupling agent or sizing reacts with the glass and plastic to form a more continuous structure. It is very important in determining the ultimate properties of the molded part. They are usually organosilanes or chrome complexes of the Werner type (167 - 169).

The glass fibers used are among the strongest commercial materials, with

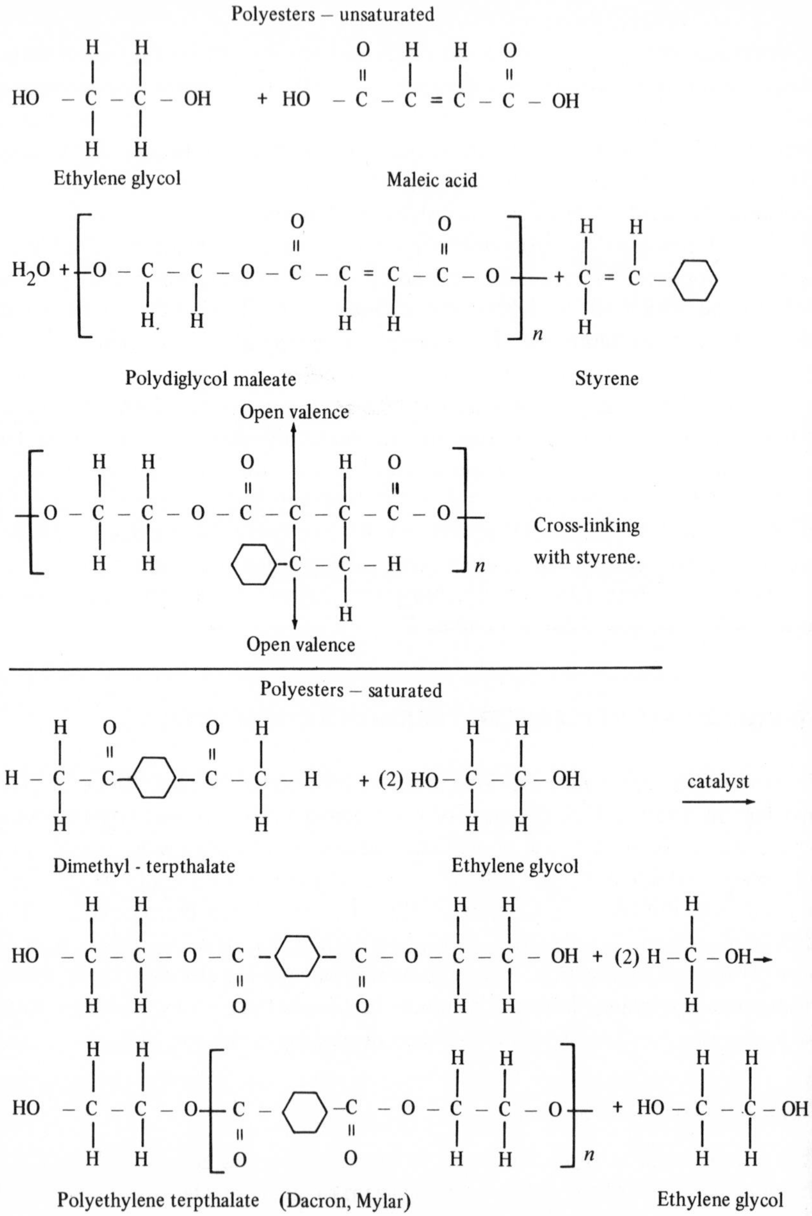

Figure 4-14 Saturated and unsaturated polyesters.

414

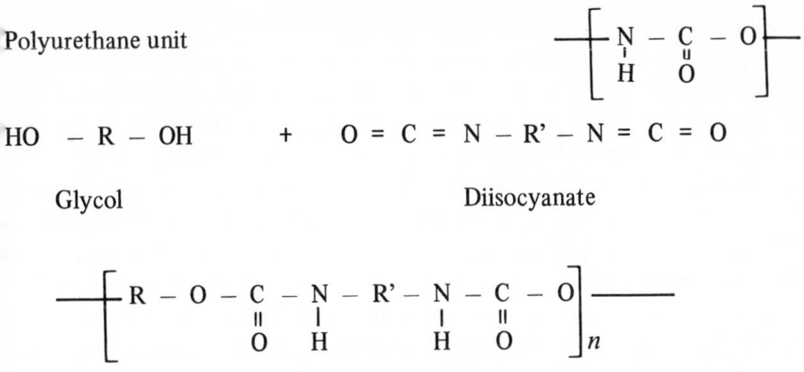

Polyurethane unit

$$\left[\begin{array}{ccc} N & - & C & - & O \\ | & & \| & \\ H & & O & \end{array}\right]$$

HO $-$ R $-$ OH $+$ O $=$ C $=$ N $-$ R' $-$ N $=$ C $=$ O

Glycol   Diisocyanate

$$\left[\begin{array}{ccccccc} R & - & O & - & C & - & N & - & R' & - & N & - & C & - & O \\ & & & & \| & & | & & & & | & & \| & \\ & & & & O & & H & & & & H & & O & \end{array}\right]_n$$

Figure 4-14.   Formation of a polyurethane polymer by the reaction of a glycol and a diiso-cyanate.

tensile strengths over 500,000 psi. Glass is dimensionally stable, resistant to chemical attack, incombustible, and softens at temperatures above 1500°F. Fibers are drawn down to diameters of 0.00037 to 0.00052 in. As they are extruded, they are coated with a lubricant, a bonding agent to hold the filaments in a strand and the coupling agent. They are available as continuous roving consisting of several strands each containing hundreds of filaments. They are wound into a cylindrical package which can be designed to unwind either from the inside or outside. They also come in chopped strands of lengths of 1/8, 1/4, 3/8, or 1/2 in. These are usually packaged 40 lb box, which is convenient weight for blending.

Spherical solid glass fillers can be used instead of glass fibers (170). The spheres also improve physical properties, though not to the extent of glass fibers. The flow properties of the sphere reinforced mix are better and the surface gloss is greatly improved.

Properties

FRTP plastics have increased mechanical properties (171 - 173) and are often used when nonreinforced plastics do not have the rigidity, strength, heat resistance, or other properties that are available in FRTP. Conversely many applications have developed replacing die casting metals where the factors of cost, minimal finishing, wear, corrosion resistance, weight savings, and inherent color are important. In some instances an expensive engineering material can be replaced with a glass filled version of a much cheaper material and save money, as shown in Table 4-9.

The tensile strength of glass reinforced thermoplastics increase dramatically, doubling or even tripling. Nylon, for example, has a tensile strength of 11,800 psi, while the 30% glass filled tensile strength is 29,000 psi. The increase in tensile strength is dependent upon the glass concentration. High density polyethylene with a tensile strength of 3700 psi increases to 6060 psi at 10% glass, 8840 at 20% glass, and 10130 at 30% glass (174). The length of the fiber does not have too much of an effect as long as it is at least 0.100 in. long.

A 33% glass reinforced nylon, after the fourth reground, lost approximately 20% of its tensile strength as the average fiber length was reduced from 0.022 to 0.012 (175). Other mechanical properties were similarly reduced. As one would expect the percentage elongation drops rapidly above 10% glass content. The yield points, however, increase and are considerably higher than the unmodified materials. The proportional limits (the range of Hookean behavior) are doubled or tripled. The modulus of elasticity increases dramatically. Unfilled polystyrene has a tensile modulus of 400,000 psi which increases to 1,200,000 at 30% glass reinforcement. Typical moduli for reinforced acetal is 800,000 psi, 6/6 Nylon 900,000 psi, and 1,700,000 psi for polycarbonate. The impact strength and toughness at room temperatures increase, although not as much as some of the other properties. At low temperatures most thermoplastics are brittle. Glass reinforcements seem to absorb the shock loading under such conditions and give distinctly higher Izod impact value. For higher temperatures (175a) the glass reinforcements raise the deformation temperature under load (DTL) for most plastics except for polycarbonates, polystyrene, SAN and ABS which show relatively little change.

Table 4-9 Comparison of some porperties of 20% glass reinforced polypropylene and 20% glass reinforced polystyrene with unreinforced "engineering" materials

	1970 Prices (¢/in.^{3})	Tensile Strength (psi)	Tensile Mod. (x10^5 psi)	Izod Impact	DTL at 264 psi °F	Coef. Exp. in./in./°F (x10^{-5})	Mold Shrink-age (in./in.)
PP-20% Glass	2.4	9,000	8.0	3	220	2.7	0.004
PS-20% Glass	1.35	12,000	10.0	1-2	200	2.2	0.0015
PC	3.85	9,000	3.6	2-15	270	3.8	0.006
Modified phenylene oxide PPO	2.87	10,000	3.8	2	265	3.3	0.007
Nylon	3.3	11,800	4.0	2	200	5.5	0.012
Acetal	3.34	9,000	4.0	2	250	4.5	0.022

Creep and Cold Flow.   These are significantly reduced with glass reinforcement. Mold shrinkage drops drastically to 0.003 in./in. for 20% polypropylene to 0.0005 in./in. for 20% filled polystyrene. This can be useful when dimensional problems exist, as changing concentration will change dimension. This can pose problems in ejection. Because there is a lower reduction in volume, there is less of a decrease in pressure caused by shrinkage. Therefore larger tapers are required. The material is more rigid which permits earlier ejection. For crystalline materials, the glass acts as a nucleating agent increasing the amount and distribution of crystallinity.

The Dielectric Constant and Dielectric Strength. These are relatively little affected. The volume resistivity is increased while the dissipation factor and arc resistance significantly decrease.

The Surface Finish.   This is affected by the cavity and material temperatures, the pressure, the percentage of glass, and the amount of dispersion. Above 20% glass content it is very difficult to get a good surface and a good surface gloss. Increasing the glass content also increases the flow length, probably because the glass holds heat. This makes the material good for thinner sections.

The Mold Design.   For FRTP the mold design is essentially the same as for other viscous materials. All sizes of gates and runner systems, including hot runners, have been used. Large runners, good venting, and large draft is desirable. Undercuts are very difficult to eject. Because of the abrasive nature of the material the mold should be at least R_c 40, preferably harder, and flash chromed.

How to Make FRTP

There are four methods for obtaining glass reinforced material. The simplest is to buy the material already compounded. It has the obvious advantage of ease of use and the elimination of the possibility of incorrect in-plant compounding. Compounded materials are more costly, require a larger inventory, require a longer lead time from order to production, and most important have already been processed. This means that the molder is essentially using reground material. This is important in that the length of the fiber has a direct bearing on physical properties. The tensile strength of a 20% by weight glass filled polypropylene precompounded with short fibers was 6000 psi, with longer fibers 7000 psi, while a dry blended material had its tensile strength at 9000 psi. Similarly for 30% nylon the figures were 18,500, 20,000 and 23,700 psi; for 25% polyethylene 6000, 6500, and 8300 psi (176).

The second method is to master batch using concentrates of 60 to 80% glass by weight. This has the usual advantages and disadvantages of a master batch procedure. It is particularly helpful in that glass fiber is not handled as such.

The third technique uses a powdered resin supply which is fed into a mixer. The glass rovings are fed to the mixer where they are chopped and mixed with the glass (177). This is a relatively clean way of handling the resin and is most applicable to large usages. It also involves a setup charge and potential royalty payment.

The most direct approach for in-plant compounding is to use chopped glass fiber as an additive and tumble blend or automatically mix it with the virgin plastic material. This is the least costly method and has similar advantages to dry coloring material. However, glass fiber is a difficult material to handle. Randomly oriented material might not be desirable. The possibilities of contamination and personal safety should be considered (178). Sizing agent technology is important and often unknown.

Effect of Molding Conditions

Deflection temperature under load and stiffness are the only important properties that are not at all affected by molding conditions. High screw back pressure and gate size show a constent effect. Raising the back pressure or decreasing the gate size seems to damage the glass fibers. An excellent discussion of molding conditions is found in Ref. 176. Other references for FRTP materials and properties are 167, 168, 174, 175, 179 to 185.

Other Fillers

Other materials are used as fillers and reinforcing agents to reduce the cost and acheive special properties. Talc, asbestos, graphite, molybdenum disulphide, TFE, and metallic powders are some examples. Reference 186 to 190 contain information about fillers and reinforcements for thermoplastics.

IDENTIFICATION OF PLASTIC MATERIALS

An experienced person can usually identify a plastic material with simple tests. A useful procedure is to do the following:

Look at and feel the plastic.
Differentiate between thermoplastic and thermosetting materials.
Find the density as it relates to water.
Burn a sample.
Determine halogen.
Check solubility.
Determine density.
Determine melting point.

Look and Feel Tests

The look and feel of a piece of plastic will very often identify it. Simple tests are used to differentiate between similar materials. For example, the "ring" of styrene when tapped with a finger nail or on a sharp object is characteristically different than that of acrylic. It can be used to differentiate between the two. In thick sections clarity would also be used. While "look and feel" is difficult to describe it is a rapid and excellent method of eliminating many materials.

Thermoplastics Versus Thermosets

It is very easy to distinguish between a thermoplastic and thermosetting material. If the thermoplastic is heated slowly it will soften and cool upon hardening. Thermosetting materials will not soften on continued heating.

Density Determinations

A plastic with a density less than one will float in water. This would indicate the polyethylenes, polypropylenes and ionomers. One must be careful not to use this test on foamed materials. If the material sinks it still could be one of the olefins, but with a filler.

A accurate determination of specific gravity (density) can be very helpful. It is not difficult to do. The larger the sample, the more accurate the determination. A one ounce sample is commonly used. Provision must be made to suspend the sample on a wire which is attached to one side of the balance. Accurate weighing, and removal of all air bubbles around the specimen while in water is important. A few drops of a wetting agent may be used. If the material is lighter than water a sinker should be used. Following procedure is suggested.

1. Weigh the wire (and sinker if used) in air. This is weight B.
2. Weigh the wire (and sinker if used) and plastic. This is weight A.
3. Weigh the wire (and sinker if used) and plastic in water. This is weight C.
4. Weigh the wire (and sinker if used) in water. Try to adjust the water container so that the level of wire immersion is about the same as in weighing C. This is weight D.

The specific gravity can be found

$$\text{Sp. gravity} = \frac{A - B}{(A - B) + (C - D)} \tag{4-4}$$

A final test would be determining the melting point which can be done on a Fisher-Johns (Fisher Scientific Company, New York, N. Y.) melting point apparatus. The sample is put on a hot plate and its changing condition watched. It is useful to put samples of known materials near by to observe them as well. This test is useful, for example, in distinguishing between FEP, melting point

554°F and TFE with a melting point of 621°F.

Burning the Sample

Applying heat is a useful, common, and practical test. Bunsen burners, propane torches, candles, and matches are good heat sources. The material is held in the flame to see whether and how it burns. The following burn with a yellow flame — rubber modified acrylic, ABS, SAN, cellulose acetate, ethyl cellulose, methylstyrene, polystyrene, styrene-acrylic copolymers, vinyl acetate, and vinyl alcohol. The following burn with a yellow flame with the characteristic green color — vinyl chlorides, vinyl acetate/vinyl chloride copolymers, and vinylidene chloride. Chlorinated ether has a green flame with a yellow top. The following have blue flames and yellow tops — acrylics, butyrate, proprionate, polypropylenes and nylons. Acetals and polyethylenes burn with a blue flame. With the addition of flame retardants, it is no longer possible to always use the self-extinguishing characteristics of a material for identification.

The second and most important characteristic from burning is the odor. Usually it can be clearly smelled after the flame is extinguished and/or the heat removed. If not, the plastic should be put in a test tube and heated there. Most plastics have a characteristic odor. For example, acetals smell like formaldehyde, styrenes like illuminating gas and acrylics have a fruity odor. The best way to use this test is to compare the odor with the burning of a known sample of the material.

Halogen Testing

A good way to detect chlorine and fluorine (halogens) in plastic is to use the Beilstein test. A copper wire is heated until the flame shows a very slight yellow color. A small piece of plastic is put on the wire and it is heated again. A characteristic green color in the flame indicates the presence of a halogen. The one difficulty with this test is its sensitivity. Flame retardants or other additives containing halogens will also give a positive test. The test will show that there are no halogens in the sample.

Solubility Tests

Sometimes chemical solubility can be used as a distinguishing medium. For example, ABS is partially soluble in cold carbon tetrachloride while SAN is not. Solubility tests form the basis for analytical procedures but are beyond the scope of this section.

BIBLIOGRAPHY

ASTM Standards, Volume 27, *Plastics-General Methods of Testing, Nomenclature,* ASTM Philadelphia, Pa., 1962.

The Properties and Testing of Plastics Materials, A.E. Lever, and J. Rhys, Chemical Publishing Co., New York, 1962.

Polymers and Resins, B. Golding, D. Van Nostrand Co., Inc., New York, 1959.

Manufacture of Plastics, Volume 1, W. Mayo Smith, (Ed.), Reinhold Publishing Co., New York, 1964.

Textbook of Polymer Science F. W. Billmeyer, Jr., John Wiley & Sons, Inc., New York, 1962.

Introduction to Polymer Chemistry, J. K. Stille, John Wiley & Sons, Inc., New York, 1962.

Whittington's Dictionary of Plastics, L. R. Whittington, Technomic Publishing Co., Inc., Stamford, Conn. 1968.

An Introduction to Polymer Chemistry, W. R. Moore, Aldine Publishing Co., Chicago, Ill., 1963.

Encyclopedia of Chemical Technology John Wiley & Sons, Inc., New York, 1968.

The Structure of Polymers, M. L. Miller, Reinhold Publishing Co., Stamford, Conn., 1966.

Engineering Properties of Thermoplastics, R. M. Ogorkiewicz, Wiley-Interscience, New York, 1970.

Plastics; Mach. Design, December 12, 1968.

REFERENCES

1. Material and Market Statistics for 1970, *MP*, January 1971, p. 65.
2. "Standard Tests on Plastics"; Bulletin G1C, *Celanese Plastics Company,* Newark, N.J.
3. *Modern Plastics Encyclopedia,* 1970/1, p. 65.
4. "Plastics Standards Digest", *SPE-J,* March 1969, p. 56; March 1971, p. 54.
4a. "Digest of Plastics Standards Issued in 1970," *SPE-J,* March 1971, p. 54,
4b. "Digest of Plastics Standards Issued in 1971." *SPE-J,* March 1972, p. 44.
5. "SPE Identification Properties List," *SPE-J,* March 1969, p. 54.
6. "Suggestions for Improved Physical Properties Tests " P. E. Finna, *SPE-J* November 1963, p. 1187.
7. "Meaningful Testing of Plastic Materials for Major Appliances," J. R. Thomas and R. S. Hagan, *SPE-J*; January 1966, p. 51.
8. "Cantilever Beam Test Measurement of Plastics' Aging," P. G. Kelleher, R. J. Miner, and D. J. Boyle, *SPE-J,* Feburary 1969, p. 53.
9. "Testing Molded and Extruded Parts," R. E. Hannah, and J. A. Blanchette, *SPE-J,* October 1968, p. 102.
10. "Evaluation Tests for Plastics Heels," E. E. Joiner, J. P. Szumski, and F. J. McGarry, *MP*, March 1960, p. 152.
11. "Evaluation of High Impact Polystyrene for Refrigerator Door Liners," D. A. Davis, J. V. Schmitz, R. S. Hagan, and R. O. Carhart, *SPE-J,* March 1961, p. 260.
12. "Specifying and Testing Plastics Materials for Automotive Applications," W. J. Simpson, *Plastics World,* August 1961.
13. "How to Predict PVC Bottle Strength," V. A. Matonis, and N. E. Aubrey, *SPE-J,* November 1970, p. 49.
14. "Mass Coloring of Plastics," P. K. Papillo, *MP*, December 1967, p. 131
15. "Measurement of Color and Other Appearance Attributes in the Plastics Industry," R. S. Hunter, *SPE-J,* February 1967, p. 51.
16. "Colouring of Thermoplastics," *Brit. Pl.,* April 1961; Part 1, p. 156; Part II, "The Dry Colouring Process," R. Mather, p. 162; Part III, "Extrusion Colouring," E. J. G. Balley, p. 167.
17. "Colorants for Plastics," J. E. Simpson, and D. P. Brush, *M P Enc.,* 1960.
18. "How to Choose the Right Colorant for the Right Resin," J. E. Simpson, *MP,* December 1962, p. 90.
19. Basic Colorants for Plastics," L. I. Weinrich, and H. G. Rotherham, *PDP*, September 1968, p. 12.
20. "Exotic Effects Through the Internal Coloring of Plastics," M. C. Felsher, and W. J. Hanau, *PDP*, October 1967, p. 19.
21. "Color in Plastics," *SPE-J,* September 1965, p. 1064,
22. "Dry Coloring Polypropylene," R. S. Charvat, *MP*, November 1959, p. 119.
23. "Choosing Colorants for Polystyrene," E. A. Wich, *MP*, June 1961, p. 86.
24. "Dry Coloring Lustrex Styrene," Product Information Bulletin 1031, Monsanto Chemical Co. Plastics Division, Springfield, Mass.
25. *Modern Plastics Encyclopedia,* 1969/70, p. 1010.

25a. "Process Profitably: Do Your Own Color Blending," L. L. Scheiner, *PT*, July 1971, p. 38.

25b. "How and When to Use Fluorescent Pigments in Plastics," T. J. Gray, *SPE-J*, August 1971, p. 53.

25c. "Fluorescent Whitening of Styrenes," K. Eschle and E. Preininger, *MP*, August 1971, p. 74.

25d. "Advances in Color Compounding Thermoplastics," R. Adams, *PDP*, June 1971, p. 12.

25e. "Is Bulk Material Handling for You?" R. J. Munns, *SPE-J*, October 1971, p. 65.

25f. "Custom Molder Grows (Material Handling)," *PDP*, December 1971, p. 10.

25g. "Buyers Guide to Granulators," *PDP*, February 1972, p. 14.

25h. "Nylon Reprocessing-Its Effects on Water Absorption and Mechanical Properties," E. A. Hazell and D. Sims, *BP.*, October 1971, p. 73.

26. "Pitfalls in Predicting the Performance Characteristics of High Density Polyethylene," V. L. Folt, and R. J. Ettinger, *SPE-J*, April 1961, p. 350.

27. "Flexural Strength of Polymethyl Methacrylate at Various Deflection Rates," R. E. Ely, *MP*, February 1960, p. 138.

28. "An Izod Specimen for the Simulation of Conditions of Plastics Use," C. G. Vinson, *PES*, November 1969, p. 388.

29. "Characterizing Impact Behavior of Thermoplastics," W. E. Wolstenholme, *J. Appl. Polymer Sci.*, VI, No. 21, 1962, p. 332.

30. "Impact Tests, Their Correlation and Significance," R. H. Schoulberg, and J. J. Gouza, *SPE-J*, December 1967, p. 32.

31. "An Appraisal of the Izod Impact Test for Toughness," C. E. Stephenson, *B. P.*, October 1961, p. 543.

32. "Assurance of Quality Molding by Energy to Break Testing," E. Heyman, *SPE-J*, May 1968, p. 49.

33. "Photoelastic Phenomena in Plastic During Impact," N. W. Hall, R. E. Wright, and W. C. Smith, *SPE-Trans.*, July 1962, p. 237.

33a. "On Relation Between Charpy Impact Strength and Izod Impact Strength According to ISO R 1163," Yukio Uemura, *Jap. Pl.*, July 1971, p. 10.

34. "High Speed Testing," L. Sandek, *PT*, February 1962, p. 26.

35. "Tensile Impact Properties of Thermoplastics," E. H. Simpson, and R. E. Polleck, *SPE Trans.*, January 1964, p. 25.

36. "An Investigation of Some Variables Affecting Impact Resistant Thermoplastics," E. G. Bobalek, and R. M. Evans, *SPE Trans.*, April 1961, p. 93.

37. "Dependence of Mechanical Properties on Molecular Motion in Polymers, R. F. Boyer, *PES*, July 1968, p. 161.

38. "On the Abrasion of Plastics," Y. Yamaguchi, *SPE Tech Pap.*, 1966, p. XIV-3.

39. "Hardness of Polymeric Materials," E. Baer, R. E. Maier, and R. N. Peterson, *SPE-J*, November 1961, p. 1203.

40. "A New Mar Resistance Tester for Plastics," A. E. Sherr, W. G. Deichert, and R. L. Webb, *PDP*, March 1970, p. 24.

41. "A Variable Speed Frictionometer," R. F. Westover, and W. I. Vroom, *SPE-J*, October 1963, p. 1093.

42. "The Frictional Properties of Thermoplastics," *Plastics*, March 1961, p. 117.

43. "Friction and Abrasion Characteristics of Plastics Materials," M. A. Marcucci, *SPE-J*, February 1958, p. 30.

44. "Understanding Plastic Wear," G. Bongiovanni, and M. Clerico, *MP*, May 1967, p. 126.

45. "Frictional Properties of Polyethylenes and Perfluorocarbon Polymers," R. C. Bowers, and W. A. Zisman, *MP*, December 1963, p. 139.

46. "Friction of Vinyl Chloride Plastics," J. B. DeCoste, *SPE-J*, October 1969, p. 67.

47. "Molybdenum Disulfide in Nylon for Wear Resistance," T. E. Powers, *MP*, June 1960, p. 148.

48. "TFE-Lubricated Thermoplastics," J. W. Lomax, and J. T. O'Rourke, *Machine Design*, June 23, 1966.

49. "Static Coefficient of Friction of Polyolefins," C. A. Friehe, *PES*, April 1966, p. 135.

50. "Plastic Bearings: An International Survey," J. C. Benedyk, *SPE-J*, April 1970, p. 78.

51. "Internally Lubricated RTP's for Gears & Bearings," J. E. Theberge, *MP*, March 1970.

52. "Low Friction Materials," A. R. Gardner, *Product Eng.* September 17, 1962, p. 111.

53. *Modern Plastics Encyclopedia,* 1969/70, p. 1026.

54. "Thermal Conductivity of Polymers," T. G. Smith, and A. R. Hoge, *SPE-J*, April 1967, p. 67.

55. "Heat Transfer During Polymerization," J. E. Funk, and J. F. Thrope, *MP*, August 1967, p. 137.

56. "Thermal Properties of Plastics," D. W. Sundstrom, *MP*, January 1970, p. 138.

57. "Relative Toxicity of Selected Polymeric Materials Due to Thermal Degradation," G. Epstein, and J. Heicklen, *SPE Tech. Papers*, 1970, p. 459.

58. "Combustion Gases Generating from Polyvinyl Chloride and Its Products," Y. Kobayashi, M. Hori, and H. Murata, *Japan Pl.*, January 1971, p. 40.

59. "A Method and Apparatus for Determining the Ignition Characteristics of Plastics," N. P. Setchkin, *J. Research National Bureau of Standards* 43,-Research paper RP 2052, December 1949.

60. "Ignition Temperatures of Plastics," G. A. Patten, *MP*, July 1961, p. 119.

61. , "New Flammability Indexes; What They Are, What They Mean," H. Reymers, *MP*, October 1970, p. 92.

62. "A New Temperature Index; Who Needs It," H. Reymers, *MP*, September 1970, p. 78.

63. "UL Temperature Index Chart," *M P Enc.,* 1970/71, p. 820.

64. "Flammability Chart," *M P Enc.,* 1970/71, p. 806.

65. "Candle-type Test for Flammability of Polymers," C. P. Fenimore, and F. J. Martin, *MP*, November 1966, p. 141.

65a. "Summary of Smoke and Flame Testing Activities with Plastic Materials in the United States," J. A. Blair and R. B. Aiken, *Constr. Specifier*, July 1971.

66. "Fireproofing of Polymers with Derivatives of Phosphines and with Halogen-Phosphorous Compounds," N. E. Boyer, and A. E. Vajda, *SPE Trans.*, January 1964, p. 45.

67. "Fire Retardant Components of Self-Extinguishing Polyesters," C. W. Roberts, *SPE Trans.*, April 1963, p. 111.

68. "The Polymer-Forming Reactions of 3,3,3, Trichloropropylene Oxide," H. C. Vogt, and P. Davis, *SPE Tech. Pap.*, 1970, p. 469.

69. "A Study of Flame Retarded ABS, Polystyrene and Polyester Systems," J. DiPietro, and H. Stepniczka, *SPE Tech. Pap.*, 1970, p. 463.

70. "Steady-State Combustion of Polyolefins," D. E. Stuetz, *SPE Tech. Pap.*, 1970, p. 369.

70a. "What Everyone Should Know About Fire Retardents for Plastics," C. D. Storrs and O. H. Lindemann, *PDP*, July 1972, p. 13.

70b. "Fire-Retarding Plasticizers-What's Happening Here and Now," S. C. Stinson, *PT*, March 1972, p. 34.

70c. "What You Should Know About Flame Retardants," I. Touval and H. H. Waddell, *PT*, July 1971, p. 29.

71. "Factors Affecting the Oxygen Index Flammability Ratings," J. DiPietro, Stepniczka, *SPE-J*, February 1971, p. 23.

72. "Flame Retardants Chart," *M P Enc.*, 1970/71, p. 854.

73. "Mechanisms Associated with Polymer Combustion and its Suppression," P. C. Warren, *SPE Tech. Pap.*, 1970, p. 362.

74. "How Plastics Burn," P. C. Warren, *SPE-J*, February 1971, p. 17.

75. "Polymer Decomposition; Thermodynamics, Mechanisms and Energetics," L. A. Wall, *SPE-J*, August 1960, p. 810.

76. "Thermal Degradation of Organic Polymers," S. L. Madorsky, *SPE-J*, July 1961, p. 665.

77. "Rates of Degradation of Organic Polymers," S. L. Madorsky, *SPE-J*, December 1962, p. 1482.

78. "Thermal Degradation of PVC," J. Stepek, Z. Vymazal, and B. Dolezel, *MP*, June 1963, p. 146.

79. "Polyvinyl Chloride Degradation and Stabilization," B. Baum, *SPE-J*, January 1961, p. 71.

80. "Progress in the Development of Heat-Resistant Organic Polymers," W. R. Dunnavant, *PDP*, April 1966, p. 11.

81. "Testing for Thermal Endurance; A Case History Based on Polysulfone Thermoplastics," T. E. Bugel, *SPE-J*, March 1968, p. 52.

81a. "A Better Way to Evaluate Arc Tracking," T. E. Steiner, *MP*, August 1971, p. 58.

82. "Chemical Structure and Stability Relationships in Polymers," B. G. Achhammer, M. Tryon, and G. M. Kline, *MP*, December 1959, p. 131.

83. "Environmental Resistance of Plastics," R. L. Bergen, Jr., *SPE-J*, July 1964, p. 630.

84. "Stress Cracking of Noncrystalline Plastics," R. L. Bergen, Jr., *SPE-J*, August 1968, p. 77.

85. "The Endurance of Polyethene under Constant Tension While Immersed in Igepal," W. A. Dukes, *B. P.* March 1961, p. 123.

86. "Environmental Stress Rupture of Polyethylene," L. I. Lander, *SPE-J*, December 1960, p. 1329.

87. "Environmental Stress and Strain Testing of HI Polystyrene and ABS," T. Stolki, *SPE-J*, October 1967, p. 48.

88. "Vapor Crazing of Polystyrene," R. W. Raetz, *MP*, September 1962, p. 153.

89. "On the Solvent Stress-Cracking of Polycarbonate," G. W. Miller, S. A. D. Visser, and A. S. Morecroft, *PES*, March 1971, p. 73.

90. "What Moisture Does to Nylons," G. W. Woodham, and D. R. Pinkston, *SPE-J*, June 1970, p. 44.

91. "Effects of Aging on ABS Plastics," B. D. Gesner, *SPE-J*, January 1969, p. 73.

92. "Photo-oxidation and Outdoor Aging of ABS," P. G. Kelleher, D. J. Boyle, and R. J. Miner, *MP*, September 1969, p. 188.

93. "Weathering of ABS," G. E. Boyce, and N. M. Jones, *B. P.* October 1970, p. 122.

94. "Dimensional Stability of Thermoplastics; Weathering Study of Nine Plastics," J. J. Scavuzzo, and C. U. McNally, *Pl. World*, March 1961, p. 26.

95. "Weathering of Plastics," R. A. Kinmonth, Jr., *SPE Trans.*, July 1964, p. 229.

96. "Fifteen Years of Weathering Results," G. R. Rugger, *SPE Trans.*, July 1964, p. 236.

97. "A Study of Weathering of an Elastomeric Polyurethane," C. S. Schollenberger, and K. Dinbergs, *SPE Trans.*, January 1961, p. 31.

98. "Natural and Artificial Weathering of Polyethylene Plastics," J. B. Howard, and H. M. Gilroy, *PES*, July 1969, p. 286.

99. "Development of an Impact Test for Evaluation of Weatherability of Rigid Plastics," R. C. Neuman, *PES*, April 1966, p. 124.

99a. "Selection Guide to Accelerated Weather-Testing Equipment," L. I. Nass, *PT*, January 1972, p. 30.

99b. "Selection Guide to Accelerated Weather-Testing Equipment," L. I. Nass, *PT*, March 1972, p. 31.

99c. "Polymers Under the Weather," F. H. Winslow, W. Matreyek, and A. M. Trozzolo, *SPE-J*, July 1972, p. 19.

99d. "Long-Term Weathering Can Change the Complexion of PVC," A. Blaga, *SPE-J*, July 1972, p. 25.

99e. "Improving Weatherability of Plastics With White Pigments," H. C. Jones, *MP*, January 1972, p. 90.

100. "UV Stabilizers for Plastics," A. L. Baseman, *PT*, April 1964, p. 30.

101. "Ultraviolet Stabilization of Plastics," F. J. Golemba, and J. E. Guillet, *SPE-J*, April 1970, p. 88.

102. "Zinc Oxide Stabilization of PP Against Weathering," D. S. Carr, B. Baum, A. Margosiak, and A. Llompart, *MP*, May 1970, p. 114.

102a. "UV Stabilization of Zinc Oxide With Thermoplastics," D. S. Carr, B. Baum, S. Margosiak, and A. Llompart, *MP*, October 1971, p. 160.

103. "How Heat and Light Affect Pigmented Polypropylene," C. W. Uzelmeier, *SPE-J*, May 1970, p. 69.

104. "Stabilization of Plastics," R. D. Deanin, *SPE-J*, September 1966, p. 12.

105. "Evaluation of Biocides for Treatment of Polyvinyl Chloride Film," A. M. Kaplan, M. Greenberger, and T. M. Wendt, *PES*, July 1970, p. 241.

106. "Brief Review of Biochemical Degradation of Polymers," R. K. Kulkarni, *PES*, October 1965, p. 227.

107. "Biodeterioration of Plastics," C. J. Wessel, *SPE Trans.*, July 1964, p. 193.

107a. "Evaluating Algistatic Activity of Films," S. Kaye, *MP*, August 1971, p. 78.

108.   "Antimicrobial Agents for Plastics," A. L. Baseman, *PT*, September 1966, p. 33.

108a.   "Thermal and Environmental (Chemical) Resistance of Glass-Reinforced Thermoplastics," J. E. Theberge, B. Arkles, and P. J. Cloud, *Mach. Des.*, February 4, 1971.

108b.   "Fillers vs. Properties of a Ductile Polyester," W. C. Jones, III and A. L. Fricke, *MP*, April 1972, p. 88.

108c.   "Organic Stabilizers for PVC Processing," C. H. Stapfer, R. D. Dworkin, and L. B. Weisfeld, *SPE-J*, May 1972, p. 22.

108d.   "Additives," A. A. Schoengood, *SPE-J*, June 1972, p. 21.

109.   "Some Characteristics of Lubricants in Rigid PVC," K. E. Andrews, C. Butters and J. Wain, *Br. Pl.*, 1970; Part 1, October, p. 97; Part II, November, p. 88.

110.   "The Aging of Plastics, P. Dubois, J. Hennicker, *Plastics;* Part 1, October 1960, p. 428; Part II, November 1960, p. 474; Part III, December 1960, p. 517; Part IV, January 1961, p. 115.

111.   "Polymers in Adverse Environments," R. E. Eshenaur, Proceedings of Conference on Polymers in High-Performance Applications, The Plastics Institute, Great Britain, p. 23.

112.   "On Aging of Plastics in Structural Applications," G. W. Ingle, *SPE Trans.*, July 1964, p. 224.

113.   "Some Physical Organic Chemistry of Aging Processes," F. R. Mayo, and K. C. Irwin, *PES*, July 1969, p. 282.

114.   "Dynamic Mechanical Testing of Polymers," G. C. Karas, and B. Warburton, *B. P.*, 1961, Part I, April, p. 189; Part II, March, p. 131.

115.   "Sonic Resonant Testing Determines Dynamic Elastic Moduli of Plastics, Nondestructively," *PDP*, December 1965, p. 18.

116.   "Nondestructive Testing of Plastics with Microwaves," R. J. Botsco, *PDP*, 1968; Part I, November, p. 12; Part II, December, p. 12.

117.   "Do-It-Yourself Polarized Light Tester", D. W. Pugh, W. F. McDonald, and W. V. Funk, *MP*, February 1960, p. 114.

118.   *Enclopedia of Chemical Technology,* Volume 16, p. 219.

119.   "Stereoregulation and Stereoregular Polymers," C. E. Schildknecht, *PES*, July 1966, p. 240.

120.   "Properties of Block Versus Random Copolymers," J. F. Kenney, *PES*, July 1968, p. 216.

121.   "Properties of a Propylene-Ethylene Block Copolymers," R. L. Tusch, *PES*, July 1966, p. 255.

122.   "New Plastic; Polyallomers," H. J. Hagemeyer, *MP*, June 1962, p. 157.

123.   "How to Modify Polyethylene with Flexible E-A Copolymers," B. H. Krevsky, S. A. Bonotto, and P. R. Junghans, *PT*, May 1963, p. 34.

124.   "Ethylene-Butene Copolymers," J. E. Pritchard, R. M. McGlamery, and P. J. Boeke, *Br. Pl.*, February 1960, p. 58; also PT, October 1960, p. 31.

125.   "Vinyl Chloride-Propylene Copolymers," R. D. Deanin, *SPE-J*, May 1967, p. 50.

126.   "Vinyl Chloride/Propylene Copolymers," M. J. R. Cantow, C. W. Cline, C. A. Heiberger, D.Th. A. Huibers, and R. Phillips, *MP*, June 1969, p. 126.

127.   "ABS Modification Improves PVC's Processibility," V.E. Malpass, *MP*, January 1970, p. 131.

128.   "Internally Plasticized Ether Copolymer Resins for Injection and Blow Molding," S. Ohtsuka, S. Yoshikawa and Y. Hoshi, *SPE-J*, May 1966, p. 75.

128a. "Acrylic Modifiers in Plasticized PVC," J. T. Lutz, Jr., *MP*, May 1971, p. 78.

129.   "Copolymers of Methylmethacrylate and Styrene for Moulding and Extrusion," T. E. Davies, *B.P.*, May 1960, p. 195.

130.   "A New Acrylic Thermoplastic," C. H. Schramm, and J. Briskin, *MP*, February 1960, p. 127.

131.   "ACS Resins", *B. P.*, September 1969, p. 129.

132.   *Engineering Properties of Thermoplastics p. 215*.

133.   *Manufacturer of Plastics p. 344.*

134.   Ref. 132, p. 150.

134a. "ASA for Aging, Weatherability, and Processability," *PDP*, July 1971, p. 20.

135.   Ref. 133, p. 437.

136.   "Properties of Nylon 11," J. G. Hawkins, *Plastics,* 1960; July, p. 257; August, p. 299.

137.   "Nylon 12," R. D. Deanin, *SPE-J*, September 1967, p. 44.

138.   "General Properties and Applications of Nylon 12," R. Kraft, *Br. Pl*, May 1969, p. 85.

139.   "Nylons from Vegetable Oils; 13, 13/13, 6/13," R. B. Perkins, J. J. Roden III, A. C. Tanquary, and I. A. Wolff, *MP* May 1969, p. 136.

140.   Ref. 118, Volume 16, p. 1-46, 88-105.

141.   Ref. 132, p. 175.

142.   Ref. 133, p. 512.

143.   Ref. 118, Volume 16, p. 106.

144.   Ref. 118, Volume 14, p. 259.

145.   "Polypropylene," Regional Technical Conference Papers, Philadelphia, Pa., *SPE* April 1962.

146.   Ref. 118, Volume 14, p. 282.

147.   Ref. 132, p. 61.

148.   Ref. 133, p. 194.

149.   Ref. 133, p. 402

150.   Ref. 118, Volume 21, p. 369-412.

151.   Ref. 132, p. 247.

152.   Ref. 133, p. 303.

153.   Ref. 133, p. 535.

154.   "Poly (4-Methylpentene-1)," R. D. Deanin, *SPE-J*, February 1967, p. 39.

155.   "TPX Methylpentene Polymers," B. S. Dyer, *Plastics,* October 1968, p. 1145.

156.   Ref. 132, p. 131.

157.   "Thermoplastic Polyarylsulfone," G. Morneau, *MP*, January 1970, p. 150.

158.   "Polysulfones-Properties and Processing Characteristics," H. M. Clayton, and A. E. Thornton, *Plastics*, 1968; January p. 76; February, p. 187; March, p. 310.

159.   Ref. 118, Volume 16, p. 272.

160. "New Elastomeric Thermoplastics," P. F. Hartman, C. L. Eddy, and G. F. Koo, *SPE-J*, May 1970, p. 62.

161. "Butadiene/Styrene Thermoplastic Elastomers," R. D. Dean, *SPE-J*, January 1967, p. 45.

162. "Thermoplastic Polyurethanes," C. L. Gable *PDP*, May 1968, p. 16.

163. "Comparison of Properties of Flexible Thermoplastics," S. Bonotto, and R. K. Walton, *MP*, May 1963, p. 143.

164. Ref. 118, Volume 21, p. 57.

165. Ref. 133, p. 456.

165a. "Fluropolymer" *SPE-J*, November 1971, p. 17.

165b. "Epoxy-Acrylate Resin, Ripoxy," *Jap. Pl.*, October 1971, p. 6.

165c. "Polytetramethylene Terepthalate (PTMT)," *SPE-J*, July 1971, p. 10.

166. U. S. Patent 2,877,501, Rexford Bradt (assigned to Fiberfil, Inc.), 1959.

167. "Silane Sizes for Glass-Reinforcement in Polyolefins," J. G. Marsden, *SPE-J*, December 2, 1970, p. 46.

168. "Chemically Coupled Glass-Reinforced Polypropylene," L. C. Cessna, J. B. Tomson, and R. D. Hanna, *SPE-J*, October 1969, p. 35.

169. "The Effect of Silane Coupling Agents in Improving the Properties of Reinforced Thermoplastics," S. Sterman, and J. G. Marsden, *PES*, April 1966, p. 97.

170. "Spherical Glass Fillers," O. R. Strauch, *SPE-J*, September 1969, p. 38.

171. "A Study of the Tensile Strength of Short Fiber Reinforced Plastics," J. K. Lees, *PES*, July 1968, p. 195.

172. "A Study of the Tensile Modulus of Short Fiber Reinforced Plastics," J. K. Lees, *PES*, July 1968, p. 186.

172a. "Are Fibre Reinforced Thermoplastics Insufficiently Promoted," *Br. Pl.*, December 1970, p. 83.

173. "Flexural Fatigue of Glass-Reinforced Thermoplastics," L. C. Cessna, J. A. Levens, and J. B. Thomson, *PES*, September 1969, p. 339.

174. "How to Get More from Glass-Fiber Reinforced HDPE," A. C. Bernardo, *SPE-J*, October 1970, p. 39.

175. "Reinforced 66 Nylon-Molding Variables vs Fiber Length vs Physical Properties," W. C. Filbert, Jr., *SPE-J*, January 1969, p. 65.

175a. "Glass Fiber Enhances High-Temperature Performance of Thermoplastics," F. C. Krautz, *SPE-J*, August 1971, p. 74.

176. "Critical Parameters for Direct Injection Molding of Glass Fiber Thermoplastic Powder Blends," W. R. Schlich, R. S. Hagan, J. R. Thomas, D. P. Thomas, and K. A. Musselman, *SPE-Tech. Pap.*, 1967, p. 929.

177. Canadian Patent 750,283, Dow Chemical Co.

178. "What's the Best Route to RTP Processing," J. E. Hauck, *MP*, July 1968, p. 80.

179. "Glass-reinforced PE vs Ethylene/acrylic Acid Copolymers," L. Lieng-Huang, and J. K. Rieke, *MP*, August 1969, p. 98.

180. "New Measures of Performance for Glass-HDPE Concentrate Blends," Y. Cornwell, and D. S. Stotz, *MP*, October 1969, p. 82.

181. "Injection Moulding of Glass Reinforced Nylon," *BP.*, May 1969, p. 89.

182. "Effect of Moisture on Reinforced Nylons," K. Hattori, G. W. Woodham, and D. R. Pinkston, *PDP*, March 1971, p. 28.

183. "The Effects of Fiber Glass Reinforcement on the Flammability Properties of Thermoplastics," K. Hattori, and D. Harris, *PDP*, August 1967, p. 28.

184. "Fiberglass-Reinforced Thermoplastics," R. D. Deanin, *SPE-J*, June 1967, p. 87.

185. "Property and Cost Advantages with Reinforced Thermoplastics," W. G. Miller, *BP.*, August 1970, p. 59.

186. "How to Select Fillers and Reinforcements for Thermoplastics," R. E. Hunt, *PT*, November 1969, p. 38.

187. "Mineral-Reinforced Thermoplastics," G. J. Fallick, H. J. Bixler, R. A. Marsella, F. R. Garner, and E. M. Fettes, January 1968, p. 143.

187a. "New Mineral Fiber Reinforcement for Thermoplastics," *SPE-J*, February 1972, p. 18.

188. "Asbestos-Reinforced Thermoplastics," E. A. Noga, and R. T. Woodhams, *SPE-J*, September 1970, p. 23.

189. "One Way to Strengthen Graphite-Polycarbonate Composite," F. S. Cheng, J. L. Kardos, T. L. Tolbert, *SPE-J*, August 1970, p. 62.

190. "Filled and Reinforced Polyolefins," R. F. Jones, *SPE-J*, August 1968, p. 71.

191. "Waxes as Lubricants in Plastics Processing," G. Illmann, *SPE-J*, June 1967, p. 71.

192. "Effect of Variables in Artificial Weathering on the Degradation of Selected Plastics," M. R. Kamal, *PES*, October 1966, p. 333.

193. "Mechanism of Ultraviolet Degradation and Stabilization in Plastics," R. D. Deanin, S. A. Orroth, R. W. Eliasen, and T. N. Greer, *PES*, July 1970, p. 228.

194. "Mechanisms of Ultraviolet Stabilization of Plastics," J. H. Chaudet, G. C. Newland, H. W. Patton, and J. W. Tamblyn, *SPE Trans.*, January 1961, p. 26.

195. "Ultraviolet Degradation of Plastics and the Use of Ultraviolet Absorbers," R. C. Hirt, N. Z. Searle, and R. G. Schmitt, *SPE Trans.*, January 1961, p. 21.

196. "A Method for Determination of Maximum Heat Resistance of Plastics," I. F. Kanavetz, and L. G. Batalov, *SPE Trans.*, April 1961, p. 63.

197. "Formulating Ethylene-Vinyl Acetate Copolymers," J. Fischer, and K. S. Tenney, *SPE-J*, December 1967, p. 65.

198. Ref. 132, p. 101.

199. Ref. 133, p. 66.

CHAPTER 5

Correcting Molding Faults

Injection molding faults may appear when starting a new mold, when changing to a new material, or during the regular operation of a mold.

INJECTION MOLDING FAULTS

The causes of these faults can be grouped as follows:

Machine.
Molds.
Operating conditions.
    Time
    Temperature
    Pressure
Material.
Part design.
Management.

Quality Control is Important

Before correcting a fault, one must find it. To find a fault good quality control is necessary. Quality control should not start when a customer returns rejects. It should be a continuing process that starts when the raw material is ordered and follows each operation until the product is shipped.

    Unless the equipment is adequate and subjected to a continual, effective, maintenance program, consistent, quality, injection molding is not possible. Molds must be kept in good operating condition. Auxiliary equipment, such as mold heating and refrigeration units, grinders, finishing tools, and gauges, have to be readily available.

Good Housekeeping Is Necessary

Good lighting, clean, quiet, comfortable working conditions, and an attitude reflecting maximum safety are also needed. Cleanliness is hard to maintain in a molding plant. Spilled powder, oil from the hydraulic system, and parts on the floor are some of the things which make good housekeeping difficult. Notwithstanding good housekeeping is very important. Lack of it is a major cause for rejected parts.

Communications Are Important

Poor communication and record keeping are an obvious source of molding problems. Injection molding permits the use of operators with minimal skills, and it is normally difficult to communicate effectively at this level. It is particularly so when the communication often involves information that is difficult to understand without adequate education. In some areas of the country many operators do not speak English. Until the industry is able to upgrade its operators, maximum attention should be paid to in-plant training.

Molding Conditions Should Be Watched

The molded part is the result of the effects of temperature, time, and pressure on the material during the molding process. Their equilibrium conditions in molding are precarious and extremely interdependent. For example, if the operator leaves the gate open too long during molding, he will change the temperature of the material, which could cause degradation or burning if it increases. The higher temperature will lower the viscosity which might flash the mold. It will also transmit pressure more effectively, possibly packing the mold causing sticking. The operator might then have to stop the machine to clean out the parts. He will then purge from the cylinder and start to mold. The material may be too cold for the first several shots, not filling out and giving different shrinkages and physical properties. This interruption might also effect the mold temperature. It is not hard to understand why the following conditions must be met before attempting to correct faults.

1. Consistent machine operation,
2. Consistent temperature control,
3. Consistent mold operation,
4. Consistent cycle time.

A single cavity mold should fill evenly. A multiple cavity mold must fill evenly. If one cavity fills first, the gate may seal off so that it will not be fully packed. Consequently, a late filling cavity might be overly packed, causing sticking and a high molded-in stress level. The warpage and shrinkages will be different. Additionally, a part might start to fill, the gate freeze, and the filling stop. Subsequent pressure might reopen the gate and start filling again. This is called interrupted filling and causes all kinds of problems.

Some of the areas for machine-caused faults as opposed to people-caused faults follow:

Clamping capacity,
Plasticizing capacity,
Plasticizing type,
Ejection system.

Unfortunately, we do not directly measure what we wish to control. We measure the temperature of the cylinder and nozzle instead of the material. We measure the hydraulic pressure of the oil behind the injection cylinder rather than the pressure on the plastic in the cylinder, nozzle, and mold. Injection speed, which is very important, is not measured at all. New equipment and techniques are becoming available to remedy these conditions.

The operating conditions which affect molding faults are time, temperature, and pressure. *Time* is influenced by

Cycle delay time.
Injection speed.
Injection forward time.
Screw rotation time.
Curing time.
Platen motion time.
Time to open and close gate.
Time to remove parts.
Condition of hydraulic equipment which is reflected in the speed of hydraulic action.

Some of the factors relating to *temperature* in the molding cycle follow:

Cylinder temperature.
Nozzle temperature.
Mold temperature.
Material temperature in the hopper.
Screw speed.
Back pressure.
Design of the screw.

When molding problems seem to be temperature related, it is good policy to check the temperature control *system* of the machine. Thermocouples should be calibrated in boiling water. Each band should be checked with an ammeter while energized. If there are relays in the heating system, they should be inspected.

The effects of *pressure* are found in:

Clamp pressure.
Injection pressure.
Boost time.
Screw torque.
Material feed.

In turn, the major variables related to *material* are

Viscosity—temperature characteristics.
Viscosity—shear characteristics.
"Set-up" temperature for ejection.
Crystallization characteristics.
Granulation.
Lubrication.
Colorant.
Additives.
Mechanical properties.

BASIC TROUBLE-SHOOTING EQUIPMENT

The following are extremely useful in trouble shooting:

1. Hand pyrometer with a flat tip for measuring mold and cylinder temperatures and a needle probe for measuring material temperatures.
2. Voltmeter.
3. Hook-on ammeter.
4. Ohmmeter.
5. Stop watch.
6. Micrometers, verniers, and scales.
7. Weighing scale to check shot and piece weights.
8. Magnifying glass.
9. Knife.
10. Polaroid sheets for checking birefringence of clear parts.
11. Marking pen, crayon, or self-adhesive labels for identifying parts.
12. Paper for recording results.

There are a number of basic rules to follow:

1. Think before you act.

2. Consult with others.

3. Discuss what has happened and what you expect to do.

4. Try one thing at a time and continue to do so until the results of the change are known. There are no short cuts in this area.

5. If possible observe the job running.

6. Take advantage of other people's experience and consult with the foreman, set-up men, quality control men, operators, or others involved in the job.

7. Carefully visually inspect the molded part, the mold, the material, and the machine.

8. If there is a choice of doing several things, change things that are easiest to revert to their original condition (molding conditions rather than mold changes) and those which are quickest to do.

9. If something *has* to be done, do it regardless of the time involved. For example, if contamination seems to be coming from the cylinder, a choice has to be made of stripping the cylinder or moving the mold to another machine. Both are time consuming. If it has to be done, the quicker it is started the quicker the fault will be corrected.

10. Finally, do not make an ill advised judgment because there is pressure to get the job done. These "short cuts" rarely work.

Before doing anything one should attempt to narrow the areas into which the problem belongs, that is, machine, mold, operating condition, material, part design, and management. Some simple suggestions follow:

1. Change the material. Try different material from the same vendor, different vendors and reground. If the problem remains the same it probably is not the material.

2. If the problem appears in about the same position of a single cavity mold, it is probably a function of the flow pattern and system from the front of the plunger through the nozzle, sprue, runner, and gate. It might also indicate a scored cylinder or something hanging up therein. In a plunger machine it may indicate trouble in a particular spot in the cylinder.

3. If the problem appears in the same cavity or cavities of a multicavity mold, it is in the cavity or the gate and runner system leading thereto.

4. If the trouble occurs at random it is probably a function of the machine, the temperature control system, or the heating bands. Changing the mold from one press to the other permits a determination of whether it is in the machine, mold and/or powder.

5. If the problem appears, disappears, or changes with the operator, observe the differences in their action.

6. Changing the type material may pinpoint the problem.

Specific molding problems may be arbitrarily divided into the following categories:

1. Short shots.
2. Parts flashing.
3. Sink marks and bubbles.
4. Poor welds.
5. Brittleness.
6. Material discoloration.
7. Splay, mica, and flow marks.
8. Surface blemishes at the gate.
9. Warpage and shrinkage.
10. Dimensional control.
11. Sticking in the mold.
12. Sticking in the sprue.
13. Nozzle drooling.
14. Faults in molding thermosets.
15. Excessive cycles.

It is not possible to present the items in each category in order of importance or frequency of occurrence. In most instances the fault will suggest both the cause and the remedy. Additional information can be had by referring to the text and manufacturer's literature. Material manufacturers also have technical service men available.

HOW TO AVOID SHORT SHOTS

Short shots are usually caused by the material solidifying before it completely fills the cavity, usually due to an insufficient effective pressure on the material in the cavity. Cure for short shots may require increasing the temperature and the pressure on the material, increasing the nozzle-sprue-runner gate system improving the mold design, and redesigning the part.

Machine Causes

1. No material in the hopper.
2. Hopper's throat partially or completely blocked.
3. Feed control set too low.
4. Feed control set too high which can cause lowering of injection pressure in a plunger machine.
5. Feed system operating incorrectly.
6. Plasticizing capacity of the machine too small for the shot.
7. Inconsistent cycles—machine-caused, operator-caused, or mold-caused.
8. Malfunctioning of the return valve on the tip of the screw. This is usually indicated by the screw turning during injection.

Molding Condition Causes

1. Injection pressure too low.
2. Loss of injection pressure during cycle.
3. Injection forward time too short.
4. Injection boost time too short.
5. Injection speed too low.
6. Unequalized filling rate in cavities.
7. Interrupted flow in cavities.
8. Inconsistent cycle—operator-caused.

Temperature Related Causes

1. Raise cylinder temperature.
2. Raise nozzle temperature.
3. Check pyrometer, thermocouple, heating bands system.
4. Raise mold temperature.
5. Check mold temperature equipment.

Mold Related Causes

1. Runners too small.
2. Gate too small.
3. Nozzle opening too small.
4. Improper gate location.
5. Insufficient number of gates.
6. Cold slug well too small.
7. Insufficient venting.
8. Inconsistent cycles-mold cause.

Materials Deficiencies.   Flow properties of the material are too low for the parts.

HOW TO AVOID PARTS FLASHING

Parts flashing are usually caused by a mold deficiency. Other causes are the injection force being greater than the clamping force, overheated material, insufficient venting, excess feed, and foreign matter left on the mold. Typical causes follow:

Mold Problems

1. Cavities and cores not sealing.
2. Cavities and cores out of line.
3. Mold plates not parallel.
4. Insufficient support for cavities and cores.

5. Mold not sealing off because of foreign material (flash) between surfaces
6. Something other than flash keeping the mold open (i.e., foreign material in leader pin bushing so that leader pin is obstructed when entering the bushing, keeping the mold open).
7. Insufficient venting.
8. Vents too large.
9. Land area round the cavities too large, reducing the sealing pressure.
10. Inconsistent cycle—mold-caused.

Machine Problems

1. Projected area of the molding parts too large for the clamping capacity of the machine.
2. Machine set incorrectly.
3. Mold put in incorrectly.
4. Clamp pressure not maintained.
5. Machine platens not parallel.
6. Tie bars unequally strained.
7. Inconsistent cycle—machine-caused.

Molding Problems

1. Clamp pressure too low.
2. Injection pressure too high.
3. Injection time too long.
4. Boost time too long.
5. Injection feed too fast.
6. Unequalized filling rate in cavities.
7. Interrupted flow into cavities.
8. Feed setting too high.
9. Inconsistent cycle—operator-caused.

Temperature Problems

1. Cylinder temperature too high.
2. Nozzle temperature too high.
3. Mold temperature too high.

Materials Problems.   Flow too soft for parts.

HOW TO PREVENT SINK MARKS—VOIDS

Sink marks are usually caused by insufficient pressure on the parts, insufficient material packed into the cavity, and piece part design. They occur when there are heavy sections, particularly adjacent to thinner ones. Such defects are predictable and the end user should be cautioned before the mold is built. Many times the solution to the sink mark problem will lie in altering the finished part to make the sink mark acceptable. This might include a design over the sink mark or hiding it from view.

Voids are caused by insufficient plastic in the cavity. The external walls are frozen solid and the shrinkage occurs internally creating a vacuum. Inadequate drying of hygroscopic materials and residual monomer and other chemicals in the original powder may cause the trouble. When voids appear immediately upon opening the mold, it is probably a material problem. When the voids form after cooling it is usually a mold or molding problem. Typical causes for voids are listed below:

Molding Causes

1. Insufficient feed (see section on *short shot*).
2. Increase the injection pressure.
3. Increase the injection forward time.
4. Increase the boost time.
5. Increase the injection speed.
6. Increase the overall cycle.
7. Change the method of molding (try intrusion).
8. Inconsistent cycles—operator-caused.

Temperature Related Causes

1. Material too hot causing excessive shrinkage.
2. Material too cold causing incomplete filling and packing.
3. Mold temperature too high so that the material on the wall does not set up quickly enough.
4. Mold temperature too low preventing incomplete filling.
5. Local hot spots on the mold.
6. Change the cooling pattern.
7. Mold temperature control system malfunctioning.

Possible Mold Changes

1. Increase the gate size.
2. Increase the runner size.
3. Increase the sprue size.
4. Increase the nozzle size.
5. Vent mold.
6. Equalize filling rate of cavity.
7. Prevent interrupted flow into the cavities.
8. Gate in thick sections.
9. Reduce uneven wall thickness where possible. Use cores, ribs, and fillets.
10. Inconsistent cycle—mold-caused.

Machine Changes

1. Increase plasticizing capacity of the machine.
2. Make cycle consistent.

Materials Changes

1. Dry material.
2. Add lubricant.
3. Reduce volatiles in material.

Changes in Cooling Conditions

1. Piece cooled too long in mold, preventing shrinking from the outside in. Shorten mold cooling time.
2. Cool part in hot water.

HOW TO PREVENT POOR WELDS (FLOW MARKS)

Poor welds and flow marks are caused by insufficient temperature and pressure at the weld location. Poor welds have the dismaying proclivity of showing up as broken parts in the field.

Flow marks are the result of welding cooler material around the projection, such as a pin. And their visibility depends on material, color, and surface. Flow marks rarely present mechanical problems, they are inherent in the design of some parts and cannot be eliminated, and this should be kept in mind before the part design is finalized.

Poor welds may be prevented by venting at the weld (where possible), adding a run-off from the weld section may help, increasing the thickness of the part at the weld, lubricating the material and/or local heating of the mold. Ways of eliminating poor welds and flow lines are listed below:

Temperature Problems

1. Too low cylinder temperature.
2. Too low nozzle temperature.
3. Too low mold temperature.
4. Too low mold temperature at spot of weld.
5. Uneven melt temperature.

Molding Problems

1. Inection pressure too low.
2. Injection feed too slow.

Mold Problems

1. Insufficient venting at the weld.

2. Insufficient venting of the piece, therefore add run-off at weld.
3. Runner system too small.
4. Gate system too small.
5. Sprue opening too small.
6. Nozzle opening too small.
7. Gate too far from the weld. Add additional gates (these might add additional welds but put them in a less objectionable location).
8. Wall section too thin causing premature freezing.
9. Core may be shifting causing one wall to be too thin.
10. Mold may be shifting causing one wall to be too thin.
11. Part is too thin at weld; thicken it.
12. Unequal filling rate, which should be equalized.
13. Interrupted filling.

Machine Problems

1. Plasticizing capacity too small for the shot.
2. Excessive loss of pressure in the cylinder (plunger machine).

Materials Problems

1. Contaminated material—this can prevent knitting properly.
2. Poor material flow. Lubricate material for better flow.

HOW TO PREVENT BRITTLENESS

Brittleness is caused by degradation of the material during molding. It may be accentuated by a part which is designed at the low limits of mechanical strength. Typical problems and solutions follow.

Molding Problems

1. Low cylinder temperature. Increase cylinder temperature.
2. Low nozzle temperature. Increase nozzle temperature.
3. If material is thermally degrading, lower cylinder and nozzle temperature
4. Increase injection speed,
5. Increase injection pressure,
6. Increase injection forward time,
7. Increase injection boost.
8. Low mold temperature. Increase mold temperature.
9. Part stressed. Mold so that part has minimum stress.
10. Weld lines. Mold to minimize weld line.
11. Screw speed too high, degrading the material.

Mold Problems

1. Part design too thin.
2. Gate too small.
3. Runner too small.
4. Add reinforcement (ribs, fillets).

Materials Problems

1. Contaminated material.
2. Wet material.
3. Volatiles in material. Use material with lower volatile content.
4. Too much reground. Reduce the amount of reground.
5. Low strength materials. Increase strength of material (i.e., add more rubber to high impact polystyrene).

Machine Problems

1. Machine plasticizing capacity too low for the machine.
2. Cylinder obstruction degrading the material.

HOW TO PREVENT MATERIAL DISCOLORATION

Material is often discolored by burning and degradation. This is caused by excessive temperatures, material hanging up somewhere in the system—usually in the cylinder—and flowing over sharp projections such as a nick in the nozzle. The other major cause for discoloration is contamination. This can come from the material itself, poor housekeeping, poor handling, and colorant floating in the air. Causes for material discoloration follow.

Materials Problems

1. Contamination.
2. Material is not dry.
3. Too many volatiles in the material.
4. Material degrading.
5. Colorant degrading.
6. Additives degrading.

Machine Problems

1. Dirty machine.
2. Dirty hopper dryer.
3. Dirty atmosphere. Colorants can float in the air and settle in the hopper and grinder.

These problems may be solved, if you do the following:

   a. Clean nozzle,
   b. Inspect nozzle and sprue bushing for burrs
   c. Purge cylinder
   d. Reseat nozzle
   e. Clean cylinder and check for burrs
   f. Check for cracked cylinder

4. Injection end of machine too large.
5. Thermocouple not functioning.
6. Temperature control system not functioning.
7. Heater band not functioning.
8. Cylinder obstruction degrading the material.

Temperature Based Problems

1. Cylinder temperature too high; decrease temperature.
2. Nozzle temperature too high; decrease temperature.

Molding Problems.

1. Decrease screw speed,
2. Decrease back pressure,
3. Reduce clamp pressure,
4. Decrease injection pressure,
5. Decrease injection forward time,
6. Decrease injection boost time,
7. Slow down injection rate,
8. Decrease cycle.

Mold Problems

Degradation and burning this may be prevented, if you do the following.

   a. Vent mold,
   b. Increase gate size,
   c. Increase runner-sprue-nozzle system,
   d. Change gating pattern,
   e. Remove lubricant and oil from mold,
   f. Investigate mold lubricant.

2. Too much stress. Check ejection system. Some rubber based materials such as high impact polystyrene and ABS will discolor on stressing.

HOW TO PREVENT SURFACE DEFECTS (Splay Marks – Mica)

Surface defects are the result of the following:

1. Contamination.
2. Wetness of hygroscopic material and water condensed on nonhygroscopic pellets.
3. Nozzle drool which is picked up by the incoming material and deposited in the mold. This material does not have time to melt.
4. Degraded and burnt material.
5. Jetting, where the material shoots into the mold and exhibits melt fracture.
6. Splay marks or blushing at the gate, caused by melt fracture.
7. Excessive lubricant.
8. Pressure defects which cause characteristic ripples and pit marks.

Splays, mica, flow marks, and surface disturbances at the gate are probably the hardest molding faults to overcome, (1). If molding conditions do not help, it is usually necessary to change the gating system and the mold temperature control. In some cases, the thermal conductivity of the mold material may initiate these surface defects (2). Sometimes localized heating at the gate will solve the problem. Heating at the gate with a propane torch will determine the advisability of adding a heat sink or a heating cartridge in the gating area. Reasons for defects and some of the solution follow.

MATERIAL PROBLEMS

1. Contamination (see "Material Discoloration" section) p. 442. Eliminate contamination as a source of the problem.
2. Wet material, which should be properly dried.
3. Nonuniform particle size materials. Therefore, do the following.

   a. Use uniform size particles,
   b. Reduce the amount of fines.

Machine Problems

1. Obstructions. Therefore, do the following:

   a. Check nozzle for partial obstruction.
   b. Check sprue-nozzle-cylinder system for restrictions and burrs.

2. Drool. Therefore, do the following:

   a. Use positive shutoff nozzle to prevent drool,
   b. Use sprue break to prevent drool,
   c. Use suckback to prevent drool.

3. Insufficient machine capacity.

Molding Problems

1. Degradation. Therefore,

   a. Reduce screw speed.
   b. Reduce back pressure

2. If jetting is the cause, reduce injection speed.
3. Alter injection speed,
4. Increase injection pressure,
5. Increase injection forward time,
6. Increase booster time,
7. Increase cycle.

Temperature Problems

1. Too low or too high cylinder temperature depending on problem. Therefore, change temperature profile of cylinder.
2. Too low mold temperature. Therefore raise mold temperature.
3. Nonuniform mold temperature. Check.
4. High nozzle heat may cause drooling. Lower if necessary.

Mold Problems

1. Increase cold slug well,
2. Increase runner extension.
3. Increase runner.
4. Polish sprue runner and gate.
5. Open gate or change gate to tab.
6. Change gate location. If jetting, flare gate or use tab or flared gate.
7. Increase venting,
8. Improve mold surface,
9. Clean mold surface.
10. Excess lubricant. Therefore, do the following:

    a. Use minimum amount of lubricant.
    b. Change type of lubricant.

11. Water caused by leaks and condensation. Remove.
12. Flow over depressions and raised section. Change the part design.
13. Try localized gate heating.

SURFACE DISTURBANCES AT THE GATE

Splay marks or blushing at the gate is usually caused by melt fracture as the material expands entering the mold. It is usually corrected by changing the gate design and localizing gate heating. The latter can be checked by using a propane

torch to temporarily heat the area. Occasionally it will be necessary to change
the gate location.

Molding Changes

1. Increase barrel temperature.
2. Increase nozzle temperature.
3. Slow down injection rate.
4. Increase injection pressure.
5. Change injection forward time.
6. Use minimum lubricant.
7. Change lubricant.

Mold Changes

1. Raise mold temperature.
2. Increase gate size.
3. Change gate shape (tab or flare gate).
4. Increase cold slug well.
5. Increase runner size.
6. Change gate location.
7. Increase venting.
8. Radius gate at cavity.

Material Changes

1. Dry material.
2. Remove contaminents from the material.

WARPAGE AND SHRINKAGE

Warpage and excessive shrinkage are usually caused by the design of the part, the
gate location, and the molding conditions. Orientation and high stress levels are
also factors. It is recommended that the sections on warpage, shrinkage,
orientation, and crystallinity be reviewed. Warpage and shrinkage can be cured
by the following techniques in the operation.

Molding Changes

1. Increase cycle time.
2. Increase injection pressure without excessive packing.
3. Increase injection forward time without excessive packing.
4. Increase injection boost time without excessive packing.
5. Increase the feed without excessive packing.
6. To reduce warpage lower the material temperature.
7. To reduce warpage keep packing at a minimum.

8. To reduce warpage minimize orientation.
9. To reduce shrinkage raise the material temperature to permit more packing.
10. To reduce shrinkage keep packing at a maximum.
11. Increase injection speed.
12. Slow down ejection mechanism.
13. Annealing parts after molding may reduce warping.
14. Cool in shrink fixture.
15. Cool in water.
16. Make cycle consistent.

Mold Changes

1. Change gate size.
2. Change gate location.
3. Add additional gates.
4. Increase knockout area.
5. Keep knockouts even.
6. Have sufficient venting, especially for deep parts.
7. Strengthen part by increasing wall thickness.
8. Strengthen part by adding ribs and fillets.
9. If differential shrinking and warping is caused by irregular wall sections core, if possible, or change the part design.
10. To reduce warpage reduce the mold temperature to stiffen the outer surface.
11. To decrease shrinkage raise the mold temperature to increase packing.
12. Check mold dimensions. Wrong mold dimensions may cause parts to appear to have shrunk excessively.

Material.   Use faster setting material.

HOW TO CONTROL DIMENSIONS

Dimensional variations are caused by inconsistent machine controls, incorrect molding conditions, poor part design, and variations in materials. Once a part has been molded and the machine conditions set, dimensional variations should maintain themselves within a small given limit. At this point, the quality control department should set these limits and designate which dimensions are to be measured to ensure acceptable parts. When parts vary from accepted standards the problem is usually in a change of molding conditions, but some of these are not readily apparent. For example, a voltage drop will affect the heat output of the cylinder band. The pyrometer will try to compensate, sometimes successfully, sometimes unsuccessfully. At the same time, this same voltage condition will have an effect on motor horsepower and the speed of solenoid operation, which are not compensated for by the machine circuit. Their effect on dimensional control is rare, but when it does occur it is most difficult to detect. Typical problems are listed below.

Mold Problems

1. Incorrect mold dimensions causing parts to appear out of tolerance.
2. Distortion during ejection. See "Sticking in the Mold," p. 449.
3. Uneven mold filling.
4. Interrupted mold filling.
5. Incorrect gate dimensions.
6. Incorrect runner dimensions.
7. Inconsistent cycle—mold-caused.

Machine Problem

1. Malfunctioning feed system in a plunger machine.
2. Inconsistent screw stop action.
3. Inconsistent screw speed.
4. Malfunctioning nonreturn valve.
5. Worn nonreturn valve.
6. Uneven back pressure adjustment.
7. Malfunctioning thermocouple.
8. Malfunctioning temperature control system.
9. Malfunctioning heater band.
10. Insufficient plasticizing capacity.
11. Inconsistent cycle—machine-caused.

Molding Problems

1. Uneven mold temperature.
2. Low injection pressure. Therefore, increase injection pressure.
3. Insufficient fill or hold time. Therefore, do the following:

   a. Increase injection forward time.
   b. Increase injection boost time.

4. Too high barrel temperature. Therefore, lower barrel temperature.
5. Too high nozzle temperature. Therefore, lower nozzle temperature.
6. Inconsistent cycle—opeator-caused.

Materials Problems

1. Batch to batch variation.
2. Irregular particle size.
3. Wet material.

HOW TO PREVENT STICKING IN MOLD

Parts stick in the mold primarily because of molding defects, insufficient knockout, packing of material into the mold, and incorrect mold design. If parts stick in the mold, it is impossible to mold correctly.

Parts stick when molding conditions cause the material to pack into the cavity. This will expand the cavity, force the plastic into microscopic depressions, and reduce the amount of normal shrinkage, which would ease ejection from the cavity. Deformation of cavity and core are common causes for mold release problems.

Theoretically, mold release agent should not be required. Practically this is not the case. The two most widely used lubricants are the heavy metallic salts of stearic acid, such as zinc stearate, applied as a powder, and silicone compounds. The silicones are mixed with the necessary solvents and propellants. They are dispensed from aerosol cans, self-propelled pressure systems, and bulk containers using air or Freon as the propellant. It is necessary to select the right solvent which does not affect the material being molded. Mold releases can cause difficulties in painting, decorating, and metallizing (3). Remember also that excess lubricant does not increase mold release. The most costly part of using lubricants is the cycle time lost by the operator during its application. A new method of permanently lubricating the cavity consists of impinging a 0.0002 to 0.0004-in. layer of graphite mixture into the metal surfaces of the mold with a patented air gun (4). The remaining graphite acts as a lubricant.

Nonetheless parts do stick. Their removal can cause severe mold damage. Steel should never be used on the mold by an operator. Hard wood, soft brass, and soft copper are materials of choice. Occasionally it is necessary to heat a screw or a broken hacksaw blade and force it into the molded part. After cooling the screw and blade are used as handles to release the molded part. Occasionally blowing air across the surface of a deep molded part will cause its release. This is particularly true if the part has been allowed to shrink for a moment. Typical solutions follow.

Mold Remedies

1. See if sticking is caused by short shot not engaging knockout system.
2. Remove undercuts.
3. Remove burrs, nicks, and similar irregularities,
4. Remove scratches and pits.
5. Improve the mold surface.
6. Restone and polish using movements only in the direction of ejection. This is called draw polishing.
7. Ascertain if the mold surface is correct for the material being molded.
8. Increase the taper.
9. Increase the effective knockout area.

10. Change the knockout location.
11. Check the operation of the knockout system (plates might not be moving in the proper sequence, cams loose, etc.)
12. See if the cams are clearing before the part is ejected.
13. Add vacuum breaks and air ejection in deep drawn moldings.
14. Check if the core is shifting during molding. This can be done by measuring wall thicknesses.
15. See if the cavities are deforming during molding.
16. See if the mold base is deforming during molding. This should be carefully checked when there is a large projected area perpendicular to the clamping force. Often there is not enough steel in the supporting pocket.
17. Check if the mold shifts when opening. Items 14 to 17 are suspects when there is scratching or rubbing on the outside of the part near the parting line. Shifting can be caused by the mold or machine.
18. Decrease the gate size.
19. Add additonal gates.
20. Relocate the gates. The purpose of items 18 to 20 is to decrease the pressure in the cavity.
21. Equalize the mold filling rate.
22. Prevent interrupted filling.
23. Determine whether the part is strong enough for ejection.
24. Radius and reinforced parts giving greater rigidity.
25. See if the parts are degrading and losing mechanical properties.
26. Add additional knockout systems such as air ejection.
27. Redesign the part.
28. If the part remains on the wrong side you can undercut the other side, change tapers, and provide mold temperature differentials.
29. Make inconsistent mold-caused cycles consistent.

Molding Remedies

1. Use more mold release agent (if insufficient).
2. Use proper mold release agent (if wrong agent).
3. Reduce material feed.
4. Reduce injection pressure.
5. Reduce injection forward time.
6. Reduce injection boost time.
7. Reduce mold temperature.
8. Increase overall cycles. This lowers the temperature making the part more rigid and increases the amount of shrinkage.
9. Make inconsistent operator-caused cycles consistent.

Materials Solutions

1. Remove contamination in the material.
2. Add lubricant to the material
3. Dry the material.

Machine Solutions

1. Repair any malfunctioning of the knockout system.
2. Lengthen insufficient knockout travel distance (if insufficient)
3. Make inconsistent machine-caused cycles consistent.
4. Check to see if platens are parallel.
5. Check the tie rod bushings.

HOW TO PREVENT STICKING IN THE SPRUE

Sprue sticking is caused by improperly fitted sprue-nozzle interface, pitted surfaces, inadequate pull-back, and packing. Occasionally the sprue diameter will be so large that it will not solidify enough for ejection at the same time as the molded parts. Here is how sticking may be prevented.

Sprue Solutions

1. Match sprue radius to nozzle radius if mismatched.
2. Correct sprue-nozzle seating.
3. Make sure that nozzle orifice is not larger than sprue orifice.
4. Polish the sprue,
5. Increase the taper of the sprue.
6. Increase the sprue diameter if it is too weak.
7. Decrease the sprue diameter if it is too large for cooling.
8. Control temperature of sprue (rare).

Mold Solutions

1. Increase the pull-out force of the spure-puller system.
2. Reduce mold temperature.

Molding Solutions

1. Use sprue break (machine moves back slightly, breaking contact between nozzle and sprue).
2. Increase suckback.
3. Reduce feed.
4. Reduce injection pressure.
5. Reduce ram forward time.
6. Reduce injection boost time.
7. Reduce material temperature.

8. Reduce cylinder temperature.
9. Reduce nozzle temperature.

Materials Solutions

1. Check for material contamination.
2. Dry material.

HOW TO PREVENT NOZZLE DROOLING

Nozzle drooling is caused by overheated material. For a material with a sharp viscosity change at molding temperature such as nylon, the use of a reverse taper nozzle or a positive-seal type of nozzle is recommended. The objection to nozzle drooling is that it introduces solidified material into the part which causes surface defects. It may also interfere with the flow and mechanical properties. Typical solutions to nozzle drooling follow.

Nozzle Solutions

1. Use positive-seal type nozzle.
2. Use reverse taper nozzle.
3. Reduce nozzle bore diameter.

Molding Solutions

1. Reduce nozzle temperature
2. Increase suckback.
3. Use sprue break.
4. Decrease material temperature.
5. Reduce injection pressure.
6. Reduce injection forward time.
7. Reduce injection boost time.

Mold Solutions

1. Increase cold slug well.
2. Increase run-off.

Materials Solutions

1. Check for contamination.
2. Dry the material.

HOW TO SOLVE THERMOSET MOLDING PROBLEMS (5)

The differences between thermoplastic and thermoset trouble shooting derive from the irreversible plasticity of thermosetting materials. The temperature in the cylinder and nozzle must be low enough to prevent premature setup. The temperature of the mold sets the material. Therefore, raising mold temperature increases viscosity and decreases pressure transmission. Listed below are those areas where the method of correcting molding faults differ.

Screw Wear, Screw Squeaking Problems

1. Decrease screw speed.
2. Decrease back pressure.
3. Decrease cylinder temperature,
4. Lubricate material.

Blistering on Material

1. Increase back pressure.
2. Increase injection pressure.
3. Increase feed.
4. Increase mold temperature.
5. Increase cycle time.
6. Increase gate size.
7. Increase runner size.
8. Increase cavity vents.

Setup Problems of Material in Cylinder

1. Reduce cylinder temperature.
2. Reduce screw speed.
3. Reduce back pressure.
4. Reduce nozzle temperature.
5. Increase nozzle opening.
6. Clean (flush) screw.
7. Maintain even cycle.

Material Setting up in Nozzle

1. Lower nozzle temperature.
2. Lower cylinder temperature.
3. Open nozzle hole.
4. Use sprue break if there is excessive heat transfer from the mold to the nozzle.
5. Clean the nozzle.
6. Decrease screw speed.

Accomplished by Changing Mold Temperature

1. Increase mold temperature to minimize or cure the following;

    a. Parts blistered.

    b. Flashed mold.

    c. Incompletely cured (flexible) parts.

2. Decrease mold temperature to minimize or cure the following:

    a. Flow lines.

    b. Dull surface.

    c. Mold stains.

    d. Sticking parts.

    e. Nonfilled parts.

    f. Porous parts.

    g. Entrapped gas.

HOW TO PREVENT EXCESSIVE CYCLES

Excessive cycles are usually caused by poor management. Proper records are not kept, standards not established, and constant monitoring of output not established. Most jobs could have their cycles reduced by an experienced molding superintendent. In practice unfortunately, these new reduced cycles will very often slowly creep back to the original cycle. Other causes of excessive cycles are insufficient plasticizing capacity, inadequate cooling, channels in the mold, insufficient cooling fluid, and erratic cycles. Problems by type follow.

Management Problems

1. Nobody in authority cares.
2. Incorrect, poor, or nonexistent records.
3. Incorrect setting of controls.
4. Unauthorized cycle changes.
5. Nonmonitoring of cycle time.
6. Too much for the operator to do.

Machine Problems

1. Malfunctioning of the machine.
2. Platen slow down time excessive.

Mold, Molding and Materials Problems

1. Insufficient cooling capacity in mold.
2. Insufficient cooling capacity in chillers.
3. Material temperature too high.
4. Material too slow in setting up.
5. Material contaminated.
6. Inconsistent cycle — mold-caused.
7. Inconsistent cycle — operator-caused.

HOW TO BREAK IN NEW MOLD

The following information is helpful in breaking in a new mold. This is a critical step in a molding plant and requires the utilization of the best talent. A plant that can get a new mold in operation in a minimum of time and with minimal or no help from the moldmaker has a definite competitive advantage. The following procedures will help in that respect and tend to eliminate or minimize mold damage.

1. If a mold is "new" to the shop but has been run before, obtain samples and as much information as possible.
2. Clean the mold carefully.
3. Visually inspect the mold. This is the time for questions, consultations, and changes. Obvious corrections, such as improving the polish or removing undercuts, should be done before the mold is put in. The object is to put a mold in that will produce satisfactory parts.
4. Understand the actions of the mold and try cams, slides, locks, unscrewing devices, and such on the bench.
5. Install safety devices. For example, if a knockout plate must be back before the mold closes, electrical interlocks should be installed. It is very easy to damage a mold this way during the break-in period.
6. Put the mold into the press and move very slowly under low pressure.
7. Open the mold and inspect it again.
8. Dry cycle the mold without injecting material. Check knockout stroke, speeds, cushions, and low pressure closing.
9. Bring mold to operating temperature and dry cycle again. Expansion or contraction of the mold parts may effect the fits.
10. Take a shot using maximum mold lubrication and under conditions least likely to cause mold damage. These are usually low material feed and pressure.
11. Build up slowly to operating conditions. Run until stabilized, at least 1 hr.
12. Record operating information.
13. Take part to quality control for approval.
14. Make required changes.
15. Repeat process until approved by quality control and/or customer.
16. Record all the information on the appropriate record forms.

REFERENCES

1. "Get Rid of Splash Marks on Injection Molded Parts," J. M. Sherlock, *PDP* June 1972, p. 18.
2. "Improved Surface of Acetal Moldings," W. C. Filbert, T.M. Roder, *SPE-J,* February 1964 p. 149. 3. "
3. "Mold Release, R. B. Bishop, *SPE-J*, May 1969, p. 151.
4. "Permanent Mold Release, *SPE-J*, January 1971, p. 17.
5. "Trouble Shooting Thermoset Molding," J. Prause, *PT*, May 1970, p. 63.

CHAPTER 6 _____

ydraulic Mechanisms
and Circuits

Hydraulics is the study of mechanical properties of liquids and their applications in engineering. Three forms of energy are found in oil hydraulic systems — heat, kinetic energy, and potential energy. Heat energy, the result of the flow of oil, is caused by the internal friction of the oil molecules flowing over themselves and over the walls of their containments. Heat energy is undesirable because it is not useful in the process and in addition must be removed from the system. A certain amount of heat is generated because of the nature of the pump and the hydraulic mechanisms. Excessive heat is caused by excessive length of flow, excessive bends, turns and obstructions, insufficient pipe diameter, inefficient or worn parts, and incorrect design. A further discussion of this topic is beyond the scope of this text.

Kinetic energy is the energy of motion. Its main effect is in terms of velocity. We attempt to minimize the kinetic energy of a hydraulic system in molding machines. Velocity, which is distance per unit time (such as feet per second), should not be confused with volume (amount of flow) which has the dimensions of quantity per time (such as gallons per minute). We are primarily concerned with the potential or pressure energy. Pressure is created by resistance to flow. It implies a closed or throttled system with a source to supply oil. This normally is a pump or an accumulator (a container of oil with a compressed gas exerting force behind it).

HYDRAULIC MECHANISMS

The three types of matter react differently to a force (Figure 6-1). A gas will compress. A solid will transmit the force in the direction of the applied force. A

457

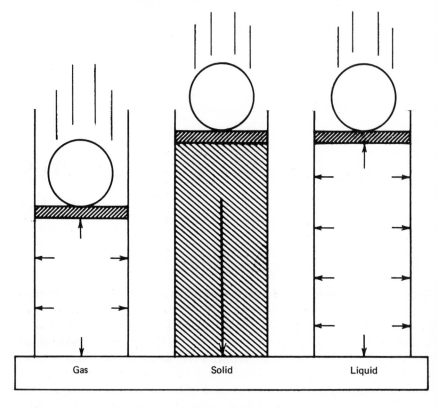

Figure 6-1   How gases, solids, and liquids react to force.

liquid obeys Pascal's law, which states that, in a static liquid, pressure acts equally in all directions and will be transmitted undiminished in all directions and act with equal force on all equal areas (Figure 6-2). An irregularly shaped container filled to the top and covered with a floating seal with an area of 1 in.^{2} has a 10-lb weight placed upon it. The liquid under the seal has a pressure of 10 lb/in.^{2} (psi). Every square inch of surface of this container has a 10-lb force acting perpendicular to it. If the surface had an area of 1000 in.^{2} then the total force on the container would be 1000 in.^{2} × 10 psi, or 10,000 lb. Thus a 10-lb force has now become a 10,000 lb force.

This is the principle of the hydraulic jack, (Figure 6-3). An hydraulic pump consists of a hand operated piston with a 1-in.^{2} area in a cylinder. When the piston is raised it lowers the pressure in the cylinder. Atmospheric pressure forces the oil from the tank through check valve B into the pump. Check valves allow flow in one direction and block flow in the other direction. (The symbols

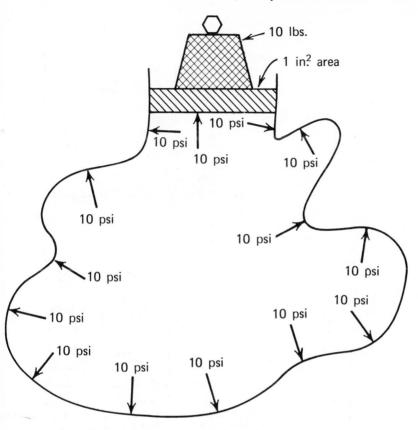

Figure 6-2 Illustration of Pascal's law – "In a static liquid, pressure acts equally in all directions and will be transmitted undiminished, in all directions. It will act with equal force on all equal areas."

are those of the American Standard Association and adopted by the Joint Industry Conference (J.I.C.) which are used in hydraulic circuitry (Figures 6-4 and 6-5). The new standards were adopted in December 1966. Since the old symbols have been used on most injection molding machine hydraulic drawings, both old and new symbols will be used in the book.) Oil can not go through check valve A because the force generated by the weight of the jack will be high enough to keep it closed. The piston in the cylinder is now pushed down with a force of 100 lb. The oil is blocked at check valve B and goes through check valve A underneath the jack raising it. The jack is another piston in a cylinder. The portion above the piston is vented to tank to remove any oil

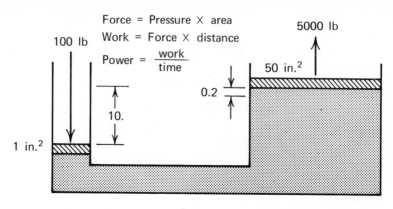

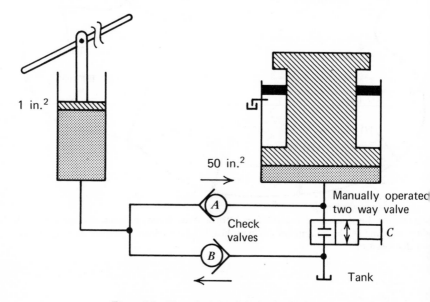

Figure 6-3   The principle of the hydraulic Jack.

leakage past the seals. The area under the piston of this jack is 50 in.^{2}. Since we are pushing with a pressure of 100 psi the total force on the top of the jack will be 50 in.^{2} × 100 psi or 5000 lb. Thus a force of 100 lb is converted into one of 5000 lb. The pumping action is repeated until the jack reaches the maximum movement desired or available. To return the jack to its starting position or to a lower position, oil is returned from underneath the jack to the tank through the manually operated two way valve. Any number of jacks, each with different surface areas, could be fed simultaneously from the same pump.

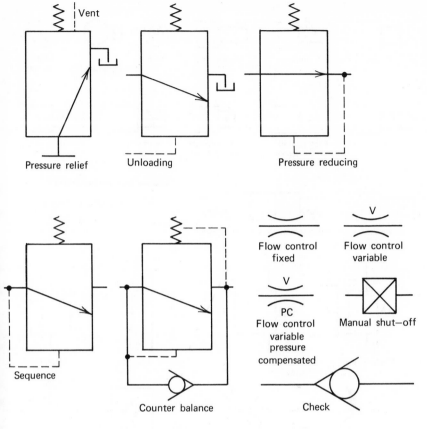

Figure 6-4 Fluid power symbols valves (old).

One should not confuse force (F), pressure (p), area (A), work (W), and power (P). They are related as follows:

$$F = pA$$
$$W = F \times (\text{distance})$$
$$P = \frac{FD}{t} = \frac{W}{t}$$
$$t = \text{time}$$

The pump side has a force of 100 lb on an area of 1 in.^{2} giving a pressure of 100 psi. The jack side has this pressure (100 psi) on an area of 50 in.^{2} for a force of 5000 lb. Obviously they are not going to move the same distance or energy would be created in the system. This is shown by the work done on each side which must be the same (neglecting heat energy losses). On the pump side the

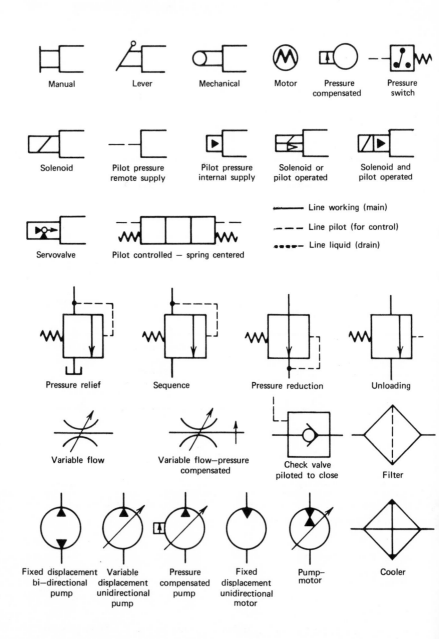

Figure 6-5 Fluid power symbols (new). USAS Y32.10 12/66.

100-lb force has moved 10 in. for work of 1000 in.-lb. On the jack side the force of 5000 lb has moved 0.2 in. for the same amount of work — 1000 in.-lb. If one pumping stroke was done each second then the power of the system would be 1000 in.-lb/sec.

Advantages and Disadvantages of Hydraulic Systems

From Pascal's law and the operation of a jack, some reasons for the advantages and disadvantages of hydraulics in a molding machine can be deduced.

Advantages of a Hydraulic Systems

1. The parts and mechanisms can be located in any place.
2. The force can be transmitted around corners and in all directions.
3. The oil is not subject to damage or breakdown in the same degree as mechanical linkages, cams, gears, and such.
4. The motion is transmitted throughout the system with little slack.
5. The power is transmitted rapidly at a long distance with small loss.
6. The speed and force can be controlled in stepless increments between limits.
7. Large forces can be applied with no motion.
8. It is relatively unaffected by variations in load and by stalling. When stalled, the volume requirements are low.
9. Large forces can be generated by small compact units and can be controlled by much smaller forces. Forces can be multiplied.
10. Action is smooth with no vibration and with an inherent cushion effect.
11. Automatic control is easily obtained.
12. Overload and breakdown protection can be built in.
13. Nonproductive energy (heat) can be absorbed and easily dissipated.
14. Because oil is the hydraulic fluid, lubrication is automatic and there is little wear.
15. Maintenance is low.
16. By means of gages there is a continuous indication of the state of the system.
17. Reciprocating motion is easily obtained.
18. Repair parts are readily available and can be made in local machine shops.
19. Since there is low power consumption, it is an economical method of operation.
20. It is a comparatively silent system.

Disadvantages of Hydraulic Systems

1. It must be a totally confined system.

2. Leaks must be prevented. This can present problems when working in the pressure ranges of modern machines.
3. Mechanical requirements of all the parts of the system are relatively large.
4. The heat-energy built up must be removed.
5. Moisture vapor and water must be kept out of the system.
6. The oil must be clean and contain the proper additives to prevent formation of sludge and gums which would impair the proper functioning of the machine
7. Oil and oil reclamation are relatively costly.
8. Hydraulic oils are a fire hazard.
9. Personnel are usually more familiar with electrical and mechanical systems.

It can be seen that for injection molding machines the advantages of an hydraulic system far outweigh the disadvantages.

HYDRAULIC OILS – REQUIREMENTS

Five basic properties are required of a good hydraulic fluid: oxidation stability, rust prevention, water separating ability, resistance to foaming, and lubrication. These require little consideration by the molder since, with normal care, they will maintain themselves (1, 2).

Oxidation stability is the resistance of an oil (hydrocarbon) to chemical reaction with oxygen (oxidation). Without it, the first products formed are unstable hydroperoxides which react to form other chemicals, including acids, all of which are soluble in oil. These will in turn react with each other and polymerize, producing soluble resins first and later, insoluble resins. Soluble resins change the viscosity of the oil, soluble acids tend to increase its corrosive action, and the insoluble resins precipitate out. This reduces clearances, plugs up small holes, clogs the filter, and reduces the operating efficiency of the system, eventually leading to a complete breakdown.

Oxidation is a chemical process whose rate approximately doubles for every 18°F increase in temperature. Because of this and viscosity changes the average temperature of the oil should not exceed 130°F. There will be parts of the system which are operating at higher temperatures. Since oxygen is supplied by air and trapped in the hydraulic system, it is present in the oil in appreciable amounts. The metals, particularly copper, catalyze the oxidation of oil. This is why copper tubing is not used, even though its other properties would make it desirable. Small particles of iron are also excellent catalysts. These come from the wearing of metallic parts as well as from unintentional contamination. A permanent magnet should be placed in every reservoir and cleaned periodically. This will not only reduce oxidation by removing the particles, but also help prevent excessive frictional wear.

Rust formation is caused by water which appears in the hydraulic system through atmospheric condensation. In addition, water used as a cooling medium in molds and on machines, may possibly leak into the hydraulic system. It does not take much water to introduce rust into the system. Corrosion not only causes damage to parts, but the abrasive action of iron oxide (rust) circulating in the system will cause additional damage. Rust is prevented by an additive which has great affinity for metal surfaces. It forms a film that resists displacement by water.

The water separating ability of oil is basically present and is obtained in the refining process rather than from additives. Water is always present in the oil because of atmospheric condensation and possible accidental addition. It emulsifies with oil and its contaminants form slurries which can foul up pumps, controls, and cylinders. The water separating or demulsifying properties of the oil cause the water to separate out and thus prevent slurry formation. This property is also useful when the oil is purified outside the hydraulic system by means of separating tanks.

Resistance to foaming is required in a good hydraulic oil. The oil usually has about 10% by volume of dissolved air, which normally remains in solution and causes no trouble. Oil can absorb a considerable amount of air when under pressure. When this pressure is released, the air will come out of solution and may produce foam. Trapped air has serious consequences in the system. Oil cannot be compressed easily, but the air can. Temperatures of 1000°F are possible through rapid compression of air such as occurs in the cylinders and pump. This is not observable on the oil temperature indicator which reads only the average oil temperature. The oil immediately surrounding the hot spot will oxidize — an undesirable condition.

Most foaming in injection molding machine systems is due to mechanical causes. These include air leaking in suction lines, too low an oil level, leaking packing, improper functioning of air bleeders, drains discharging above the oil level, and certain deficiencies in the design of the hydraulic system. The latter should not be true of original equipment, but develops through modifications in the molding plant.

The lubricating properties are normally taken for granted. It should be noted that hydraulic oil has excellent antiwear properties compared with other oils of similar viscosity. The higher the viscosity, the better the film strength or antiwear properties. This is the reason for selecting an oil of high viscosity and is another important reason for keeping the temperature of the system low. The higher the temperature, the lower the viscosity.

VISCOSITY

Viscosity is the most important single property of hydraulic oils. It is a measure of the shearing stress or the resistance to flow of sliding layers of molecules.

It is equal to the tangential or shearing force on two parallel plates of unit area at a unit distance apart moving at a unit velocity in relation to each other. The viscosity, therefore, is the shear stress divided by the shear rate. The shear stress is shown as force per unit area. The shear rate is the volumetric flow rate divided by the volume.

$$\text{Absolute viscosity } (\mu) = \frac{\text{shear stress } (\tau)}{\text{shear stress } (\gamma)} = \frac{F/L^2}{(V|t)|V} = \frac{F\,t}{L^2}$$

where F = force
   L = length
   V = volume
   t = time.

The units of absolute viscosity in the English and metric systems are shown below. The viscosity of water at 20°C equals 1 cp.

English system		Metric system		
$\dfrac{\text{lb-sec}}{\text{in.}^2}$ = reyn		$\dfrac{g}{(cm)(sec)}$ =	$\dfrac{(dyne)(sec)}{cm^2}$ =	P

$$1 \quad \frac{1b-sec}{in.^2} = 6.89 \times 10^4 \ P = 6.89 \times 10^6 \ CP$$

The kinematic viscosity (ν) is the absolute viscosity divided by the mass density (ρ). Mass density is the weight per unit volume divided by the acceleration of gravity. In metric units the mass density is the same as the specific gravity. The units for the kinematic viscosity follow.

English system		metric system		
$\dfrac{(lb)(sec)/ft^2}{(lb/ft^3)/(ft/sec^2)}$ =	$\dfrac{ft^2}{sec}$	$\dfrac{[g/(cm)(sec)]}{(gm/cm^3)}$ =	$\dfrac{cm^2}{sec}$ = St	

$$1 \ \frac{ft^2}{sec} = 929 \ St \ = \ 9.29 \times 10^4 \ cSt$$

$$1 \ \frac{in.^2}{sec} = 6.45 \ St \ = \ 645 \ cSt$$

The English system has no unit for kinematic viscosity. The conversion is made by changing from cm^2/sec to $in.^2/sec$ (dividing centistokes by 645) or ft^2/sec to cm^2/sec (dividing centistokes by 9.29×10^4).

The viscosity of oil is measured in Saybolt universal seconds (SSU or SUS). It is reported in the number of seconds it takes for 60 cc of oil at $100°F$ to pass through a standard orifice 0.0695 in. diameter and 0.483 in. long. Viscosity of oil used in molding machines ranges from 150 to 300 SUS. It can be converted into kinematic viscosity by the following empirical formula:

$$\nu = 0.226t - \frac{195}{t} \qquad t < \quad 100 \text{ SUS}$$

(6-2)

$$\nu = 0.220t - \frac{135}{t} \qquad t > \quad 100 \text{ SUS}$$

where

ν = kinematic viscosity (c St)

t = seconds (SUS).

As oil is heated the distance between the molecules expands lowering their resistance to flow, or viscosity. The viscosity index is a measure of the effect of temperature on viscosity. It is an empirical number based on the viscosity/temperature relationship of two specified oils. The viscosity index of hydraulic oils for molding systems should not be less than 78.

The factors that make for inefficiency in an hydraulic system are mainly slippage in the pump, internal leakage in the control components, cavitation, oil leakage, fluid friction, mechanical friction, and the rate of response. These are all affected by viscosity.

Effects of Too High Viscosity

If the viscosity of the oil is too high, more force will be required to move the same volume of fluid. Since this energy is nonproductive, it will reduce the efficiency of the system. Too viscous an oil can also cause cavitation.

Furthermore, viscous oil is not sucked into the pump in sufficient quantities so that a relatively high vacuum is produced within the pump. This phenomenon lowers the vaporization point of the oil which then begins to vaporize. Air dissolved in the oil is released, air pockets form and then collapse when the vacuum is reduced. When this happens the pump becomes noisy and vibrates during pump operation. The noise and vibration shortens the life of the pump. Plugged intakes, clogged filters, improper design such as intake connections that are too small, or any oil starvation of the pump will cause the same thing.) Finally, high viscosity oils, which on the one hand increase the lubricity of the oil, may cause sluggish response in the valves on the other hand.

Effects of Too Low Viscosity

The undesirable effect of excessively low viscosity are excessive internal and external leakage, slippage in the pump, increased rate of wear on moving parts, larger pressure drops in the system, increased temperature, and a lowering of efficiency. The lower viscosity gives fast control action and trouble free cold starting. The selection of the viscosity and viscosity index of the oil is a compromise.

Operating Range

The maximum operating temperature range for molding machines is 40 to 150°F. The viscosities of the oils outside of this range will cause improper operation of the machine and possible damage. Since leakage in a cylinderical part increases as the cube of the clearance between the parts, these clearances are held to close tolerance. If the temperature is not properly controlled, clearances will change because of the differential expansions of the piston and the spool. This, and a lowered lubricity of the oil at high temperatures, can cause severe wear and even galling.

Fire Resistant Fluids

Fire resistant hydraulic fluids (3) have been developed for those locations where there is danger of ignition of the hydraulic fluid caused by failure of a pressure line or components. There are two types —water based and chlorinated or phosphate ester based. Neither are as desirable as hydraulic oils. Should conversion be required consultation is needed with the machinery manufacturer. Seals, packing, paint, and other components may have to be changed.

MAINTENANCE OF HYDRAULIC OILS

Experience has shown that hydraulic systems treated with a reasonable amount of care will seldom cause trouble. When trouble does occur, it is usually very costly in terms of down time and repair. Luckily, difficulties in the system occur before a total breakdown and the experienced molder will take advantage of these warnings to remedy the faults in the system. For example, dirty oil can cause improper functioning of valves which can lead to erratic molding. Overheating of the oil can be detected by increased power consumption which is also costly.

It has been estimated that 70% of hydraulic trouble is caused by the improper condition of the hydraulic oil.

Contamination

There are two types of contaminations:

1. Contaminants resulting from degradation of the hydraulic oil itself,
2. Contaminants resulting from the addition of foreign materials (which include plastic, water, packing and gasketing material, metal particles, and rust).

The deleterious effects of high operating temperatures on oil must be reemphasized. The decomposition of oil is rapidly accelerated, once the process starts, by the decomposition products formed.

The major source of accidental contamination is during the transfer of the oil from the drum to the machine. All drums should be stored indoors and on their side. The best means of transfer is by blowing from the drum to the machine. Needless to say, the top of the drum, the bung, the filler hole of the machine, and the connecting system must be scrupulously clean.

Water accumulates in the hydraulic system because of atmospheric conditions. When a machine is shut down the heated air above the reservoir can contain more water than does air at room temperature. Upon cooling under certain conditions, water will condense. Continual repetition of this process will introduce an appreciable amount of water into the oil. One of the major contaminants to be checked, therefore, is water.

Oil will also receive contamination from the dusty atmosphere. Dust settles on the exposed parts of the hydraulic system, and is washed back through the packing and into the oil. Fine particles of iron are a normal result of wear. These will be removed by a magnet placed in the reservoir.

The final source of outside contamination is dirt which enters the system during machine repairs. Before the hydraulic system or any part of it is exposed, the working areas should be thoroughly cleaned and all tools wiped. If the job cannot be completed at once, the opening and surrounding areas should be covered.

Despite every effort, contaminants will get into the system. The purpose of maintenance is to keep these under control so that the machine can function properly. The better method of controlling oil purity is through straining and filtration (4). A strainer is a device for the removal of solids from a fluid, where resistance to the motion of the solid is in a straight line, such as a wire mesh screen. A filter is a device for the removal of solids from a fluid where resistance to motion of such solids is in a tortuous path, such as cellulose or fibers.

A strainer should be put before every pump. They are constructed to bypass the strainer when it is dirty. This will prevent cavitation of the pump, but not damage by contaminants. There is an indicator telling when the strainer should be cleaned.

Continuous filtration should be provided on every machine. A larger capacity filter of the same type, which should be able to heat the oil before filtration, should be available to be installed on a particular machine if required.

Oil can be reclaimed by putting it in a settling tank for at least 24 hrs. The sludge and water is drained from the bottom. The remaining oil is recirculated through a filter until it looks clean and nothing precipitates out upon standing. A sample should be sent to the manufacturer to check the condition of the oil and see if any additives are required. Periodic samples of oil from each machine should be taken and allowed to settle. Regular inspection of any precipitated sludge or water and the "sparkle" of the oil is a good indication of the state of the system.

Figure 6-38 shows an oil filtration circuit with its own motor and pump. Figure 6-6 shows an oil filtration circuit where the supply of oil is from one of the main return or tank lines. It can be from the main pressure relief valve, an unloading valve, a cylinder, or a four way valve. If the filter is clogged, and the pressure is over 60 psi, the overload check valve will open bypassing the return of load to tank. The filter can be changed while the machine is in operation by closing the shut-off valve. The oil will then return over the check valve. The gauge reading will be a measure of the resistance in the filter which is a function of the amount of dirt.

Gauges

Gauges are used in hydraulic machines to measure the pressure at various point in the system. There are at least two gauges on the injection machine; one measures the injection pressure and the second the clamping pressure. The pilot system will have a gauge if it is in a different pressure range.

It is essential that gauges function properly and be kept in good condition since many molding problems can be traced to malfunctioning gauges. For example, a particular molding job is known to require a 1500-psi gauge reading on the clamping circuit. If the gauge is incorrect and there are actually only 900 psi the pressure may not be enough to prevent flashing of the mold. The gauge is usually the last thing checked. Gauges usually fail because of pressure surges. A snubber should always be used with a gauge. Its purpose is to cushion the shock of the oil. Cut-off valves for pressure gauges are especially designed with slow opening speeds to prevent sudden pressure surges.

The gauge most commonly used for measuring pressure in the injection machine is a Bourden-type gauge which works on the principle that oil pressure in a curved tube will cause the tube to straighten out, and that the amount of straightening is directly proportional to the pressure. They are accurate to approximately 0.3% of the full scale reading. They are relatively fragile but can often be repaired by maintenance personnel. When this is done the reading should be compared with a gauge of known accuracy or a dead weight tester.

The plunger type gauge is a relatively rugged, reliable gauge, slightly less accurate than is the Bourden-type. The liquid whose pressure is being measured

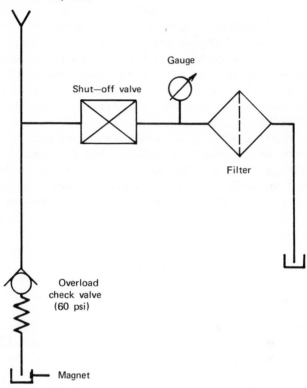

Figure 6-6   Oil filtration circuit.

actuates a piston to which is attached an indicator. The piston is held in place by a spring, and the force of the oil compresses the spring, lifting the valve and the indicator out of the body. It is very useful in the maintenance department, particularly when checking out pumps and other hydraulic components where sudden surges of pressure occur. It is not used as an operating pressure indicator, because of the higher accuracy of the Bourden-type gauge.

The pressure indicated by either gauge is the pressure of the system, excluding atmospheric pressure. Thus it is called gauge pressure. Absolute pressure is a system's pressure plus atmospheric pressure and is never used on molding equipment. Gauges in the hydraulic system are comparable to meters in the electrical system. They are continuing indicators of the operation, and are invaluable in hydraulic trouble shooting.

CONNECTORS

The power to operate the molding machine, generated in the pump, must be transmitted to the cylinders and hydraulic motors which are the working components of the machine. The hydraulic power transmission lines in the injection molding machine are steel pipes, steel tubing, and rubber covered flexible braided wire hose (5). The choice of these transmission units depends on the flow and pressure requirements of that portion of the system for which they are designed (6).

The proper selection of the power transmitting units or piping and their connectors are of primary importance. This is adequately done on new equipment. Many of the older machines in use today can benefit from the improved knowledge and experience that has been accumulated. Leaks in transmission lines and connectors often develop in the same place, and are detected by a good maintenance program. The reasons for the breakdowns should be analyzed and the condition corrected.

Transmission lines and their connectors are fully treated in many references and in manufacturers' bulletins. A brief discussion follows.

Steel Pipe Tubing, Hose

Steel pipe has a minimum tensile strength of 60,000 psi, a minimum yield point of 35,000 psi, and an allowable stress in the steel (S value) of 15,000 psi. The inside diameter will be determined by flow considerations. The wall thickness is generally determined by various codes. For molding machines Barlow's formula is used. Pipes are sized by schedule numbers which are related to the ratio of the allowable stress in the pipe over the allowable stress in the pipe material. Table 6-1 shows some dimensions and properties of pipe used in molding machines.

Hydraulic piping is connected by threading and welding. Threading alone is not completely satisfactory. Pipe compound can be used over the threads however, the preferable method is to use a Teflon tape which is wound once around the thread and overlapped by a 1/2 in., after which the thread is screwed tight. Threaded fittings have certain disadvantages. Aside from weakening of the wall, there are flow turbulences generated because of the enlargement and reduction of the cross-sectional areas. These disadvantages can be overcome by welding. Forged steel fittings of the socket welding type are available in Schedule 80 and 160 for use on molding machines. They are easily installed by slipping the pipe into the fitting and then welding. This technique provides a strong joint resistant to shocks and vibration without the turbulence and

Table 6-1 Dimensions and properties of ASTM A106 grade B pipe (working pressures calculated from ASA B31.1 where $y = 0$ (Barlow's formula), $S = 15,000$ psi C plain end value, and manufacturer's tolerance are 12½%.

Nominal Pipe Size (in.)	Threads (per inch)	O. D. (in.)	Length of Thread Screwed Into Fitting	I. D. (in.)	Schedule 80 Nominal Wall Thickness	Schedule 80 g/min at 10 ft/sec velocity	Schedule 80 Allowable Working Pressure (psi)	Schedule 160 I.D. (in.)	Schedule 160 Nominal Wall Thickness	Schedule 160 g/min at 10 ft/sec velocity	Schedule 160 Allowable Working Pressure (psi)
1/8	27	0.405	1/4	0.215	0.095	1.1	2460				
1/4	18	0.540	3/8	0.302	0.119	2.2	3000				
3/8	18	0.675	3/8	0.423	0.126	4.4	2680	USE SCHEDULE 80			
1/2	14	0.840	1/2	0.546	0.147	7.3	2810				
3/4	14	1.050	9/16	0.742	0.154	13.5	2410				
1	11 1/2	1.315	11/16	0.957	0.179	22.4	2440	0.815	0.250	16.3	3880
1 1/4	11 1/2	1.660	11/16	1.278	0.191	40.0	1950	1.160	0.250	32.9	2880
1 1/2	11 1/2	1.990	3/4	1.500	0.200	55.0	1740	1.338	0.281	43.8	2860
2	11 1/2	2.375	3/4	1.939	0.218	92.0	1590	1.689	0.343	69.8	2990

473

weaknesses associated with threaded pipe.

Means must be provided for assembling and disassembling of the piping by means of hand tools. The standard methods are using forged steel unions and flanges. The unions have steel to steel seats with spherical to angle mating surfaces to provide positive seating. The nut is tightened to increase the tension of the thread.

The second method for assembling the piping system is to use flanges. This method has gained wide acceptance. The pipe is threaded or welded to the flanges, which are held together by two of four bolts. The flanges are sealed with O rings. Older equipment used copper, aluminum, leather, and synthetic seals. Flanges are made to connect pipe, but not to correct misalignment. This is a serious condition which will lead to premature piping and sealing failure.

In designing hydraulic piping and tubing for molding machines it should be assumed they will be used as a stepladder, if at all possible. They should also be supported to prevent excessive vibration and corresponding joint and seal failure. Failure of piping is relatively rare. If it appears in the same place, the system should be evaluated.

Tubing   The tubing used in the hydraulic system of molding machines is a low carbon steel similar to grades 1010 and 1015. It is fabricated by cold drawing and annealing to give seamless tubing, or by cold working and annealing electric resistance welded steel. Steel tubing is identified by its outside diameter and its wall thickness. There are many thicknesses available. Table 6-2 shows dimensions and properties of some of the steel tubings which can be used on molding machines. The strength of the tubing is determined by its wall thicknesses, which are usually designated by Birmingham wire gauge (BWG) numbers.

It should be noted again that copper tubing is not desirable for hydraulic use. Copper promotes catalytic decomposition of oil. It also work hardens by vibration, causing failure. Galvanized steel is never used because it will react with additives to create insoluble zinc metallic soaps that will cause malfunctioning.

Tubing is selected because of good flow characteristics (the usual velocity of oil in the tube is between 10 to 15 ft/sec) and strength. One of the superior characteristics of tubing is that it can be bent, permitting fewer fittings in the system. Since a fitting is costly in terms of material and labor, and a source of leakage and maintenance, this is a significant advantage. In addition, properly bent tubing will keep turbulence to a minimum and increase the efficiency of the system. Bench type tube benders which can handle tubing through 1 in. O.D. should be available in the plant. The radius of bending is approximately 2 1/2 to 3 times the inside diameter of the tube. Larger diameter tubing can be hand bent by filling the tube with sand, sealing off both ends, heating it to cherry red heat, and bending it around pins placed on a steel plate. A properly bent tube has a maximum flattening of between 5 and 10%.

Table 6-2 also indicates the approximate recommended minimum distance between supports for tubing. If insufficiently supported, vibration, leakage, and premature failure will occur. When oil under pressure flows around bends, it will tend to straighten the tubing. This, of course, is the principle of the Bourdon gauge. If tubing failure reoccurs at a given spot, additional support should be considered.

Unlike pipe, which is threaded, tubing is always connected by flared, flareless, or welding type fittings. Much progress has been made in recent years in the development and design of such connectors, with the result that maintenance of hydraulic tubing can be kept at a minimum. Flared fittings are rarely used on molding machines. The flareless type gives at least equivalent performance and is much simpler to install.

A flareless fitting consists of three parts; the body, a case-hardened sleeve, and a nut. The tubing is cut square, and the internal and external burrs removed. The nut is put on the tube, then the sleeve is slipped over the tube, the tube inserted into the fitting, and the nut tightened. The nut causes the heavy edge of the sleeve to shear a groove on the outer surface of the tube, making a tight joint between the fitting and the tube. The nut presses on the bevel at the rear of the sleeve, causing it to clamp tightly to the tube. When fully tight, the sleeve is bowed at the midsection and acts as a spring to maintain constant tension between the body and the nut, preventing the nut from leaking.

There are elastomeric seals using 0 rings, rubber bushings, or teflon rings, which provide soft seals with a cushion, preventing leakage from vibration. A sleeve is inserted over the tubing into which is placed the seal. A nut tightening the sleeve compresses the elastomeric substance causing the seal. Larger diameter tubing is usually welded into socket fittings. This has the same advantages as welding pipe. Tube connectors are available in many different configurations.

Tubing gives a small amount of flexibility to the system which helps dampen and absorb vibration. Therefore, it is desirable to use bends rather than straight tubing to enable one to maximize its vibration and leak resistant properties. Proper terminal fittings should be selected to minimize the number of bends. Maintenance of tubing almost always centers around leakage at end of the connectors.

Hose.   The third major hydraulic transmission mechanism is the flexible reinforced rubber hose. The inner core is made of oil resistant synthetic rubber. The outer core for higher pressures is synthetic rubber, and for lower pressures can be synthetic rubber impregnated with cotton braid. Low pressure hose uses single wire braid reinforcement, medium pressure hose multiple wire braid, and high pressure multiple spiral wound wire braid. Table 6-3 gives typical dimensions and properties of flexible rubber hose used on injection molding machines.

Rubber hose is usually the connection of choice between two movable parts. Examples are the movable carriage on the injection end, and hydraulic cylinders

Table 6-2 Dimensions and Properties of steel tubing

Size	O.D. (in.)	Length of Tubing Between Supports (ft)	0 to 1000 psi (8-1 Safety Factor)				1000 to 2500 psi (6-1 Safety Factor)				Thread Description for Connector
			BWG No.	Wall Thickness (in.)	I.D. (in.)	g/min 10 ft/sec velocity	BWG No.	Wall Thickness (in.)	I.D. (in.)	g/min at 10 ft/sec velocity	
1/4	0.250	3	20	0.035	0.180	0.8	20	0.035	0.180	0.8	7/16 X 20 UNF
5/16	0.312	3	20	0.035	0.242	1.4	18	0.049	0.215	1.2	1/2 X 20 UNF
3/8	0.375	3	20	0.035	0.305	2.3	17	0.058	0.259	1.7	9/16 X 18 UNF
1/2	0.500	3	19	0.042	0.416	4.2	15	0.072	0.356	3.1	3/4 X 16 UNF
5/8	0.625	4	18	0.049	0.527	6.8	13	0.095	0.435	4.6	7/8 X 14 UNF
3/4	0.750	4	16	0.065	0.620	9.4	11	0.120	0.510	6.4	1-1/16 X 12 UN
7/8	0.875	4	15	0.072	0.731	13.1	10	0.134	0.607	9.0	1-3/16 X 12 UN
1	1.000	5	13	0.095	0.810	16.1	9	0.148	0.704	12.2	1-5/16 X 12 UN
11/4	1.250	7	12	0.109	1.032	26.1	7	0.890	0.890	19.4	1-5/8 X 12 N
11/2	1.500	7	11	0.120	1.260	38.9	5	0.220	1.060	27.3	1-7/8 X 12 N

Table 6-3   Typical dimensions and properties of synthetic rubber inner tube and outer core, wire reenforced flexible hose used on injection molding machine

Low Pressure – Single Wire Braid

Hose I.D. (in.)	Hose O.D. (in.)	Recom. Working Pressure (psi)	Min. Burst Pressure (psi)	Min. Bend Radius (in.)
0.188	0.516	3,000	12,000	3
0.250	0.578	3,000	10,000	3 3/8
0.313	0.672	2,250	9,000	4
0.406	0.766	2,000	8,000	4 5/8
0.500	0.922	1,750	7,000	5 1/2
0.625	1.078	1,500	6,000	6 1/2
0.875	1.234	800	3,200	7 1/2
1.125	1.500	600	2,500	9
1.375	1.750	500	2,000	10 1/2
1.813	2.219	350	1,400	13 1/4
2.375	2.875	350	1,400	24

Medium Pressure – Multiple Braid

Hose I.D. (in.)	Hose O.D. (in.)	Recom. Working Pressure (psi)	Min. Burst Pressure (psi)	Min. Bend Radius (in.)
0.250	0.688	5,000	20,000	4
0.375	0.844	4,000	16,000	5
0.500	0.969	3,500	14,000	7
0.625	1.094	2,750	11,000	8
0.750	1.250	2,250	9,000	9 1/2
1.000	1.562	2,000	8,000	12
1.250	2.000	1,625	6,500	16 1/2
1.500	2.250	1,250	5,000	20
2.000	2.750	1,125	4,500	25

High Pressure – Multiple Spiral Wire

Hose I.D. (in.)	Hose O.D. (in.)	Recom. Working Pressure (psi)	Min. Burst Pressure (psi)	Min. Bend Radius (in.)
0.375	0.844	5,500	22,000	7
0.500	0.969	5,000	20,000	9
0.750	1.266	4,000	16,000	12
1.000	1.562	3,500	14,000	13 1/2
1.250	2.000	3,000	12,000	18
1.500	2.250	2,500	10,000	22
2.000	2.750	2,250	8,000	26

attached to molds. Hose is easier to install than permanent connections, and because of its ability to expand slightly with pressure, it exerts an accumulator action which helps dampen pressure surges and vibration in the system.

Hose and fittings come in many shapes and configurations. They are either permanently assembled or reusable. The reusable ends have the advantage of being able to be fabricated in the plant. This significantly reduces the stock of hose assemblies that would be required. The disadvantage is the possibility of poor workmanship leading to failure.

When hose is installed, there should be no twisting or torsion at any time during installation or service. The full minimum radius requirements should be met. When the hose is used between two points in a straight line, enough slack should be left to allow for the bending radius. Hosing will change in length when pressurized, and the slack compensates for this change. Elbows are very useful in designing for flexible hose. Hoses should be kept free so that there is no abrasive action on the outside. They should be away from hot parts, such as heating cylinders. Hose manufacturers have manuals for installation which should be consulted.

Maintenance of Connectors

Maintenance of hydraulic fluid power connectors is almost always repairing leakage around the connectors and their joints. It is important to have records which will indicate the frequency of occurrences of failure at each location. A leak at the rate of one drop per second on a three shift, 6 day basis, loses approximately 350 gal of oil per year. The cost of reclaiming oil, its loss of properties, the labor in refilling machines, and the gathering up of spilled oil should be held to a minimum.

Not the least disadvantage of a leaking machine is the possibility of the oil level falling below operating requirements, without detection, causing damage to the equipment. A leaky machine (and the consequent dirt around it) does not make for good housekeeping and clean work. In those applications where vacuum metallizing, spraying, and painting are to be post molding operations, oil is a major hazard. Obviously, spilled oil increases the potential for fire.

The common causes of leaks in connectors and their fittings on molding machines follow:

1. Mechanical vibration and stress.
2. Improper selection of the connector.
3. Improper installation of the connector.
4. Water hammer.

Leak Prevention. A method of maintenance to prevent leakage will include the following:

1. Clear management policies stating that leakage is not desirable.

2. A maintenance reporting schedule that would allow identification of the offending units.

3. A complete supply of necessary piping, tubing, hoses, and fittings to meet maintenance problems as they occur. A good portion of improper maintenance is due to emergency and make-shift repairs because parts were not available. These tend to remain permanently.

4. Adequate machinery for fabricating tubing, piping, and rubber hose.

5. A continuing evaluation of new methods of connecting.

6. A continuing review of the hydraulic breakdowns and their solutions.

Water Hammer; What Is It and How To Avoid It

When there is a sudden change of velocity, in a fluid moving in a closed container, such as when there is a rapid shifting of valves, a phenomenon occurs called water hammer. The kinetic energy of the moving oil is converted suddenly into pressure energy. This pressure energy coupled with the inertial effect of the oil behind it causes a sudden surge of pressure in the pipe. This can result in pressures as much as 1000 psi higher than encountered in the normal operation of the system. It will expand the pipe and travel to the end. The shock wave will vibrate back and forth, with a clanking and banging noise, until its energy is dissipated by friction. Sometimes the noise is obscured by the normal noise level of the machine. The expansion of the walls of the pipe will cause a shortening in its length, thus putting a strain on the welds and connectors. One of the causes of water hammer can be traced to dirty oil.

Dirty oil can cause valve spindles to hold tight, requiring a much greater than normal pressure to move them. When the spindle starts to move, it clears itself and jumps rapidly, shifting the direction of flow. Water hammer effect can be eliminated by the following techniques:

1. Keep the velocity of the fluid low.

2. Slow down the valve closing speed either by built-in chokes or by the use of surge-damping valves which allow controlled, gradual acceleration of flow in one direction and free flow in the other direction.

3. Add a damper accumulator to the system.

4. Clean or replace dirty oil.

PACKING AND SEALS

Hydraulic machinery would not be possible without seals and packing (7). The requirement of 100% reliability in aircraft and the space program has led to major improvements in this area. The plastics engineer should be alert to the possibility of modifying existing conditions to utilize these improvements.

The three systems which use packing and seals in the palnt are the hydraulic pneumatic, and water. Each has different requirements, although the engineering principles are the same.

There are two types of seals — the static seal and the dynamic seal. The static seal is used to prevent the leakage of liquids and gases between parts that are not moving. Gaskets are examples of static seals. The term "gasket" is usually used in referring to sealing of flat plates and surfaces, such as covers on an hydraulic tank. Materials used for flat gasketing are vegetable fiber sheets, synthetic rubber, duck, and asbestos. The second type of static seal is used where diametral clearances are involved, such as sealing a stationary pin in a hole or a valve seat. The 0 rings which make superior gaskets are used for these applications.

A dynamic seal is a seal of liquid or gas between two members which are moving with respect to one another. There is a reciprocating type seal where the motion is reciprocal, such as in a piston in a cylinder, and the rotary type of seal where the motion is rotary, such as the shaft of a pump or occasionally a combination of the two.

Packing can also be classified in terms of shape. Six major types found in molding plants follow:

1. The V-type packing, used on the injection and clamping cylinders to seal between the piston and the housing.

2. The U-type packing, used as pneumatic and certain low-pressure hydraulic applications.

3. Cup packing, primarily used in air cylinders and air applications.

4. Flange packing, used in low-pressure applications as a shaft seal and also for a wiper.

5. Metallic piston rings, used on the injection and clamping rams and smaller hydraulic cylinders.

6. The 0 rings, used in almost all components and applications.

The materials that are used in packings and seals are numerous, with the main ones as follows:

1. Metal (copper, steel, stainless steel, bronze, and iron).

2. Natural and synthetic rubber.

3. Fluorocarbon plastics.

4. Asbestos.

5. Cloth impregnated with synthetic rubber.

6. Leather impregnated with wax or synthetic rubber.

The two classes of packing most widely used in molding are the fabricated and the homogeneous rubber-type packings. The fabricated packings use various cloth or asbestos and impregnate them with one of the many types of synthetic rubber or polymers. Their main use is in the V rings which seal the reciprocating plungers on the clamp and injection ram. The homogeneous packings are synthetic rubber. Their main use is in cups and 0 rings.

Fabricated packings require less tolerances, and lower concentricity requirements. They can be used with more metals. For example, cast iron (used for many rams) is only usable with fabricated packings. They will take high pressures up to 10,000 psi without excessive leakage or wear. This is well above the required operative range for molding machines. Fabricated V rings can be cut for easy installation while the homogeneous variety cannot.

The homogeneous type of packing is elastomeric by nature. The essential requirements of a dynamic seal is a surface finish finer than 16 microinches rms. Pistons should be hardened, tolerances held, and excentricity at a minimum. Homogeneous packing will take temperatures up to 300°F, and maximum pressures to 5000 psi in dynamic applications with the use of antiextrusion devices. In static homogeneous seals pressures can exceed 20,000 psi.

V-Type Packing

The V-type packing (Figure 6-7) consists of three parts — the male adapter (which is placed in the packing gland opening first, with the flat side down); V-rings (the number determined by the pressure, and the cylinder and plunger diameters); and the female adapter (which is the last packing to go in and on which the packing gland seats).

The principle of the packing is that the pressure of the fluid forces the V-ring apart, forming the seal. The higher the pressure of the fluid, the tighter the packing seals against thy plunger. At zero and low pressure the seal is kept because the cross section of the V-ring is slightly larger than the packing box, causing a slight compression of the packing. It is desirable, when possible, to preload the packing by means of either a conical spring or individual springs spaced on the packing. When refacing or replacing pistons care should be taken to maintain the slight compression, if not there will be leaks under low pressure. Overtightening this packing will not increase the sealing. On the contrary, excessive pressure will cause rapid wear and deterioration.

In fabricating packing, the cuts should be staggered at 90 and 180° intervals. They are usually prelubricated with graphite making for easier inserting. If there is any difficulty, particularly on larger glands, soaking the packing in oil or generously lubricating the packing box will help. Machine oil pressure can be used to remove the packing.

In applications where pressure exists in both directions, such as on a cylinder piston, two sets of the packing can be installed, with the V's facing each other. They must be separated by a metal separator firmly attached to the piston, or the pressures will be transmitted to each other, nullifying the effect of the packing. This type seal has very low leakage, good wear and has found increasing application.

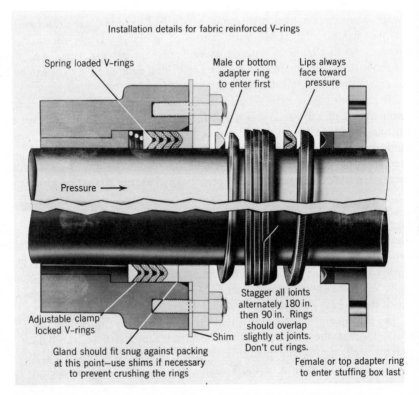

Figure 6-7 Installation details for fabric reinforced V-rings (Crane Packing Company).

U-Rings

The U-ring packings (so called because the cross sections resemble the shape of a U) works on a similar principle to V-packings; the pressure causes the packing to expand and hug the shaft. The pressure of the seal is proportional to the pressure of the fluid. Homogeneous U-rings have low frictional characteristics and are used with pressures up to 3000 psi. They are single installations and are never stacked. A popular use of U-rings is on the piston of an air cylinder. On new equipment U-rings have been largely supplanted by V-rings or 0-rings.

Cup Packing

Cup packings are widely used on the ends of pistons in pneumatic applications. Fabricated and rubber impregnated leather cups are used. The homogeneous type is not often found since other packings are less critical and perform equally well. The cup packing is usually mounted by means of a boss with a metal to metal tightening seal. This centers the cup and prevents excessive tightening of the packing.

Metallic Piston Rings

Metallic piston rings of the hydraulic-step type are used extensively on the injection and clamping cylinders for the internal piston seal, where they have the tremendous advantage of reliability and long trouble free life. The leakage past a series of piston rings is remarkably low and can be readily tolerated in molding machines. It has been found that the sealing is done by the first one or two rings. The additional rings increase the service life of the seal and guard against maladjustments or malfunctioning of a single ring.

Malfunctioning of piston rings will cause a hot spot to develop on the cylinder wall where the oil rushes by. It is also suspect when the cylinder moves toward the rod end when pressure is applied to both ends. If the occasion arises for replacing piston rings or reinserting plungers with piston rings on them, it is well to obtain standard automobile type piston ring contractors. If not available, thin spring steel or shim stock can be wrapped around the piston rings and held tightly while the assembly is inserted.

0-Rings

One of the great technological breakthroughs in sealing and packing was the discovery of the elastomeric, homogeneous 0-ring (Figure 6-8). This, coupled with synthetic rubber, and the new material discoveries that came after World War II, has led to a versatile, useful, simple, long lasting, inexpensive type of seal. While the 0-ring will not solve every packing and sealing problem, its intelligent use is a great help in the injection molding field. For static applications, the 0-ring seal can hold all pressures that will be encountered in the molding machines. In dynamic applications, seals of 5000 psi are obtainable with the proper antiextrusion devices. The seals can be placed either in a groove in the piston or cylinder wall. The preferred location is on the piston, because it is easier to machine the groove there than in the cylinder. They are very easily maintained and replaced, are readily available, and have small space requirements. The seal can usually be reused when taken apart for inspection.

The retaining grooves for 0-rings are very economical to make and do not require any packing glands or other paraphernalia associated with some packing types. They can be installed quickly and with no special tools on both regular and irregular parts. However, they do have definite limitations, such as in high speed reciprocal and rotary actions the tendency to spiral and roll, the requirements for close tolerances and clearances in the mating parts, and like all homogeneous packings the requirement of a fine finish on the moving parts.

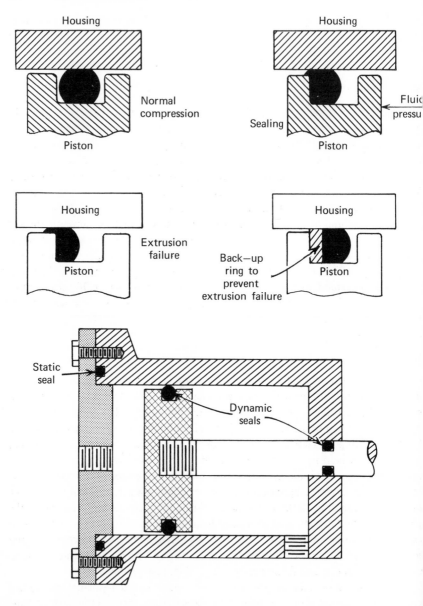

Figure 6-8   Use of O rings.

Spiral failure can be reduced by using an X shaped 0-ring which fits in the same packing gland as a standard 0-ring.

The 0-rings are made of many synthetic rubber compounds. When used on plastics care should be taken to see that they are compatible. The O-ring seals because the fluid pressure forces the precompressed O-ring in the direction of the flow. The pressure of the fluid squeezes the O-ring either against the piston and the cylinder in dynamic applications or the mating surfaces in static applications. When the pressure is reversed the O-ring moves in the other direction with the same results. When the pressure is too high or the clearance too low, the condition of extrusion occurs and causes rapid seal failure. To overcome this, leather backup rings are installed for pressures of 1500 psi and over. A durometer hardness of 70 is recommended for pressures to 1500 psi; 80 durometer to 2500 psi; and 90 durometer, for higher pressures. The extra hardness acts to limit the extrusion of the seal.

The O-rings can be used in noncircular sealing providing there is a minimum radius for bending. Square, rectangular, and irregular grooves can be machined. The O-rings of the proper circumference can be inserted and a very effective seal had. This is particularly useful in mold cooling applications. Premachined plates with O-rings attached are commercially available.

Manufacturers of O-rings have excellent manuals describing the applications and engineering properties of their product. These are readily available and it is strongly recommended they be obtained and read.

Maintenance of Packing and Seals

Maintenance of packing and seals does not present any difficulty. Packing failure, aside from normal wear, is most often caused by scoring on the plunger. If this is the case, the plunger should be either welded or brazed and then resurfaced. When packing consistently gives trouble, the system in point should be examined to see if there is extensive wear, whether the cylinder I.D. is out of round and whether the packing is acting as a bearing instead of just a packing. Dirty hyraulic fluid is a cause of packing failure. The decomposition products get on the moving pistons and initiate scoring. Once this starts, leakage increases rapidly. At the first sign of scoring, the machine should be stopped and the cause of the trouble found and rectified.

At least one spare should be in stock for every packing used in the plant. Invariably a cost of a repair is greater than the cost of the packing. If new packing is available, maintenance men will change it depending on conditions rather than availability.

To summarize, the major factors that cause packing failure follow:

1. Improper original design.
2. Wrong clearance.
3. Improper support.
4. Wrong metal on sealing parts.
5. Improper finish on metal surfaces.

6. Overheating of system.
7. Contaminated hydraulic fluids.
8. Improper lubrication.

FLUID POWER CALCULATIONS

The following formulae are useful for calculating connector sizes, volume, speed, and time in various parts of the hydraulic system.

$$R = \frac{7740 U D}{\nu} = \frac{3160 Q}{\nu D} \tag{6-3}$$

where
R = Reynold's number
U = velocity (ft/sec)
D = inside diameter (in.)
Q = flow rate (gal/min)
ν = kinematic viscosity (cSt).

$$\Delta P = \frac{0.0808 f L U^2 S}{D} \tag{6-4}$$

ΔP = pressure loss (psi)
f = friction factor
L = length (in.)
U = velocity (ft/sec)
S = specific gravity
D = inside diameter (in.) .

$$f_{laminar} = \frac{64}{R}$$

$$f_{turbulent} = \frac{0.316}{R^{0.25}}$$

R = Reynolds number

Flow rate in circular tubes

$$Q = 2.45 \, D^2 \, U \tag{6-5}$$

where
Q = flow rate (gal/min)
D = inside diameter (in.)
U = velocity (ft/sec) .

The volume of oil to move a cylinder is

Back end **Rod end**

$$V = 0.785 \, D_c^2 \, L \qquad\qquad V = 0.785 \, (D_c^2 - D_R^2) \, L \qquad\qquad (6\text{-}6)$$

where
- V = volume (in.³)
- D_c = diameter cylinder (in.)
- D_e = diameter rod (in.)
- L = length (stroke) (in.) .

The cylinder speed is

Back end **Rod end**

$$U = 1.273 \, \frac{Q}{D_c^2} \qquad\qquad U = \frac{1.273 \, Q}{(D_c^2 - D_R^2)} \qquad\qquad (6\text{-}7)$$

where
- U = speed (in./min)
- Q = flow rate (in.³/min)
- D_c = diameter cylinder (in.)
- D_R = diameter rod (in.) .

Back end **Rod end**

$$Q = \frac{U D_c^2}{294} \qquad\qquad Q = \frac{U \, (D_c^2 - D_R^2)}{294} \qquad\qquad (6\text{-}8)$$

where
- Q = flow rate (gal/min)
- U = speed (in./min)
- D_c = diameter cylinder (in.)
- D_R = diameter rod (in.)

The time to move the cylinder is

Back end **Rod end**

$$T = \frac{0.204 \, D_c^2 \, L}{Q} \qquad\qquad T = \frac{0.204 \, (D_c^2 - D_R^2) \, L}{Q} \qquad\qquad (6\text{-}9)$$

where
- T = time (sec)
- L = length stroke (in.)
- Q = flow rate (gal/min)
- D_c = diameter cylinder (in.)
- D_R = diameter rod (in.)

It is best to design for laminar flow (Reynolds number below 2000). To accomplish this larger connectors, fittings, and valves are required than if turbulent flow occurred. The result is often a compromise between cost and optimum design. It is interesting to note that the friction factor for R (Reynolds number) 2000 (laminar flow) is 0.032 (eq. 6-4). At R equal 4000, it is 0.040, at R 8900 0.032. Above this value the friction factor is lower than the lowest found amoung laminar flow condictions. There are other factors, which cause problems aside from turbulent flow.

One of these is the velocity of the oil. If it is too high there are significant pressure drops in the system (eq. 6-4). When rapidly moving oil is stopped surge pressures occur. In extreme cases they cause "water hammer." It has been estimated that rapidly stopped oil generates a surge pressure of approximately 40 psi per ft/sec of velocity. Velocities that are too low increase the cost of valving and piping. The maximum velocity generally recommended is 15 ft/sec. Higher velocities in certain parts of the equipment can be found.

To illustrate the use of these formulae let us suppose the following condition. An hydraulically activated cam has to be returned to its position in one second to prevent interference with the mold closing. It is activated by a 3 in. cylinder with a 10 in. stroke. The oil has a specific gravity of 0.85 and a viscosity at the operating temperature of 300 SUS. The problem is to determine the hydraulic and piping requirements.

From (6-9) we can determine the flow rate needed.

$$1 = \frac{0.204\,(3)^2\,(10)}{Q}$$

Q = 18.4 gal/min

To determine the size connector for laminar flow we must find the Reynolds number. From (6-2) we can determine the kinematic viscosity.

$$v = 0.220\,(300) - \frac{135}{300}$$

v = 65 cSt

From (6-3) we can determine the diameter of the pipe for laminar flow with a Reynolds number of 2000.

$$2000 = \frac{(3160)(18.4)}{65D}$$

D = 0.45 or ½ in. I.D. connector

The diameter of the connector calculates to 0.45 in.. A 1/2 in. connector would be used.

The pressure loss can now be calculated. Equation 6-5 will give the velocity.

$18.4 = (2.45)(0.5)^2 U$

$U = 30 \text{ ft/sec}$

Even though this velocity is high it is acceptable in this part of the circuit. If it were not, the diameter of the connector would have to be increased. The friction factor is obtained from (6-4).

$f = \dfrac{64}{2000} = 0.032$

The pressure loss is calculated from (6-4)

$$\Delta P = \frac{(0.0808(0.032)(10)(30)^2(0.85)}{0.5}$$

$\Delta P = 39 \text{ psi}$

The pressure loss is 39 lb psi, which is acceptable for this application.

If 18.4 gal/min were not available at that time in the cycle, a smaller source could be used in conjunction with an accumulator. The accumulator can be charged at a much lower rate so that smaller diameter hose could be used from the pressure source to the accumulator. The connection from the accumulator to the cylinder would still be 1/2 in., though much shorter in length. An accumulator circuit is shown in Figure 6-29.

HYDRAULIC VALVES

Three types of valves are used in hydraulic systems. Directional control valves change or control the direction of the flow of oil. Examples are two way, four way, and check valves. Pressure control valves control the pressure in the system. The consist of relief, unloading, pressure reducing, sequence, and counter balance valves. The third type is flow control valves which control the amount of flow.

Two Way Valves

The simplest directional control valve is a two way valve. It is either open or closed permitting or stopping oil flow. Figure 6-9 shows a schematic diagram of a normally open, manually controlled two way valve. Housing of directional control valves are usually an iron or steel casting with a hole machined through. The ports are drilled and tapped or adapted for flanges. The end plates are held in with screws and sealed with O rings (figure 6-10). The spool or spindle is a

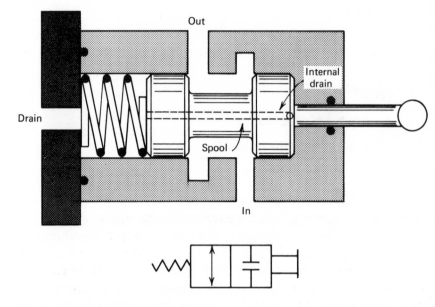

Figure 6-9 Schematic representation of two way, spring returned, manually operated, normally open valve with symbol.

turned piece of hardened, ground steel. A second type of construction, where size and weight considerations are important, uses a hollowed-out spool with radial holes drilled for proper flow paths for the valve's function. In this particular valve (Figure 6-9) a spring is used to maintain the valve in its normal, or open position. When the manual side is pushed, the spool shifts over, compressing the spring. A drain is provided to remove the oil from behind the spool caused by leakage between the spool and the housing. Without it, it would be difficult and perhaps impossible to move the spool. When the spool is shifted, the land of the spool covers the inlet opening. When the manual force holding the spool is released, the spring will push the spool back to its original position, opening the valve. A hole drilled the length of the spool and exiting at right

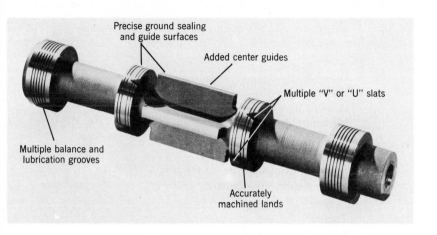

Precise ground sealing
and guide surfaces

Added center guides

Multiple "V" or "U" slats

Multiple balance and
lubrication grooves

Accurately
machined lands

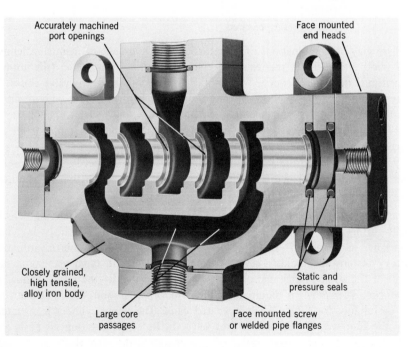

Accurately machined
port openings

Face mounted
end heads

Closely grained,
high tensile,
alloy iron body

Static and
pressure seals

Large core
passages

Face mounted screw
or welded pipe flanges

Figure 6-10 Four-way valve showing machined plunger and cored body (The Oilgear Company).

angles at the front end of the spool acts as the drain for the other side of the valve. Beneath the valve is its JIC symbol. It is always shown so that when the spring is activated, the valve will assume the configuration shown in the box adjacent to the symbol. The following are the most common methods of activating valves:

Spring.
Manual.
Push-button
Lever.
Pedal or treadle.
Mechanical.
Solenoid.
Reversing motor.
Pilot pressure (supplying or releasing).
Pilot pressure, differential.
Solenoid pilot.
Thermal.
Servo.
Detent.
Pressure compensated.

There may be combinations of actuators such as hydraulic or manual, solenoid or pilot, and such. In this example, when the spring is activated (the normal position), the valve is open. When the manual is operated, the valve closes. If there were no spring in the valve and a manual control was used on both sides, both sides would have the "manual" symbol.

Some of the major uses for two way valves in injection molding are in deceleration circuits, pump separation circuits, the selection of different hydraulic pressures, the selection of different hydraulic speeds, and in prefill circuits. Figures 6-11 and 6-12 illustrate some of these applications. The others will be subsequently illustrated.

Deceleration Circuits. Figure 6-11 shows a deceleration circuit which will bring a cylinder to a smooth stop. Speed control is important in injection molding. Because of the high inertia of the heavy platens and molds a severe jarring can occur with rapid opening at the end of the stroke. Since many knockouts are activated by the return action of the ram, slow down is required for their smooth operation. A circuit like this is useful too, for hydraulically operated cams. The clamp cylinder is operated by the four way directional control valve. In its center position all ports are blocked. When the forward solenoid F is energized the valve shifts so that the pressure is ported to the rear of the clamp cylinder and the tank connection to the front. Even though the two-way valve is energized (open) (because the

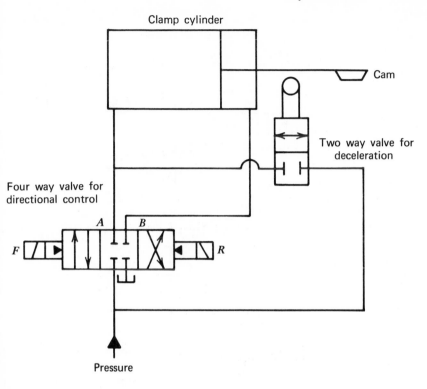

Figure 6-11 Deceleration circuit to bring cylinder to a smooth stop using a two-way valve.

cam has depressed the roller) both sides are connected to pressure, thus not affecting the forward motion. The clamp cylinder now moves forward, releasing the mechanical plunger of the two way valve. For return, solenoid R is energized. This send pressure to port B and port A to tank. The cylinder begins to retract. When the cam hits the plunger of the two way valve it starts to shift it to the open position. Since port A is connected to tank it will start to bypass the pump pressure to tank. This will slow the ram down to a gentle stop. The four way valve is now shifted to its neutral position so that port A is blocked and the pressure of the system will no longer go to tank. While this is a simple way to decelerate, it is good in only one direction at a time and is difficult to adjust.

Figure 6-12 shows an improved deceleration circuit. The circuit consists of a limit switch (LS-1) which controls a two-way valve used for speed control (No. 1) and therefore a flow control valve (No. 1). They are attached to the

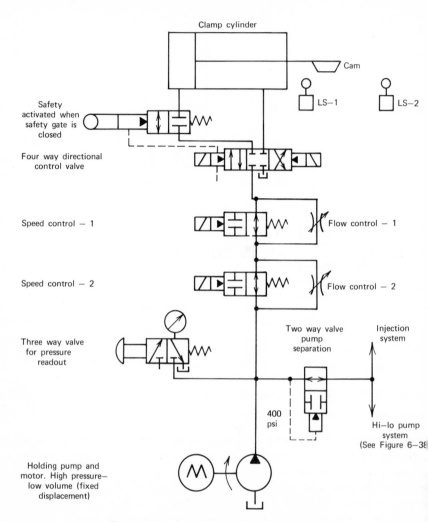

Figure 6-12 Uses ot two and three way valves.

pressure inlet before the directional control valve. The speed control valve is
normally open. As long as LS-1 is not contacted the flow in the system will
be maintained as such and the clamp cylinder will move accordingly. (All
other elements of the circuit not pertaining to the use of two way valves have
been omitted for clarity.) When LS-1 is contacted the solenoid controlled,
pilot operated speed control valve No. 1 is activated, closing the valve and
forcing the oil to go through the flow control valve No. 1. The speed of the
cylinder will depend upon the setting of this valve. This will happen at any

time when the cam activates LS-1. Therefore speed control is available at any time of the forward or return stroke of the clamp ram, whenever the cam hits the switch. Should slow-down be desired in only one direction, a one way dog type of arm can be used on the limit switch. Otherwise the limit switch can be taken out of the circuit by means of relays or timers. Different speeds can be had by adding an identical control circuit (No. 2). As many can be added as required. Another way of accomplishing the same thing would be to use a pilot operated flow control valve which would be placed into or removed from the circuit by another two way valve.

Holding Pumps.   In most hydraulic and many toggle operated presses a high pressure-low volume holding pump is used to maintain the pressure in the clamp system while the injection and cooling take place. The output of the pump should be slightly more than enough to overcome the internal leakage of the system. It is necessary to isolate the clamping circuit at that time, from the injection end. An excellent way to do it is by an hydraulically piloted two way valve (bottom of Figure 6-12). When the clamping pressure is below 400 psi, all pumps in the system are used to move the clamp. As soon as resistance is met the pressure begins to build. When it reaches the valve setting (in this instance 400 psi) it will shift the spool of the valve, closing it and isolating the main pump circuits, which can be used for the injection circuits. When the four way directional control valve shifts to open the clamp cylinder pressure will drop until resistance is met when the cylinder opens up. The pressure will drop below 400 psi and open the pump separation valve and permit all pumps to move the clamp cylinder back rapidly.

Safety.   Safety is important in a molding machine. The two way valve illustrated in the top of Figure 6-12 is used as an hydraulic interlock. Its normal position prevents oil from the four way valve reaching the clamp forward position. To activate the safety valve two things are required; (*a*) the safety gate must mechanically depress the valve and (*b*) pilot pressure must be available. Pilot pressure is initiated by energizing the four way directional control valve, which must be done through the electrical safety. Without pilot pressure in the four-way valve there is no possibility of a malfunctioning of any part of the clamp circuit. The cycle, therefore, cannot start unless the pilot pressure is available.

Three Way Valves

A three way valve is one where the outlet can be connected either to pressure or to tank. Its three main uses are for pressure readouts, safety circuits, and activating spring returned cylinders. Figure 6-12 shows a three way valve used for reading pressure. The normal (spring controlled) position has the pressure port blocked and the gauge ported to tank. When the button is pushed, the

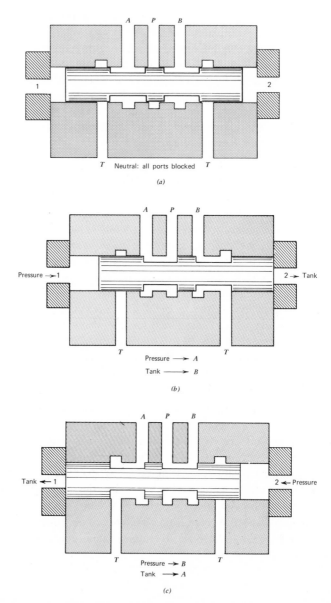

Figure 6-13. (*a*) Hydraulically operated, three position, spring centered, four way valve i▮ *neutral* or *central* position. All ports are blocked. (*b*) Pilot pressure is applied at port 1, por▮ 2 goes to tank. Pressure port (*P*) of the main valve is connected to port *A* and tank port (*T*▮ is connected to port *B*. (*c*) Pilot pressure is applied to port 2, port 1 goes to tank. Pressur▮ port (*P*) is connected to port *B* and tank port to *A*.

gauge is attached to the pressure system. This prevents the gauge from being attached to the system at all times. There are times when the same hydraulic line will have different pressures, such as the back pressure on a reciprocating screw. Then it is desirable to use a low range gauge for accuracy instead of one which will read maximum pump pressure. This type circuit makes it possible.

If in the previous illustration, the clamp forward connection was attached where the gauge is, no oil would reach it in the normal spring position. If the activator was a mechanical cam moved by the safety gate, it would serve as an hydraulic safety.

If the clamp forward port of a spring returned cylinder was attached to the gauge port, nothing would happen until the valve was activated. Then pressure would force the cylinder forward. When the activator was removed, the clamp forward port would go to tank and the spring would retract the cylinder.

Four Way Valves

Four way directional control valves, as the name implies, have four ports: pressure, tank, and two working ports A and B. Their main use is to extend and retract cylinders; to start, stop, and reverse the direction of hydraulic motors; and to sequence machine operations.

Hydraulic cylinders used on molding machines consist of a housing with a cylindrical bore in which is placed a piston or ram. The piston has a shaft on one side which is connected to the platen or member to be moved. The large diameter end of the piston is put into the cylindrical hole and that side is sealed. The smaller diameter piston shaft extends outward through V shaped hydraulic packing which prevents leakage (Figure 6-7). It is supported by a bushing. When oil under pressure is applied to the large or head end, and the piston end is vented to tank, the piston will move forward. When the pressure is reversed and applied to the shaft side the head end is vented to tank, the piston will return. To accomplish this, a four way valve is used to direct the flow of oil (Figure 6-13). The valve, has three positions, though a two position valve will also reverse the direction. Figure 6-13A shows the spool in the central or neutral position held there by springs which are not shown. This spool configuration has all ports blocked in the central position, hence is called a closed center valve. The valve is hydraulically operated. If oil under pressure is applied to port 1 and port 2 is directed to tank, the spool will shift to the right (Figure 6-13B). The pressure port is now connected to A and B to the tank. If the pilot pressure is reversed with pressure at port 2, and 1 going to tank, the spool will shift to the left (Figure 6-13C), allowing the pressure to go to B and A to tank.

Assume a different spool configuration so that the area presented to the pilot pressure at port 1 was larger than the area presented to the pilot pressure at port 2. If oil at the same pressure were applied to both sides the larger area at port 1 would cause the valve to shift to the right. If port 1 was ported to tank, the pressure at port 2 would then shift the valve to the left. This is called a differential operated valve.

Other Configurations. The valve in Figure 6-13 which blocks all ports in the central position must be used with accumulators. It will hold the cylinder or fluid motor in a stopped position and close the pressure and tank ports. It can be used in parallel with other circuits. A second possible configuration is the open-center type, where all four ports are connected together and hence to tank. This permits the cylinder and hydraulic motor to move. More important it unloads the pump from the pressure port to tank. This type configuration can be used to unload the pump while the safety gate is open or cylinders are at rest. Another configuration is with the pressure port blocked and A and B connected to tank. This drains both sides of the cylinder and the hydraulic motor. It can be used with other closed center valves. It is almost invariably used as the pilot operator of a four way valve. A fourth configuration blocks the tank port and attaches A and B to pressure. When moving a single acting cylinder it will act as a regenerative circuit permitting oil to go from the rod end to the back end without going through the pump. This gives fast cylinder speed. The tandem center blocks A and B and connects pressure to tank. It will lock a cylinder and a motor when the valve centers. The pressure flows through the center valve and can be directed to additional tandem centered valves. Other configurations are blocking port A and connecting port B, pressure and tank; and blocking port B and connecting A, pressure and tank.

Solenoid Operation.   While direct solenoid operation can be satisfactory for valves up to 1 in. pipe size, contemporary hydraulic design has limited their use to a maximum size of 1/2 in.. A solenoid is an electrical device consisting of a coil with a steel rod through it. When the coil is energized, the magnetic field causes the rod to move. Since the force of a solenoid varies with the square of the voltage, solenoids should be selected so that they operate properly at 85% of the nominal voltage. Low voltage can cause sluggish solenoid operation which can cause machine malfunctions that are difficult to locate. On larger valves the solenoid is impractical as a direct drive. Instead it is used to operate a small four way valve which acts as a control for pilot pressure oil supply and provides oil to move the spindle or spool as previously described. These pilot operated, solenoid controlled, four way valves are usually built as one unit utilizing the pressure and drain connections of the large valve. They can, however, be operated at different pressures and with a completely different oil system. Figure 6-14 shows the new JIC configuration for a solenoid operated, pilot controlled, spring centered, four way, three position valve. Figure 6-14B shows the condensed symbols. In the event of an electrical failure a spring centered solenoid will return to its central position.

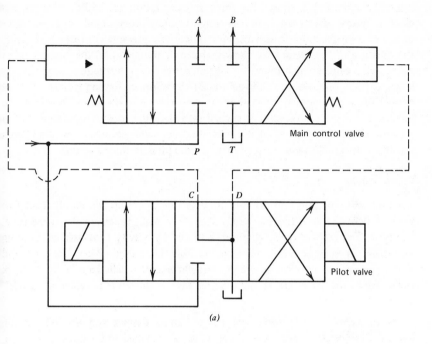

Main control valve

Pilot valve

(a)

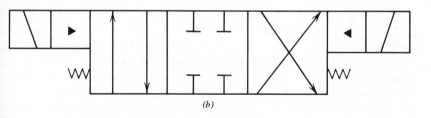

(b)

Figure 6-14 *(a)* Full JIC symbols for solenoid operated, hydraulically piloted, with internal pilot supply, spring centered three position, four way valve. *(b)* condensed symbol of *a*.

Figure 6-15*a* shows a pictorial representation of the method of moving ar hydraulic cylinder; *b* shows the same representation using JIC symbols. Ar electric motor drives an hydraulic pump which provides oil to a manually operated, two position, four way valve. There is a pressure control which limit the pressure of the system. As shown, the oil is pumped past the pressure relief valve, and a gauge, which shows the system's pressure, into the four way valve. It is set so that oil goes into the rod end of the cylinder forcing it back. The oil behind the piston goes through the four way valve into the tank. To reverse the direction, the four way valve is shifted into the other position. Should the oil pressure exceed the setting of the pressure relief valve, enough oil would bypass through it to tank, to lower the pressure to the predetermined setting.

Check Valves

A check valve permits unrestricted flow in one direction and is self-closing to prevent flow in the other direction. The three main types are the swing-type ball-type, and poppet-type. They are actuated by gravity (not used in molding machines) or springs. Pilot operation is sometimes used. When the inlet and outlet ports of the check valve are in a straight line, it is called an "in-line" check valve. When the ports are at right angles (as an elbow) it is called an "angle" check valve.

Swing Type.   The swing-type has a hinged flapper and is rarely spring loaded. Gravity closes the valve which must be installed in the correct position It is not used on molding machines and is found primarily in low pressure water applications.

Ball-Type.   The ball-type check valve is basically a ball with a spring behind it sealing off on a drilled inlet port. When oil comes in this direction (unrestricted flow) with more force than the spring, the ball moves off the seat and oil flows around it. When oil comes from the other direction, the spring has already sealed the ball onto the seat and the higher the pressure of the oil the tighter the seal. The ball-type, while simple and relatively inexpensive, has a number of disadvantages: (*a*) the design is such that it produces large turbulence and consequent pressure drop and heat; and (*b*) it has poor seal characteristics because the ball never reseats in exactly the same position. Ball-type checks however, are preferred with the more viscous liquids such as molten plastic, and are used (without springs) in preplasticizer systems between the injection and shooting cylinders.

To overcome the deficiencies of a ball in the hydraulic system, a poppet is used. Figure 6-16 shows a poppet-type, in-line check valve in the open position with oil flowing from the free flow port. It flows into chamber *A* through hole *B* and out the right-hand port. When oil stops flowing, the spring expands sealing the poppet on the seat. The higher the pressure from the blocked port, the

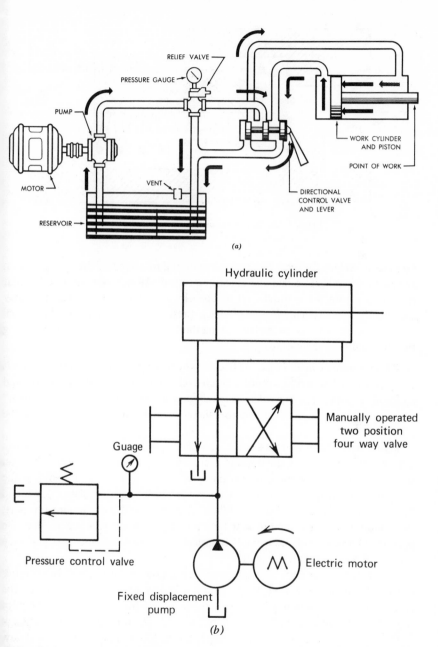

Figure 6-15 (a) A standard method of moving an hydraulic cylinder. (NAVPERS 16193). Pictorial representation. (b) Simple control of an hydraulic cylinder, as shown in using JIC symbols.

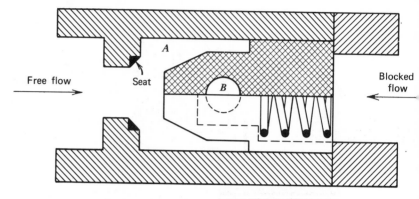

Figure 6-16   In-line check valve in open position.

tighter the seal on the seat. The pressure required to overcome the spring tension
and open the valve is called the cracking pressure. Most valves use a spring that
requires 5 psi on the hydraulic oil to open it; 65 psi springs are available. They
are usually used to create enough back pressure to pilot operate directional
control valves. This is an inefficient method because of the loss of pumping
capacity and heat generation.

Pilot Operated Check Valve.   Because the principle of using pilot pressure is
so common, we describe a pilot operated check valve. As with all valves and
hydraulic mechanisms, it works on the application of "force = pressure x area."
Areas and pressures are selected to generate enough force to accomplish the
purpose.

Figure 6-17 shows a cross-section of a pilot operated, right angle check valve.
Spring 5 held in place by the housing exerts an upward force on piston 2 held on
spool 1 by nut 3. The spring keeps the spool on seat 4. In normal operation the
oil coming from the free flow port on the left will force the piston down and
flow out the port on the bottom. When oil flow comes from the bottom port
the piston is initially on the seat because of the spring pressure so that the
pressure of the oil will keep the valve sealed. The pilot pressure port is on the
top. The drain port is on the right.

Figure 6-18 illustrates the schematic operation of this valve. In *A* oil is
flowing from the free flow port out the blocked port. In *B* oil goes to the
blocked port and cannot pass. Figure 6-18*c* shows the operation of the pilot
valve. Pilot pressure is applied on top of piston 3. The force generated is enough
to overcome the force of the oil coming from the blocked port. The oil will now
be able to flow in either direction in the valve and will remain so as long as there
is pilot pressure. To move the spool, the pilot pressure times the pilot piston area

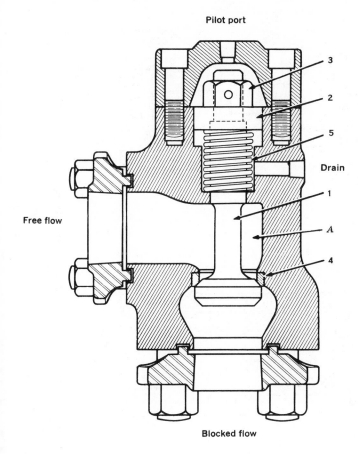

Figure 6-17   Pilot operated, right angle, check valve (Vickers, Inc.).

...ust be greater than the spring force and the product of the valve sealing area ...mes the maximum pressure applied from the blocked port. The rate of pilot oil ...ow will determine the speed of the spool movement. Particularly in directional ...ontrolled valves, a choke is used to control the spool speed to prevent shock in ...he system.

   A main use for pilot operated check valves is in the prefill circuits of fully ...ydraulic clamp molding machines, which are very large and permit the back of ...he clamping ram to be filled by gravity or atmospheric pressure instead of going ...hrough the pumping system. These valves are used with jack rams (Figure 6-19). ...A jack ram is a cylinder within the main ram.) The jack ram is much smaller in ...iameter and will therefore cause the main ram to move much faster with the ...naximum pump output than if the oil had to fill the larger volume behind the

504

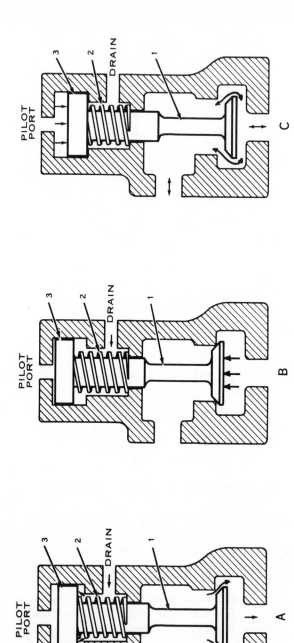

Figure 6-18 Schematic operation of valve in Figure 6-17 (Vickers, Inc.)

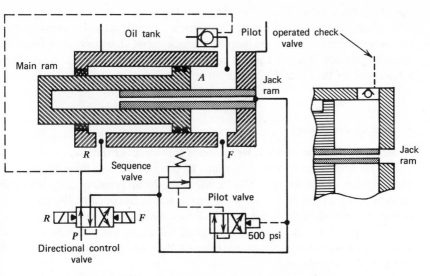

Figure 6-19   Prefill circuit using pilot operated check valve.

main ram. When the forward solenoid (*F*) of the directional control valve is energized, oil under pressure goes into the jack ram. The movement of the main ram will not have enough resistance to generate a 500-psi back pressure so that the pilot valve will not shift to operate the sequence valve. The forward motion of the main ram will cause the check valve to open permitting oil flow from the tank to the main ram area. The front of the ram is ported to tank. When the main ram is clamped and meets resistance the pressure builds up to 500 lb which operates the pilot valve, closes the sequence valve, and brings main pump pressure behind the main ram. The check valve will close sealing off the chamber *A* from the oil tank. Full pressure will be built up. When the ram is to return the directional control valve is shifted to the *R* position. This ports the jack ram to tank closing the pilot valve and the sequence valve. Pressure is applied to port *R* and to the pilot operated check valve. As the main ram moves back, the oil in chamber *A* returns to the tank through the check valve which is held open. There are decompression valves built into the pilot operated check valve.

Decompression Valves.   When a mold is clamped or an injection ram moved forward the resistance build-up is gradual because of the yield of the metal machine parts, and the slight compressibility of the oil. However, when this oil thus compressed is decompressed rapidly, the same factors act in reverse and send damaging shock waves through the system or wear circuit components. The higher the system pressure, the greater the danger of decompression. To overcome this, a decompression valve is used which is a variation of a pilot operated check valve. The tip of the pilot piston hits a small secondary check valve spool, before it hits the main check valve. This permits a very small amount of oil at a high velocity, to pass the check valve, drastically dropping the system's pressure. The pilot piston continues to move and opens the main check valve spool. In the event that this is not enough, a third decompression stage can be added.

Pressure Control Valves

If we take an in-line check valve and vent the restricted flow side to tank, we have a pressure control valve. If the area of the exposed free flow side is 1 in.^{2} and the force of the spring is 500 lb, whenever the oil pressure on the inlet side exceeds 500 psi (500 lb $\times$ 1 in.^{2}) the piston will move, opening the valve, and permit enough oil to flow past it to tank to reduce the system pressure to less than 500 psi. Then the force of the spring will move the piston to close the valve. Figure 6-20 shows a schematic drawing of an internally drained relief valve operating on this principle.

A valve of this kind can chatter, causing pressure fluctuations in the system. Damping arrangements and other techniques overcome this. This type valve will be fail safe, as wear will reduce the controlled pressure rather than raise it. It is very often used as a maximum pressure safety valve on pumps and in systems.

Balanced Piston-Type Valve.   A balanced piston type valve overcomes most of the limitations of the spring loaded relief valve (Figure 6-21). The pressure to be controlled can be placed in either of the "out/in" ports. The bottom port goes to tank. The hollow piston 1 is held down by spring 5 sealing the tank outlet from the pressure side. There is a small hole (Y) in the piston, connecting the pressure side (Z) and the chamber above the piston (X). The oil pressure in both chambers Z and X are normally the same. Since the area of the piston on the X side is larger than the area on the Z side and assuming them both at the same pressure, the area differential will help spring 5 keep the piston seated.

Assume, for the moment, that the vent (W) is plugged. Nothing will happen until the pressure reaches the point where it exerts more force than the compressed spring 3 exerts on the poppet seal area at point 2. Then the seal at point 2 will open and oil in chamber X will flow past it through the center of the hollow piston into the tank. This reduces the pressure in chamber X (the top side of the piston) which will now be lower than the pressure in chamber Z. The piston moves up permitting oil to flow past the main seal to tank. When the oil pressure in chamber Z is low enough so that the poppet 2 reseats, the original conditions prevail, and the valve closes. Such pressure control is more gentle than the direct spring operated valve.

If the vent was connected to another poppet valve assembly, whose spring chamber behind the poppet had a direct connection to tank, the identical type

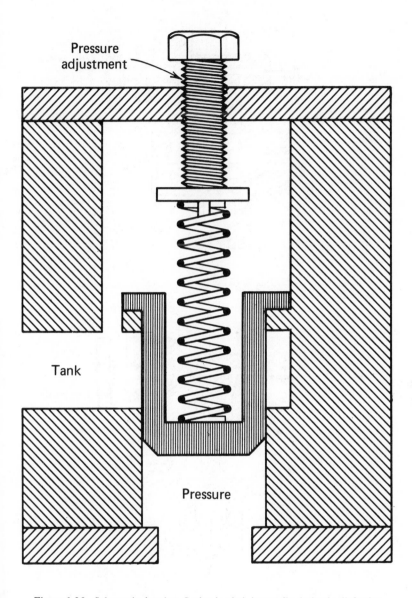

Pressure
adjustment

Tank

Pressure

Figure 6-20   Schematic drawing. Spring loaded, internally drained relief valve.

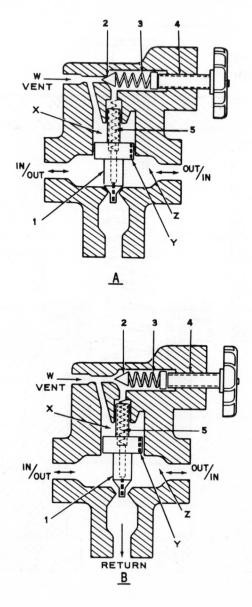

Figure 6-21   Balanced piston type relief valve (Vickers, Inc.).

f control occurs. The only limitation is that the secondary control must be set
t a lower pressure than the primary control. If not, the primary control would
perate first.

How To Vary The Pressures.   By using a number of remote control valves
which are simpler, less expensive, use smaller connectors, and are easier to
nstall than the main valves), any number of different pressures become
vailable. Figure 6-22 shows such a circuit. When the directional control valve is
n its neutral position the main relief valve operates on the setting of the primary
ilot (*C*). This will be the highest setting in the system. When solenoid *A* is
nergized, pilot valve *A* will control the pressure and when solenoid *B* is
nergized pilot valve *B* will control the pressure. The *A* and *B* can be set at any
ressure lower than *C*. If the vent is connected directly to tank by energizing
olenoid *D*, then the oil in chamber *X* of Figure 6-21 will be ported to tank
ermanently raising the piston *I* and dumping all the oil in the system to tank.
The relief valve will now be acting as an unloading valve. This is acceptable for
nfrequent operation. Otherwise, because of the internal construction of the
alve, it will overheat. For full scale operation unloading valves are used. This
ype of circuit can be used for installing several pressures and speeds in the
clamp or injection circuits.

Unloading Valves

When the flow of a pump is not needed during the molding cycle, it is
uneconomical and impractical to stop the pump from rotating. Sending the oil
over relief valves is wasteful because of the heat energy lost by compressing and
decompressing. An unloading valve is used. It sends the pump output to tank at
close to atmospheric pressure, after a preset pressure has been reached in another
part of the system which is applied to its pilot port. Unloading valves are used in
accumulator circuits to unload the pump after the accumulator has been
charged. They are also used to remove large volumes of oil rapidly from the head
end of a cylinder when it is retracting. These valves are pilot operated, spring
returned, normally closed two way valves.

Unloading valves are made with poppet type pilot operation similar to relief
valves. This type of valve requires a somewhat higher pressure during unloading,
which adds to the heat and horsepower loss. For this reason, most larger
unloading valves are manufactured with direct spring action. An example of the
use of an unloading valve in a hi-lo pump circuit is shown in Figure 6-38.

Sequence Valves

A sequence valve blocks the flow of oil in a secondary circuit until a set pressure
is reached. The valve then opens, permitting this primary system to be connected

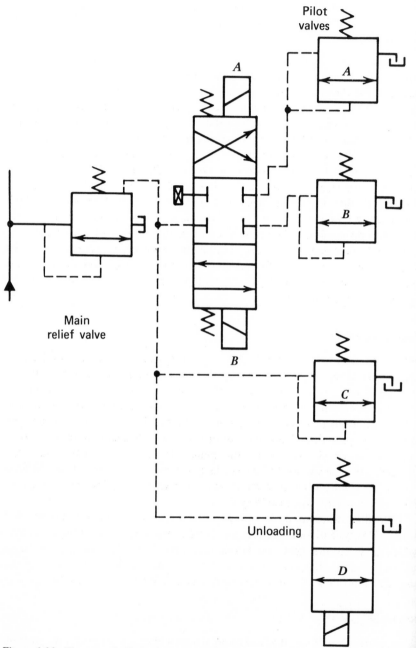

Figure 6-22 The use of pilot valves to get different pressures from the same main relief valve.

510

ɔ a secondary system (Figure 6-23). The secondary system pressure can vary ᴚom zero to that of the primary system before the valve opens. Therefore, the ᴚain must be externally connected. The sequence valve, in essence, is a normally ᴚosed, spring-returned, piloted, two way valve.

The pilot is usually internal. It is connected from the primary system to the ᴚottom part of the spool. When enough force is developed to overcome that of ᴚhe spring the valve opens. It will remain open as long as the primary pressure is ᴚaintained at that or higher value. If the pilot is external it can be used as a ᴚafety interlock or to operate the valve when other conditions of the circuit are ᴚet (Figure 6-19 and 6-40). In many circuits it is desirable to have unrestricted ᴚeverse flow from the secondary to the primary port. This is done with a check ᴚalve either integrally built or externally connected.

Sequence valves are used in regenerative circuits (Figure 6-40), where a ᴚinimum pressure is required for a control circuit, and in operations where it is ᴚssential to have one part of the system pressurized before the other.

Ꞓounterbalance Valves

Ꞁ counterbalance valve (foot valve, back pressure valve) has free flow in one ᴚirection (through an integral check valve) and blocked flow in the other ᴚirection, until a preset pressure is reached. Counterbalance valves are often ᴚonnected to the outlet ports of vertical cylinders to counterbalance their ᴚeight. They are also used on the jack ram end of horizontal clamping cylinders ᴚo counteract the inertia of the platen and cylinder and prevent lunging or ᴚamming of the piston. They provide the resistance for the screw while ᴚlasticizing and are commonly known in that capacity as the back pressure valve.

Figure 6-24 shows a schematic of a counterbalance valve with oil flowing past ᴚhe check valve in the free flow direction. When oil comes from the restricted ᴚow port, nothing will happen until there is enough force generated under the ᴚiston to counterbalance the spring. The valve will then open and remain open as ᴚong as that minimum pressure is maintained.

Ꞓressure Switches

Ꞁ pressure switch will activate an electrical switch when the pressure rises (or ᴚalls) to a preset value. These switches are used on molding machines to initiate ᴚhe injection forward cycle when the clamp pressure has been built up. At the ᴚame time, they can start the overall cycle timing. They are used in accumulator ᴚircuits for controlling the accumulator pressure. They are used for sequence ᴚontrol, initiating an action when the pressure indicates a previous action has ᴚeen completed.

The simplest type consists of a piston with one side exposed to pressure and ᴚhe other side spring loaded. The piston depresses an electrical switch whose

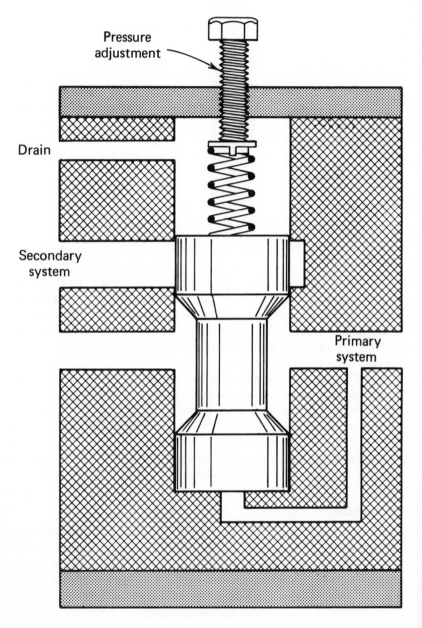

Figure 6-23 Sequence valve. When pressure in primary system reaches control point, plunger moves upward, connecting primary and secondary systems.

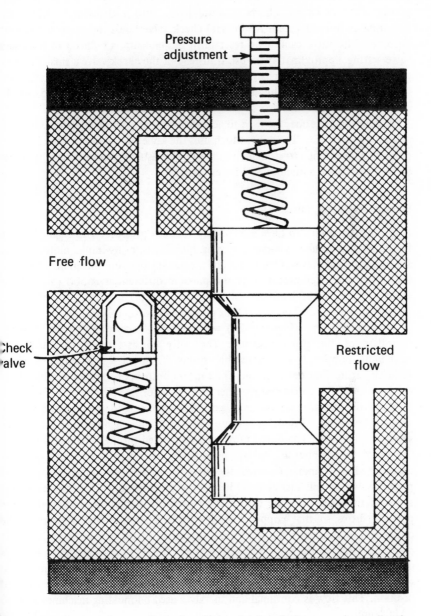

Figure 6-24 Counterbalance or foot valve which maintains a back pressure in one direction and free flow in the other direction.

position can be varied to adjust the pressure. A second type has a poppet valve which, after unseating, allows pressure to go into a chamber in which the actuating piston is located. An integral check valve holds the oil in the control chamber. The valve cannot be reset until the pressure is reduced to the check valve setting, approximately 25 psi. If instead of the check valve a small orifice is used to drain the control chamber to tank, the size of the orifice will control the time to reset. This can be used as a timer, from milliseconds to 5 sec. It accuracy depends on the viscosity of the oil.

A Bourdon tube may be used to actuate a pressure switch adjustable to rising and falling pressures. The tube, which uses the same principle as a Bourdon type pressure gauge, tends to straighten out with increasing pressure. This motion can be used to actuate a number of different switches.

Pressure Reducing Valves

It is often necessary to have a secondary reduced pressure. This might be used for running hydraulic cylinders for cams, knockouts, or hydraulic motors. There are two types of valves. One maintains a constant pressure differential between the outlet and inlet pressures. If the pressure differential is 500 psi, and the input pressure is 1800 psi the valve will put out oil at 1300 psi. If the input pressure drops to 1500 psi the output pressure will drop to 1000 psi.

The second type (Figure 6-25) will maintain a constant output pressure regardless of the input pressure. The valve consists of a pilot assembly containing poppet 1 and adjustable spring 2. Spool 4 has spring 5 which maintains it in the open position. The top and bottom of the spools, which have the same areas, are connected together through an orifice (E).

When the pressure at the inlet port does not exceed the pressure setting spring 5 keeps the valve open because the pressures in chamber F and under the spool are the same. The outlet pressure is connected through channel D through the spool to chamber F. As the pressure builds up in the outlet port, it is transmitted to chamber F. When the pressure reaches the spring setting of the poppet valve, it will unseat it, reducing the pressure in F. This will cause the spool to rise restricting the orifice (C) between the inlet and outlet, until the pressure drops so that poppet 1 reseats itself. Therefore, the pressure at the outlet will be limited to the sum of the equivalent forces of the two springs. The control position of the valve is shown in Figure 6-25b. In normal operation (C) is never entirely closed. It must pass sufficient oil to permit the functions of the secondary pressure system and enough flow to hold the spool at the control position. This flow is continual and in valves used on molding machines is 60 to 90 in.^{3}/min.

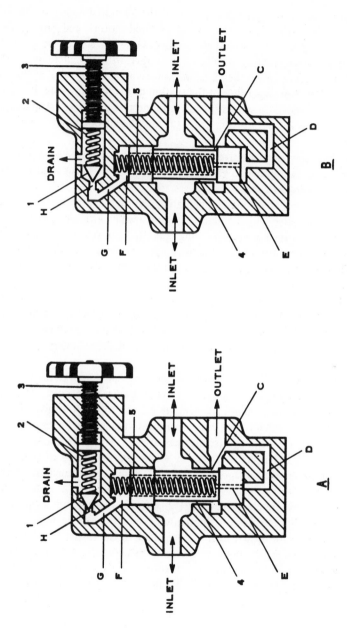

Figure 6-25 Pressure reducing valve with constant pressure output (Vickers, Inc.).

Flow Control Valves

Flow control valves control the rate of flow of the oil in hydraulic circuits. They are used to control the velocity of cylinders, fluid motors, and valve shifting. They can be of fixed orifice or variable orifice. Once the size of an orifice has been set, the volume passing through it will vary with the pressure. A flow control valve of this type is called a nonpressure compensated valve. If the valve delivers a constant flow regardless of fluctuation of the inlet pressure, it is called a pressure compensated flow control valve. In circuits with fixed displacement pumps, the only way to vary the speed of component operation is with flow control valves.

The simplest flow control valve is a hole drilled in a plug in a pipe. These are usually small and used either to generate back pressure or to act as a snubber to prevent pressure surges to gauges.

There are two types of noncompensated, adjustable orifice, flow control valves—the needle valve and the globe valve. A needle valve has a long tapered point which seats against a corresponding cone. It is moved up and down by a threaded stem. It requires many turns to completely close. This allows for a fine control. They are primarily used as chokes in the pilot systems of valves and as a two way valve for gauges.

The globe valve has a sharply beveled disk which seats on a cone. It requires few turns to open fully and gives a coarse control. Its primary use in molding machines is as an "inching" valve (Figure 6-38). It is used to bypass the main pressure line to tank, mainly to permit slow motion of the cylinders during setup or repair.

When a fixed rate of flow is required regardless of pressure fluctuations (such as for an hydraulic motor in a screw machine), the valve must be pressure compensated. Figure 6-26 shows a schematic diagram of such a valve. The Y is an adjustable orifice which is set by the operator to determine the flow rate. The outlet is connected to the spring chamber area C. The inlet is connected to chambers A and B. The sum of the areas of the spool exposed to chambers A and B is equal to the area in spring chamber C. A pressure change in either the inlet or the outlet port will unbalance the forces on the spool causing the spool to move and change the clearance (X). This changes the pressure drop through (X) compensating for the pressure fluctuation. The spool always seeks a position to maintain a pressure drop which is equivalent to the spring force. Since the spring force is constant, the pressure drop will be constant. If the pressure drop between the inlet and outlet ports of a fixed orifice are constant, the flow rate will be constant. Therefore, in this valve the flow rate will be unaffected by pressure.

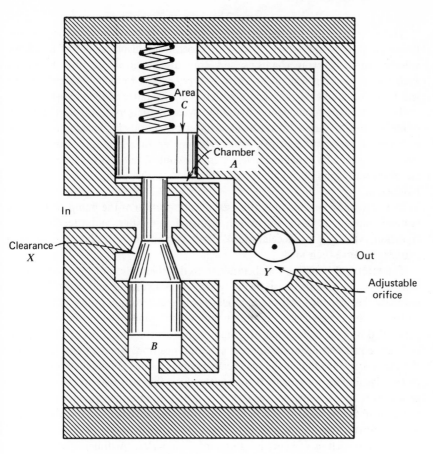

Figure 6-26   Flow control valve, pressure compensated.

Servovalves

The type of fluid control devices we have discussed has no feedback as the result of their action. Their input is not controlled by the devices (cylinders, hydraulic motors) which they actuate. This is known as an open-loop system. One of the major faults of molding machine design is that we are measuring and controlling secondary actions. For example, we measure and control the pressure and flow of oil behind the injection plunger. We are really concerned with the velocity, acceleration and final position of the injection ram and with the pressure on the molded material. To accomplish this a closed-loop system or servocontrol is needed. In this way, the controller is continually aware of the momentary condition of the function to be controlled, compares this with the programmed function, signals to a servovalve or servopump which makes the adjustments required to reduce the error to zero. This process is continuous.

The only servomechanism that has been used on molding machines for any length of time is one which is used to control the displacement of a variable volume pump. Servomechanisms and computer control are now becoming available in molding machines. There is every reason to believe that their cost will be reduced so that they come within the price range of the average molder.

There are two fundamental hydraulic systems for servocontrol. The servo-pump output is varied by feedback from a potentiometer (position control), tachometer (velocity control), or accelerometer (acceleration control). The system consists of a variable delivery pump, a control valve with a torque motor, an electronic amplifier, and the feedback. It is the pump output which is modulated.

In a servovalve system, only the used output portion of the pump changes the system's output. It has a servovalve, electronic amplifier, and a feedback unit (position, velocity, or acceleration). Both systems have a fluid motor or a cylinder to drive their load.

The electrical impulses which result from the comparison of the programmed and actual movements must be converted into hydraulic action (8). The two methods that can be used are a moving coil and a torque motor. The torque motor is used in servovalves found on molding machines. The motor armature is controlled by two coils. When the input signal is zero, the currents through the coils are equal and the armature is in a neutral position. When a signal is applied, the current is different in the coils, causing an armature movement proportional to the signal. This movement is transmitted by a thin stiff wire to the servovalve spool. This spool, which moves 0.006 to 0.010 in. in either direction, controls the control pressure oil which in turn controls the movement of the main spool, which controls the supply pressure. There are other methods of controlling the main spool such as flapper-nozzles and jet-nozzles. A fuller discussion of servovalves and circuits is beyond the scope of this book. Further information is available in *Oil Hydraulic Power and Its Industrial Applications and Fluid Power Directory* (see Bibliography, p.   549) and from manufacturer's literature.

PRESSURE INTENSIFIERS

Very high pressure pumps have a high initial cost and maintenance. When relatively small amounts of oil are needed, such as maintaining 5000 psi in a clamp holding circuit, the initial cost and maintenance suggest the use of a pressure intensifier (booster). Pressure is generated by means of using the difference in areas of two pistons, such as in an hydraulic jack. The increase in pressure will depend on the ratio of the piston areas. The flow will vary inversely as their ratio.

For example, if the ratio of the large to the small piston area is 7 : 1, and the oil pressure behind the large cylinder is 500 psi, then the small cylinder will produce oil at 3500 psi. If the flow rate to the large end of the cylinder is 35 gal/min, the flow from the small end would be 5 gal/min. The pressure intensifier, therefore, can also be used as a volume amplifier by applying the pressure on the small end and is used as a prefill system on some molding machines instead of a prefill check valve.

Figure 6-27 shows a single stage pressure intensifier. When A of the four way valve is energized pressure is blocked to the back of the intensifier, port E, by the check valve and the sequence valve. Oil goes into port C through the front end of the intensifier, out port D into the back of the work cylinder. When the work cylinder meets resistance and pressure builds up, the sequence valve opens permitting oil to go into the back of the intensifier, port E. As the intensifier plunger moves forward it seals off port C and applies the intensified pressure through port D into the work cylinder.

When the cycle is complete, B is energized shifting the four way valve. Oil under pressure enters the rod end of the work cylinder and the front of the intensifier through port F; C and the sequence valve are ported to tank. This closes the sequence valve. Oil from behind the intensifier flows out through port E and the check valve. The oil behind the work cylinder rapidly forces the intensifier plunger back and goes to tank through port C.

Obviously the system will work as long as there is enough oil in the chamber in front of the intensifier to overcome leakage. Otherwise a double stage pressure

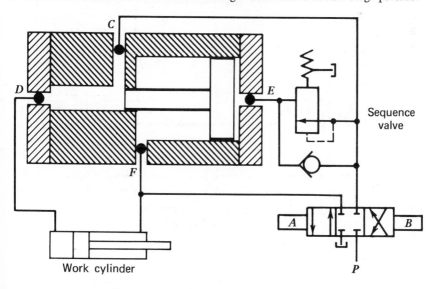

Figure 6-27   Single stage pressure intensifier.

intensifier can be used (Figure 6-28). It consists of a reciprocating double-ended piston. When the piston completes its stroke in one direction it mechanically shifts over a three way valve, which is shifted in the opposite position hydraulically. Let us consider the action in the diagram with the intensifier moving toward the right. Low pressure is applied behind the intensifier through port F and through the check valve (5) to port E. The previous cycle has pushed plunger A so that pressure has piloted A - 1 and A - 2. The pilot pressure at A - 2 has extended the plunger B, and has exhausted B - 1 and B - 2 to tank, so that these valves could shift. As the plunger moves, oil in front of the large piston area is exhausted through port H to tank. Intensified, high pressure oil leaves at port G and flows through check valve 7 to the main circuit, through J. It is blocked by check valve 8. Because the pressure at J is much higher than at port E check valve 6 will remain closed. When the piston hits plunger B the valve shifts. Low pressure oil now activates B-2 which resets plunger A, and connects A-1 and A-2 to tank. It also shifts B-1 which moves the four way directional control valve and reverses the procedure. High pressure oil will now come from port E past check valve 6 to J. The booster operates automatically and does not raise the horsepower requirement. For example, using a 5-1 booster ratio, 20 gal/min entering at 500 psi will leave at approximately 4 gal/min and 2500 psi.

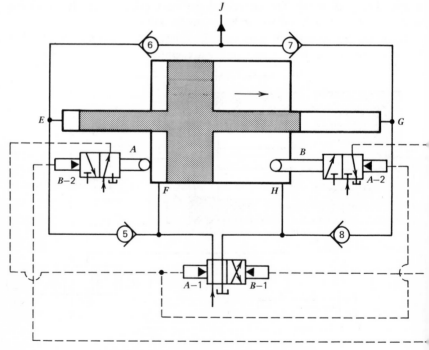

Figure 6-28   Double stage pressure intensifier.

Aside from their use in the molding press, they are valued also in tools for punching, pressing, riveting, shearing, and such. Pressure intensifiers can be powered with compressed air instead of hydraulic oil and, as such, are commonly used for other tools, but not on molding machines.

ACCUMULATORS

An hydraulic accumulator is a devise which allows the oil to store energy for later release (9,10,10a). The energy can be stored by raising weights (gravitational), compressing springs (mechanical), or compressing gases (pneumatic). They are analagous to batteries in electrical circuits, and springs in mechanical circuits.

Gravity Type

The simplest accumulator is a gravity type which consists of a cylinder in a vertical position on whose piston heavy weights are placed. Oil is pumped into the bottom of the cylinder raising the weight. When the oil is released, gravity will force the oil out at a constant speed and constant pressure. This is the only hydraulic accumulator which will do so. The disadvantages of this type (large bulk and large weight) has precluded its use for molding machines.

Spring Loaded Types

Spring loaded accumulators work by having the oil pressure compress springs. It consists essentially of a cylinder with a sliding piston with suitable packing. Springs are placed on one side of the piston and the oil pumped on the other side. They are occasionally used in molding machine circuits; for example, when it is necessary to maintain pressure during the shifting of a valve.

Gas Loaded Types

Gas loaded accumulators are used on injection molding machines. Hydraulic fluid will compress less than 2% at 5000 psi. This means that when 2% of the compressed fluid has been released, the pressure in the system is 0. On the other hand, gases are very compressible and easily store energy. They follow Boyle's law; which says that the product of the pressure and volume of the same amount of gas is constant at a given temperature. In other words, if the volume is decreased by a factor of 10, the pressure increases by the same factor 10; that is, 10 gal at 200 psi has the energy equivalent of 1 gal at 2000 psi. The gas accumulator is a device which allows the hydraulic fluid to compress the gas in charging, and allows the expansion of gas in discharging, to provide energy to the fluid.

Nonseparator Accumulators

A nonseparator type accumulator consists of a shell with the gas pressure on top and the oil on the bottom. The gas acts directly on the oil without a separator between them. Its main advantage is its ability to accommodate large volumes of oil. Because of the possibility of dissolving gas in the oil and its sluggish operation it is not used on molding machines.

Separator Accumulators

The separator type keeps the oil and gas separate. It is accomplished either by a piston, diaphragm, or bladder. The piston type consists of a cylinder with a piston containing the appropriate seals, which separates the oil from the gas. They are relatively costly to build and to maintain. The diaphragm type consists of two hemispheres with a synthetic rubber diaphragm between them. Because of its small weight to volume ratio it is mainly used in airborne applications.

Bladder Accumulators

The bladder type consists of a shell with a synthetic rubber bladder into which is charged the gas under pressure. The other end receives the oil. Its main advantage is a positive seal between the gas and oil with a light weight separation. This provides for quick pressure response for pressure regulating, and shock absorbing uses. Air is never used to charge oil accumulators as a failure could result in an explosive mixture. Nitrogen is the gas of choice.

Uses of Accumulators

Accumulators are used to dispense lubricants to various parts of the machine. The lubricant is pumped into the bottom part and is dispensed by either supplying pressure to the top part or having continual pressure metering out the lubricants. Another application is as a shock absorber. Hydraulic shock (water hammer) can be effectively dampened by using an accumulator near the source, usually a directional control valve. Persistent leakage problems are often solved this way.

A major use of accumulators in injection molding is to provide an additional source of volume and pressure when needed. Hydraulic cylinders are commonly used to move cams on a mold. To achieve rapid motion large pumps would be required. Putting a properly sized accumulator, connected to the pump output, will give the volume of oil required for rapid motion. The accumulator is recharged between cycles.

There is considerable time during the molding cycle when the high volume, low pressure pump is not used. By using an accumulator, the speed of the

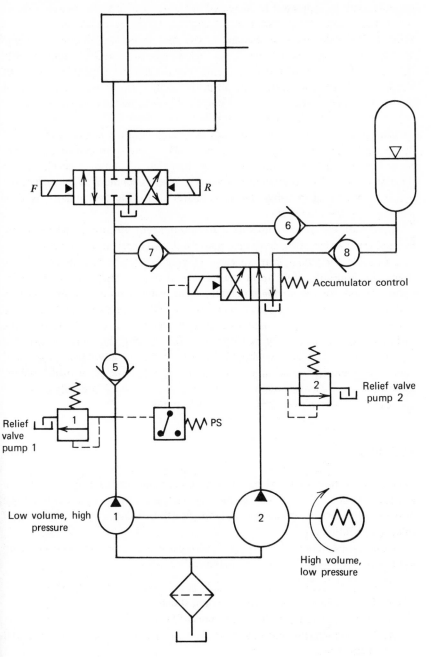

Figure 6-29 Use of accumulator to increase speed of existing circuit.

Low volume, high
pressure

High volume,
low pressure

Relief
valve
pump 1

Relief valve
pump 2

Accumulator control

PS

F

R

cylinder will be appreciably increased without changing the pumps (Figure 6-29). Each pump has its own pressure control. Assuming the accumulator to be fully loaded, solenoid F is energized. The low volume, high pressure pump supplies oil through check valve 5, the high volume, low pressure pump through the accumulator control valve and check valve 7, and the accumulator through check valve 6. In other words, the oil in the accumulator is being rapidly added to the combined output of the two pumps. When resistance is met, the pressure switch is activated, shifting the accumulator control valve. The low volume, high pressure pump builds up and maintains the clamping force in the cylinder. The high volume, low pressure pump charges the accumulator through check valve 8. Since the pressure of pump 1 is higher than that of pump 2, check valve 6 remains closed. The accumulator recharges. When solenoid R is energized, the cylinder returns again using the accumulator's oil. The accumulator is then recharged for the next cycle.

A common use of accumulators is for intermittent operations, where its use will reduce the size of the hydraulic power unit. An excellent example is the use of a cylinder for pulling cores or cams, (Figure 6-30). The accumulator slowly stores the hydraulic fluid from the clamp-end pumps, and releases it rapidly upon demand, for the auxiliary operation. The example on p. 534, needing 18.4 g/min, would require a 12½-hp power unit. With an accumulator, a 5-hp unit is more than sufficient. It is used to provide power for manual operation, if the main system is bypassed, or not operating (i.e., during the mold set-up). It is controlled by a pressure switch which starts the motor only when there is not enough pressure in the accumulator. Check valve *1* prevents oil from reaching the clamp circuit, which might be dangerous. Check valve *2* prevents the clamp oil from flowing into tank through the auxiliary pump.

FLUIDICS

Fluidics is the technology of liquid or gas streams and their reactions to control jets (11). As such, they contain no moving mechanical parts. Their use is in logic or decision-making functions and potentially can be used in place of relays and solid state controls.

Because they have no moving parts and replace electronic and electromechanical devices they have a number of distinct advantages.

1. For practical purposes they should never wear out.
2. They are insensitive to vibration and shock.
3. They generate insignificant amounts of heat.
4. The temperature and nature of the fluid is limited only by the material of the device.

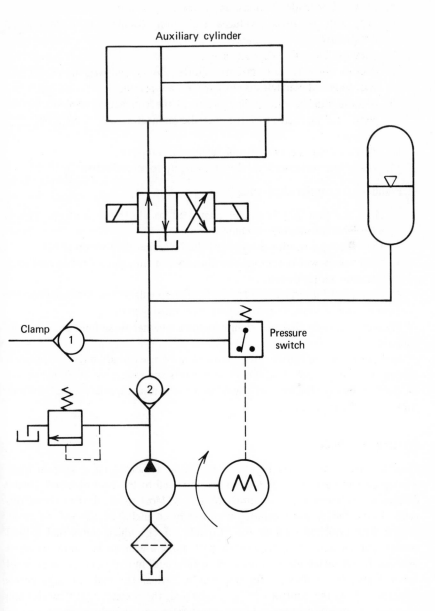

Figure 6-30 Use of accumulator to operate auxiliary cylinder.

5. They can be made of almost any material.
6. They can be mass produced (injection, compression molding) and inexpensive.
7. They can be made extremely small.
8. Because there is no electricity, shorts, shocks, and wire are eliminated; also, coils and contacts do not have to be replaced.
9. Because fluidics run at low pressures the connecting elements can be plastic tubing instead of the metallic connectors with their problem of leakage.
10. Plastic tubing is easy to install, change, and replace.
11. The power consumption is low ranging up to several watts.

Fluidics have certain disadvantages:

1. They are slow (milliseconds) compared to electronic controls. This is acceptable for molding machines.
2. They require a continual supply of fluid whether they are on or off.
3. Their low power output, at this time, is not enough to drive a cylinder or hydraulic motor directly.
4. They must "interface" with an amplifier to enlarge the signal and operate the power devices. They of course, have moving parts.

Fluidics is so new that one cannot predict whether these limitations will be overcome.

In addition to control circuits, fluidics can be used in accurate low pressure closing; they also check if all mold components and inserts are in place; and can be used to control the flow of plastic in a mold to encapsulate a part without supporting it mechanically.

Principles of Fluidics

Fluidic devices, which can be analog or digital are based on three principles. Analog devices have an output signal proportional to the control signal. Digital devices are switches which are either on or off. Most fluidic devices are of this type. As the theory and technique improves more analog devices will be made.

The first principle used in the "Coanda effect" which states that a fluid stream will continue along its existing path along a wall surface until deflected by a jet. It will then remain on the next surface until moved again by an external force. Figure 6-31a shows a flip-flop fluidic device. The main pressure enters from port P. If the control jet $C\text{-}1$ is activated, the pressure will flow through port $0\text{-}1$. If control pressure is now applied from port $0\text{-}2$, the main stream will switch and exit through port $0\text{-}2$. By adding a bias port and/or combining units such logic devices, among others, can be made: OR, AND, NOT, NOR, NAND. Other controls of this kind include binary flip-flop, half adder and summing

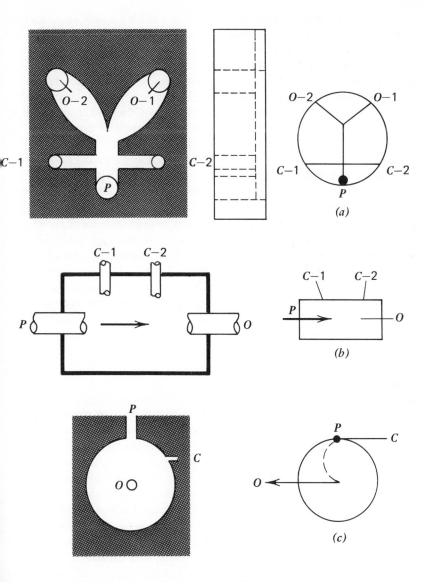

Figure 6-31 Fluidic devices: (*a*) flip-flop switch, (*b*) turbulence amplifier, and (*c*) vortex amplifier.

amplifiers. If the control flow of ports *C-1* and *C-2* were divided by another control device the outputs would be at *0-1* and *0-2*, and would be proportional to the respective inputs. This is a proportional amplifier.

The second principle involves the transition from laminar to turbulent flow by a jet (Figure 6-30b). The turbulence amplifier has the pressure entering at port *P*, maintaining its laminar flow so that the same pressure will be delivered at port *0*. If a jet with enough energy is introduced at *C*-1 or *C*-2 it will break up the laminar flow, cause turbulence so that neglible pressure is built up at the output.

The third type of fluidic device is called a vortex amplifier and works on the principle of conservation of momentum (Figure 6-30 *c*). In its normal state the pressure will flow directly through the valve and out the outlet port *0*. When a controlled jet impinges itself tangentially on the main pressure jet flow, it causes it to rotate, creating a vortex, with a high pressure drop from the circumference to the outlet. This back pressure stops the flow of the power jet. The output will vary, depending on the strength of the control jet, making the vortex amplifier one of the better analog fluidic devices. While this valve operates at low fluidic pressures, it can also operate at the hydraulic pressures of a molding machine.

Fluidics is just developing and will probably have considerable applications in the molding machine and plants. It is safe to assume that as these developments occur they will be reported in the literature (12).

PUMPS

A pump is a mechanism through which an external source of power is used to apply force to a liquid. Molding machines use electric motors as the power source. A pump does not pump "pressure." It creates flow. Pressure is resistance to flow. The pump cannot produce pressure in itself since it cannot provide resistance to its own flow. A pump is not a compressor as such, because at low pressures oil is virtually imcompressible.

Pumps can be classified according to the manner by which they discharge their liquids (such as rotary pumps, centrifugal pumps, and gear pumps). They are also classified as positive or nonpositive displacement pumps. A positive displacement pump (piston pump) is one where at a constant shaft speed a specific volume of liquid is delivered regardless of the resistance of the circuit. A nonpositive displacement pump (centrifugal pump) is one in which the volume delivered per shaft rotation is dependent upon the resistance of the circuit to flow. A positive displacement pump gives constant delivery. If the mechanism can be arranged to alter the volume delivered, it is called a variable displacement pump. Nonpositive displacement pumps are not found in the hydraulic systems of molding machines. They are used in the water system, particularly on water cooling towers. The following discussion relates to positive displacement pumps.

There are a number of specifications that apply to pumps. The *volume* of fluid pumped per revolution of the shaft is a function of the geometry of the pump. It is given either in cubic inches per shaft revolution or gallons per minute at a given shaft speed. In variable displacement pumps the maximum and minimum are given.

The *volumetric efficiency* is the ratio of the theoretical volumetric output at zero pressure over the actual output at a given pressure. It is a measure of the fluid leakage inside the pump and depends on the operating speed, pressure, and construction.

The *mechanical efficiency* is the ratio of the output horse power over the theoretical (geometric) input horsepower at a specific volume output and pressure. It is a measure of the frictional loss. The *overall efficiency* is the product of the volumetric and mechanical efficiencies.

The *pressure range* is the maximum and minimum pressures at which the pump is designed to operate.

The *speed range,* given in revolutions per minute, is the maximum and minimum limits for the efficient operation of the pump. Usually pumps run at a synchronous motor speed.

The *fluid* specification gives the recommended oil, its viscosity, its viscosity index, filtration methods, and cleanliness level.

The *temperature* range is the maximum and minimum safe operating temperatures that will give an efficient pump operation with the recommended oil and not adversely effect the seals and packing

The three types of positive displacement pumps found on injection machines are gear pumps, vane pumps, and piston pumps.

Gear Pumps

An external gear pump consists of two gears which mesh together in a closed container. One gear is attached to the motor and the other rotates freely on a shaft. As the gears rotate, the volume between the gear teeth will increase on the inlet side lowering the pressure and drawing up oil from the reservoir. Conversely, on the outlet side the volume will be reduced causing the oil to flow. The gears are usually of the spur type. Helical gears are used which are quieter but develop an end thrust. This is eliminated by using closed center herringbone gears.

They have high volumetric efficiencies up to 93% and general overall efficiencies of 75%. The pressure range is 250 to 3000 psi, and a volume of 0.2 to 150 gal/min. They are inexpensive, have few moving parts, and are relatively insensitive to dirty fluid. They are used on pilot pressure systems and as priming pumps for large volume, high pressure pumps.

External gear pumps have the teeth projecting outward from the center of the gears. In internal gear pumps, one gear has its teeth projecting outward and the other, on the perifery of the casing, has its teeth projecting inward towards the center of the pump. The inner gear always has fewer teeth than the outer gear. They are not ordinarily used on molding machines.

Vane Pumps

The vane pump is widely used on injection machines. The pump (Figure 6-32) consists of a drive shaft on which is attached a rotor that is not concentric with the ring. The rotor has slots in which the vanes slide. Centrifugal force or springs keep the vanes touching the ring. As the pump rotates, the volume of a segment bounded by the two vanes, ring, and rotor increases on the suction side, creating a vacuum. Air pressure on the oil in the tank or oil from a priming pump fills this volume through the inlet port while on the suction side. The inlet and outlet ports continue for almost the full length on each side. On the pressure side of the pump, the volume of the segment between vanes decreases, forcing the oil out of the outlet port, creating flow.

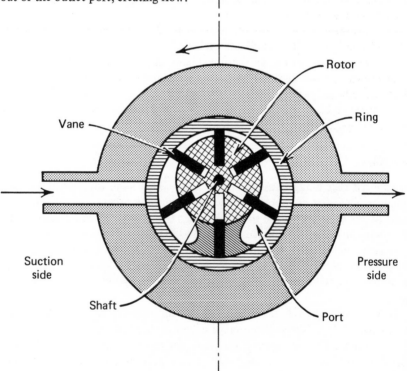

Figure 6-32 Vane pump with unbalanced shaft.

The vane pump illustrated is unbalanced, in that there is pressure on the ight-hand side and suction on the left-hand. This would tend to push the whole .haft toward the left and give relatively rapid wear. Vane pumps designed to >revent this are called balanced pumps which are used on molding machines. It is iccomplished by shaping the cam ring with two lobes. Variable displacement /ane pumps shift the cam ring. Therefore, they are inherently unbalanced •ecause they use a single lobe cam.

Vane pumps are rugged and not too susceptible to dirty oil. Their efficiency s high, volumetric ranging from 85 to 90% and overall 82 to 87%. Additionally hey keep their high efficiency for a long period because the wear of the vane ·nds upon the ring is automatically compensated. They are easy to repair with tandard kits to replace the moving parts. The pressure range is from 250 to ;000 psi and delivery from 0.5 to 250 gal/min. The delivery can be altered by :hanging the cam ring.

Vane pumps, like other types, also come with two units on one shaft to be un by one motor. They may be of any size or combination and are ıydraulically independent. For the high pressure ranges one is used to uperchange the other.

Rotary Piston Pumps

Rotary piston pumps are either axial, where the pistons are parallel to the axis of he shaft, or radial, where the pistons are perpendicular to the shaft. The three ypes of internal valving are flat or plate type (Figure 6-33), check valves, or >intles (Figure 6-34).

A piston pump uses a piston in a hole to decrease or increase the volume of a :hamber (Figure 6-33). The action is analogous to a vane pump. The shaft turns he cylinder block. The pistons are anchored to the rotating cam ring which is et at an angle. As it rotates toward the top position in the illustration, the >iston pulls back increasing its volume and creating a vacuum which is filled with >il through the inlet port. As it rotates downward, the piston is forced in, ·xpelling the oil through the pressure port, until it reaches the bottom position, vhere the cylinder has its minimum volume. The port plate is shown on the left.

If the cam plate and port plate are nonadjustable the output per shaft ·evolution will be constant. This is called a positive displacement, fixed volume >ump. Supposing the cam plate angle was changed so that the cam ring was >erpendicular to the shaft. There would then be no motion of the pistons ·elative to the block. The pump would not deliver any oil. By tilting the cam ·ing slightly, a small displacement would occur and the volume would be small. \s we vary the cam plate angle, we vary the displacement of the pump. The >ump would now be called a positive displacement, variable volume pump.

There is another way of accomplishing the same result. If the cam plates were

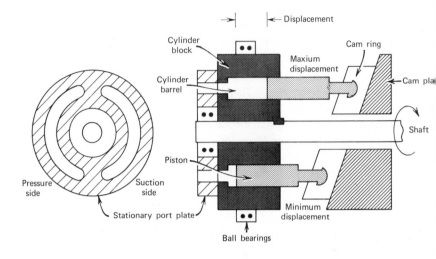

Figure 6-33   Positive displacement axial piston pump.

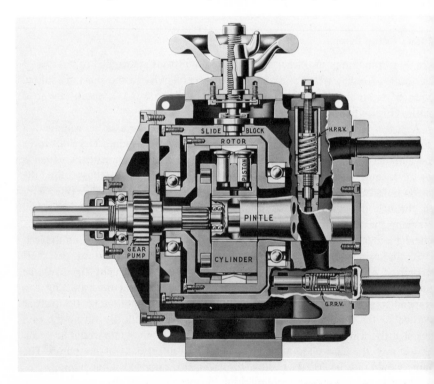

Figure 6-34   Variable displacement radial piston pump. (The Oilgear Company).

at the angle to produce maximum displacement, and the port plates rotated 90°, the pump would deliver no oil, because an equal volume of inlet and outlet ports are connected to each other. By altering the amount of rotation of the port plate, the volume per shaft rotation can be infinitely varied. This, too, would be called positive displacement, variable output pump.

Figure 6-34 shows a variable displacement radial piston pump. It consists of a stationary pintle upon which is mounted the rotating cylinder containing the pistons. The pintle has four holes drilled parallel to the shaft axis. It is milled under the piston holes so that two act as inlets and two as outlets. The cylinder assembly is driven by the shaft. Centrifugal force and oil pressure keep the beveled heads of the pistons against the thrust ring of the rotor. This causes the rotor to rotate with the cylinder. The rotor is attached to the slide block. The illustration is a top view of the pump looking down, an horizontal view. The slide block slides back and forth. In the illustration it is fully moved to the hand wheel or control side. The pistons have their maximum movement on the hand wheel side, and decrease in volume as they rotate 180°, sending oil through the outlet ports of the pintle. If the slide block is central, the pistons will have no radial movement, and no oil will be delivered. If the slide block moves toward the bottom of the illustration the ports will reverse, the pressure ports becoming tank ports and the tank ports becoming pressure ports. This type pump has been used very successfully on injection molding machines, eliminating the four way directional control valve for the injection and clamp cylinder (13). The shock transmitted is so small that the author has seen a nickel balanced on its end on the stationery platen remain so during dry cycling of the machine. The main pump is supercharged with a gear pump. Both pumps have their own built-in relief valves.

The location of the slide block is controlled by a hand wheel. In any given position the pump acts like a positive displacement unit. The control need not be a hand wheel. It can be an electrically controlled motor. This controls the movement of the slide block very accurately, so that the volume of fluid can be varied smoothly over a stepless range from zero to maximum output in either direction. This characteristic is also true of the other methods of control. A pressure unloading control will hold a preset pressure and shift the slide block so that only enough oil is delivered to maintain this pressure. This eliminates excess heating, power loss, relief valves, unloading valves, and bypass valve. Hydraulic or electric controls are available to change the volume of the pump at any given time to a specific preset delivery. This type of pump is ideally suited for electrohydraulic servo control.

Some of the advantages of variable displacement pumps are infinitely variable pressure and volume without external valving and piping; lower power consumption; lower heat development; more flexibility in molding; use of servocontrol; and longer life.

Some of the disadvantages are higher initial cost; more expensive repair longer downtime, because repairs cannot usually be done in the plant; and larger weight and size.

Piston pumps are the most efficient, with volumetric efficiencies from 90 to 98% and overall efficiences about 90%. Pressure ratings are from 600 to 5000 psi and volume from 0.5 to 2200 g/min.

Horsepower Requirements for Pumps

The input horsepower required for an hydraulic systems is

$$HP = 5.82 \times 10^{-4} \, P \, Q \qquad (6\text{-}10)$$

where HP = input horsepower
          P = pressure (psi)
          Q = flow rate (g/min).

Let us determine the horsepower requirements for the hydraulic system described on p. 488. Assume that a 7000-lb force is required to seal the cam. The 3-in. diameter cylinder has a piston area of 7.07 in^2, the volume required is 18.4 gal/min. Substituting (1-18)

$$P = \frac{F}{A} = \frac{7000}{7.07} = 1000 \text{ psi}$$

Using this in (6-10)

$$HP = (5.82 \times 10^{-4})(1000)(18.4) = 10.7 \text{ hp}$$

Assuming an 85% overall efficiency

$$HP = \frac{10.7}{0.85} = 12\tfrac{1}{2} \text{ hp}$$

This is a very large power supply and can be substantially reduced by using a small accumulator, (Figure 6-30).

Pump Maintenance and Trouble Shooting

The maintenance of pumps essentially is one of keeping the oil clean. Normal wear is taken care of by replacing the bearings and moving parts. The following suggestions do not cover the control mechanisms of variable displacement pumps.

Pump Not Delivering Oil.   This can be readily checked by opening a connection on the pressure side of the pump. The problem could be one or more of the following:

1. Not enough oil in the tank.
2. Oil intake line, filters, or strainers are clogged.
3. Air entering in the suction line. This can be indicated by unusual loud noises.
4. Pump shaft turning in the wrong direction. Maintenance might have inadvertently reversed the three phase motor input during repair.
5. Pump shaft speed too low. This can be caused by single phasing of a three motor or a loose or slipping coupling.
6. Mechanical trouble. Usually this is accompanied by noise in the pump. The usual mechanical troubles are worn bearings, broken shafts, rotors, pistons, or vanes.

Pump Not Delivering Full Pressure and Volume.   This can be checked by collecting the unrestricted flow of the pump during a timed interval. A restricting valve and pressure gauge is put over the pump output and the output similarly measured at a specific pressure. These figures are compared to the rating of the pump. If they are acceptable then the problem is downstream in the system.

1. The internal relief valve setting of the pump (if there is one) is too low or acting incorrectly.
2. The wrong viscosity oil has been used permitting too much internal leakage, or the oil is too hot, lowering its viscosity.
3. Broken, worn or stuck pump parts. This requires disassembly of the pump.

Noisy Pump.   This condition might be due to one of the several causes listed below:

1. Air leaking into the system.
2. Cavitation, which is oil starvation of the pump. The intake filter system should be checked.
3. Air leaks due to a worn shaft packing.
4. The pump might be out of line with the motor.
5. The coupling can cause noise from wear or might require lubrication.
6. The noise might come from chattering relief valves in the pump.
7. Internal causes within the pump.

Overheating.   Common reasons for overheating follow:

1. The heat exchanger might be dirty, has insufficient water or the inlet water has too high a temperature.
2. Oil viscosity is too high.

3. The discharge pressure is too high for the design of the system.

4. Internal leakage in the pump is too high. This is caused by wear.

5. Leakage in the system. If a valve is not functioning properly, or a piston ring is worn, the pump will operate above the system design conditions to make up for the leakage. Additionally, oil flowing with high velocity through the leakage points which act like small orifices will overheat.

6. The oil in the reservoir is low. The reservoir acts as an heat exchanger and insufficient residence time in the reservoir will reduce the heat dissipation of the system.

HYDRAULIC MOTORS

Hydraulic motors are the reverse of hydraulic pumps. In a pump, power is applied to the shaft which turns the pump causing oil to flow. In an hydraulic motor, oil is forced through the motor which causes the shaft to turn. A pump is designed to deliver the maximum volume per shaft rotation. Therefore, the design emphasis is on volumetric efficiency. The hydraulic motor converts energy in a fluid to rotary motion and torque. The emphasis is on mechanical efficiency. For this reason there is a slight difference in design between a motor and pump, though they operate using the same principles. The three types of hydraulic motors are gear-type, vane-type, and piston-type. They can have fixed or variable displacements.

Displacement

The displacement is given as the amount of fluid required for one shaft revolution. Units are usually cubic inches per revolution. A fixed displacement motor provides constant torque and variable speed (at a given pressure). The speed is controlled by changing the flow rate to the motor. A variable displacement motor provides variable torque and speed. With constant pressure and input flow, the ratio between torque and speed can be varied to meet the load requirements by changing the displacements. Varying combinations of torque, speed and power can be had by combining pumps and motors with fixed or variable displacements.

Speed

The speed of an hydraulic motor is a function of the flow rate and displacement less the internal leakage. Volumetric efficiency is a ratio between the theoretical speed of the motor for a given flow and the actual speed developed.

Torque Output

Torque output, which is expressed in inch-pounds or foot-pounds, is a function of the motor displacement and the pressure. Torque is usually reported for a specific pressure drop across the motor. Mechanical efficiency is the ratio of actual torque delivered to the theoretical torque. In a fixed displacement motor torque varies directly with the pressure.

The relationship between horsepower, torque, speed, and displacement is given in (1-11) (1-12), (6-10), (6-11), and (6-12):

$$T = \frac{D \Delta P}{75.5} \tag{6-11}$$

where T = torque (ft-lb)
  D = displacement (in^3/rev.)
  ΔP = pressure drop across motor (psi).

$$HP = \frac{FL}{6600T} \tag{6-12}$$

where HP = horsepower
  F = force (lb)
  L = length moved (in.)
  T = time (sec)

Motor in Use

Hydraulic motors are primarily used to turn screws in the plasticing chamber. With the proper combination of hydraulic pump and motor, stepless variations in speed and torque are available. This is of great value in molding. In the event they are overloaded they will stall without mechanical damage. For these reasons, too, they are well adapted for auxiliary operations, such as turning cores in an automatically unscrewing mold.

Figure 6-35 shows an hydraulic motor circuit with slow deceleration and braking controls. It consists of a motor, pump, pressure compensated flow control valve, pressure relief valve, four way, three position, directional control valve. The pressure compensated flow control valve regulates the speed of the motor by metering part of the pump flow to tank. This is known as a *meter to waste* installation. The flow control valve could have been put in series with the inlet or outlet of the hydraulic motor. The illustrated method is more efficient because the pump output pressure will only be enough to overcome the work resistance.

When solenoid R is energized the motor is in its normal operating position. Oil from the pump goes through the motor whose output goes directly to tank.

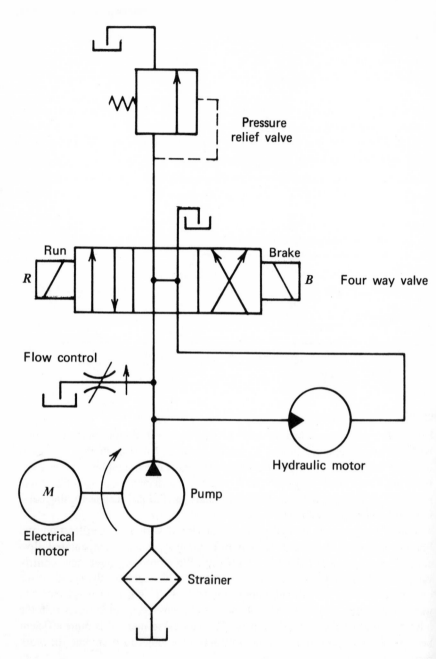

Pressure
relief valve

Run   Brake

R   B   Four way valve

Flow control

Hydraulic motor

M   Pump

Electrical
motor

Strainer

Figure 6-35 Hydraulic motor circuit with slow deceleration and braking.

The maximum torque is controlled by the pressure relief valve. The speed is controlled by the pressure compensated flow control valve. To slow down normally, the four way valve is shifted to its neutral position. The pump and motor are ported to tank and the motor coasts slowly to a stop. If it is desirable to stop the motor quickly, solenoid *B* is energized, the pump is dumped to tank, and the motor exhaust goes to the pressure relief valve where it is blocked, quickly braking the hydraulic motor.

Plasticizing the more viscous materials requires high torque at low speeds. This can be accomplished by attaching two hydraulic motors of equal displacement to the gear reducing transmission (Figure 6-36). When the three position, four way valve is in the position shown on the drawing oil goes to both hydraulic motors. Since the oil flow is divided between them they will go at a slow speed. The torque, however, is constant and depends on the pressure control setting. Therefore, the torque will be additive and the circuit will have low speed and high torque. If the valve is shifted into detent position 3, oil will only flow through motor 3. The return oil exhausts through check valve 2. Check valve 1 permits oil to be on both sides of motor 1 so that it does not act as a brake. The screw will now have one-half the torque but double the speed as compared to both motors operating. This is the high speed-low torque range. With the directional control in detent position 1 motor 1 will run instead of motor 3. The motors need not be of equal displacement allowing three speed-torque ranges.

HYDRAULIC CIRCUITS

Hydraulic circuits used in injection molding machines are relatively simple to follow, when broken into their component parts (Figure 6-37). There is a pump circuit with its related oil cooling and filtering. Cylinder plungers have safety controls, directional controls, speed controls, pressure controls, regenerative controls, and prefill controls. Hydraulic motors have pressure and speed controls. Examples of these circuits can be found throughout this chapter. Figure 6-38 shows an oil filtration circuit with its independent pump and motor. This can come with the machine or be added afterwards. It has its own electric motor and pump *F*. Oil is drawn from the tank through check valve *F* which prevents drainage of the system when the pump is off. It goes through the filter and back to tank. As the filter clogs up, more pressure will be required to force the oil through. In most filter systems, the cartridges should be replaced at about 60 psi. The unloading valve *G* is set at 60 psi. This prevents a pressure build-up beyond the capacity of the filtering element. Should the element be destroyed, it would go into the oil tank. A magnet is put into the oil tank to attract small iron particles from erosion or breakage in the system, or any larger particles which might get in accidentally.

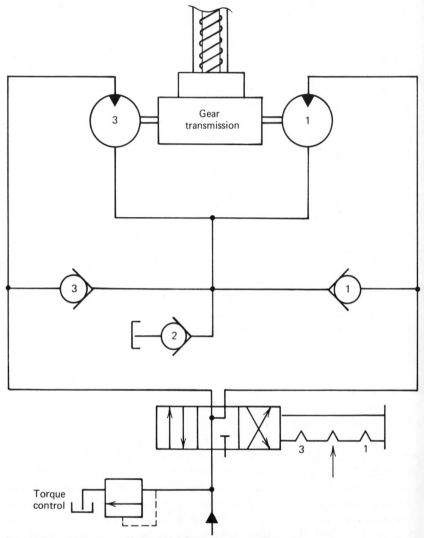

Figure 6-36   Two speed-torque ranges using two identical hydraulic motors.

High-Low Pump Control Circuit

Variable volume pumps will deliver just enough oil to maintain the pressure required by the load. Fixed displacement pumps use their full volume for moving a cylinder and then have to dump most of the pumped volume over a relief valve while the system is being held under pressure. This is wasteful of power and generates heat which is expensive to remove. To overcome this, two

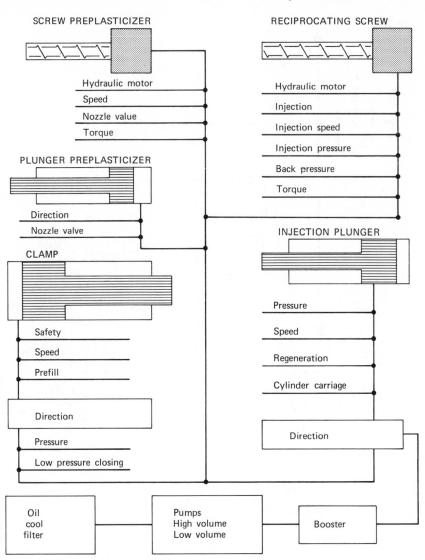

Figure 6-37 Component functions of hydraulic circuit.

pumps are used. They are used together to provide the volume for the rapid motion. The smaller of the two is then used to maintain the pressure while the larger is unloaded to the tank at low power consumption. The other section of Figure 6-38 shows this. Low volume-high pressure pump 1 and high volume-low

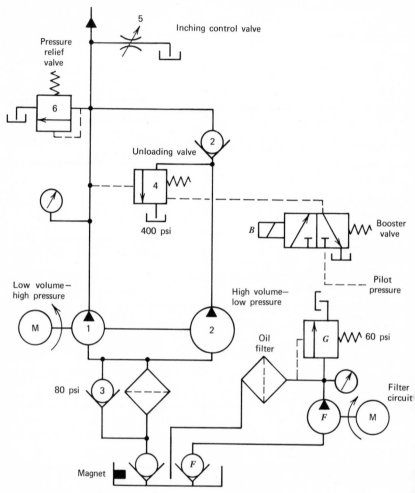

Figure 6-38   Hi-lo pump and filter circuits.

pressure pump 2 are turned on a common shaft by the electric motor. Oil is supplied from the tank through a strainer. If the strainer is clogged, it will be bypassed by check valve 3, which will open at 60 psi. This prevents cavitation or oil starvation of the pump. When a directional control valve shifts to move a cylinder, the output of both pumps will flow into the system. When the cylinder meets resistance, the pressure will rise. When it reaches the 400 psi setting of the unloading valve 4, high volume low pressure pump 2 will unload to tank. The pressure will continue to rise with oil supplied from low volume-high pressure pump 1 until it reaches the setting of the relief valve 6. This relief valve 6 will bypass that flow of pump 1 not required to maintain the leakage through the valve cylinders and other parts of the hydraulic system. Oil from pump 1 is blocked from pump 2 by check valve 2 because pump 2 is at a higher pressure than pump 2.

There are times during the molding cycle when the output of the low pressure pump is required above the 400 psi unloading pressure. This might happen during injection forward. If solenoid B is energized, pilot oil will enter the control chamber of the unloading valve 4, preventing it from unloading regardless of the pressure of the system. Solenoid B it usually controlled by a timer, or it can be positionally controlled with a limit switch mostly commonly activated by the injection plunger motion.

Valve 5 is a hand-operated globe valve which bypasses the output of both pumps to tank. The inching valve is used to move the cylinders slowly during setup and maintenance.

Let us see how this circuit is used in a molding machine. Most fixed displacement power supplies have an additional low pressure high volume pump which is used as a holding pump for the clamp circuit. After the gate is closed all three pumps deliver oil to the clamp circuit. As the pressure begins to rise in the clamp circuit, the third low volume, high pressure pump is separated from the other two pumps (1 and 2) which are illustrated in Figure 6-38. (One way of separating them is show in Figure 6-12). The circuit (Figure 6-38) is timed so that at this moment, the injection forward control valve will be shifted to permit oil to go behind the injection ram. This will drop the pressure in the lines connected to pumps 1 and 2 and will close the unloading valve 4. They will both send their oil there until the pressure builds up to 400 psi when pump 2 will unload through unloading valve 4. It can be prevented from unloading by energizing solenoid B. The injection pressure will be controlled by pressure relief valve 6. More often there will be two pressure controls with a circuit similar to Figure 6-22.

When the injection four way valve shifts, as it would in a plunger system, the pressure will drop in the lines from pumps 1 and 2 and both pumps will deliver oil to return the plunger. The unloading operation will be repeated. On clamp return, the shift of the four way valve will open the pump separation two way valve (Figure 6-12), dropping the pressure and bringing the three pumps into the circuit to open the machine.

CLAMP CONTROL

A typical clamp control circuit for a toggle machine is shown in Figure 6-39. System pressure flows through the flow control valve 1 and check valve 2, into

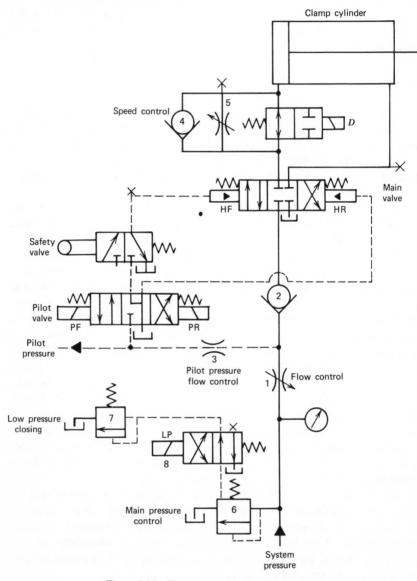

Figure 6-39 Clamp control hydraulic circuit.

the hydraulically operated, spring centered directional four way clamping valve. Pilot pressure is supplied through a flow control valve 3, to the solenoid operated pilot valves. When the pilot forward solenoid *PF* is operated, the valve shifts and pilot pressure oil goes into the safety valve. This is mechanically operated and only closes when the gate is closed. Only when the gate is closed does pilot pressure oil flow into the clamp forward side (*HF*) of the four way valve. It shifts so that pressure is sent to the back of the clamp cylinder.

Regardless of the position of deceleration valve *D*, oil will flow without restriction through check valve 4 during clamp forward. The clamp return movement of the platen can be slowed down by energizing solenoid *D*. This closes the oil path through that valve. Since it cannot flow back through the check valve 4 it must flow through the restricted opening of the deceleration flow control valve 5, and move at a speed determined by that valve setting.

The clamping system's main pressure control valve 6 has a secondary pressure control 7, which is connected through the solenoid operated low pressure relief valve 8. When the mold closes and there is no obstruction, a limit switch is closed which energizes solenoid LP, and blocks out the low pressure relief valve 7. Full system pressure is now obtained in the clamping system. If there is any obstruction on the mold, the limit switch will not be contacted. Solenoid *LP* will not be energized, and the clamping pressure will never be higher than the setting of the low pressure relief valve 7. This is set low enough to prevent damage to the mold. Electrically, the circuit will prevent the injection ram from coming forward and may also set off an alarm. This is called the low pressure, mold safety circuit.

INJECTION CONTROL CIRCUITS

The injection control system (Figure 6-40) is fed by the system oil pressure through the solenoid operated, hydraulically piloted, spring returned, four way valve. When the pilot solenoid for injection forward (*PF*) is energized, the pilot valve shifts, sending oil into the injection forward side (*F*) of the main valve. This shifts the spool sending oil into the back of the injection ram. The regenerative circuit blocks the return flow at check valve 1, and sends it through check valve 2, back into the pressure line, combining it with the pump flow. Normally it would go to tank and have to be repumped. This continues as long as the sequence valve 3 remains closed. When the injection pressure rises to the sequence valve setting, 400 psi, it opens, permitting the small amount of oil in front of the injection ram to go to tank. On injection return, (solenoid *PR* energized sending oil to *R* side of main valve) the oil returning the injection ram goes through check valve 1. The purpose of the regenerative circuit is to increase the injection forward speed at low pressure.

When the pilot valve shifts for injection forward, the pilot oil branches off to be used as the booster oil supply. In this circuit, then, the unloading valve for Figure 6-38, can be blocked only during injection forward.

The pressure reducing valve 6, in Figure 6-40 will supply oil to secondary

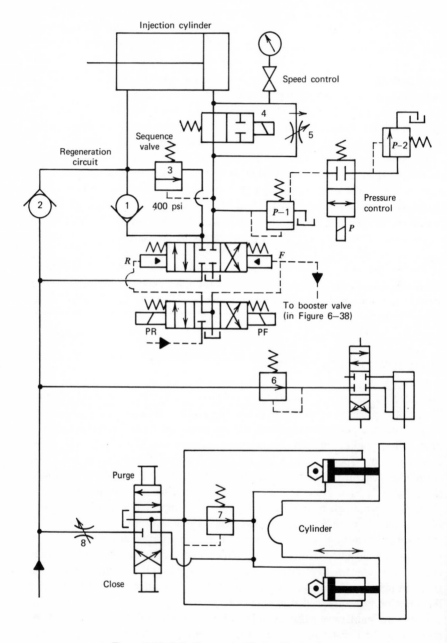

Figure 6-40 Injection control hydraulic circuit.

hydraulic mechanisms such as hydraulic knockouts and cams.

The carriage on which the injection cylinder is mounted is controlled by a manually operated, three position, four way valve. The speed is controlled by a needle valve 8. The pressure reducing valve 7, is adjusted to limit the seating pressure of the nozzle against the sprue. Too much pressure might push the mold from the platen.

RECIPROCATING SCREW CIRCUIT

The hydraulic system for a circuit to operate a reciprocating screw. (Figure 6-41) is fed through a solenoid piloted, hydraulically operated, four way valve. When the hydraulic motor solenoid M is energized, the oil goes into the "screw on" solenoid S. In its normal position the pressure outlet port is blocked. When S is energized it sends oil through the hydraulic motor, which turns the screw. The torque of the motor is controlled by the pressure relief valve T. A bleed-off circuit controls the speed of the hydraulic motor, using a pressure compensated flow control valve 1. It can be blocked by a manual shut off valve 2 to give maximum motor speed. The two check valves 3 and 4 keep the ram end of the injection cylinder full of oil at all times.

As the screw turns the plastic is melted and collects in front of the screw. This forces back both the injection screw, transmission, and the injection cylinder ram. As the injection cylinder ram moves back, the oil return is blocked by check valve 5. It is also blocked by the gauge valve, whose back pressure solenoid BP was energized in parallel with solenoid S. The pressure gauge reads the back pressure of the oil. In molding, the back pressure rarely exceeds 300 psi, while the injection pressure might be over 2000 psi. To set the back pressure more accurately, a low range pressure gauge is used. The back pressure sequence valve 6 remains closed until the preset pressure is reached. Then the valve opens and excess oil dumps to tank. This controls the pressure on the material while it is being plasticized. When the carriage has been pushed back by the plastic to a preset location, a limit switch is contacted deenergizing the "screw-on" solenoid S and stopping the hydraulic motor, and deenergizing solenoid BP.

The plastic material could be under enough pressure to drool out of the nozzle. To prevent this the "suckback" solenoid SB is energized. This sends pilot pressure oil to the return side of the injection cylinder backing off the screw and decompressing the material. The amount of decompression is controlled by the "suckback" timer. The suckback valve is also used to return the screw without preplasticizing material. Its maximum pressure is controlled by the spring of check valve 4.

When the cycle calls for injection, the main control valve is shifted to "inject" sending oil into the injection forward end of the cylinder through check valve 5.

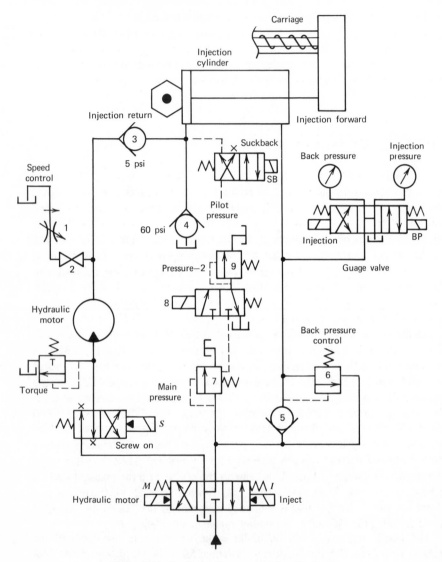

Figure 6-41   Reciprocating screw hydraulic circuit.

This brings the screw forward forcing the material out of the nozzle. The gauge valve has also shifted to the injection forward side connecting the injection pressure gauge to the circuit. Two injection pressures are obtained with valves 7, 8, and 9. With solenoid 8 deenergized relief valve 7 controls the maximum pressure. When solenoid 8 is energized by a timer or a limit switch relief valve 9 is attached to the circuit and now controls the pressure at a value less than relief valve 7. When the injection is complete all solenoids are deenergized, preparing for the next cycle.

BIBLIOGRAPHY

Oil Hydraulic Power and Its Industrial Applications 2nd ed. W. Ernest, McGraw-Hill Book Co., New York, 1960

Fluid Power Controls, J. J. Pippenger, and R. M. Koff, McGraw-Hill Book Co., New York, 1959.

Hydraulic and Pneumatic Power for Production, H. L. Stewart, Industrial Press, New York, 1955.

Simplified Hydraulics, D. S. McNickle, Jr., McGraw-Hill Book Co., New York, 1966.

Fluid Power Directory, Penton Publishing Co., Cleveland, Ohio,

Industrial Hydraulics Manual-935100, Vickers, Inc., Troy, Mich., 1965.

Basic Course in Hydraulic Systems, Machine Design, Penton Publishing Co., Cleveland, Ohio, 1968.

Basic Hydraulics, NAVPRS 16193, Supt. of Documents, Washington, D. C.

"Bookshelf for Fluid Power," *Hydraul. Pneum.*, August 1968, p. 96.

REFERENCES

1. *Hydraulic Systems for Industrial Machines*, Socony Mobil Oil Co., New York, N. Y.

2. "Hydraulic Fluids, Fact and Fancy," G. R. Arbocus, *SPE-J*, August 1961, p. 767.

3. "Fire Resistant Hydraulic Fluids for the Plastics Industry," J. Mathe, *SPE-J*, July 1967, p. 17.

4. "Evaluating Fluid Filtering Media," C. H. Hacker, *Product Engineering*, November 28, 1960, p. 29.

5. "Piping Fluid Power Systems," *Hydraul. & Pneum.*, May 1963, p. 95.

6. "Sizing Components for Fluid Power Systems," *Hydraul. & Pneum.*, April 1969, p. 119.

7. *"Seals; Machine Design,"* Penton Publications, Cleveland, Ohio.

8. "Industrial Hydraulic Control Techniques," J. D. Rowe, *Plastics,* July 1967, p. 852.

9. "Accumulators and their Applications, A. Zahid, Greer Olaer Products, Los Angeles, Calif.

10. "The Hydraulic Accumulator Circuit," W. R. Groves, *IPE* March 1963, p. 90.

10a. "Accumulators on Injection Machines," A. Kliene-Albers and H. Heyden, *IPE*; Part I – August 1962, p. 349; Part II – September 1962, p. 412.

11. "Let's Look at Fluidics," *Hydraul. & Pneum.*, Industrial Publishing Co., reprint, 1966.

12. "Fluidics in Plastic Processing," R. Henke, *PDP*, Feb. 1969, p. 28; March 1969, p. 33.

13. *B.P.*, December, 1967, p. 89.

_____

CHAPTER 7

Electrical Mechanisms and Circuits

Alternating current is supplied in a three-phase system; AC generators are wound with three armature circuits spaced 120 electrical degrees apart, producing currents and voltages that are separated by this amount. They are transmitted at high voltage and reduced to the plant voltage by transformers (1).

BASIC INSTALLATION DATA FOR ELECTRIC POWER

The two types of power connections are the delta and wye. The delta connection has the three transformers connected one to the other to form a triangle. Power is taken from each of the three connections. It is a three phase, three wire system where the voltage between each pair of line wires is the actual transformer voltage. The line voltage is in phase with the voltage across any one winding. The line current is either 30 or 150° out of phase. By adding vectors, we find that the line current is the $\sqrt{3}$, or 1.73 times the individual or single phase current.

The wye connection (Figure 7-1) is almost always found in molding plants. It is a three phase, four wire system with a neutral grounded line. The current rather than the voltage is in phase. The single phase voltage $(A, B,$ or $C)$ to neutral is equal to the three-phase voltage $(A - B, B - C,$ or $A - C)$ divided by 1.73. This provides two voltages from the same system. Any unbalance in the system (such as caused by resistive heating) is carried by the grounded neutral. The 208Y/120-V system supplies 120-V single phase for lighting and 208-V three phase for small motors. Where possible, this voltage system should not be

550

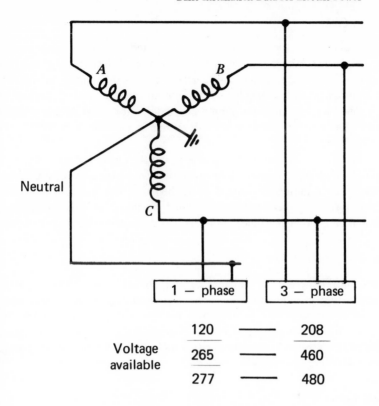

Figure 7-1   WYE – connection for power supply.

used for molding equipment as the higher voltages are more economical. The 460Y/265-V and 480Y/277-V are virtually the same, the difference being in the internal connections of the standard single phase 240-V transformers. Since the 277-V single phase is directly usable with fluorescent lighting loads, and the 460/480-V for three-phase motors, it fits well for molding plant operation; 120-V is provided by transformers.

The power in three-phase systems is

$$P = \frac{1.73 \, E \, I \, \cos\theta}{1000} \tag{7-1}$$

Where
P = power (killivolt-amperes, KVA)
E = volts
I = current
$\cos \theta$ = power factor.

Installation Costs

The higher the voltage the lower the current required for a given power output (see Eq. 7-1). The size of the copper wire is a function of the number of amperes to be carried. Therefore, going from a 208-V to a 480-V circuit reduces the copper requirement to less than half. It also requires smaller conduit. Because of this and lower equipment costs, motor installations for lower voltage systems cost more. The cost of installing a 208-V system is approximately 55% higher than a 480-V System.

For other reasons as well, it is desirable to use the higher voltage system (2). If a plant has a 208-V system and needs additional wiring capacity, consideration should be given to raising the voltage. If the higher voltage was used in place of the 208-V system no new installation would be necessary. Almost all large motors used on molding machines can operate at the higher voltage.

Ratings

The difference between the nominal system voltage rating and equipment nameplate rating can be confusing. The system voltage is the transmission voltage under no load and is decreased by the voltage drop through the transformer and transmission system. For example, a single phase 120-V system will have a 115-V motor and control rating and a 120-V heating device rating. Similarly a 480Y/277 system will have a 440-V motor nameplate rating.

Power Factor

The cosine θ of (7-1) is the power factor. It is an indication of the efficiency of the electrical system. Induction equipment requires magnetizing power to produce the flux needed for its operation. This reactive power does no useful work and is measured in kilovars (kvar). The real or working power is measured in kilowatts (kw). The power factor is the ratio of the power actually used (kw) as measured by a watt meter, divided by the apparent power or product of the volts and amperes. In a resistive circuit (heating bands, ovens, lights) there is no inductive load so the power factor is unity. A 40-hp induction motor has a full load power factor of 0.89 and a half load power factor of 0.83. Since motors on injection molding machines often run at less than half a load, the power loss is significant. It requires the unnecessary transportation of electrical energy. It is counterbalanced by the unity power factor of the heating loads. The power factor of an inductive load can be improved by using a capacitor. It is normally uneconomic to raise the power factor above 90 to 95%. Improving the power factor has the following advantages:

Lowers demand charges by the utility company.

Increases system capacity.

Lower $I^2 R$ losses.

Lowers voltage drop.

Lessens transformer size requirements.

Low Voltage

Low voltage in a system that originally had a permissible voltage drop (2% at full load in feeder circuits and 1% at full load in branch circuits) can be caused by the addition of too much equipment, decrease in power factor, and a drop in voltage from the utility. Unfortunately the latter is quite common in major industrial areas such as New York City. Extra equipment requires more current in the distribution line. Since the conductor's resistance is constant, the larger the current the higher the voltage drop. These $I^2 R$ losses cause the conductor to heat, increasing the line resistance and aggravating the voltage drop. Some cures follow:

Run more copper. A machine or group of equipment can be wired directly to the power source.

Increase the power factor. This will reduce the reactive current in the lines.

If possible, raise the voltage from 208 to 480. This will almost double the current capacity of a given conductor.

Change the transformer taps to offset the drop in voltage.

Low voltage adversely affects the performance of induction motors, heating devices, and solenoids. The torque of the motor, the output of the heater, and the pull of the solenoid vary as the square of the voltage. A 10% voltage drop will therefore decrease the torque 19%, the wattage 19%, and the pull of the solenoid. With a 10% voltage drop the temperature of the motor rises 6 to 7°C. Most solenoids are designed to operate successfully at 15% under voltage. When equipment is working close to capacity, voltage drops may prevent molding. For example, a 19% drop in torque might prevent an electrically powered screw from turning. A 10,000-W heating system will deliver 8100 W, which might not be enough to operate an oven. Similar problems could occur with the heating bands of a cylinder and with hopper dryers. An inexpensive recording voltmeter is useful in locating this problem.

PROTECTIVE DEVICES

The purpose of the electrical system is to provide uninterrupted power to the operating devices. The system is designed so that short circuits and overloads are isolated, allowing the rest of the system to remain in operation. Fuses, circuit breakers, safety switches, or their combinations provide this protection. The protection of the main feed system and the secondary distribution lines are beyond the scope of this book. The following discussion relates to the protection of machines, individual devices, and circuits.

Fuses

A fuse is a device which protects a circuit by melting open its current responsive element when an overcurrent or short circuit passes through it. The plug type is rated up to 30 A and not used on systems over 125 V. They screw into standard sockets or special coded sockets which prevent insertion of the wrong size fuse.

Cartridge fuses up to 60A have the ferrule-type connection. Above 60 A, they have knife blade-type contacts. They are made either for one time use or have provisions for replacing the fuse links. While renewable fuses are initially more expensive, the low cost of the links make them economical. Several different link ratings fit in the same cartridge. Both the outside of the cartridge and the panel should be clearly marked with the proper link rating.

Both type fuses are made either with single elements for short circuit protection or double elements for overload protection. A motor winding may draw enough current to damage its insulation, but is not enough to open the short circuit element of the fuse. The overload protection consists of a copper heat sink, which will melt an alloy, permitting a spring to pull a connection which opens the circuit. This type unit will permit the normal high starting currents of a motor but will open on a 35% continual current overload.

Some types of electrical equipment, such as meters, transistors, diodes, and semiconductors are damaged by even slight overloads. A current limiting fuse which interrupts in milliseconds is used to stop or limit the buildup of short circuit currents that can damage these devices.

Circuit Breakers

A circuit breaker is a device which will automatically interrupt the circuit when abnormal amounts of current in excess of its rating are applied. It can be reset and used again as contrasted with fuses which have to be replaced. The circuit breaker has a trip unit which is activated by the overload; contacts which break, opening the circuit; arc chutes which contain the arc caused by the circuit breaking; and an operating mechanism energized by the trip unit. A thermal trip unit has a bimetallic element which heats up as current flows through it. It bends or deforms unlatching the mechanism which opens the main contacts. A thermal-magnetic trip has the same bimetallic unit to which is added a magnetic mechanism activated by the magnetic field created by the current. An hydraulic-magnetic trip has a solenoid with a dashpot time delay element.

Changes in the coil current change the flux which causes the iron core to move within the coil. The iron core is in a silicone fluid which acts as a dashpot. As the core moves into the coil, it will reach a point where the magnetic force is strong enough to activate the tripping element. The time delay is a result of the dashpot effect. On extreme overloads or short circuits the coil itself develops enough magnetic field to activate the trip.

As with fuses, circuit breakers can have both instantaneous and time delay characteristics. The tripping curve can be adjusted to the application. Rapid acting circuit breakers are available for sensitive equipment although they do not open as quickly as special fuses.

Switches are used to disconnect a circuit element so that it can be handled safely. They are used as on-off devices with noninductive loads such as ovens and heating cylinders. Motors are controlled by starters. Switches in motor lines are best used when the motor is not in operation.

CONTROL RELAYS

The purpose of a relay is to activate control devices in their proper sequence. It is done by opening or closing contacts. A control relay is a device that is operated by a change in the condition of one electric circuit. This causes the operation of other devices in the same or other circuits. They do not control power consuming devices such as ovens, motors, and lights, except for solenoids and motors that use less than 2 A. There are many different types of relays including solid-state, reed, mercury wetted, and armature. The electromechanical relay (Figure 7-2) is found on molding machines and is discussed here.

Control relays contacts are made of silver or silver alloys and should only be cleaned with fine emery cloth and never with a file. The contact arrangements follow:

1. Normally open (NO),
2. Normally closed (NC),
3. Overlapping.

Item 3 is a combination of two sets of contacts actuated together and arranged so that the contacts of one set open (or close) after the contacts of the other set have closed (or opened).

The poles on which the contacts are mounted can be:

1. Fixed, cannot be changed from NO to NC or NC to NO.
2. Convertible, contacts can be readily changed from NO to NC, or NC to NO.
3. Universal, which has one each NO and NC contacts on a pole.

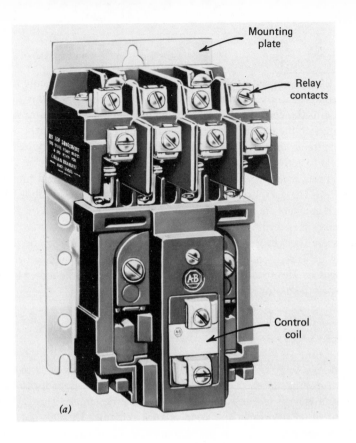

(a)

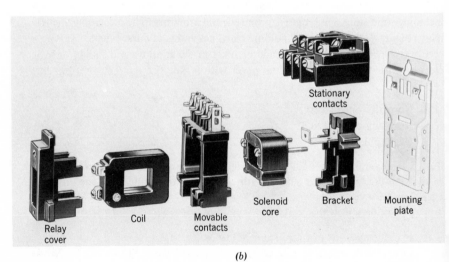

(b)

Figure 7 2   Four pole electromechanical contact relay (Courtesy of Allen Bradly Co.).

For practical reasons there are rarely more than six poles on a relay. There are many different types of relays. For example, when two large motors are to be started together it is desirable to start one slightly after the other to reduce the feeder in-rush current. This can be done with a time delay relay. The three ways to do this are pneumatic (dashpot), thermal (bimetallic strip), and electronic (resistor and capacitor). Latching relays remain activated until the latch (mechanical or magnetic) is released. Overload relays are used on motor contactors. Relays are also available that are monitored by frequency phase changes and voltage conditions.

Relay Activation

The poles are activated by an electromagnetic device consisting of a stationary iron core and coil and a movable relay magnet or armature, that is composed of layers of laminated high-permeable steel. The armature is attached to the moving contacts of the relay. When the coil is energized the armature will move up activating the relay. When the coil is deenergized gravity and possibly a spring will return it to its original position, deactivating the contacts. In 60-Hz circuits, the magnetic field will be zero 120 times each second as the current alternates. This will create hum and chatter which is objectionable mechanically and esthetically. A shading coil is used to establish a lagging magnetic field to keep the armature in position all the time. If a relay chatters or buzzes the shading coil should be checked.

Thus it is obvious from the construction of the relay that the coil and each contact are separate circuits and can use different voltages. For safety reasons, the control circuits on molding machines should be 120 V with one side grounded. Figure 7-3 shows the JIC symbols for electrical devices. The relay coil, for example, is a circle, and the normally open contacts are two vertical lines with an oblique line added for the normally closed designation.

Relay Failure

The major cause of relay failure is dirt which causes improper seating of the armature or plunger, with corresponding overheating of the coil and its eventual burn-out. Dirt also prevents the proper seating when closing and opening of the contacts causing excess arcing and rapid contact failure. If a relay does not function it should be mechanically checked for loose or broken wires. The armature should be moved by hand to see if the circuit functions. If it does, the fault lies in the coil. This can be checked by disconnecting it and applying an outside source of current. If the fault is mechanical and is not cleared up by an air blast, the relay should be disassembled and inspected.

SWITCHES

DISCONNECT	CIRCUIT INTERRUPTER	CIRCUIT BREAKER	LIMIT				MAINTAINED POSITION
			SPRING RETURN			NEUTRAL POSITION	
			NORMALLY OPEN	NORMALLY CLOSED			
					HELD CLOSED	HELD OPEN	NP	

LIQUID LEVEL		VACUUM & PRESSURE		TEMPERATURE ACTUATED		FLOW (AIR, WATER, ETC)	
NORMALLY OPEN	NORMALLY CLOSED	NORMALLY OPEN	NORMALLY CLOSED	NORMALLY OPEN	NORMALLY CLOSED	NORMALLY OPEN	NORMALLY CLOSED

PUSH BUTTONS

SINGLE CIRCUIT		DOUBLE CIRCUIT	MUSHROOM HEAD	MAINTAINED CONTACT
NORMALLY OPEN	NORMALLY CLOSED			

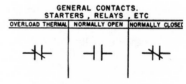

COILS				CONDUCTORS	
RELAYS TIMERS, ETC.	OVERLOAD THERMAL	SOLENOID	CONTROL TRANSFORMER	NOT CONNECTED	CONNECTED
			H1 H3 H2 H4 / X1 X2		

GENERAL CONTACTS. STARTERS , RELAYS , ETC		
OVERLOAD THERMAL	NORMALLY OPEN	NORMALLY CLOSED

Figure 7-3   JIC electrical symbols.

CONTACTORS

Contactors are devices for repeatedly establishing and interrupting an electric power circuit. They are used for controlling motors, ovens, heating cylinders, and lights. They are essentially similar in construction to relays excepting they have heavier power ratings. They are combined with various overload protective devices selected for the appropriate use.

Overload Protectors for Contactors

For example, contactors for ovens or heaters only require fuses or circuit breakers. Overload protection is required for motors, which prevents overheating of the motor caused by excessive current.

Magnetic-type overload protectors have a coil connected in series with the motor load and a plunger which will trip when the motor current flowing through the coil develops a strong enough magnetic field to move the plunger and open the motor contactor coil. This operates instantaneously when the tripping current is reached. Since the inlet starting current of the motor is much higher than the full load running current, a dashpot type time delay is included. This type motor protection is inferior to the thermal and only used when ambient temperatures make thermal controls impractical.

Thermal overload protectors use either bimetallic materials or low melting eutectic alloys. A heater coil in series with the motor winding is placed near a bimetallic disk. When the coil overheats it will cause the bimetallic disk to pop, mechanically opening the motor contactor coil, stopping the motor. When the disk cools, it will pop back closing the coil and restarting the motor. This automatic reset makes it undesirable for molding machine control.

For totally enclosed motors where a locked rotor condition causes rapid heat buildup, thermal overload protectors are installed in the motor housing. It is evident that this type of thermal overload protection will not protect the motor from short circuits. This is done by adding fuses or circuit breakers in the line.

Mercury type contactors are used in heating-load applications. They consist of hydrogen filled glass tubes, containing contacts; a ceramic-lined, stainless steel plunger; and mercury. The plunger (which is weighted in the NC unit and floating in the NO one) acts to raise or lower the level of the mercury, making or breaking the circuit. It is energized by a coil surrounding the tube.

The limitations are that they must be mounted vertically, cannot be mounted on a vibrating machine, and are subject to the breaking of glass. The advantages are no moving parts other than the plunger; no contact pitting; contacts unaffected by water or atmospheric contaminants; no mechanical banging; no 60-Hz hum; and little maintenance.

ELECTRIC MOTORS

Information on the theory and operation of motors is readily available elsewhere. We briefly discuss them from an operation and maintenance (3) point of view.

Motor Failure

Motor failure should be thoroughly investigated, since a modern motor, properly maintained, should outlast the useful life of the machine. If a motor does not function electrically the fault is in the power circuit, the control circuit, or the motor. The first thing to do is to measure the surface temperature of the motor with a pyrometer. Previous maintenance records should indicate its normal operating temperature. If it is within this range, the indications are the motor was not overloaded. Mechanical or electrical faults will cause overheated motors

Mechanical Faults.   Mechanical causes would include the bearings; a failure of the fan (if there is one in the motor); dust, dirt, or plastics inside the windings; and an overload caused externally, such as a jammed grinder or hydraulic overload. Mechanical breakdowns are usually heralded with motor noise. Although a noisy motor can also be caused by single- phasing (one of the phases is not receiving current). This condition can be checked by stopping the motor; if it will not start again the problem is electrical, not mechanical. The motor can also be checked by using a clip-on ammeter on each leg.

Electrical Faults.   The first step in locating an electrical fault is to use a voltmeter on the line side of the fuses or a circuit breaker. If power is available, the fault is not in the main feeder. Voltage is checked on the line side of the motor contactor. If there is no voltage, the fuse or circuit breaker has opened. The disconnect switch between the feeder lines and the motor is opened isolating the motor power circuit. The fuse or circuit breaker faults are corrected. An ohmmeter is then connected to each side of the contactor and ground. The ohmmeter is also placed between each pair of legs. The short circuit will be readily located. If it is in the motor it is usually a burned insulation on the stator winding and is readily smelled. This should not happen with proper motor protection. After the motor is repaired, the overload control system must be thoroughly investigated. If no grounding is located, it pays to start the motor again under no load. Occasionally fuses and circuit breakers go with no apparent fault. One should be certain the circuit breaker is functioning correctly.

If voltage is available on the line side of the contactor, the fault is in the contactor, control circuit, or motor. If the contactor is momentarily closed manually with the power on, and the motor starts, the fault is in the contactor or control circuit. The control circuit fuses are checked. The quickest way to check the contactor-coil is to disconnect the two leads. Then another source of current with the same voltage is connected to the coil through a switch. If the contactor does not operate the coil is replaced. If it operates the fault is in the remainder of the control circuit. The thermal overloads may be faulty and can be easily replaced for checking. The electrical circuit diagram of the machine is followed and the offending unit or wire will be found. Once the motor is running a clip-on ammeter on each leg will tell if each winding of the motor is drawing approximately the same current. The variation should not exceed 5%. If not, the motor should be stopped and checked.

If the motor is faulty it is wise to call in an experienced electrician to confirm the diagnosis and locate the reason for failure.

Maintenance Prevents Motor Failure.   Dirt is the prime cause of motor trouble. It acts as an abrasive and breaks down the insulation of the stators. Because of the oily and dusty nature of the molding plant, it will settle in the motor and tend to act as a heat insulator preventing full removal of heat as designed into the motor. Cleanliness of the motor and its surrounding area is important and part of the daily maintenance schedule.

The motor should be checked weekly by feeling the bearing housing and the motor itself for excessive heat and vibration. While the hand is not an accurate temperature indicator, experience will soon show when a motor is hot. At the same time, visual inspection of the couplings is made. Any unusual noise, smell, vibration, or any unusual sign requires the motor be shut down immediately and the cause determined.

The current in each winding should be checked with an ammeter monthly. Special ohmmeters are available to measure the insulation resistance between the field windings and ground. The severe service of motors in the injection plant require lubrication of the bearings. Motor manufacturers have excellent service manuals which should be read and followed.

CONTROL SWITCHES

Control switches are devices for making, disconnecting, or redirecting an electric current. They are distinguished from power switches such as circuit breakers by their function and small size. They are designated both by what they do and how they operate. If a switch has one pole or contact it is a single-pole switch. "Contact" means the ability to make or break a circuit. If it has two sets of contacts, it is a double-pole switch. If it makes and breaks only one contact, it is called a single-throw switch. If in making one contact, it breaks another, or in breaking one contact makes another, it is called a double-throw switch. Therefore a double-pole double-throw (DPDT) switch makes and breaks two independent circuits. Each pole has a normally open and normally closed contact. The switch could have a center or off position where the input lines are isolated. Switches may have the following actions.

1. *Momentary action.* When an operator (human, mechanical, or electrical) activates the switch, it remains activated only for the time the force is applied. When the force is removed it returns to its original position.

2. *Maintained action.* When activated they transfer the circuit from one set of contacts to another. They maintain this position regardless of whether the activating force is removed, even though the activator may return to its original position. When the switch is activated again the contacts will transfer the circuit to another circuit or back to its original position.

3. *Sequential action.* Two or more sets of contacts are switched in a predetermined sequential order. Some sets might be operated simultaneously and others may actuate more than once in any sequence.

Switch Activation

Switches are activated in many ways.

Push-Buttons. Push-buttons are perhaps the simplest type. They are used on molding machines to stop and start the motor, and in manual operations. The toggle switch is basically a push-button switch with a maintained contact. They are general purpose switches used for lighting, in molding machines and all over the plant. A variation of this switch uses a key instead of a lever to open and close the circuit. It is very useful where it is desired to limit the number of people who can set the circuit which it controls. An example might be switching from semiautomatic to automatic cycle. This can be an important safety measure.

Rotary Type Switches. Rotary-type switches are used to transfer circuits. They are primarily used in the molding machine for changes to automatic, semiautomatic, and manual operations.

Limit Switches. A limit switch (4) is a means of interfacing position or mechanical motion with the electrical circuit. The contacts are mounted in rugged enclosures and usually consist of one set of normally opened and one set of normally closed contacts. They can be had in up to four-pole configurations. The most common operator is a rotating lever with a wheel at the end. It travels approximately $100°$ with a $2°$ movement required for operation. Other operators are top push-buttons, top push-rollers, side push-buttons, and side push-rollers. Oil tight enclosures should be used on molding machines. They are widely used on molding machines, being found in safety control circuits, low pressure closing, platen speed, injection control, and mold safety circuits.

Proximity Switches. Proximity switches sense and indicate the presence or absence of a moving or stationary object without physical contact. The types primarily used on molding machines operate on the principle of changing the balance of magnetic fields causing contacts to operate. Their primary use is in the low pressure closing systems of toggle operated machines.

Snap-Acting Switches. Snap-acting switches work on a cantilever beryllium-copper spring system. The action is positive, regardless of the speed of the operating force. Once the operating position is reached it snaps over rapidly in less than 0.005 sec. The contacts are usually single-pole double-throw, small in size, rugged, dependable, and exceptionally versatile. Actuated by plunger, pin, lever, push-button, toggle, one-way dog, and wire, they come in all types of enclosures and are used as limit switches. These devices are of great assistance in designing automatic fixtures and equipment and safety devices in molds.

Pressure and Temperature Actuated Switches.   These switches work on the principle of mechanical action induced by a movement resulting from temperature or pressure changes. In temperature switches, the switch member might be opened by the action of a bimetallic element, the movement of a bellows, the movement of a capillary, or a change in pressure due to the temperature of certain gases. In pressure switches, the pressure usually operates a bellows. A transducer is a pressure switch where the pressure changes the electrical resistance of a circuit.

TIMERS

All industrial processes are related in some way to time. A timing device will measure a predetermined interval of time and, at its conclusion, operate a device (5).

Types of Timer Elements

The timing elements can be classified into six groups.

1. Mechanical.
2. Thermal.
3. Dashpot.
4. Electronic.
5. Solid state.
6. Electric motor-driven.

Mechanical Timers.   These timers consist of a spring, escapement and actuator, and switch. The spring is wound and its unwinding rate is controlled by the escapement, similar to a watch. At the end of a preset interval the actuator controls a switch. They are low in cost, unaffected by electrical conditions, noisy, and have relatively limited operating life. They are occasionally used for auxiliary operations such as timing a tumbling barrel.

Thermal Timers.   These are of the bimetallic and expansion type. The latter uses the principle that metal expands with heat. It mechanically amplifies this motion to throw a switch. Aside from bimetallic elements used on motor controls, thermal timers are not used in the molding plant.

Dashpot Timers. The timing of dashpot timers is controlled by the rate of flow of air or a fluid through a fixed or adjustable orifice. When the coil of a solenoid is energized, it moves at a rate controlled by the pneumatic or hydraulic flow. They are not often used on molding machines.

Electronic Timers. Electronic timers use R-C (resistance-capacitor) circuits for timing. The controls element can be on a transistor, a transistor and a silicone-controlled rectifier (SCR) combination, or a unijunction transistor (UTJ) and SCR combination. The R-C timing is based on the discharging action of a capacitor through a resistor. The time of discharge will depend on the size of the resistor. The timing range is usually limited to 5 min. This timing device can drive an electromechanical relay to operate the timed device, which has the limitation of any mechanical device. Solid state switching can be used, which makes the timing much more expensive. Electronic timers are more accurate than motor-driven ones, but voltage and temperature changes can affect them adversely.

Solid state timers count the pulses of the 60 cycle alternations of the current. Even though they are more expensive their extreme accuracy and reliability has resulted in their growing use in molding machines.

Electric Motor Timers. Almost all timers used on molding machines are driven by a small synchronous electric motor attached to a gear train. Because of the remarkably accurate frequency control of modern power systems, such timers are exceptionally accurate. They can be classified into cycle timers and reset timers. They are activated by the result of another physical set of conditions, such as a limit switch, push-button, pressure or temperature switch, or another timer. Certainly, not every part of the injection molding process is or should be controlled by timers. Untimed motions are those which are variable, and whose usefulness is determined by the completion of a motion or action. This can either start a timed action or another similar untimed action. An example of an untimed action is the return of a plasticizing reciprocating screw, which is stopped by a limit switch at a given position. Another example is the closing of the platen which may start the overall clamp timer and the injection forward motion.

Cycle Timers

A repeating cycle timer, once the program is set and the motor circuit closed, will continually repeat itself without interruption. If it is stopped, it will start again at precisely the same point where it is stopped. The mechanism of such a timer is a synchronous motor to which is attached a cam or cams, which mechanically open and close switches. The cams can be adjustable. When the adjustment is readily made by a dial they are called percentage timers. They keep the switch on for a given percentage of the cycle.

One of the most useful repeating cycle timers is a time-switch clock. It has a motor which operates a series of adjustable cams that trip an "on-off" switch. It has a 24-hr cycle, but provision for skipping days. These timers are used, for example, to turn on the heat for heating cylinders and ovens for start-up at the beginning of a week.

A time meter is not a timer in the sense that it does not perform the function of activating a circuit, but instead is a meter which totalizes the length of time electric current is flowing in a given circuit. It is basically a synchronous motor attached through a gear train to an indicating dial. These dials read in minutes or hours and are useful in recording interval operations such as the length of time a heating element of a cylinder is in operation.

Automatic Reset Timers

Reset timers start from a zero "reference point," measure time for a predetermined time interval, trip its contacts, deactivate the motor and clutch, and reset itself to its original zero "reference point."

A clutching mechanism is required. A clutch motor has a gear on its rotor which is separated from the gear train by spring action when the motor is deenergized. When the motor coil is energized, the magnetic field brings the rotor gear into contact with the gear train starting the timer. The gear train must be reset by a spring. The reset time is small for timers used on molding machines. Timers utilizing clutch motors use them to activate a switch. A relay is usually used.

As an interval timer (which limits the duration of an operation), the relay starts the motor and energizes the load at the same time. When it times out, the switch opens the relay coil deenergizing the load and the time motor and resetting the timer.

To be used as a delay timer (which varies the delay between two operations) the relay is activated at the end of the delay period, starting the load. It can also be accomplished by permitting the motor to stall at the end of the cycle.

The second way to operate a clutch is to use a solenoid which is external from the motor. This has the advantage of adding additional contacts energized by the solenoid. A type of timer using this system is a variation of a program timer. Any number of contacts are located from a zero time position. When the clutch is energized the motor will drive the cam timing the contacts. At the end of the cycle the clutch is disengaged and a spring returns the timer to its original position.

Solenoid-Clutch Timer.   It is impossible to follow the electrical circuits of molding machines without fully understanding the operation of the solenoid-clutch operated timer. There are three separate positions for the timer. The three switch positions are indicated under the timer contact symbol in the order shown in Figure 7-4.

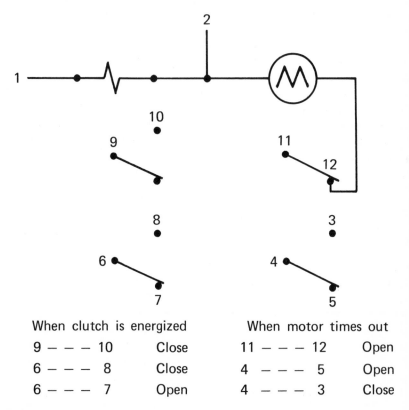

When clutch is energized

9 – – – 10	Close
6 – – – 8	Close
6 – – – 7	Open

When motor times out

11 – – – 12	Open
4 – – – 5	Open
4 – – – 3	Close

When timer resets—all contacts revert to the diagramed positions

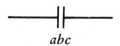

abc

Figure 7-4 A solenoid-clutch type timer in reset position. The three timer positions are (*a*) timer reset, (*b*) clutch energized, and (*c*) timer motor times out.

Figure 7-4 shows the reset position which occurs when the clutch circuit is deenergized. This is the position at the beginning of the operation. When the clutch is energized, the two switches underneath are activated. The timer motor is an independent circuit. When the timer motor times out, the two switches beneath the motor in the diagram are activated.

These contact arrangements permit flexibility in circuit design. Figure 7-5*a* illustrates how a timer might be connected. In the reset position (shown on the diagram) contacts 3 and 5 are open because 9 and 10 are open. Contact 8 is open, but contact 7 has a completed circuit. Circuit diagrams use two vertical

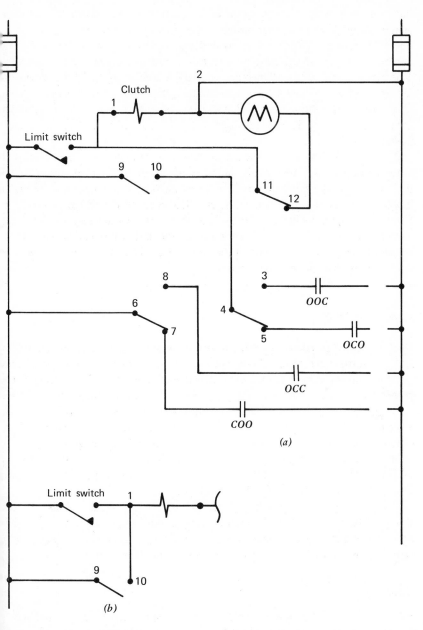

Figure 7-5 A solenoid-clutch type timer connected for maintained contact (a) and momentary contact (b).

lines on each side of the diagram for the power supply. When the clutch i
energized by closing the limit switch, 9 and 10 close, 6 and 8 close, and 6 and 7
open. Thus in the clutch energized position, contacts 5 and 8 are closed and
contact 7 is open. It will have no effect on contact 3, which is motor actuated
The motor is started simultaneously with the clutch being energized. When the
motor times out 11 and 12 will open stopping the motor; 5 will open and 3 will
close. The motor timing out will not affect circuits 7 and 8 which are clutch
activated. When the limit switch opens, the circuit will be reset to its original
position.

If the limit switch is opened during the timing cycle the timer will
automatically reset. This circuit therefore requires a maintained closing of the
limit switch. If the starting impulse was a momentary contact, such as through a
limit switch, push-button and such the timer can be connected as shown in
Figure 7-5b. A momentary closing of the limit switch would energize the clutch
and close contacts 9 and 10. They would interlock the clutch keeping the timer
activated regardless of the state of the limit switch. The circuit would have to be
opened at another place to reset the timer.

HEAT CONTROLS

Heat is used to plasticize material, to dry materials in ovens and hopper dryers
and to control mold temperature. They all use resistant type heaters which are
usually monitored by a sensing element which is amplified in a controller. The
controller energizes and deenergizes the heaters to maintain the set temperature.
Temperature is a fundamental measurement analogous to voltage in the
electrical system and pressure in the hydraulic system. It is a measure of the
potential that determines a body's ability to transfer heat energy from itself to
another body by radiation, conduction, and/or convection. It cannot be
measured directly and is indicated by its results on the physical properties of
other materials that are in equilibrium with it.

These properties are the expansion and contraction of solids, liquids, or gases;
vapor pressure of liquids; radiation from a hot body; amount of voltage
generated when two dissimilar metals are heated (as in thermocouples); and a
change in resistance of conductors with temperature (thermistors).

Temperature

Determining a temperature scale is not simple. Lord Kelvin devised a
thermodynamic scale which uses the temperature of absolute zero as $-273.6°C$
based on Carnot's reversible heat engine, temperature increments being in direct
proportion to increments in the engine's efficiency. International Temperature

Scale in use today is defined by means of seven reproducible basic temperature points and 12 reproducible secondary points. Two basic points are the equilibrium states of ice and water, and water-steam, at atmospheric pressure. The Fahrenheit (F) scale has the ice point at 32° and the steam point at 212°. The Celsius (C) uses 0° and 100°. The lowest theoretical temperature, where there is minimum molecular motion, is given in the Fahrenheit scale as 0°. Rankine, (R) and in the Celsius scale as 0° Kelvin (K). Table 7-1 shows the relationships and provides other useful information. A college or high school physics review book will refresh and provide other information.

The molder should use the boiling point of water to calibrate heating systems. It is not unusual for thermocouple-pyrometer systems to drift 15°F. When heats are set from molding records this difference will cause trouble. The systems should be checked at least monthly.

Table 7-1. Some useful electrical and temperature information

$$1 \text{ W-h} = 3600 \text{ J}$$
$$= 2655 \text{ ft-lb}$$
$$= 3.413 \text{ Btu}$$
$$= 1.3 \times 10^{-3} \text{ hp/h}$$

$$1 \text{ hp} = 33,000 \text{ ft-lb/min}$$
$$= 746 \text{ W}$$
$$= \text{about } 1 \text{ KVA } (3\phi)$$

$$E = IR = \frac{P}{I} = \sqrt{PR}$$

$$I = \frac{E}{R} = \frac{P}{E} : \sqrt{\frac{P}{R}}$$

$$R = \frac{E}{I} : \frac{E^2}{P} = \frac{P}{I^2}$$

$$P = EI = I^2R = \frac{E^2}{R}$$

$$C = \frac{5}{9}(F - 32)$$

$$F = \frac{9}{5}(C + 32)$$

$$K = C + 273.2$$

$$R = F + 459.7$$

ELECTRICAL HEATING

All cylinders for injection molding thermoplastics are heated electrically, with resistance heaters. Cylinders for injection molding thermosets are usually heated by circulating controlled temperature hot water in a jacket around the barrel. The nozzle is heated electrically. There are several types of resistance heaters such as tubular resistance heaters, cartridge heaters and band type heaters.

Tubular Heaters

Tubular resistance heaters are made by suspending a coiled resistance heating element made of nichrome in a metal tube or sheath. Magnesium oxide powder is vibration packed into this tube, and mica disks on each side seal the powder under pressure and act as insulators for the heating element. The tube is swedged to size to increase the packing density of the magnesium oxide. An induction heater tempers the outside so it can be bent and shaped to form. Tubular heaters come in different diameters with approximately 20 to 40 $W/in.^2$ (The watt density is determined by dividing the total number of watts over the available heating surface.)

Tubular heaters on injection cylinders are assembled by bending and forcing them into machined grooves with a soft mallet, and securing them permanently by peening the grooves with a prick punch. When properly applied with good contact surfaces, and not overrated, these heaters can last the life of a cylinder. They are considerably more expensive to install because machining of the cylinder is necessary, and slightly more steel is required to compensate for the grooves. Because of their construction they are unaffected by molten plastic, accidental physical damage, and mechanical separation at the interface. They require no maintenance aside from keeping the terminals tight.

Tubular heaters are cast in aluminum shaped to the outside of the heating cylinder. They are also formed into grooves machined in aluminum bands that fit over the heating cylinder. These have good heat characteristics and low maintenance. Tubular heaters are very useful on the nozzle where there is good chance of hot material accumulating over the heater. This would burn out a band heater.

Cartridge Heaters

These heaters are made with nichrome wire wound on forms with a magnesium oxide type of cement. The heaters are sheathed with copper (primarily used for heating water), stainless steel, and other alloys. Approximately 40 $W/in.^2$ are delivered with stainless steel sheaths. The introduction of an iron-chromium-nickel metal has raised the allowable watt density to 400, and these type units are extensively used for mold heating applications.

The successful use of cartridge heaters depends on an accurate fit between it and the hole. The maximum diametrical clearance is 0.004 in. Molybdenum-based compounds which are liquid when cold and solid when hot are used to lubricate their insertion and reduce the effect of the air gap. The IMS Company (Cleveland) (which has pioneered in the development of heating cylinders and associated equipment) has introduced a nozzle with cartridge heaters which give double the heat and much less maintenance than a standard heating band.

Band-Type Heaters

These heaters are made of nichrome wire wound on a form and insulated. The outside of the band is made of a high grade chromium steel. Mica bands use mica insulation and ceramic bands use ceramic. Compound bands have an insulation of a high heat paste inside a formed shell, which is bent to shape before hardening the compound.

Most injection cylinders are heated with band heaters. The maximum heat input which can be used in plastic processing is 30 to 40 W/in.^{2}. Higher watt densities tend to give poorer control and shortened band life; therefore, band voltages should be specified carefully. A 208-V band on a 220-V line increases the wattage by 21%. It also substantially decreases the heater life. Conversely, using a 220-V band on a 208 line will reduce the wattage but give much longer band life.

In connecting bands copper wire is not used. Special heat resistant alloys which resist oxidation but carry about a third as much current as the equivalent copper wire are used. The current carrying capacity of the wire rather than its size is specified.

A main cause for band heater failure is loose bands. The life of the heater depends on the ability of the system to remove the heat provided. This is best done by good physical contact with the metal to be heated, since air is an excellent insulator and hinders the drawing off of heat from the metal of the heater. The contact of heater bands should be periodically checked, particularly after the installation of a new band. At this time the electrical connectors to the band are tightened. A major cause for band failure is contact with plastic. The plastic melts and gets inside the heater, carbonizing and shorting the nichrome. A loose cover is provided with the cylinder which protects it from plastic and allows enough ventilation to prevent overheating of the band. A totally enclosed heating cylinder, while economical in electricity, is very expensive in short heater life.

The 440-V bands are not often used because the thin nichrome wire does not give good service life. Pairs of 230-V heaters in series are used instead.

The proper way to check resistance heater operation is to use a clip-on ammeter. A voltmeter will indicate voltage to the terminal but not a break in the wire. Together they can be used to check the wattage of the band.

CYLINDER TEMPERATURE MEASUREMENT

Temperature measurement of cylinders requires the generation of a control signal to a controlling device which maintains the barrel temperature at predetermined level (5a). Devices used are thermistors, wire wound element and thermocouples.

Thermistors

Thermistors are solid state semiconductors that have a high negative coefficient of resistance change with temperature. They have a quick response and can have long probe leads without compensation. Relatively unstable, thermistors are subject to thermal shock. Consequently, they are not often found on molding machines.

Wire Wound Elements

A second sensor uses the principle of the change of resistance with temperature in a wire wound element. This sensor is considerably more sensitive than thermocouples (discussed below). It does not require cold-junction compensation or thermocouple-break protection and eliminates the need for special lead and protection. It is used in a Wheatstone bridge (5b).

The Thermocouple

The universal temperature sensing device for injection molding heating cylinder is the thermocouple (6). When two dissimilar metals are welded together, they will convert heat energy into electrical energy. The amount of energy converted depends upon the metal chosen and the temperature. The two materials used for thermocouples on heating cylinders are made of iron and constantan (60% copper, 40% nickel) or I/C. The wire has been standardized at 20 gauge (0.03 in.). The J-type (used in the molding industry) has $32°F$ as the reference junction. The emf generated by an I/C thermocouple at $300°$ is 7.94 mV, at $400°$ 11.03 mV, and at $500°$ 14.12 mV. In this range, the increase per $°F$ is approximately 0.0307 mV. Since the reference junction is at room temperature thermocouple sensing circuits compensate for the change. A DC current is generated, with the positive wire being iron (white insulation) and the negative wire constantan (red colored insulation) and the indicating instrument is a D'Arsonval galvanometer. If the thermocouple is connected with the wrong polarity on the meter, the pointer will go to the left and off the scale instead of its normal direction to the right.

The thermocouple is protected by steel tubing whose O.D. has been standardized at 3/16 in. The fitting which screws into the heating cylinder has been standardized at 1/8 in. NPT. The two main ways of holding the thermocouple tightly into the fitting are compression types and quick disconnect bayonet types. The standard shapes of thermocouples are straight, 45° and 90° bends. Thermocouple extension wire is 14 or 16 gauge and must match the temperature-emf characteristics of the thermocouple. The wire, in effect, transfers the reference junction from the thermocouple to the meter or controller. These meters should be protected by iron conduit as stray electrical fields can affect their accuracy. In time, thermocouples will deteriorate and spares should be available. Unfortunately the need for replacement will not be indicated until there is a processing failure.

How to Check the Thermocouple

An excellent way to check the system is to immerse the thermocouple in boiling water. An inexpensive testing unit can be built consisting of a small pot and a heater for boiling water in a glass, permanently mounted. These units can be brought up to the machine and a thermocouple inserted, without disconnecting it from the circuit. If the system is accurate at 212°F it is most likely good throughout its range. Other standards are pure tin in equilibrium at its melting point (449.4°F) and pure lead in equilibrium with its melting point (621.2°F).

The system can be checked at different temperatures by putting the thermocouple in a calibration bath on a hot plate. A stirrer keeps the liquid in equilibrium. An accurate mercury thermometer next to the thermocouple will give true temperature readings.

The most accurate way to check a thermocouple (through not the control system) is to use a potentiometer connected to the thermocouple. The millivolt output is measured and should agree with the value given on standard charts for that temperature.

TEMPERATURE CONTROLLERS

There are two ways of using the current generated by the thermocouple to indicate temperature. One method is the electronic way which uses a Wheatstone-bridge circuit. The second uses a millivolt meter.

Wheatstone Bridge Circuit Indicator

A known voltage determined by the desired temperature is applied to one side of the bridge to balance the unknown voltage from the thermocouple. When the bridge is balanced, no current flows from the points of the bridge and the system has reached its set temperature.

This system does not always use a meter for control, although sometime meters are used to show the operator the barrel temperature. Other instruments just have an on-off light indicator. Still others have a meter which indicates the number of degrees deviation above or below the set temperature.

The Millivolt Meter Indicator

The second method of using thermocouple output is to drive a D'Arsonva millivolt meter. Attached to the pointer is a small lightweight vane. One manufacturer uses the vane to interrupt the light in a photoelectric system. Another uses the vane to change the inductance of an oscillating circuit when it moves between two coils. The phototube assembly, or oscillator coils, is mounted on the indicating pointer which is manually set for the desired temperature.

Temperature Control

A thermocouple attached to a portable galvanometer is called a surface pyrometer. It is indispensable for measuring mold temperatures, motor temperatures, and other solid surfaces. Probes can be had to be inserted into the molten plastic to measure its temperature.

By means then of the thermocouple, galvanometer, and vane on the indicator the controller can be so activated that heat will be called for when the temperature is below the point set manually on the pyrometer and turned off when the indicator is above the set point. This on-off type controller is not satisfactory for molding. If we examine the nature of cylinder and resistance heating and temperature measurement, control methodology will become apparent.

How the Cylinder is Heated

The resistance heater applies heat on the outside of the cylinder. The cylinder of necessity, has considerable steel to contain the pressures of the plastic material. Therefore, heat transfer from the resistance heaters on the cylinder surface to the plastic material inside the cylinder is not instantaneous. It takes a significant length of time for the heat energy to travel by conduction to the inside wall in contact with the plastic material.

The thermocouple will indicate the temperature at some point in the cylinder wall, probably near the middle. This introduces considerable error too. It can be significantly decreased by using two thermocouples in parallel at each location, one being as near as possible to the inside wall of the cylinder and the other to the outside wall (7, 8). When the thermocouple(s) indicates that the required temperature is reached, the heat is shut off. However, the heat energy in transit in the cylinder wall will travel through to the plastic and raise the plastic to a temperature significantly above the controlled temperature point. The reverse will happen on cooling.

One of the major advantages of screw plasticizing is the elimination of this cyclic heat input. The screw mechanically shears the plastic and applies heat "inside" the polymer. As soon as the screw stops this thermal input stops preventing temperature override. Since resistance heaters provide a significant portion of the heat, the "heat sink" effect of the barrel is still significant.

Proportioning Systems.   To improve the on-off type controller an electronically generated proportioning system is used. One way to do this is to feed a small control millivoltage into the thermocouple circuit which will cause the galvanometer to read higher than the actual temperature when the system is calling for heat, and slightly less when the system has been satisfied. Below the level of the proportioning band the heat is 100% on. As it reaches the set point it decreases to 50% on, and until at the far end of the proportioning zone, it is completely off. This is, in effect, averaging the power output, proportional to the deviation from the set point.

In this type of controller, the equilibrium temperature might not be the same as the set point. Automatic reset circuits are used to correct this.

Extruders sometimes require cooling during their operation. It is not inconceivable that this might be required with some of the newer heat sensitive injection molded materials. A three position controller, which has a proportioning heat control band, a neutral band, and a proportioning cooling band, is used for such control.

The galvonometer type instruments use a mechanical relay in the pyrometer to operate the contractor coil of the heat control relay. The closer the temperature is controlled the more often they operate. They are a major source of maintenance. A stuck contactor or pyrometer relay fault will keep the heating bands on continually. This may cause severe degradation of the material and even fires. In many instances the only way to remove the burned plastic is to take the cylinder apart and clean it mechanically. These faults can be very difficult to detect. A useful device is to put one pilot light across the power side of the contactors and another one activated by the contactor coil current. They are mounted side by side under the pyrometer. They should go on and off together. If not the cause should be immediately ascertained. High limit control systems are available which sound an alarm and cut off the heating power when the cylinder barrel reaches a given temperature increment over the set temperature.

Solid State Devices.   A much better way of temperature control is to use solid state devices totally eliminating the electromagnetic relays (9-11). The output of a galvonometer or potentiometric type control is essentially proportional to the temperature of the barrel. They can be used to drive a saturable core reactor or a silicon-controlled rectifier (SCR). Both these devices will provide a continual amount of current proportional to the temperature o. the barrel rather than the on-off type of the proportional controllers. Thi increases the life of the bands and gives a much better melt temperature control

Saturable Core Reactors.   The saturable core reactor works on the principle that AC current flowing through a coil is much less than would be indicated by Ohm's law. Counter currents are induced in the coil to impede the flow o currents (Lenz's law). This effect is tremendously increased if an iron core is used. As the current and voltage increase in the coil surrounding the iron core the flux builds up until the core is magnetically saturated. Then an increase in applied voltage will cause the current flow to increase rapidly. A saturable reactor is one in which the degree of saturation can be controlled independently This is done with a control winding ultimately driven by the instrument output Saturable core reactors give excellent results but are being replaced by silicon controlled rectifier units because of certain limitations. They are bulky, have to be matched to the heater load, provide a maximum of 90% power to the heaters and do not shut off completely when in the circuit.

Silicon Controlled Rectifier.   Silicon controlled rectifiers solid state con trols use the rectifier as a gate, switching from a nonconducting to a conducting state. This can be done in two ways. It is to be recalled that an alternating current changes its polarity with each cycle. A plot of the amount of curren versus time shows a sine curve, half of which is above and half below the zero line. One method of gating cuts off the same portion of the current in each cycle. The amount of cut-off depends on the input voltage from the controller For example, if half the power was required half the cycle would be cut off While this controls the process very well it also produces transient curren interference in the power line which can interfere with other solid state controls In addition, it cuts off current for part of the cycle and causes a surge in the remaining part. The power source must supply enough current for the whole cycle. This in effect lowers the power factor.

A much better way of controlling the SCR gate is to turn it on and off only when the current output of the power system is zero (when the sine curve reverses). The current is cut off for a given number of cycles depending on the control requirements. Since each SCR picks the "off" cycles at random, the average current drawn from the lines will have a normal 100% power factor.

Basic Limitation of the Control System

All the controllers we have discussed measure the temperature of the steel barrel. We are really concerned with the viscosity of the polymer. Polymer temperatures and pressures in the nozzle are measurable, and will be used as part of the controls in the more sophisticated equipment being developed (12-15).

ELECTRICAL CIRCUITS

Knowledge of the electrical components combined with a thorough understanding of the hydraulic and mechanical action of the molding machine will facilitate following the electrical control circuits. Like the hydraulic circuit, the electrical circuit can be divided functionally (Figure 7-6). A good instruction manual can be very helpful (16).

An electrical schematic drawing will certainly be furnished, although not always conforming to JIC specifications. It is particularly important to have each line numbered on the left and corresponding relay and timer contact identification on the right. Tables listing all of the switches, limit switches, timers, and relays should be drawn as has been done for Figure 7-14. The information should include their location, function, contact location, manufacturer, and part identification. The latter is useful in checking individual components and ordering spares parts and units.

Figure 7-7 shows some methods of controlling solenoids electrically, drawn using JIC symbols (16a). Convention keeps solenoids, coils, and activating devices on the right. The right vertical line is one side of the circuit and the left vertical line the other. The numbers on the left identify each line in the circuit. The numbers on the right show the location of the contacts activated by the coil, switch, timer, and so on. If the number is not underlined the contact is NO. If it is underlined, the contact is NC. In a large drawing it would be otherwise difficult to quickly and completely locate them.

The lower section, lines 1 and 2, show a standard interlocking circuit. Once the NO start contacts are closed, even momentarily, the control relay will remain energized until the NC stop contacts are momentarily opened. Closing the start contacts energizes the coil of the control relay and closes its contacts on line 2. Once this is done there is a continuous circuit through the control relay coil, its NO contact line 2 (now closed), and the NC stop contacts. Momentarily opening the NC stop contacts on line 1 will deenergize the control relay and open the contacts on line 2. These cannot be closed again until the stop contacts are closed again and the start contacts energized. Interlocks are very commonly found in control circuits.

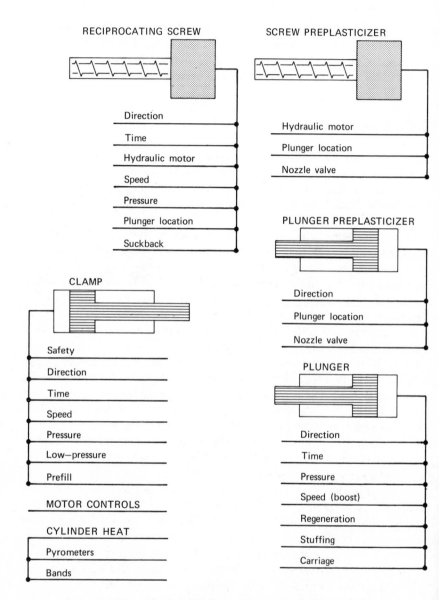

Figure 7-6 Component functions of electrical circuit.

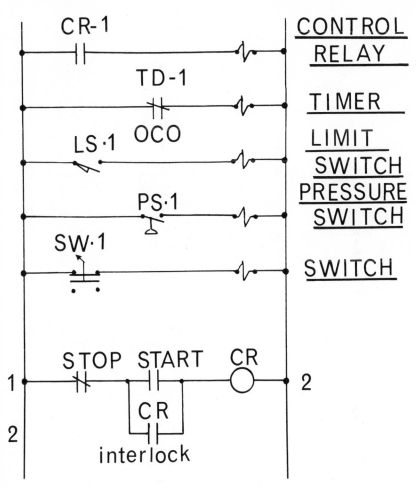

Figure 7-7 Methods of controlling solenoids and an interlock circuit.

Heat Control Circuit

Figure 7-8 shows a typical circuit for heat control. A timing circuit has been added which permits energizing the heat control circuit so that machines will be ready to operate at the beginning of the first shift of the week.

The timing circuit consists of 24-hr clock 3.* It is normally left running continuously but can be turned off by switch 3. Pyrometer 4 is controlled by the NO contacts 4 of control relay 1. The pyrometer is manually controlled by switch 1. For the timed operation, manual switch 1 is opened and night switch 2

*numbers refer to the lines on the left-hand side of Figure 7-8.

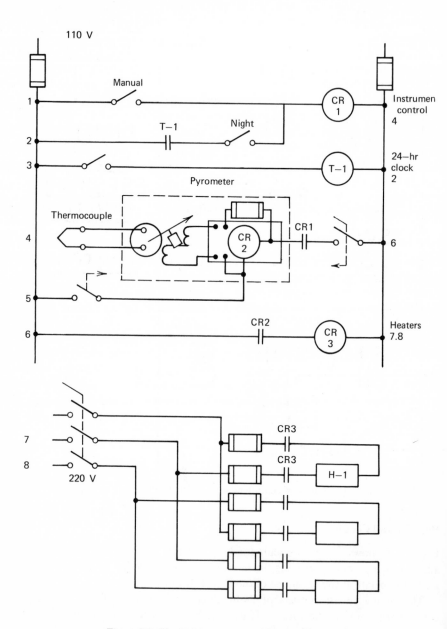

Figure 7-8 Typical pyrometer and heating band circuit.

is closed. When the timer times out its contacts *T-1 2* will close energizing the instrument control relay, CR-1; closing NO CR-1 contacts 4 will energize the pyrometer.

Because the cylinder is below the set temperature, the thermocouple will cause instrument relay CR-2 4 to close. This energizes heater control relay CR-3 6 (for heater H1). The contactors of CR-3 on line 7 and 8 will close sending 220-V power to the heating bands assuming the disconnect switch 7 is closed. When the cylinder is up to heat the thermocouple will cause the pyrometer to deenergize, CR-2, 4 which deenergizes CR-3, 6, and opens the contacts CR-3, 7, 8, deenergizing the heater band. It is to be noted that contactor coil relay 6 operates on 110 V, while the contacts of contactor 7, 8, operate on a different voltage (220 V). The heaters are connected to roughly balance the amount of current drawn on each of the three phases. Unbalanced phase demand is costly in power charges.

Nozzles can be controlled by thermocouples and pyrometers. More often they are controlled by autotransformers. These are very susceptible to overloads and should be protected by special fast acting circuit interruptors. They should never be started at their full load. When the press is shut down they should be turned to zero and turned up gradually. It is not good practice to tie the nozzle band into the control circuit of the first heating bands, as the heat requirements of the two areas are different.

Many machines have the heat control and heating circuits wired into the main machine circuit. This has a serious disadvantage and should be changed. When machines are down for repair and occasionally for mold change, the machines are disconnected. It is desirable to have the heat ready when the repairs are done. The only way to do this, if safety regulations are to be followed, is to have the heating power and control circuits completely independent with their own disconnect switches.

Motor Control Circuits

Motor control circuits are shown in Figure 7-9. The overload protectors (OL) will open if there is a sustained "over-current" situation. The fuses will protect from instantaneous faults. The contactor coils close the power contacts that connect the motor to the 220 V or higher power lines and also close a set of relay contacts (1-M, 2-M) which are used in the control circuit.

Lines 1 and 2 show the control circuit for a single motor. When the start button is pushed, the contactor coil is energized. It interlocks through contact M, 2. The motor will run until either the stop button is pushed, deenergizing the contactor coil; overload (OL-1 or OL-2) opens; a fuse blows in the power circuit; or the disconnect switch is opened. The latter method is not recommended. The stop button should be used first.

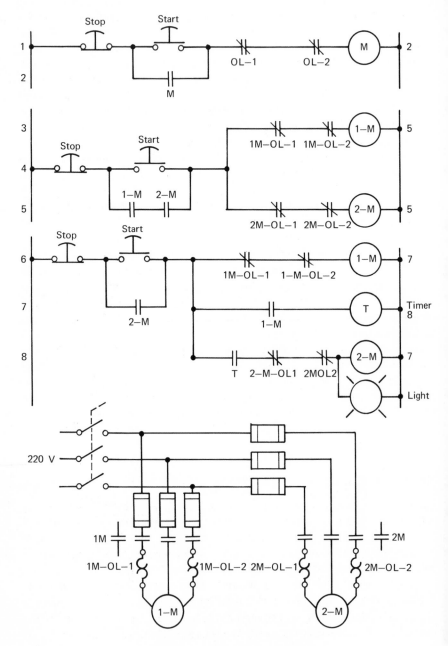

Figure 7-9 Motor control circuits.

Lines 3, 4, and 5 show a control circuit for two motors operating together. The motor circuits are shown at the bottom of the drawing. When the start button is pushed, each motor contactor coil is energized through its overloads. The start button is interlocked through a series connection of the two interlock contacts (1-M, 2-M). If either motor contactor coil opens because one of its overload contacts open, the interlock of the start button will be broken, deenergizing both motor contactor coils.

Lines 6, 7, and 8 show a circuit for starting two large motors (which have to operate together) in sequence. This is to prevent excessive inrush currents in the feeder line. When the motor start button is closed, motor contactor coil 1-M, 6 is closed. The start button is not interlocked. Instead the interlock contact of 1-M is used to start a timer T,7. When the timer times out, its contacts T,8 energize motor contactor coil 2-M,8. This interlocks the start button through contact 2-M,7. The light goes on showing that the start button can be released.

If either overload contact of 1-M opens, contactor coil 1-M opens, opening in turn contact 1-M,7. This resets the timer, opening timer contacts T,8, opening the second motor contactor coil 2-M,8. If this or either of the 2-M overload contacts opens, the contactor coil 2-M,8 opens breaking the interlock of the start button, 2-M,7, stopping both motors.

Low Pressure Closing Circuit.   See hydraulic diagram, Figure 6-39 and Figure 7-10. When the mold closes without any obstruction on the mold

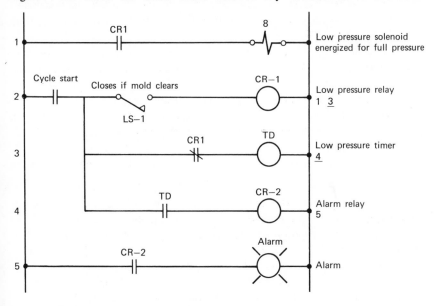

Figure 7-10 Low pressure circuit with alarm for hydraulic circuit. (See figure 6-39.)

584 Electrical Mechanisms and Circuits

surfaces, a limit switch closes which energizes the low pressure solenoid. This blocks the low pressure relief valve out of the circuit and allows full clamp pressure. If there is an obstruction, the limit switch does not close and the low pressure timer times out. This activates an alarm (bell, light), stops the control circuit, and can be used to shut off the machine.

Analysis of Circuit. Machine closes, at which time the cycle start contacts, 2, close. Low pressure timer TD,3 starts timing.

If there is no obstruction A- LS1,2 closes energizing the low pressure control relay CR-1,2, which does the following:

1. Closes CR1,1, contacts energizing low pressure solenoid 8,1. This permits full system pressure.
2. Opens CR1,3 contacts stopping and resetting timer,3.

If mold has obstruction, then the following is true:

1. LS1,2, does not close and timer TD,3, times out, closing timer contacts TD,4, energizing alarm control relay CR2,4.
2. CR2,5 contacts close starting alarm, 5. Other sets of contacts on this relay could be used to stop the cycle or machine.

Control Circuit for Automatic Cycling. Figure 7-11 shows the following:

1. With the selector switch, 1 set at automatic, the returning clamp contacts limit switch LS1,1, within 2 in. of full clamp return, starting cycle delay timer T,1.
2. Machine opens fully, contacting limit switch LS2,2, which deenergizes control relay CR,2. This deenergizes relay contacts CR,4, which will interlock the timer contacts T,3, while machine is forward of LS2. It also resets the control circuit for the next cycle (contacts not shown).
3. Timer T,1, times out, closing contacts T,3, which starts the machine cycle.
4. Clamp ram moves forward permitting LS2,2, to close. This interlocks the control circuit.
5. Machine platen moves more than 2 in., opening LS1,1 and resetting the timer.

Circuit for Semiautomatic and Manual Control of a Clamping Cylinder Using a Solenoid, Piloted, Three Position, Four Way Hydraulic Valve with all Ports to Tank in the Neutral Position and Two Solenoids Controlled by a Timer. This circuit is used to operate another cylinder (see Figure 7-12). When the gate is closed, a limit switch is contacted, starting the clamp timer and energizing the clamp relay which starts the machine closing. When the clamp timer times out, its contacts deenergize the clamp relay which sends the clamp back. At full return, a limit switch opens so that both clamp forward and return solenoids are

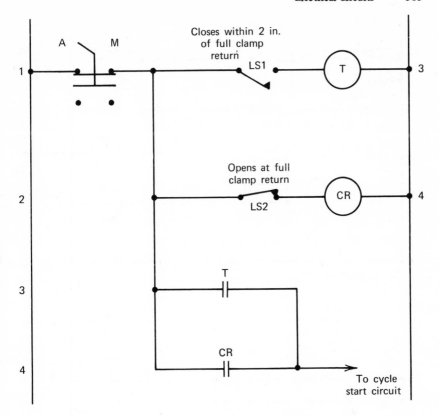

Figure 7-11 Circuit for automatic cycle control with time delay.

deenergized, shifting the hydraulic valve to the neutral position. This unloads the pump through the valve to tank.

An additional timing circuit is shown. It is started at the same time as the clamp timer and controls a pair of solenoids, energizing only one of the pair at the same time. This is the standard way of extending and retracting cylinders.

Some cylinders are built so that they must be stopped before full forward stroke. Otherwise the packing bearing and glands may be damaged. An overstroke limit switch LS3,6, is located so that it will open, stopping clamp forward motion before that position is reached.

Analysis of Circuit – Semiautomatic Cycle.   First turn the three operating switches to semiautomatic.

1. Gate is closed, closing LS1,3. Its safety contacts 6 and cycle start contacts 3 close. This starts clamp timer T1,4, closing its contacts OCO,3, and energizing control relay CR,1.

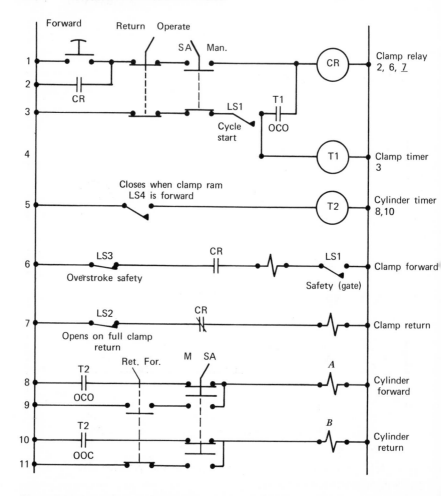

Figure 7-12 Manual and semiautomatic control of a clamping cylinder – using a solenoid piloted, three position, four-way hydraulic valve with all ports to tank in neutral. Included is another cylinder controlled by a timer and two solenoids, with all ports blocked in neutral position.

   a. CR,6, contacts close, energizing clamp forward solenoid 6.
   b. CR,7, contacts open deenergizing clamp return solenoid 7.
   The clamp moves forward.

2. When clamp ram is fully forward it closes LS4,5, starting cylinder timer T2,5.

   a. This closes T2 contacts, 8.
   b. Opens T2 contacts 10. The cylinder moves forward.

3. Timer T2,5, times out. (It must complete its cycle before clamp ram returns, because that unloads the pump output to tank through its the open ports in the neutral position).

    a. OOC contacts 10 close energizing solenoid B 10

    b. OCO contacts 8 open deenergizing solenoid A8. The cylinder returns.

4. Clamp timer T1,4, times out, opening timer contacts OCO, 3. This deenergizes clamp relay coil 1 reversing steps 1-a and 1-b. The clamp returns.

5. On full clamp return limit switch LS-2, 7, opens deenergizing clamp return solenoid 7. Solenoids 6 and 7 of the four way valve are deenergized, so the valve shifts into its neutral position which unloads the pump.

6. The gate is opened, opening LS1, 3 which resets timer T1, 4, deenergizing clamp forward solenoid 6. When the clamp returns LS2, 7, opens deenergizing clamp forward solenoid 8.

7. The clamp return also opens LS4, 5, resetting timer T2, 5.

8. Opening the return switch 1 at any time will reset timer T1, 4, opening its OCO, 3, contacts, deenergizing CR, 1, which opens the clamp.

Manual Operation

9. Set the switches to manual operation. This removes timer T1, 4 and the contacts of timer T2, 8 and 10, from the circuit.

10. The clamp control relay CR, 1, is controlled by the forward, 1, and return, 1, switches. Closing clamp forward button 1 energizes CR, 1. It interlocks through contacts CR, 2. When the safety gate is closed, LS1, 6, contacts will close. *The clamp closes.*

11. When the clamp return switch 1 is opened, the interlock contacts CR, 2, opens deenergizing the clamp relay coil, 1, reversing steps A1 and A2. This causes the *clamp to open.*

12. Similarly, the cylinder solenoids A, 8, and B, 10, are operated by switch contacts 9 and 11.

Electrical Circuit for Controlling Three Hydraulic Pressures. See Figure 7-13 and hydraulic diagram, Figure 6-22.

1. With selector switch 1 in "3" position, cycle start contacts 1 are closed by an external source. This starts timers T1, 1, and T2, 2. Since both solenoids A, 4, and B, 5, are deenergized, pilot valve C controls the pressure.

2. Timer T1, 1, times out closing contacts T1, 3, energizing control relay CR, 3, which does the following:

    a. Closes CR, 4, contacts energizing solenoid A, 4.

    b. Opens CR, 5, contacts deenergizing solenoid B, 5. Pilot valve A controls the pressure.

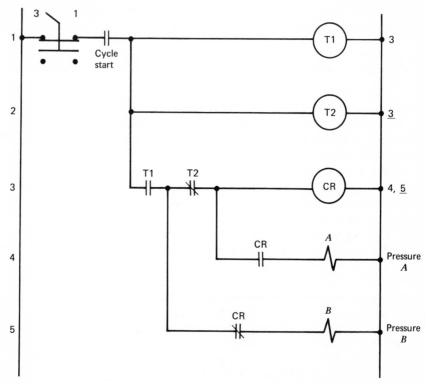

Figure 7-13 Electrical circuit for three different hydraulic pressures. See hydraulic circuit, figure 6-22.

3. Timer T2, 2, times out, opening contacts T2, 3, deenergizing control relay CR, 3, which does the following:

    a. Opens contacts CR, 4, deenergizing solenoid A, 4.

    b. Closes CR, 5, contacts energizing solenoid B, 5 through T1, 3, contacts which are still closed. Pilot valve B controls the pressure.

4. Cycle start contacts 1 are opened (from external source), resetting the timers T1 and T2.

Simplified Electrical Circuit for a Reciprocating Screw Machine.   See Figure 7-14. Refer to hydraulic diagrams, Figures 6-39 and 6-41. The balance of the circuits can be located in the other hydraulic drawings. The low pressure safety (Figure 7-10) and automatic cycle (Figure 7-11) circuits have been eliminated. The following refers to Figure 7-14.

Control Relays

Location	CR	Function
1	1	Starts clamp forward.
3	2	Starts injection. Connects injection gauge. Energizes controlled speed.
7	3	Energizes extruder run and hydraulic motor solenoids. Energizes full speed solenoid during purge.
10	4	Starts the extruder and clamp timer.
11	5	Stops extruder and starts suckback and suckback timer.
15	6	Starts clamp return. Resets clamp forward relay. Energizes full speed solenoid.
18	7	Controlled speed circuit.

Timers

Location		Function
4	1	Injection timer.
9	2	Clamp timer.
5	3	Second injection pressure timer.
13	SB	Suckback timer.

Solenoids

Location		Function	Location	No.	Function
20	A	Clamp forward	8	G1	Back pressure gauge
19	SEP	Separates pumps	29	HM	Starts hydraulic motor.
26	B	Starts injection forward	13	SB	Starts suckback.
27	G2	Injection gauge	30	D	Clamp return.
6	8	Second injection pressure	21	E	Full speed.
28	S	Extruder fun	25	F	Controlled speed.

Semiautomatic Manual Purge- Switch

Location	Switch	Purge	Manual	Semiautomatic
1	1	+	+	+
3	2			+
4	3	+	+	
4	4			+

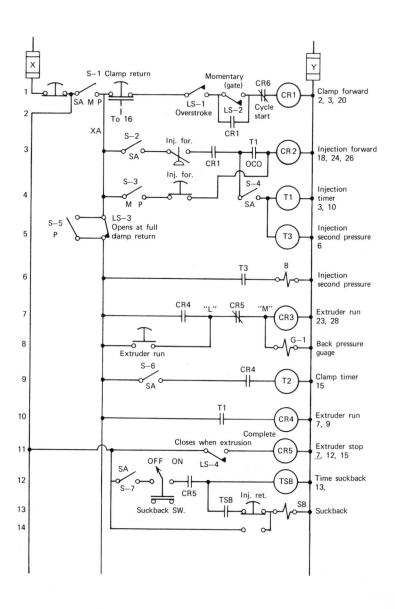

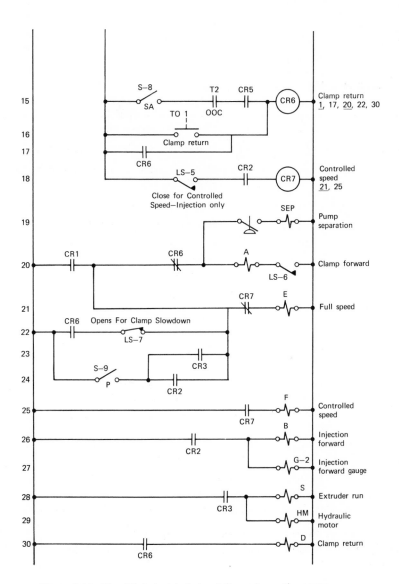

Figure 7-14 Simplified electrical circuit for reciprocating screw.

Location	Switch	Purge	Manual	Semiautomatic
5	5	+		
9	6			+
12	7			+
15	8			+
24	9	+		

Limit Switches

Location	Switch	Function
1	1	Overstroke protection.
1	2	Momentary gate switch. Starts cycle.
5	3	Opens at clamp return. Resets timers and relays.
11	4	Closes when extrusion is complete. Stops screw turning. Starts suckback.
18	5	Closes for controlled speed. Only on during injection.
20	6	Clamp forward safety. Machine cannot be closed unless gate closes this switch.
22	7	Clamp speed slowdown.

A description of the circuit follows:

1. Gate closes starting cycle.
2. Machine clamps; pressure builds up, closing pressure switch.
3. Injection timer starts and injection ram goes forward.
4. Injection timer times out starting clamp timer and hydraulic motor which drives screw.
5. Screw turning plasticizes material, forcing screw back until it hits limit switch, stopping the screw, and starting the suck back timer.
6. Suck back starts until timer times out.
7. Clamp timer times out, opening clamp.
8. Fully open clamp opens limit switch which resets timers and machine.

Analysis of Circuit on Semiautomatic Operation.   Selector switch set to semiautomatic. Gate is open and all devices are deenergized as diagrammed. Gate is closed.

1. Limit switch LS6, 20, closes safety contacts setting clamp forward solenoid A, 20.
2. Momentary contacts of LS2, 1, are hit by gate closing, energizing clamp forward control relay CR1, 1 which does the following:

a. Closes CR1, 2 contacts interlocking CR1, 1.
b. Closes CR1, 3, contacts setting injection forward relay CR2, 3.
c. Closes CR1, 20, contacts setting pump separation solenoid 19 and energizing clamp forward solenoid A, 20. *Clamp closes.*

3. Pressure builds up, closing pressure switch contacts 19 which energize the clamp separation solenoid 19 and separate the small holding pump from the large volume pump(s).

4. Injection pressure forward switch contacts 3 close, starts second injection pressure timer T3, 5 and starts injection forward timer T1, 4 closing T1, (OCO), contacts to energize CR2,3 *which does the following:*

a. Closes CR2, 18, setting controlled speed relay CR7, 18.*
b. Closes CR2, 26, energizing injection forward gauge solenoid G2, 27, and injection forward solenoid B, 26. Injection starts, which does the following:
c. Opens LS-4, 11, deenergizing CR5, 11, which opens contacts CR5, 12, resetting suck back timer T-SB 12.

5. (Concurrently with Step 6) At appropriate location of injection forward travel, limit switch LS5, 18, closes, energizing controlled speed solenoid CR7, 18, which does the following:

a. Opens contacts CR7, 21, deenergizing full speed solenoid E, 21.
b. Closes contacts CR7, 25, energizing controlled speed solenoid F, 25. *Injection forward moves at controlled speed.*

6. Concurrently with Step 5. Second injection pressure timer T3, 5, times out, which does the following: closes contacts T3, 6, energizing secondary pressure solenoid 8, 6. *Injection proceeds at secondary injection pressure setting.*

7a. Injection timer T1, 4, times out closing contacts T1, 10, energizing extruder run relay, CR4, 10, which does the following:

a. Closes CR4, 7, contacts energizing back pressure gauge solenoid G-1, 8, connecting back pressure gauge.
b. Energizes extruder run relay CR3, 7. This closes CR3, 28, contacts energizing extruder run solenoid 28 and hydraulic motor solenoid 29. (Note: CR3, 23, contacts are only used during purge.) *Screw turns.*
c. Close contacts CR4, 9, starting clamp timer T2, 9.

*The large pump is a variable volume, pressure compensated type, whose speed is controlled by solenoids E and F. When E is energized, full volume is delivered by the pump. When F is energized the controlled volume (previously set by the operator) is delivered. When neither are energized, the pump short strokes delivering no volume. At that time in the circuit, the only oil comes from the small high pressure holding pump.

7b. Injection timer contacts T1, 3, open deenergizing injection forward relay CR2, 3, which does the following:

    a. Opens contacts CR2, 26, deenergizing injection forward solenoid B, 26, and injection forward gauge solenoid G2, 27.

    b. Opens contacts CR2, 18, deenergizing controlled speed relay CR7, 18, which opens contacts CR7, 25 deenergizing controlled speed solenoid F, 25 and closes contacts CR7, 21, energizing full speed solenoid E, 21. *Pump delivers full volume.*

8. Extruder runs until its backward movement contacts limit switch LS4, 11. The location of this limit switch determines the amount of plastic in front of the screw. Extruder stop relay CR5, 11, is energized, which does the following:

    a. Open contacts CR5, 7, deenergizing extruder run relay CR3, 7, reversing steps 7(a) a, and b. *Screw stops turning.*

    b. Closes contacts CR5, 15, setting clamp return relay CR6, 15. This is a safety to prevent mold opening while extruder is running. This can happen if extruder slows down (material too viscous, hydraulic pressure drops, etc.), or clamp timer setting is incorrect.

    c. If suck back is used contacts CR5, 12, close, which does the following:

        (1) Start suck back timer T-SB, 12,

        (2) Energize suck back solenoid SB, 13, starting suck back

        (3) Suck back timer T-SB, 12, times out opening suck back timer contacts T-SB, 13, deenergizing suck back solenoid SB, 13, stopping suck back.

9. Clamp timer T2, 9, times out closing timer contacts T2, 15, energizing clamp return relay CR6, 15 which does the following:

    a. Opens contacts CR6, 1, deenergizing clamp forward relay CR1, 1. This reverses steps 2a-c. \It also resets injection timer T1, 4, and second injection pressure timer T3, 5. This would happen in any event, because injection pressure switch contacts 3 would open.

    b. Opens contacts CR6, 20, deenergizing clamp forward solenoid A, 20.

    c. Closes CR6, 30, energizing clamp return solenoid D, 30. *Machine opens at full speed.*

    d. Closes CR6, 22, setting clamp opening slow down limit switch LS7, 22. Note that CR1, 20, is open (step 9a); CR2, 18 open (step 9a); CR6, 22, is closed (step 9), therfore, the full speed solenoid E, 21, is controlled by LS7, 22. Controlled speed solenoid F, 25, is deenergized by open contacts CR7, 25.

    e. As platen moves back, LS7, 22, is contacted, deenergizing full speed solenoid E, 21. *Machine slows down* moving only on volume of small holding pump.

10. Machine opens fully opening limit switch LS3, 5, which does the ollowing:

    a. Resets clamp timer T2, 9.

    b. Deenergizes relays CR3, 7; CR4, 10; CR6, 15; CR7, 18. Gate is opened and machine is ready for the next cycle.

Analysis of Manual Operation.   Selector switch turned to manual.

1. Clamp forward is started by closing the gate which energizes clamp forward relay CR1, 1. Contacts CR1, 20, close energizing clamp forward solenoid A, 20. CR1, 1, is interlocked through contacts CR1, 1.

2. Extruder run manual button 8 is closed to turn screw. It must be held. The screw will stop when the carriage contacts limit switch LS4, 11, following the same circuit used in semiautomatic operation.

3. Injection forward button 4 energizes injection forward relay CR2, 3. When button is released, injection forward stops.

4. Injection return button 13 energizes suck back solenoid SB, 13. Screw will return (without turning) under low pilot pressure, to prevent damage. Normally the screw is returned by plasticizing.

5. To open clamp, clamp return button is pushed. Contacts 1 open, breaking the interlock and deenergizing clamp forward relay CR1, 1. Contacts 16 close, energizing clamp return relay CR6, 15, opening the clamp and interlocking through CR6, 17. Relay CR6, 22, also energizes the full speed solenoid E, 21.

Purge.   Selector switch turned to purge.

1. Switch S5, 5, permits purging while clamp is open, by allowing extruder run button 8 to be energized even if LS3, 5, is open.

2. Switch S9, 24, permits full speed when the injection forward relay CR2, 3, or extruder run relay CR3, 7, are energized.

TROUBLE SHOOTING

A plastics engineer dealing with thermoplastics must fully and completely understand the function of machines. Many failures of plastic parts are caused by improper settings or erratic action of the processing equipment. It may be necessary to alter the mechanical, hydraulic, or electrical functions to achieve a certain result. The limitation of the equipment often determines the applicability of the parts.

Most often he is asked to "trouble shoot" a malfunctioning machine. This should not be difficult (17). The speed and efficiency of this recommendations have a lot to do with his status with the operating personnel.

Complete, useful drawings of the hydraulic, electrical, and mechanical parts of the machine should be immediately available. They should be properly labeled and have sufficient tables, pressure information, and mechanical data logically and conveniently displayed.

Preparatory to a breakdown, the engineer must understand thoroughly the machine. Probably the best way is to copy the hydraulic diagram. The electrical circuit should be redrawn and a detailed step by step analysis be written. This can be similar to that used for Figure 7-14, or it may be graphed or made in any manner useful to the engineer.

The instruments needed are a voltmeter, hook-on ammeter, pressure gauge and surface pyrometer. The fault is either electrical, hydraulic, mechanical, or a combination. One should determine what the trouble is by asking the foreman and operators. Any other attempts to find and correct the fault should be ascertained. If possible one should watch the operation until the "trouble" occurs.

Operation of the manual controls, hand operating solenoids, relays, and limit switches can give valuable information. Circuits can be traced for faults by using the voltmeter or ammeter. Electrical and hydraulic components can be visually examined and, if required, disassembled, checked, and repaired. Timers, limit switches, relays, and other parts can be replaced.

All of these things can be done "haphazardly." This is the procedure of the operating personnel not familiar with or unable to use electrical and hydraulic circuit diagrams. The experienced engineer may do some of these things intuitively and rapidly. However, if they do not bring prompt results he will methodically and logically investigate the fault, usually starting from the malfunctioning part.

It is difficult to postulate a universal trouble shooting procedure. We describe tracing the possible causes of a stopped screw in a machine described by the hydraulic circuit Figure 6-41 and the electrical circuit Figure 7-14.

Stopped Screw Example

Set the machine controls at OFF and press the motor start button. If the electric motor is operative, set the controls at purge, make sure LS4, 11, is not contacted, and push the extruder run button 8. Check the output shaft of the hydraulic motor and the input and output shafts of the gear reducer. (We eliminate obvious conclusions, i.e., if only the input shaft turns, the trouble is in the gear reducer.) If relay CR3, 7, does not energize, check the control fuses and use a voltmeter across XA and Y. "No" voltage with good fuses suggests trouble in either stop button 1, S1, 1, or S5, 5.

This would be a good time to look at the heat. If the heat has dropped, the viscosity of the plastic might rise enough to prevent the screw turning. A reading of barrel temperatures with a surface pyrometer will check for incorrect readings or temperature sensing by the thermocouples. As far fetched as it might seem, check the material in the hopper to be sure it is the one specified. Errors, particularly when using reground, are conceivable. If the cylinder heats and material are right, the problem is in the electrical circuit, hydraulic circuit, or mechanical.

Close relay CR3, 7, by hand. If the screw turns, the fault is in the relay coil, the NC CR5 contacts 7, or the extruder run button 8. The offender is easily found by connecting one voltmeter probe to Y, and touching the other probe successively to the XA side of CR3, 7, NC CR5, 7, and extruder run button, 8. If, for example, there is no voltage at "M" and voltage at "L", the NC CR5, 7, contacts, connections, or wire in that part of the circuit are open.

If relay CR3, 7, is energized, check the extruder run solenoid S, 28, and the hydraulic motor solenoid HM, 29. If they are energized the fault is not electrical. If they are not, the circuits on lines 28 and 29 are checked.

Solenoids S and HM could have been operated by hand before checking the electrical circuit. This is not good practice since there are situations when manually operating a solenoid with electrical faults will cause damage.

If the manual extruder run button 8 started the extruder, the trouble is in the initiator of the automatic circuit, NO CR4, 7, contacts. Relay CR4, 10 is closed manually and the contacts are checked. The following would then be checked in order:

CR4, 10, relay coil.
T1, 10, contacts.
T1, 4, timer.
CR1, 3, contacts.
Pressure switch 3.
S4, 4, switch.
S2, 3, switch.

If the fault is suspected to be hydraulic, the system pressure (and pilot pressure) is checked at the inlet of the inject-hydraulic motor valve. A pressure guage is attached to the inlet port of the hydraulic motor. If no pressure is available, a guage at the top blocked port of the "screw on" valve will pinpoint whether the trouble is there, in the main valve, or in the pressure control valve.

If there is pressure going into the hydraulic motor, and unusually high pressure at the outlet port, check valve 3 is blocked. A broken hydraulic motor can be detected by the inability to maintain pressure on the inlet side with the motor stalled. The motor manufacturer will describe test procedures for determing wear (slippage).

If the back pressure valve 6 is stuck closed, there is no place for the returning oil from the injection cylinder to go. The screw may not have enough torque to overcome this 100% back pressure by "churning" the material. Even a low back pressure may be enough to stop screw rotation. Similarly, a mechanical seizing or stopping of injection return carriage will give the same results.

The final reason for screw stoppage is mechanical, where a "foreign" substance wedges itself between the screw and the barrel. Fortunately, this i rare. The unit must be carefully disassembled and any damage repaired.

Solid State Controls

Solid state controls are being used at an increasing rate in the molding machine particularly for temperature control. Some of the advantages of solid state control, as compared to electromechanical control, follow:

1. No moving parts.
2. Not affected by the environment (vibration and temperature).
3. More accurate.
4. More reliable.
5. Faster.
6. Longer life for heating bands.
7. Take up less space.
8. Susceptible to computer control.
9. Allow safety to be built in at lower cost.

Some disadvantages follow:

1. Increased initial cost.
2. Unfamiliar to operating personnel.
3. Susceptible to "spikes" and other interferences in the electrical system.

Solid state controls are those which switch without any moving parts. This i done with certain crystalline materials which are electrical insulators o conductors, depending on the input signal. A review of electronics is found i Ref. 17A.

The heating cylinder can be controlled by a silicone controlled rectifie (SCR), which provides stepless control by changing the current to the heating units based on the input of the thermistor or thermocouple. There are two methods of SCR temperature control. The SCR can be fired when the voltage i zero and held for a given number of cycles (zero crossover), or the SCR can be fired for a given part of each cycle (phase angle). The first method is preferred because it will not generate "spikes" in the control system which can cause a malfunction. SCR control of the cylinder increases the accuracy of the contro and increases the life of the heater.

Solid state timers are more accurate than the electromechanical timers and are used in sequential switching circuits. NOR/Nand logic systems can be used to program the sequential switching. With electronic control of valves, solid state switching can provide stepless pressure and volume control of the hydraulic

ystem. The machine can be programmed for gradual pressure changes, instead
of the sudden changes found in conventional circuits; this process reduces or
eliminates shock in the hydraulic system. Many times reed switches (which are
not solid state but are highly reliable) are used to interface with solid state
circuits.

Trouble shooting is aided by having lights show which part of the circuit is
operating. When the malfunction is located, the defective electronic circuit
board or plug-in unit can be replaced. The use of solid state controls is relatively
new, with each machine manufacturer using his own system. General discussions
of solid state circuits are found in Ref. 18 to 24.

REFERENCES

1. "Electric Power Distribution for Industrial Plants," *Electrified Industry*.

2. "The Right Electrical System Can Save You a Lot of Money," R. F. Coleman, *MP* May 1969, p. 118.

3. *How to Maintain Motors and Generators*, General Electric Co., Schenectady, N.Y.

4. "Guide to Limit Switches," E. L. Rudisill, and F. D. Yeaple, *Prod. Eng.,* November 12, 1962, p. 84.

5. "How to Choose a Timer," J. Proctor and F. Hall, *Prod. Eng.*, February 19, 1962, p. 96.

5a. "Temperature Controllers for Plastics Machinery," H. E. Harris, *PDP*, April 1970, p. 36.

5b. "Resistance Temperature Sensors for Injection Molding Machines," H. E. Harris, *MP*, November 1969, p. 94.

6. "Thermocouple Pyrometry," G. A. Hormuth, *Mach. Des.*, August 21, 1969, p. 128.

7. "Advantages of Dual Thermocouples in Injection Cylinders and Extruder Barrels," R. K. West, *SPE Tech. Pap,* III, 1957, p. 417.

8. "Effects of Controlled Wattage on Extruder Barrel Temperature," H. H. Ramsey, *SPE-J*, October 1960, p. 1111.

9. "Development of a Solid State Process Controller," N. Hambley, *B. P.,* August 1968, p. 85.

10. "What's New in Temperature Measurement and Control," L. L. Scheiner, *PT*, June 1968, p. 35.

11. "Fundamental Analysis of Extruder Temperature Control," H. E. Harris, *MP*, August 1967, p. 115.

12. "Measurement of Screw and Plastic Temperature Profiles in Extruders," D. I. Marshall, I. Klein, and R. H. Uhl, *SPE-J*, April 1964, p. 329.

13. "Temperature Profile of Molten Plastic Flowing in a Cylindrical Duct," H. Schott, and W. S. Kaghan, *SPE-J*, February 1964, p. 139.

14. "Extruder Melt Temperature Control," G. A. Petit, and P. E. Ahlers, *SPE-J*, November 1968, p. 59.

15. "Extruder Instrumentation," H. B. Kessler, et al, *SPE-J*, March 1960, p. 267.

16. "What's Wrong with Machine Instruction Manuals," I. I. Rubin, *PT* August 1961, p. 33.

16a. "Electrical Circuits for Solenoid Valves," M. G. Saake, *Hydraul. & Pneum.;* Part I – February 1970, p. 67; Part II – March 1970, p. 101.

17.   "How to Trouble Shoot Your Injection Molding Machine," J. Newlove, *PT*, November 1966, p. 39.

17a. "Electronics", *SPE-J*, December 1970, P. 22.

18.   "Temperature Controlers,' *B. P.,* December 1970, p. 103.

19.   "What's Available in Solid State Controls for Molding Machines," K. B. Rexford, *M. P.,* October 1968, p. 131.

20.   "An Injection Molder's Guide to Solid-State Control Systems," J. Carlson, *P. T.,* November 1968, p. 39.

21.   "Process Control," *SPE-J*, October 1969, p. 24.

22.   "Nor/Nand Logic;" Product Engineering Reprint R-119, McGraw-Hill Book Company, New York, N. Y.

23.   "Static Switching," F. Gruner, *Mach. Des.*, January 22, 1970, p. 121.

24.   "Do Solid State Controls Turn You On?", G. A. Tanner, *SPE-J*, December 1971, p. 34.

CHAPTER 8 _____

Examples of Molded Parts

A number of parts, most of which were injection molded by the author, have been selected to illustrate some of the principles and ideas previously discussed.

Filter Disk

Figure 8-1 shows a 12-in.-diameter filter grid molded in high impact polystyrene, over which is later sewn a polypropylene fiber filter bag. A number of these are stacked on a center tube and enclosed in the filter container. The filtering media, usually clay, is put in the container outside of the fiber bag. Dirty water flows through the filtering media, through the bags, on to the grids. This filtered water is carried off through the center hollow tube. A significant number broke radially while in operation. More broke when the user dismantled and cleaned the unit.

The original end user and molder must have decided to economize. The thickness was minimal. The part being center gated was oriented in the direction of flow. The strength was therefore much stronger radially than circumferentially. Since the ribs (which were used to keep the polypropylene bag off the surface of the filter grid) were in the direction of flow, they did not contribute to the strength of the piece.

A second supplier decided to build a disk engineered to overcome these faults (Figure 8-2). The part is much thicker and reinforced circumferentially with ribs where it is weakest. The radial slots permit the water to run into the center tube. The other side of the piece is identical except that the slots have been moved to the midpoint of the ribs on the other side. Therefore there is always at least one rib supporting the flat part of the filter disk. The part was overdesigned because

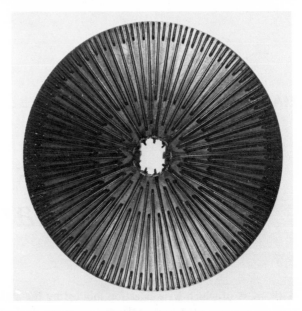

Figure 8-1   A 12-in. high-impact polystyrene filter grid with radial ribs.

of consumer resistance caused by the "cheaper" unit. After consumer acceptance was established another unit was made (Figure 8-3), which was sufficiently strong and considerably less expensive.

Four Way Valve

One of the advantages of injection molding is that parts can be produced requiring little or no finishing. An inherent advantage of plastic is its corrosion resistance, particularly for water applications. Both these properties were used to replace a four way valve for swimming pool applications originally made in an aluminum die casting (Figure 8-4). The diameter of the valve body is 4 3/8–in. and its height 2½–in. The rough casting had to be machined and the valve assembled. After one season's use the chemicals in the pool began to corrode the valve causing it to jam. When leverage was used on the handle, the "freeze" was so severe that the metal handle bent or broke.

The part was converted into polycarbonate (Figure 8-5). The design chosen was very close to the die cast part for two reasons. Die casting is similar to injection molding and there were a large number of valves in the field so that the outside dimensions had to be similar. There were also many internal parts that would be obsolete if they could not be used in the polycarbonate version.

The main problem was the mold design. There were four parts – the valve

Figure 8-2 A 12-in. high-impact polystyrene filter grid with circumferential ribs (Courtesy of American Machine Products Inc.).

body, the valve cover, the spool body, and the spool cover. For economic reasons they had to be molded in one shot. For engineering reasons the body had to be center gated. This required a three plate mold with seven cams and four irregular seal-offs for the four parts of the valve body.

Complicating the decision was that at that time the raw material supplier did not know of any three plate molds with pin point gates for parts of this complexity molded in polycarbonate. Nonetheless the mold was built. All parts were of air-hardening tool steel, highly polished. The hardened cams moved on hardened ways. The mold base was of a superior grade steel. Precision fitting and excellent workmanship were required. Since the cams and molds would be smashed if the mold closed when they were not in place, electrical interlocks were designed, so that the machine could not close unless the cams were in the correct position.

The valve's cover was attached to the valve body by a combination of solvent cementing and ultrasonic welding. The only machining required was to tap the eight holes that hold the cover on to the body, tap the one body hole for a pipe

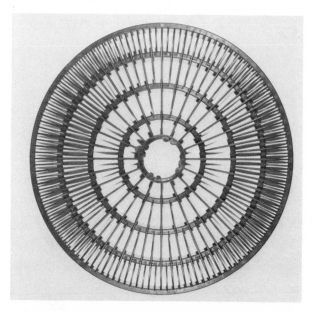

Figure 8-3 A 12-in. high-impact polystyrene filter grid with circumferential ribs, but cored to remove extra material.

connection, and tap the spool cover for the handle. The cost of the valve was reduced by the approximately two-thirds. The tooling cost for the plastic part was approximately double that of the die cast part. Rejection in the field because of corrosion was eliminated.

Mortar Training Device

Figure 8-6 shows a mortar training device used by the Armed Forces. Instead of using live shells, a special tubular device is inserted in the mortar. The plastic projectile is dropped into the tube and is ejected by an air-blast. When it hits the ground, an inertial device in the body explodes a "22" blank cartridge. Special air gauges and tables simulate the calculations and aiming, as if live ammunition were fired. The performance specifications are close, requiring the missile to hit a target about the size of a garbage can cover, one half a city block away.

Many nonplastic materials were tried unsuccessfully. Their main problem, aside from cost, was that they were nicked when they fell on pebbles or rocks. The nicks had to be removed so that the projectile would slide through the tube. Plastic was tried as a last resort and ABS proved to be satisfactory. Usually the indentations did not interfere with the reusability. If they did, rubbing them on

Figure 8-4 Aluminum die cast four way valve. The body is 4-3/8 in. in diameter and 2½ in. deep. (Courtesy of American Machine Products, Inc.)

a rock or using a pen knife or bayonet would solve the problem.

Prototypes were made and it became evident that the exploding blank cartridge would eventually affect the plastic. It was decided to mold a steel insert A into the plastic body F. The firing pin E, which is held in its back position by the spring D, is retained in the molded insert by metal ring C and snap fastener B. The firing pin has its bearing surface on the plastic. There are four rings on the outside which act as a seal between the projectile and the barrel. The compressed air expands, imparting velocity to the projectile. The fit had to be very close and consistent. A slight oversize would cause binding and an undersize permit too much air to escape. The required accuracy could not be maintained by molding but was achieved by centerless grinding. The tail section had to be molded without significant voids. If not, it caused an irregular trajectory which was too inaccurate.

The main problem was caused by the design limitations of the piece. The part was 4 3/8— in. long and 1—in. in diameter. The internal parts did not permit a

Figure 8-5   Aluminum valve of Figure 8-4 molded in polycarbonate (Courtesy of American Machine Products, Inc.).

thick enough wall around the steel insert A. Inspection of the piece will show that it must be gated into the center of the fin section. The part above the fin is cammed. The greatest strength of the plastic would be in the direction of flow. This is longitudinal in relation to the insert rather than circumferential which would be desired. Since this is an orientation effect it was felt that it might work if the bushings were kept very hot. This would permit relaxation of the plastic around the bushing and increase its hoop strength. It was subsequently shown that those parts molded with a cold insert failed in the field and those with hot inserts were satisfactory. Today the problem would not be so severe, as stronger materials are available.

Lamp Housing

A lamp part, 8 in. high and 5 in. in diameter, molded out of general purpose heat resistant polystyrene, is shown in Figure 8-7. The bottom cap is solvent

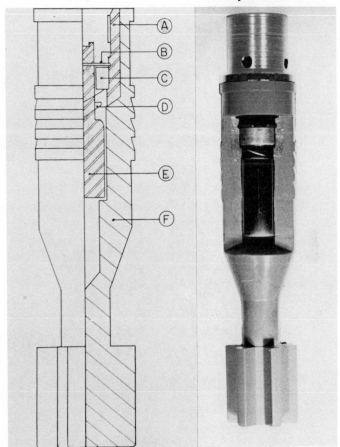

Figure 8-6 ABS mortar training projectile. It is 4 3/8 in. high and 1 in. in diameter (Courtesy of Anchor Precision Products, Inc.).

cemented after molding. The material was selected for low cost, beauty, and ease of molding. The main problem was the tool design. To achieve esthetically pleasing triangular cutouts, four cams were required. The cutout effect could have been achieved with two cams, but the triangles close to the parting line would be severely distorted. The cam blocks were massive and held together by a large ring which acted as a cam lock. To prevent tearing, the cams move at right angles to the core. The cams are on the ejection side and are activated by the pins that are attached to the injection side. The cores must be maintained accurately in their position when the machine closes. If not, the core pins will not locate. If one visualizes the mold in a horizontal press one can see that the top cam must be supported against the pull of gravity. This is done with springs and detents. The cams and knockout plate are electrically interlocked. If the

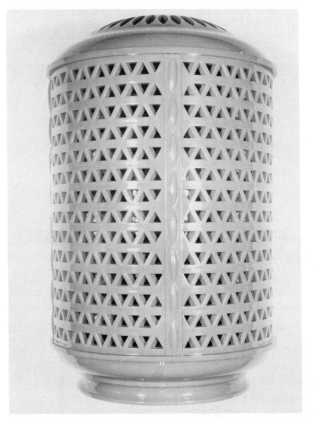

Figure 8-7 Lamp housing molded in high heat polystyrene, 5 in. diameter and 8 in. high (Courtesy of Robinson Plastics Corp.).

knock-out plate is not completely back, the cams would hit the knockout pins, damaging the mold.

Skimmer

Figure 8-8 is the front view of a stationary skimmer that removes surface debris from a swimming pool. It is 10 in. long by 8 in. wide by 7 in. high. The top and bottom housings are molded in ABS and the hinged weir (flap) is made of polypropylene. The hinge is kept at water level by a styrofoam panel which is attached to the underside of the flap, with three stainless steel clips that snap on over three tits molded on the polypropylene (Figure 8-9). The debris is caught in a polypropylene strainer (Figure 8-10).

Figure 8-8 Stationary skimmer for removing debris from the surface of swimming pools. It is 10 in. long, 8 in. wide, and 7 in. deep. The housing is molded of ABS and the hinged weir of polypropylene (Courtesy of American Machine Products, Inc.).

Figure 8-9 Underside view of the skimmer cover, showing the foamed polystyrene pad held on to the weir by stainless steel clips.

Figure 8-10   View of bottom part of skimmer showing polypropylene strainer.

There are some interesting applications of plastic in this part. The cover has a 3/4-in.-long undercut hook on each side molded with a "jiggler" pin. They snap into the corresponding slots in the base (Figure 8-8). To separate the cover, the flexible property of the ABS is used. The sides are pushed in to relieve the undercut hooks.

The hinged weir makes use of the orienting characteristics of polypropylene. Its size is 7 3/4 in. long by 5½ in. wide by 0.075 in. thick. The part is molded so that all the plastic must flow through the hinge section. Because of the direction of flow and the thinness of the section the molecules are oriented in the direction of flow. As soon as the flow stops, they freeze. The hinge must be flexed right after molding. This further orients the molecules by stretching them. It probably also freezes them in place. The hinge works because the C-C linkages are strong and flexible. Such hinges have been experimentally flexed over a million times without failure. If the hinge section is made thicker, 0.030 in. for example, the hinge will not work. The extra thickness will make it impossible to orient sufficiently the molecules.

The strainer appears to have a large cross hatched design. Its purpose is not esthetic. The hinged weir and the strainer are molded on the same mold. It was impossible to fill the strainer without packing the hinge. Therefore, the hatched design was added as additional runners to provide more flow to fill the part easily. They also provide mechanical reinforcement. It is multigated on the a parting line along the long side. This gives no distortion. However, if the part is packed and too much material is forced in at the gate sections, the side by the gate will subsequently bow, as the excess material must go somewhere when it cools.

Vacuum Metallized Lamp Parts

Figure 8-11 shows vacuum metalized lamp parts molded in general purpose or heat resistant polystyrene. The part on the left is 5 1/2 in. long and 1 1/4 in. in diameter. It is slipped over a pipe separating other components of the lamp. The finial (lamp top) in the upper right is 3 1/2 in. long and 1 1/8 in. in diameter, and the other finial 1 1/4 in. long by 1 in. in diameter. Both finials have metal threads inserted after molding. Plastic threads were unsatisfactory because of the high heat developed by the incandescent bulb.

With the start of the Korean war, it became evident that the nonmilitary use of brass would be restricted. At this time vacuum metallizing of plastic was beginning to become commercially available. Molds were built and an extensive molding and metallizing experimental program undertaken. The main problems were adhesion and quality. If the part had the slightest flaw, the housewife would attempt to polish it off with brass polish, removing the metalized finish. Although the parts were less expensive than brass, weighed less, and had a surface that did not tarnish, their acceptance by lamp buyers in stores was poor. The attitude of the lamp buyers was why try something new, even if better, when the old item was accepted. When brass was no longer available they accepted vacuum metallized parts which today are the standard of the industry. This was one of the first large scale commercial uses of metallizing. Fortunately, the attitude toward plastic today is one of acceptance for its own merits and it is no longer considered a substitute.

The parts are produced in large multicavity cammed molds. The spindle is cammed from both sides to reduce part weight. The large finial uses an air operated cam and the small one a mechanical cam. Since these parts were copies of brass turnings with large smooth surfaces, quality metallizing is difficult. Exceptional attention is paid to molding clean shiny surfaces which are protected from mold release, oil, and dust. Special spinning and filtering techniques are used in the lacquering operations of the metallizing.

Venturi Tube

Figure 8-12 shows a venturi tube used in a swimming pool skimmer system. It is 9 1/2 in. long and 1 3/8 in. diameter, with a 1-in. molded pipe thread on the outside. An adjustable plastic flap controls the flow of water and is held with a brass knurled screw threaded into a brass knurled nut.

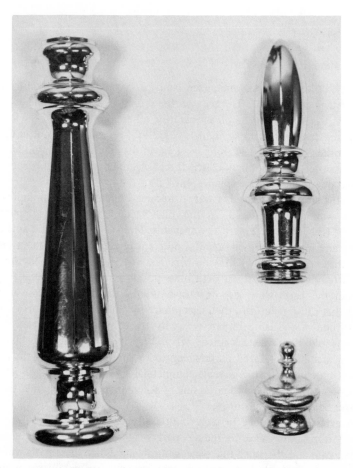

Figure 8-11 Vacuum metalized polystyrene lamp parts. The spindle on the left is 5½ in. long and 1¼ in. in diameter. It has a ½ in. hole. The finial on the top right is 3½ in. long and 1 1/8 in. in diameter. The finial on the bottom right is 1¼ in. long and 1 in. in diameter. Both finials have a metal thread inserted after molding (Courtesy of Robinson Plastics Corp.).

Molding the part in one piece would require a complicated mold with three cams, one having a pull of 10 in. The wall thickness would be uneven, reducing the water flow, and an expensive brass turning would have to be inserted.

The part was made in two halves with a tongue and groove joint. It was solvent cemented using a specially devised jig with four clamps. This was necessary for the mechanical strength of the bond. An inexpensive brass nut was

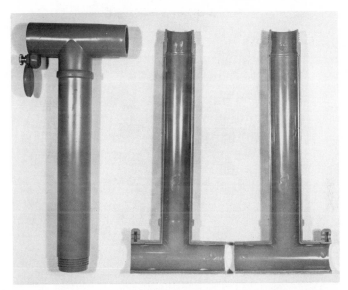

Figure 8-12   Venturi tube molded in high impact polystyrene. It is 9½ in. long and 1 3/8 in. in diameter. The tab is held in place by a brass knurled screw which is tightened against a brass nut held in place by the two cemented halves (Courtesy of American Machine Products, Inc.).

inserted in an hexagonal recess and held in place by the cementing of the part. The total cost for mold and part for the quantities contemplated was considerably less this way.

Tissue Box

Figure 8-13 shows a tissue dispenser molded in general purpose crystal styrene. It is 10 in. long, 5 1/2 in. wide and 2 1/2 in. high with an average wall thickness of 0.200 in. The box was designed to simulate the quality and brilliance of hand carved and polished acrylic. The design of the carvings was based on the angle of refraction of polystyrene, and selected to give maximum brilliance.

For economic reasons the base plate and body had to be run together. The box was assembled and packed at the molding machine. The major problem was designing a cam for the 3/16-in. deep slots, in which the base plate slid. Pulling a 10-in. core, with a minimum of draft, would require that the base plate go in in one way only. This was not acceptable. Two internal cams, one for each side, were designed. The total movement was small, 1/4 in. Even though the internal

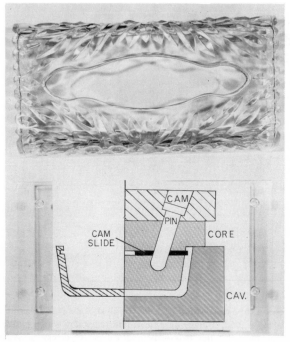

Figure 8-13 A tissue dispenser molded in crystal polystyrene. A schematic drawing of the internal cam is laid upon the base plate (Courtesy of Rialto Products Corp.).

cam looked perfectly feasible on paper the author looked for "moral" support before starting the mold. At that time (1952) neither he nor the many people asked were able to discover a similarly built mold. Nonetheless the mold was built and ran over 750,000 shots without replacement or repair of the cams. They were lubricated externally once every 8 hr shift with a silicone mold release spray. Today, of course, complexity or novelty of a mold construction is no longer a bar to the production of a plastic part.

Flashlight Head Assembly

Figure 8-14 shows the head assembly for a rechargeable flashlight. The assembled part is 2 1/2 in. long, 1 1/2 in. wide, and 1 1/2 in. deep. It consists of a vacuum metallized reflector into which is eyeletted two metal parts, which serve as the bulb holder and contact. The bezel is molded in polystyrene with a protrusion on each side that snaps the unit into the main housing. These

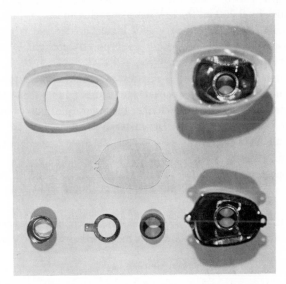

Figure 8-14 Head assembly for a rechargeable flashlight. The bezel is made of polystyrene, the reflector of vacuum metallized polystyrene, and the clear shield of polycarbonate sheet. The metal parts are staked together and the whole assembly held by heat "smearing" the tips on the bezel. The head assembly is 2½ in. long, 1½ in. wide, and 1½ in. deep (Courtesy of Gulton Industries, Inc.).

protusions are molded with jiggler pins. The lens protector is stamped out of clear polycarbonate sheet. This was the only material which had the clarity, scratch resistance, and strength in the thin section needed for the assembly. The back of the bezel has two round tits which correspond to the holes at the ends of the reflector and the cutouts in the polycarbonate shield. They are assembled together and the tits are smeared over with a hot punch, holding the assembly together. This method was chosen because there was not enough room for a metallic fastener. It is not possible to cement vacuum metallized polystyrene to polycarbonate. The beryllium molds presented no problem. Beryllium pressure castings were used because of the ease in duplicating the irregularly contoured elliptical surface of the bezel and the optical requirements of the curvature of the reflector. The product illustrates the use of different materials – polystyrene, polycarbonate, steel, and beryllium stamping – and different processes – injection molding, vacuum metalizing, plastic stamping, metal stamping, and assembly – to produce an economical, well engineered unit.

Combs

Figure 8-15 shows part of a shot for an eight cavity standard comb 8 1/4 in. long by 1 1/4 in. high by 0.200 in. thick. What makes this unusual is that it is molded in polycarbonate. It is inserted in a gold plated holder and is part of a very expensive boudoir set. The previously used polystyrene comb broke readily which was not acceptable for this high cost item.

Molding combs is not simple. The slightest short shot immediately shows by a curvature on the line of teeth. A packed shot causes minute flash which will cut the scalp. There was real question whether the viscosity of polycarbonate would permit proper filling of the teeth. The material supplier suggested borrowing a comb mold and trying it. With the customer's acquiescence it was decided to build the mold.

Since no one was sure whether to gate at the thin tooth end, thick tooth end, or center, the mold was designed to gate in all locations. The initial attempt was made at the thick tooth end.

The mold filled very nicely except there were severe blemishes along the root of the comb. Because the mold was designed for easy runner changes, it was

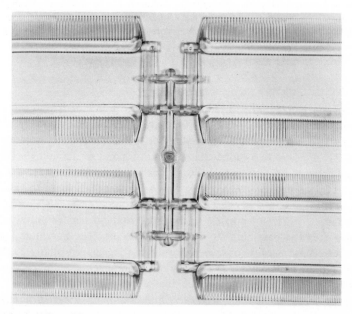

Figure 8-15 Gate section of eight cavity polycarbonate 8 in. comb mold (Courtesy of Globe Silver Co., Inc.).

decided to attempt to eliminate this even though it was covered by the gold plated holder. Several different types of runners, including an H runner, had no effect on the blemish. Initially a restricted gate was used which was very satisfactory from the point of view of mold filling. Many different gates were tried. The final solution was a tab gate, with the tab being the width and height of the root.

Car Tray

Figure 8-16 shows a car tray of low density polyethylene for holding 12 MATCHBOX® miniature cars. It is 9 3/8 in. long by 6 in. wide by 1 3/8 in. high with an average wall thickness of 0.050 in. It was originally a one cavity mold. Production estimates were very large. For this reason and the cost of the part it was decided to attempt a four cavity, three plate, back gated mold.

The mold was large, 19 in. by 31 in. and had to run automatically. Great rigidity was required for any shifting would thin out one wall preventing filling. Each cavity and core had its individual water channels. The sprue was brought right up to the secondary runner. To obtain even flow, minimize packing (which would cause sticking) and prevent orientation warping, it was decided to gate at six different places (1-6).

The mold filled and ran beautifully. One problem developed. The separating ribs or dividers above gates 3 and 4 warped. This was shown to be caused by packing too much material in the gate area, similar to what happened in Figure 3-29. Investigation showed that ribs over gates 1, 2, 5, and 6 were also packed. Since they had one end terminating on the outside they stretched the rim (which was not esthetically visible) and looked straight. The dividers over gates 3 and 4 were constrained by the structure and therefore warped. The solution was simple. Gates 3 and 4 were blocked from the back at the secondary sprue bushing. After this was done, one looking at the part would be "sure" that there were six gates. If he decided to copy exactly the mold he might wonder how the original part was molded with a straight divider.

Subsequently three other molds were built. The runner system weighed almost as much as the parts. Its handling and regrinding were difficult and costly. It was decided to build one of the molds using an insulated runner, with heated nozzle tips. The runner ran very well unless a nozzle plugged up. When that happened the insulated runner plate opened, plastic covering the whole plate. This gave an effect of projected area of about 400 in. square which was too much for the 425-ton clamp. This would blow the plate causing one-half hour loss in production. For a long time this was a rare occurrence, happening perhaps once or twice a week. Suddenly it became intermittent, happening three or four times a day and then not appearing for a week. It was eventually traced to very fine particles of Texas sand which plugged up the nozzles and which was introduced in the bagging operation of the raw material supplier.

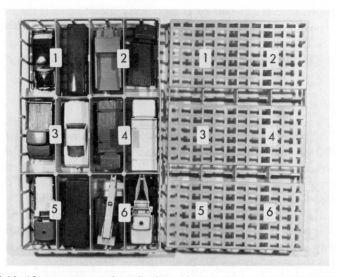

Figure 8-16 12 compartment low density polyethylene tray to hold MATCHBOX®
miniature cars. The tray is 9 3/8 in. long, 6 in. wide, and 1 3/8 in. deep, with an average wall
thickness of 0.050 in. (Courtesy of Lesney Products Corp.)

Apollo Bust

It is fitting to end this book with the first custom molded product in which
the author was the "engineer." Figure 8-17 shows the bust of Apollo, molded
in acrylic, 4 in. high, 2 1/2 in. across the shoulders, and 1 in. thick, and
molded in 1940. Incidentally, the parts still retain their original clarity and
brillance. The part was put on an alabaster base, lighted from underneath by a
battery operated bulb and used as bookends. The customer was having them
hand carved from blocks of acrylic. He could not turn them out fast enough
to satisfy his demand.

He was sent to my employer who asked me (on the basis of my 4 months'
experience) if the part could be molded. I suggested that if the customer
could engrave the reverse of the part in steel I would build a mold around it
and try it. The customer asked for the steel. Knowing nothing about steel, I
bought two pieces of boiler plate about the right size from the local junk
yard.

Some weeks later he returned with the steel, a mold was built around it,
and the part was molded. Initially there was considerable flash which was
eliminated by hand fitting in the machine. The customer was exceptionally
well pleased; he took the parts, buffed the seam, and gave the bust the
appearance of a handmade part.

Figure 8-17. Bust of Apollo, 4 in. high 2½ in. across the shoulders and 1 in. thick molded in acrylic in 1940. (Courtesy of Robinson Plastics Corp.)

There was an interesting sequel. Some months later a very knowledgeable and experienced person in the plastic business saw the polished sample on my employer's desk and suggested that it could be molded if the undercuts were removed. My employer, who had a fine sense of humor, called me in and repeated this suggestion. I asked him what I should do with the 20,000 parts already molded. After duly chastising me for being disrespectful to an older person, he suggested that I learn something. I then found out what an undercut was. It tied in with an observation about the part that, until then, was not clear.

If one waited too long to open the machine, it was relatively difficult to get Apollo out of the mold. The longer one waited, the harder it got. This new knowledge of "undercuts" explained why. The parts were molded with a 1-minute cycle and ejected relatively soft. As it knocked out, the head bent forward, because of the undercut, to clear it from the cavity. This did cause a small tear on the shoulder, but fortunately the tear looked like muscles. Obviously the longer one waited, the harder Apollo became and the more difficult it was to eject.

Appendix

1. Commonly used abbreviations
2. Cycle time (in seconds) into shots per hour and hours for molding 1000 shots
   Weight (in grams) into pounds per 1000 pieces and into ounces
3. Fractional diameters, decimal diameters, millimeter diameters, areas of circles, circumferences of circles, surfaces of spheres, and volumes of spheres
4. Draft angles per side
5. Conversion of Rockwell C and Brinell hardness scales
6. Specific gravity into grams per cubic inch and ounces per cubic inch
7. Temperature conversions
8. Millimeters into inches
9. Mensuration formulae
10. Metric conversions
11. Pounds into kilograms and kilograms into pounds.
12. Pounds per square inches into kilograms per square centimeter
13. Kilograms per square centimeter into pounds per square inch
14. Conversion factors

1. Commonly used abbreviations

ABS	Acrylonitrile-butadiene-styrene	CS	Casein
ACS	Acrylonitrile-styrene-chlorinated polyethylene	CTFE	Monochlorotrifluoroethylene
ASA	Acrylic-styrene-acrylonitrile	DAP	Diallyl phthalate
CA	Cellulose acetate	DMC	Dough molding compound
CAB	Cellulose acetate butyrate	EP	Epoxy
CAP	Cellulose acetate propionate	EC	Ethyl cellulose
CP	Cellulose propionate		
CPE	Chlorinated polyether	EVA	Ethylene vinyl acetate
CPVC	Chlorinated polyvinyl chloride	FEP	Fluorinated ethylene-propylene

1. (Cont.)

FRTP	Fiberglass reinforced thermo-plastics	PP	Polypropylene
GP	General purpose	PS	Polystyrene
GR	Glass reinforced	PTMT	Polytetramethylene tere-phthalate
HI	High impact	PUR	Polyurethane
HD	High density		
LD	Low density	PVAc	Polyvinyl acetate
MD	Medium density	PVAl	Polyvinyl alcohol
MF	Melamine formaldehyde	PVC	Polyvinyl chloride
MI	Melt index		
M_W	Molecular weight	PVDC	Polyvinylidene chloride (saran)
MWD	Molecular weight distribution	PVDF	Polyvinylidene Fluoride
PA	Polyamide (nylon)	PVCAc	Polyvinyl chloride-acetate
PAN	Polyacrylonitrile	PVF	Polyvinyl fluoride
PBS	Polybutadiene-styrene	RP	Reinforced plastics
PC	Polycarbonate	SAN	Styrene-acrylonitrile
PE	Polyethylene	SBP	Styrene-butadiene plastics
PETP	Polyethylene terepthalate		
PF	Phenol-formaldehyde	SI	Silicones
		TFE	Polytetrafluoroethylene
PMMA	Polymethyl methacrylate (acrylic)	UF	Urea-formaldehyde
POM	Polyoxymethylene (acetal)	VCP	Vinyl Chloride-propylene

Table 2. Cycle time (in seconds) into shots per hour and hours for molding 1000 shots; grams into pounds per 1000 pieces; grams into ounces

Shots per Hour	Hours per 1000 Shots	GRAMS or Cycle time (sec)	Pounds per 1000 Pieces	Grams to Ounces	Shots per Hour	Hours per 1000 Shots	GRAMS or Cycle time (sec)	Pounds per 1000 Pieces	Grams to Ounces
3600	0.2778	1	2.205	0.035	88	11.39	41	90.39	1.446
1800	0.5556	2	4.409	0.071	86	11.67	42	92.59	1.481
1200	0.8333	3	6.614	0.106	83	11.94	43	94.80	1.517
900	1.111	4	8.819	0.141	82	12.22	44	97.00	1.552
720	1.389	5	11.02	0.176	80	12.50	45	99.21	1.588
600	1.667	6	13.23	0.212	78	12.78	46	101.4	1.623
514	1.944	7	15.43	0.247	77	13.06	47	103.6	1.658
450	2.222	8	17.64	0.282	75	13.33	48	105.8	1.693
400	2.500	9	19.84	0.318	73	13.61	49	108.0	1.729
360	2.778	10	22.05	0.353	72	13.89	50	110.2	1.764
327	3.056	11	24.25	0.388	71	14.17	51	112.4	1.804
300	3.333	12	26.46	0.423	69	14.44	52	114.6	1.839
277	3.611	13	28.66	0.459	68	14.72	53	116.9	1.875
257	3.889	14	30.87	0.494	67	15.00	54	119.1	1.910
240	4.167	15	33.07	0.530	65	15.28	55	121.3	1.946
225	4.444	16	35.27	0.564	64	15.56	56	123.5	1.980
212	4.722	17	37.48	0.600	63	15.83	57	125.7	2.016
200	5.000	18	39.68	0.635	62	16.11	58	127.9	2.051
189	5.278	19	41.89	0.670	61	16.39	59	130.1	2.086
180	5.556	20	44.09	0.701	60	16.67	60	132.3	2.117
171	5.833	21	46.30	0.741	59	16.94	61	134.5	2.152
164	6.111	22	48.50	0.776	58	17.22	62	136.7	2.187
157	6.389	23	50.71	0.811	57	17.50	63	138.9	2.223
150	6.667	24	52.91	0.847	56	17.78	64	141.1	2.258
144	6.944	25	55.12	0.882	55	18.06	65	143.3	2.293
138	7.222	26	57.32	0.917	55	18.33	66	145.5	2.328
133	7.500	27	59.53	0.952	54	18.61	67	147.7	2.363
129	7.778	28	61.73	0.988	53	18.89	68	149.9	2.399
124	8.056	29	63.93	1.023	52	19.17	69	152.1	2.434
120	8.333	30	66.14	1.058	51	19.44	70	154.3	2.469
116	8.611	31	68.34	1.094	51	19.72	71	156.5	2.505
112	8.889	32	70.55	1.129	50	20.00	72	158.7	2.540
109	9.167	33	72.75	1.164	49	20.28	73	160.9	2.575
106	9.444	34	74.96	1.199	48	20.56	74	163.1	2.610
103	9.722	35	77.16	1.235	48	20.83	75	165.4	2.646
100	10.00	36	79.37	1.270	47	21.11	76	167.6	2.681
97	10.28	37	81.57	1.305	46	21.39	77	169.8	2.716
95	10.56	38	83.78	1.340	46	21.67	78	172.0	2.751
92	10.83	39	85.98	1.378	45	21.94	79	174.2	2.787
90	11.11	40	88.19	1.411	45	22.22	80	176.4	2.822

2. (Cont.)

Shots per Hour	Hours per 1000 Shots	GRAMS or Cycle time (sec)	Pounds per 1000 Pieces	Grams to Ounces	Shots per Hour	Hours per 1000 Shots	GRAMS or Cycle time (sec)	Pounds per 1000 Pieces	Grams to Ounces
44	22.50	81	178.6	2.857	40	25.28	91	200.6	3.210
44	22.78	82	180.8	2.893	39	25.56	92	202.8	3.245
43	23.06	83	183.0	2.928	39	25.83	93	205.0	3.281
43	23.33	84	185.2	2.963	38	26.11	94	207.2	3.316
42	23.61	85	187.4	3.000	38	26.39	95	209.4	3.352
42	23.89	86	189.6	3.034	37	26.67	96	211.6	3.386
42	24.17	87	191.8	3.069	37	26.94	97	213.9	3.422
41	24.44	88	194.0	3.104	37	27.22	98	216.1	3.457
41	24.72	89	196.2	3.140	36	27.50	99	218.3	3.492
40	25.00	90	198.4	3.175	36	27.78	100	220.5	3.528

Tables for converting:

    A. Cycle time (seconds) into shots per hour.
    B. Cycle time (seconds) into hours of molding for 1000 shots.
    C. Weight (grams) into pounds per 1000 pieces.
    D. Weight (grams) into ounces.
        1 g = 0.0353 oz   1 lb = 454 g
        1 oz = 28.3 g   1 g = 0.0022 lb

Table 3. Fractional diameters, decimal diameters, millimeter diameters, areas of circles, circumferences of circles, surfaces of spheres, and volumes of spheres

Fractional Diameter	Decimal Diameter (in.)	Millimeter Diameter	Area of Circle (in.²)	Circum. of Circle (in.)	Surface of Sphere (in.²)	Volume of Sphere (in.³)
1/64	0.015625	0.397	0.00019	0.04909	0.00076	
1/32	0.031250	0.794	0.00077	0.09818	0.00308	0.00002
3/64	0.046875	1.191	0.00173	0.14726	0.00692	0.00006
1/16	0.062500	1.588	0.00307	0.19635	0.01228	0.00013
5/64	0.078125	1.984	0.00479	0.24544	0.01916	0.00025
3/32	0.093750	2.381	0.00690	0.29452	0.02761	0.00043
7/64	0.109375	2.778	0.00939	0.34361	0.03756	0.00068
1/8	0.125000	3.175	0.01227	0.39270	0.04908	0.00102
9/64	0.140625	3.572	0.01553	0.44179	0.06212	0.00145
5/32	0.156250	3.969	0.01917	0.49087	0.07668	0.00200
11/64	0.171875	4.366	0.02320	0.53996	0.09280	0.00266
3/16	0.187500	4.763	0.02761	0.58905	0.11044	0.00345
13/64	0.203125	5.159	0.03240	0.63814	0.12960	0.00439
7/32	0.218750	5.556	0.03758	0.68722	0.15032	0.00548
15/64	0.234375	5.953	0.04314	0.73631	0.17256	0.00674
1/4	0.250000	6.350	0.04909	0.78540	0.19636	0.00818
17/64	0.265625	6.747	0.05541	0.83449	0.22164	0.00981
9/32	0.281250	7.144	0.06213	0.88357	0.24852	0.01165
19/64	0.296875	7.541	0.06922	0.93266	0.27688	0.01369
5/16	0.312500	7.938	0.07670	0.98175	0.30680	0.01598
21/64	0.328125	8.334	0.08456	1.0308	0.33824	0.01849
11/32	0.343750	8.731	0.09281	1.0799	0.37124	0.02127
23/64	0.359375	9.128	0.10143	1.1290	0.40572	0.02430
3/8	0.375000	9.525	0.11045	1.1781	0.44180	0.02761
25/64	0.390625	9.922	0.11984	1.2272	0.47936	0.03120
13/32	0.406250	10.319	0.12962	1.2763	0.51848	0.03511
27/64	0.421875	10.716	0.13978	1.3254	0.55912	0.03931
7/16	0.437500	11.113	0.15033	1.3744	0.60132	0.04385
29/64	0.453125	11.509	0.16125	1.4235	0.64500	0.04870
15/32	0.468750	11.906	0.17257	1.4726	0.69028	0.05393
31/64	0.484375	12.303	0.18426	1.5217	0.73704	0.05949
1/2	0.500000	12.700	0.19635	1.5708	0.78540	0.06545
33/64	0.515625	13.097	0.20881	1.6199	0.83524	0.07177
17/32	0.531250	13.494	0.22165	1.6690	0.88660	0.07848
35/64	0.546875	13.891	0.23489	1.7181	0.93956	0.08562
9/16	0.562500	14.288	0.24850	1.7671	0.99400	0.09318

3. (Cont.)

Fractional Diameter	Decimal Diameter	Millimeter Diameter	Area of Circle	Circum. of Circle	Surface of Sphere	Volume of Sphere
37/64	0.578125	14.684	0.26250	1.8162	1.05000	0.10115
19/32	0.593750	15.081	0.27688	1.8653	1.10752	0.10958
39/64	0.609375	15.478	0.29164	1.9144	1.16656	0.11846
5/8	0.625000	15.875	0.30680	1.9635	1.22720	0.12783
41/64	0.640625	16.272	0.32232	2.0126	1.28928	0.13765
21/32	0.656250	16.669	0.33824	2.0617	1.35296	0.14798
43/64	0.671875	17.066	0.35454	2.1108	1.41816	0.15880
11/16	0.687500	17.463	0.37122	2.1598	1.48488	0.17013
45/64	0.703125	17.859	0.38829	2.2089	1.55316	0.18200
23/32	0.718750	18.256	0.40574	2.2580	1.62296	0.19442
47/64	0.734375	18.653	0.42357	2.3071	1.69428	0.20737
3/4	0.750000	19.050	0.44179	2.3562	1.76716	0.22089
49/64	0.765625	19.447	0.46038	2.4053	1.84152	0.23496
25/32	0.781250	19.844	0.47937	2.4544	1.91748	0.24967
51/64	0.796875	20.241	0.49873	2.5035	1.99492	0.26495
13/16	0.812500	20.638	0.51849	2.5525	2.07396	0.28084
53/64	0.828125	21.034	0.53862	2.6016	2.15448	0.29736
27/32	0.843750	21.431	0.55914	2.6507	2.23656	0.31451
55/64	0.859375	21.828	0.58003	2.6998	2.32012	0.33230
7/8	0.875000	22.225	0.60132	2.7489	2.40528	0.35077
57/64	0.890625	22.622	0.62298	2.7980	2.49192	0.36989
29/32	0.906250	23.019	0.64504	2,8471	2.58016	0.38971
59/64	0.921875	23.416	0.66747	2.8962	2.66988	0.41021
15/16	0.937500	23.813	0.69029	2.9452	2.76116	0.43143
61/64	0.953125	24.209	0.71349	2.9943	2.85396	0.45335
31/32	0.968750	24.606	0.73708	3.0434	2.94832	0.47603
63/64	0.984375	25.003	0.76104	3.0925	3.04416	0.49943
1	1.000000	25.400	0.78540	3.1416	3.14160	0.52360
1/32	1.03125	26.193	0.83525	3.2398	3.34100	0.57418
1/16	1.06250	26.988	0.88664	3.3379	3.54656	0.62804
3/32	1.09375	27.781	0.93956	3.4361	3.75824	0.68509
1/8	1.12500	28.575	0.99403	3.5343	3.97612	0.74551
5/32	1.15625	29.369	1.05001	3.6325	4.20004	0.80937
3/16	1.18750	30.163	1.10754	3.7306	4.43016	0.87681
7/32	1.21875	30.956	1.16659	3.8288	4.66636	0.94785
1/4	1.25000	31.750	1.22719	3.9270	4.90876	1.0227
9/32	1.28125	32.544	1.28931	4.0252	5.15724	1.1013

3. (Cont)

Fractional Diameter	Decimal Diameter	Millimeter Diameter	Area of Circle	Circum. of Circle	Surface of Sphere	Volume of Sphere
5/16	1.31250	33.338	1.35297	4.1233	5.41188	1.1839
11/32	1.34375	34.131	1.41817	4.2215	5.67268	1.2705
3/8	1.37500	34.925	1.48490	4.3197	5.93960	1.3611
13/32	1.40625	35.719	1.55316	4.4179	6.21264	1.4560
7/16	1.43750	36.513	1.62296	4.5160	6.49184	1.5553
15/32	1.46875	37.306	1.69429	4.6142	6.77716	1.6589
1/2	1.50000	38.100	1.76715	4.7124	7.06860	1.7671
17/32	1.53125	38.894	1.84155	4.8106	7.36620	1.8798
9/16	1.56250	39.688	1.91748	4.9087	7.66992	1.9974
19/32	1.59375	40.481	1.99495	5.0069	7.97980	2.1195
5/8	1.62500	41.275	2.07395	5.1051	8.29580	2.2468
21/32	1.65625	42.069	2.15448	5.2033	8.61792	2.3788
11/16	1.68750	42.863	2.23655	5.3014	8.94620	2.5161
23/32	1.71875	43.656	2.32015	5.3996	9.28060	2.6584
3/4	1.75000	44.450	2.40530	5.4978	9.62120	2.8062
25/32	1.78125	45.244	2.49196	5.5960	9.96784	2.9589
13/16	1.81250	46.038	2.58016	5.6941	10.32064	3.1177
27/32	1.84375	46.831	2.66990	5.7923	10.67960	3.2817
7/8	1.87500	47.625	2.76117	5.8905	11.04468	3.4514
29/32	1.90625	48.419	2.85398	5.9887	11.41592	3.6268
15/16	1.93750	49.213	2.94832	6.0868	11.79320	3.8083
31/32	1.96875	50.006	3.04419	6.1850	12.17676	3.9954
2	2.00000	50.800	3.14160	6.2832	12.56640	4.1888
1/32	2.03125	51.594	3.2406	6.3814	12.9624	4.3879
1/16	2.06250	52.388	3.3410	6.4795	13.3640	4.5939
3/32	2.09375	53.181	3.4430	6.5777	13.7720	4.8054
1/8	2.12500	53.975	3.5466	6.6759	14.1864	5.0243
5/32	2.15625	54.769	3.6516	6.7741	14.6064	5.2486
3/16	2.18750	55.563	3.7583	6.8722	15.0332	5.4809
7/32	2.21875	56.356	3.8664	6.9704	15.4656	5.7185
1/4	2.25000	57.150	3.9761	7.0686	15.9044	5.9641
9/32	2.28125	57.944	4.0873	7.1668	16.3492	6.2155
5/16	2.31250	58.738	4.2000	7.2649	16.8000	6.4751
11/32	2.34375	59.531	4.3143	7.3631	17.2572	6.7410
3/8	2.37500	60.325	4.4301	7.4613	17.7204	7.0144
13/32	2.40625	61.119	4.5475	7.5595	18.1900	7.2942

Fractional Diameter	Decimal Diameter	Millimeter Diameter	Area of Circle	Circum. of Circle	Surface of Sphere	Volume of Sphere
7/16	2.43750	61.913	4.6664	7.6576	18.6656	7.5829
15/32	2.46875	62.706	4.7868	7.7558	19.1472	7.8775
1/2	2.50000	63.500	4.9087	7.8540	19.6348	8.1813
17/32	2.53125	64.294	5.0322	7.9522	20.1288	8.4910
9/16	2.56250	65.088	5.1572	8.0503	20.6288	8.8103
19/32	2.59375	65.881	5.2838	8.1485	21.1352	9.1357
5/8	2.62500	66.675	5.4119	8.2467	21.6476	9.4708
21/32	2.65625	67.469	5.5415	8.3449	22.1660	9.8121
11/16	2.68750	68.263	5.6727	8.4430	22.6908	10.164
23/32	2.71875	69.056	5.8054	8.5412	23.2216	10.521
3/4	2.75000	69.850	5.9396	8.6394	23.7584	10.889
25/32	2.78125	70.644	6.0754	8.7376	24.3016	11.264
13/16	2.81250	71.438	6.2126	8.8357	24.8504	11.649
27/32	2.84375	72.231	6.3515	8.9339	25.4060	12.040
7/8	2.87500	73.025	6.4918	9.0321	25.9672	12.443
29/32	2.90625	73.819	6.6337	9.1303	26.5348	12.852
15/16	2.93750	74.613	6.7771	9.2284	27.1084	13.272
31/32	2.96875	75.406	6.9221	9.3266	27.6884	13.699
3	3.00000	76.200	7.0686	9.4248	28.2744	14.137
1/16	3.06250	77.788	7.3662	9.6211	29.465	15.039
1/8	3.12500	79.375	7.6699	9.8175	30.680	15.979
3/16	3.18750	80.963	7.9798	10.014	31.919	16.957
1/4	3.25000	82.550	8.2958	10.210	33.183	17.974
5/16	3.31250	84.138	8.6179	10.407	34.472	19.031
3/8	3.37500	85.725	8.9462	10.603	35.784	20.129
7/16	3.43750	87.313	9.2806	10.799	37.122	21.268
1/2	3.50000	88.900	9.6211	10.996	38.484	22.449
9/16	3.56250	90.488	9.9678	11.192	39.872	23.674
5/8	3.62500	92.075	10.321	11.388	41.284	24.942
11/16	3.68750	93.663	10.680	11.585	42.720	26.254
3/4	3.75000	95.250	11.045	11.781	44.180	27.611
13/16	3.81250	96.838	11.416	11.977	45.664	29.016
7/8	3.87500	98.425	11.793	12.174	47.172	30.466
15/16	3.93750	100.013	12.177	12.370	48.708	31.965
4	4.00000	101.600	12.566	12.566	50.264	33.510
1/16	4.06250	103.188	12.962	12.763	51.848	35.102

3. (Cont.)

Fractional Diameter	Decimal Diameter	Millimeter Diameter	Area of Circle	Circum. of Circle	Surface of Sphere	Volume of Sphere
1/8	4.12500	104.775	13.364	12.959	53.456	36.751
3/16	4.18750	106.363	13.772	13.155	55.088	38.443
1/4	4.25000	107.950	14.186	13.352	56.744	40.195
5/16	4.31250	109.538	14.607	13.548	58.428	41.991
3/8	4.37500	111.125	15.033	13.744	60.132	43.847
7/16	4.43750	112.713	15.466	13.941	61.864	45.749
1/2	4.50000	114.300	15.904	14.137	63.616	47.713
9/16	4.56250	115.888	16.349	14.334	65.396	49.723
5/8	4.62500	117.475	16.800	14.530	67.200	51.801
11/16	4.68750	119.063	17.257	14.726	69.028	53.923
3/4	4.75000	120.650	17.721	14.923	70.884	56.116
13/16	4.81250	122.238	18.190	15.119	72.760	58.354
7/8	4.87500	123.825	18.665	15.315	74.660	60.663
15/16	4.93750	125.413	19.147	15.512	76.588	63.019
5	5.00000	127.000	19.635	15.708	78.540	65.450
1/16	5.0625	128.588	20.129	15.904	80.516	67.929
1/8	5.1250	130.175	20.629	16.101	82.516	70.482
3/16	5.1875	131.763	21.135	16.297	84.540	73.085
1/4	5.2500	133.350	21.648	16.493	86.592	75.767
5/16	5.3125	134.938	22.166	16.690	88.664	78.497
3/8	5.3750	136.525	22.691	16.886	90.764	81.308
7/16	5.4375	138.113	23.221	17.082	92.884	84.168
1/2	5.5000	139.700	23.758	17.279	95.032	87.113
9/16	5.5625	141.288	24.301	17.475	97.204	90.107
5/8	5.6250	142.875	24.850	17.671	99.400	93.180
11/16	5.6875	144.463	25.406	17.868	101.624	96.322
3/4	5.7500	146.050	25.967	18.064	103.868	99.549
13/16	5.8125	147.638	26.535	18.261	106.140	102.813
7/8	5.8750	149.225	27.109	18.457	108.436	106.181
15/16	5.9375	150.813	27.688	18.653	110.752	109.588
6	6.0000	152.400	28.274	18.850	113.096	113.096
1/16	6.0625	153.988	28.867	19.046	115.468	116.659
1/8	6.1250	155.575	29.465	19.242	117.860	120.303
3/16	6.1875	157.163	30.069	19.439	120.276	124.022
1/4	6.2500	158.750	30.680	19.635	122.720	127.832
5/16	6.3125	160.338	31.297	19.831	125.188	131.695
3/8	6.3750	161.925	31.919	20.028	127.676	135.656

3. (Cont.)

Fractional Diameter	Decimal Diameter	Millimeter Diameter	Area of Circle	Circum. of Circle	Surface of Sphere	Volume of Sphere
7/16	6.4375	163.513	32.548	20.224	130.192	139.671
1/2	6.5000	165.100	33.183	20.420	132.729	143.791
9/16	6.5625	166.688	33.824	20.617	135.296	147.965
5/8	6.6250	168.275	34.472	20.813	137.888	152.250
11/16	6.6875	169.863	35.125	21.009	140.500	156.583
3/4	6.7500	171.450	35.785	21.206	143.140	161.031
13/16	6.8125	173.038	36.450	21.402	145.800	165.527
7/8	6.8750	174.625	37.122	21.598	148.488	170.141
15/16	6.9375	176.213	37.801	21.795	151.204	174.812
7	7.0000	177.800	38.485	21.991		
1/8	7.1250	180.975	39.871	22.384		
1/4	7.25.00	184.150	41.282	22.776		
3/8	7.3750	187.325	42.718	23.169		
1/2	7.5000	190.500	44.179	23.562		
5/8	7.6250	193.675	45.664	23.955		
3/4	7.7500	196.850	47.173	24.347		
7/8	7.8750	200.025	48.707	24.740		
8	8.0000	203.200	50.265	25.133		
1/8	8.1250	206.375	51.849	25.525		
1/4	8.2500	209.550	53.456	25.918		
3/8	8.3750	212.725	55.088	26.311		
1/2	8.5000	215.900	56.745	26.704		
5/8	8.6250	219.075	58.426	27.096		
3/4	8.7500	222.250	60.132	27.489		
7/8	8.8750	225.425	61.862	27.882		
9	9.0000	228.600	63.617	28.274		
1/8	9.125	231.775	65.397	28.667		
1/4	9.250	234.950	67.201	29.060		
3/8	9.375	238.125	69.029	29.452		
1/2	9.500	241.300	70.882	29.845		
5/8	9.625	244.475	72.760	30.238		
3/4	9.750	247.650	74.662	30.631		
7/8	9.875	250.825	76.589	31.023		
0	10.000	254.001	78.540	31.416		
1/8	10.125	257.176	80.516	31.809		

3. (Cont.)

Fractional Diameter	Decimal Diameter	Millimeter Diameter	Area of Circle	Circum. of Circle
1/4	10.250	260.351	82.516	32.201
3/8	10.375	263.526	84.541	32.594
1/2	10.500	266.701	86.590	32.987
5/8	10.625	269.876	88.664	33.379
3/4	10.750	273.051	90.763	33.772
7/8	10.875	276.226	92.886	34.165
11	11.000	279.401	95.033	34.558
1/8	11.125	282.576	97.205	34.950
1/4	11.250	285.751	99.402	35.343
3/8	11.375	288.926	101.62	35.736
1/2	11.500	292.101	103.87	36.128
5/8	11.625	295.276	106.14	36.521
3/4	11.750	298.451	108.43	36.914
7/8	11.875	301.626	110.75	37.306
12	12.00	304.801	113.10	37.699
1/4	12.25	311.150	117.86	38.485
1/2	12.50	317.500	122.72	39.270
3/4	12.75	323.850	127.68	40.055
13	13.00	330.201	132.73	40.841
1/4	13.25	336.550	137.89	41.626
1/2	13.50	342.900	143.14	42.412
3/4	13.75	349.250	148.49	43.197
14	14.00	355.601	153.94	43.982
1/4	14.25	361.950	159.48	44.768
1/2	14.50	374.650	165.13	45.553
3/4	14.75	374.650	170.87	46.338
15	15.00	381.001	176.71	47.124
1/4	15.25	387.350	182.65	47.909
1/2	15.50	393.700	188.69	48.695
3/4	15.75	400.050	194.83	49.480
16	16.00	406.401	201.06	50.265
1/4	16.25	412.750	207.39	51.051
1/2	16.50	419.100	213.82	51.836
3/4	16.75	425.450	220.35	52.622
17	17.00	431.801	226.98	53.407

3. (Cont.)

Fractional Diameter	Decimal Diameter	Millimeter Diameter	Area of Circle	Circum. of Circle
1/4	17.25	438.150	233.71	54.192
1/2	17.50	444.500	240.53	54.978
3/4	17.75	450.850	247.45	55.763
18	18.00	457.201	254.47	56.549
1/4	18.25	463.550	261.59	57.334
1/2	18.50	469.900	268.80	58.119
3/4	18.75	476.250	276.12	58.905
19	19.00	482.601	283.53	59.690
1/4	19.25	488.950	291.04	60.476
1/2	19.50	495.300	298.65	61.261
3/4	19.75	501.650	306.35	62.046
20	20.00	508.001	314.16	62.832
1/4	20.25	514.350	322.06	63.617
1/2	20.50	520.700	330.06	64.403
3/4	20.75	527.050	338.16	65.188

Table 4. Draft angles per side[a]

Depth	1/4°	1/2°	1°	1 1/2°	2°	2 1/2°	3°	4°	5°	7°	10°	
1/32	0.0001	0.0003	0.0005	0.0008	0.0011	0.0014	0.0016	0.0022	0.0027	0.0038	0.0055	1/32
1/16	0.0003	0.0006	0.0011	0.0016	0.0022	0.0027	0.0033	0.0044	0.0055	0.0077	0.0110	1/16
3/32	0.0004	0.0008	0.0016	0.0025	0.0033	0.0041	0.0049	0.0066	0.0082	0.0115	0.0165	3/32
1/8	0.0005	0.0010	0.0022	0.0033	0.0044	0.0055	0.0066	0.0088	0.0109	0.0153	0.0220	1/8
3/16	0.0008	0.0016	0.0033	0.0049	0.0065	0.0082	0.0098	0.0130	0.0164	0.0263	0.0331	3/16
1/4	0.0011	0.0022	0.0044	0.0066	0.0087	0.0109	0.0131	0.0174	0.0219	0.0351	0.0441	1/4
5/16	0.0014	0.0027	0.0055	0.0082	0.0109	0.0137	0.0164	0.0218	0.0273	0.0384	0.0551	5/16
3/8	0.0016	0.0033	0.0065	0.0098	0.0131	0.0164	0.0197	0.0262	0.0328	0.0460	0.0661	3/8
7/16	0.0019	0.0038	0.0076	0.0115	0.0153	0.0191	0.0229	0.0306	0.0383	0.0537	0.0771	7/16
1/2	0.0022	0.0044	0.0087	0.0131	0.0175	0.0218	0.0262	0.0350	0.0438	0.0614	0.0882	1/2
9/16	0.0023	0.0050	0.0098	0.0147	0.0197	0.0245	0.0295	0.0394	0.0493	0.0691	0.0992	9/16
5/8	0.0027	0.0054	0.0109	0.0164	0.0218	0.0273	0.0328	0.0436	0.0547	0.0767	0.1102	5/8
11/16	0.0030	0.0060	0.0110	0.0180	0.0230	0.0300	0.0361	0.0480	0.0602	0.0844	0.1212	11/16
3/4	0.0033	0.0065	0.0131	0.0196	0.0262	0.0328	0.0393	0.0524	0.0656	0.0921	0.1322	3/4
13/16	0.0036	0.0071	0.0142	0.0212	0.0284	0.0355	0.0426	0.0568	0.0711	0.0998	0.1432	13/16
7/8	0.0038	0.0076	0.0153	0.0229	0.0306	0.0382	0.0459	0.0612	0.0766	0.1074	0.1543	7/8
15/16	0.0041	0.0082	0.0164	0.0245	0.0328	0.0409	0.0492	0.0656	0.0821	0.1151	0.1653	15/16
1	0.0044	0.0087	0.0175	0.0262	0.0349	0.0437	0.0524	0.0698	0.0875	0.1228	0.1763	1

[a]The table's use is illustrated in the following example. Find the difference in diameter from top to bottom of a round cavity 3/4 in. deep with a taper of 2 1/2° per side. From the table the taper per side is 0.0328 in., the difference being $2 \times 0.0328 = 0.0656$ in. Since the tables are additive, the same results could have been obtained by multiplying the 2 1/2° per side by 2 (one for each side) or 5° and read the results in the 5° column for 3/4 in. = 0.0656 in. Similarly, the taper for 1 1/4° can be found by adding the 1/4° and 1° columns.

Table 5. Conversion of Rockwell C and Brinell hardness scales

Rockwell C. Scale 120° Cone 150 kg Load	Approximate Brinell Number	Rockwell C. Sale 120° Cone 150 kg Load	Approximate Brinell Number
15	200	43	405
16	205	44	415
17	210	45	427
18	214	46	440
19	218	47	452
20	223	48	463
21	227	49	475
22	233	50	487
23	239	51	503
24	245	52	514
25	250	53	526
26	255	54	538
27	263	55	552
28	268	56	565
29	274	57	578
30	280	58	590
31	287	59	607
32	295	60	620
33	304	61	635
34	313	62	650
35	322	63	665
36	330	64	680
37	340	65	695
38	350	66	710
39	360	67	727
40	370	68	743
41	380	69	760
42	390	70	775

Table 6. Specific gravity into grams per cubic inch and ounces per cubic inch

Specific Gravity	g/in.^{3}	oz/in.^{3}	Specific Gravity	g/in.^{3}	oz/in.^{3}
0.90	14.75	0.522	1.20	19.66	0.693
0.91	14.91	0.526	1.21	19.83	0.699
0.92	15.08	0.532	1.22	19.99	0.705
0.93	15.24	0.537	1.23	20.16	0.711
0.94	15.40	0.543	1.24	20.32	0.717
0.95	15.57	0.549	1.25	20.48	0.722
0.96	15.73	0.545	1.26	20.65	0.728
0.97	15.90	0.561	1.27	20.81	0.734
0.98	16.06	0.566	1.28	20.98	0.740
0.99	16.22	0.572	1.29	21.14	0.745
1.00	16.39	0.578	1.30	21.30	0.751
1.01	16.55	0.584	1.35	22.12	0.780
1.02	16.72	0.589	1.40	22.94	0.809
1.03	16.88	0.595	1.45	23.76	0.838
1.04	17.04	0.601	1.50	24.58	0.867
1.05	17.21	0.607	1.55	25.40	0.896
1.06	17.37	0.613	1.60	26.22	0.925
1.07	17.53	0.618	1.65	27.04	0.953
1.08	17.70	0.624	1.70	27.86	0.982
1.09	17.86	0.630	1.75	28.68	1.011
1.10	18.03	0.636	1.80	29.50	1.040
1.11	18.19	0.641	1.85	30.32	1.069
1.12	18.35	0.647	1.90	31.14	1.098
1.13	18.52	0.653	1.95	31.96	1.127
1.14	18.68	0.659	2.00	32.77	1.156
1.15	18.85	0.665			
1.16	19.01	0.670			
1.17	19.17	0.676			
1.18	19.34	0.682			
1.19	19.50	0.688			

$$1 \ oz = 28.3 \ g$$
$$1 \ g \ = \ 0.0353 \ oz$$
Specific gravity $\times$ 16.387 $\ = \ $ g/in.^{3}
Specific gravity $\times$ 0.5778 $\ = \ $ oz/in.^{3}
Specific gravity $\times$ 0.0361 $\ = \ $ lb/in.^{3}
Specific gravity $\times$ 62.4 $\ = \ $ lb/ft^3
lb/ft^3 $\times$ 0.01604 $= $ specific gravity

Table 7 Temperature conversions [a]

°C		°F	°C		°F	°C		°F	°C		°F
−40	−40	−40	60	140	284	160	320	608	260	500	932
−34	−30	−22	66	150	302	166	330	626	266	510	950
−29	−20	−4	71	160	320	171	340	644	271	520	968
−23	−10	14	77	170	338	177	350	662	277	530	986
−18	0	32	82	180	356	182	360	680	282	540	1004
−12	10	50	88	190	374	188	370	698	288	550	1022
−7	20	68	93	200	392	193	380	716	293	560	1040
−1	30	86[	99	210	410	199	390	734	299	570	1058
4	40	104	104	220	428	204	400	752	304	580	1076
10	50	122	110	230	446	210	410	770	310	590	1094
16	60	140	116	240	464	216	420	788	316	600	1112
21	70	158	121	250	482	221	430	806	321	610	1130
27	80	176	127	260	500	227	440	824	327	620	1148
32	90	194	132	270	518	232	450	842	332	630	1166
38	100	212	138	280	536	238	460	860	338	640	1184
43	110	230	143	290	554	243	470	878	343	650	1202
49	120	248	149	300	572	249	480	896			
54	130	266	154	310	590	254	490	914			

°C		°F
0.56	1	1.8
1.1	2	3.6
1.7	3	5.4
2.2	4	7.2
2.8	5	9.0
3.4	6	10.8
3.9	7	12.6
4.5	8	14.4
5.0	9	16.2

$$°F = 9/5 \, °C + 32$$
$$°C = 5/9 \, (°F − 32)$$
$$°K \text{ (Kelvin)} = °C + 273.16$$
$$°R \text{ (Rankine)} = °F + 459.69$$

[a]
Temperature conversions. The center column is the temperature to be converted. Read the degrees Celsius on the left or the degrees Farenheit on the right. For intermediate temperatures use the 1 degrees to 9° table and add to the nearest lower temperature.

Table 8. Millimeters into inches

Millimeters	Inches	Millimeters	Inches	Millimeters	Inches
1	0.03937	34	1.35858	67	2.63779
2	0.07874	35	1.37795	68	2.67716
3	0.11811	36	1.41732	69	2.71653
4	0.15748	37	1.45669	70	2.75590
5	0.19685	38	1.49606	71	2.79527
6	0.23622	39	1.53543	72	2.83464
7	0.27559	40	1.57480	73	2.87401
8	0.31496	41	1.61417	74	2.91338
9	0.35433	42	1.65354	75	2.95275
10	0.39370	43	1.69291	76	2.99212
11	0.43307	44	1.73228	77	3.03149
12	0.47244	45	1.77165	78	3.07086
13	0.51181	46	1.81102	79	3.11023
14	0.55118	47	1.85039	80	3.14960
15	0.59055	48	1.88976	81	3.18897
16	0.62992	49	1.92913	82	3.22834
17	0.66929	50	1.96850	83	3.26771
18	0.70866	51	2.00787	84	3.30708
19	0.74803	52	2.04724	85	3.34645
20	0.78740	53	2.08661	86	3.38582
21	0.82677	54	2.12598	87	3.42519
22	0.86614	55	2.16535	88	3.46456
23	0.90551	56	2.20472	89	3.50393
24	0.94488	57	2.24409	90	3.54330
25	0.98425	58	2.28346	91	3.58267
26	1.02362	59	2.32283	92	3.62204
27	1.06299	60	2.36220	93	3.66141
28	1.10236	61	2.40157	94	3.70078
29	1.14173	62	2.44094	95	3.74015
30	1.18110	63	2.48031	96	3.77952
31	1.22047	64	2.51968	97	3.81889
32	1.25984	65	2.55905	98	3.85826
33	1.29921	66	2.59842	99	3.89763
				100	3.93700

Table 9 Mensuration formulae

Annulus	—	area	= difference of the square of the diameters $\times$ 0.785
Circle	—	area	= diameter^2 $\times$ 0.785
Circle	—	circumference	= diameter $\times$ 3.14
Cone	—	surface area	= area of base + 3.14 [(radius)(slant height)]
Cone	—	volume	= 1/3 (area of base)(vertical height)
Conic	—	frustrum	= 0.2618 (height)$(D^2 + Dd + d^2)$
Cube	—	surface area	= 6(length of side)^2
Cube	—	volume	= (length of side)^3
Cube	—	diagonal	= 1.732 (length of side)
Ellipse	—	area	= 0.785 [Axis (A) $\times$ axis (B)]
Hexagon	—	area	= 0.866 (diameter inscribed circle)^2
Octagon	—	area	= 0.828 (diameter inscribed circle)^2
Pyramid (regular)		area of sides	= 1/2 (area of base) (vertical height)
Pyramid	—	volume	= 1/3 (area of base) (vertical height)
Sphere	— area		= 3.14 (diameter)^2
Sphere	—	volume	= .524 (diameter)^3

Table 10 Metric Conversions

Length

1 m	=	39.37 in.			
1 m	=	3.281 ft	1 yard	=	0.9144 m
1 m	=	1.094 yards	1 ft	=	0.3048 m
1 cm	=	0.3937 in.	1 in.	=	2.54 cm
1 mm	=	0.03937 in.	1 in.	=	25.4 mm
1 km	=	0.62137 mile	1 mile	=	1.609 km

Area

			1 mile^2	=	2.590 km^2
1 m^2	=	10.764 ft^2	1 yard^2	=	0.836 m^2
1 m^2	=	1.196 yard^2	1 ft.^{2}	=	0.0929 m^2
1 cm^2	=	0.155 in^2	1 in.^{2}	=	6.452 cm^2

Volume

			1 yard^3	=	0.7645 m^3
1 m^3	=	35.314 ft^3	1 ft^3	=	0.02832 m^3
1 m^3	=	1.308 yards^3	1 ft^3	=	28.317 l
1 cc	=	0.061 in.^{3}	1 in.^{3}	=	16.387 cc
1 l	=	61.023 in.^{3}	1 gal (US)	=	3.785 l
1 l	=	0.0353 ft^3	1 quart	=	0.9463 l
1 l	=	1.760 pints	1 l	=	2.202 lb of water
1 l	=	0.2642 US gal			

639

10. (Cont.)

Weight

1 gm	=	0.0353 oz	1 oz	=	28.35
1 kg	=	2.2046 lb	1 lb	=	454 g
1 metric ton.	=	2204.6 lb	1 lb	=	0.454 kg

Pressure

$1 \ kg/cm^2 = 14.223 \ psi$

$1 \ lb/in.^2 \ (psi) = 0.0703 \ kg/cm^2$

$1 \ lb/ft^2 \qquad = 0.0049 \ kg/cm^2$

Other

1 ft-lb	=	0.1383 kg-m
1 km/hr	=	0.9113 ft/sec

Table 11. Pounds into kilograms and kilograms into pounds

Pounds into Kilograms 1 lb (av.) = 0.45359 kg				Kilograms into Pounds (1 kg = 2.0462 avoirdupois lb)			
lb	kg	lb	kg	kg	lb	kg	lb
1	0.45359	25	11.340	1	2.2046	25	55.116
2	0.90718	30	13.608	2	4.4092	30	66.139
3	1.3608	35	15.876	3	6.6139	35	77.162
4	1.8144	40	18.144	4	8.8185	40	88.185
5	2.2680	45	20.412	5	11.023	45	99.208
6	2.7216	50	22.680	6	13.278	50	110.23
7	3.1752	55	24.948	7	15.432	55	121.25
8	3.6287	60	27.216	8	17.637	60	132.28
9	4.0823	65	29.484	9	19.842	65	143.30
10	4.5359	70	31.751	10	22.046	70	154.32
11	4.9895	75	34.019	11	24.251	75	165.35
12	5.4431	80	36.287	12	26.456	80	176.37
13	5.8967	85	38.555	13	28.660	85	187.39
14	6.3503	90	40.823	14	30.865	90	198.42
15	6.8039	95	43.091	15	33.069	95	209.44
16	7.2575	100	45.359	16	35.274	100	220.46
17	7.7111	125	56.699	17	37.479	125	275.58
18	8.1647	150	68.039	18	39.683	150	330.69
19	8.6183			19	41.888		
20	9.0719			20	44.092		

Table 12. Pounds per square inch into Kilograms per square centimeter

PSI to $(KG)/(CM^2)$
1 psi = 0.0703 kg/cm^2

psi	kg/cm^2	psi	kg/cm^2	psi	kg/cm^2	psi	kg/cm^2	psi	kg/cm^2	psi	kg/cm^2
1	0.07	25	1.75	49	3.43	73	5.11	97	6.79	7500	527
2	0.14	26	1.82	50	3.50	74	5.18	98	6.86	8000	562
3	0.21	27	1.89	51	3.57	75	5.25	99	6.93	8500	598
4	0.28	28	1.96	52	3.64	76	5.32	100	7.00	9000	633
5	0.35	29	2.03	53	3.71	77	5.39	500	35	9500	668
6	0.42	30	2.10	54	3.78	78	5.48	1000	70	10000	703
7	0.49	31	2.17	55	3.85	79	5.53	1250	888		
8	0.56	32	2.24	56	3.92	80	5.60	1500	150	11000	773
9	0.63	33	2.31	57	3.99	81	5.67	1750	123	12000	844
10	0.70	34	2.38	58	4.06	82	5.74	2000	141	13000	914
11	0.77	35	2.45	59	4.13	83	5.81	2250	158	14000	984
12	0.84	36	2.52	60	4.20	84	5.88	2500	176	15000	1055
13	0.91	37	2.59	61	4.27	85	5.95	2750	193	16000	1125
14	0.98	38	2.66	62	4.34	86	6.02	3000	211	17000	1195
15	1.05	39	2.73	63	4.41	87	6.09	3250	229	18000	1266
16	1.12	40	2.80	64	4.48	88	6.16	3500	246	19000	1336
17	1.19	41	2.87	65	4.55	89	6.23	3750	264	20000	1406
18	1.26	42	2.94	66	4.62	90	6.30	4000	281		
19	1.33	43	3.01	67	4.69	91	6.37	4500	316		
20	1.40	44	3.08	68	4.76	92	6.44	5000	352	psi	kg/cm^2
21	1.47	45	3.15	69	4.83	93	6.51	5500	387		
22	1.54	46	3.22	70	4.90	94	6.58	6000	422	¼	0.0175
23	1.61	47	3.29	71	4.97	95	6.65	6500	457	½	0.035
24	1.68	48	3.36	72	5.04	96	6.72	7000	492	¾	0.0525

Table 13. Kilograms per square centimeter into pounds per square inch

$(KG)/(CM^2)$ to PSI
1 kg/cm^2 = 14.2233 psi

kg/cm^2	psi	kg/cm^2	psi	kg/cm^2	psi	kg/cm^2	psi	kg/cm^2	psi	kg/cm^2	psi
1	14.	8	114.	15	213.	22	313.	29	412.	36	512.
2	28.	9	128.	16	227.	23	327.	30	427.	37	526.
3	43.	10	142.	17	242.	24	341.	31	441.	38	540.
4	57.	11	156.	18	256.	25	355.	32	455.	39	554.
5	71.	12	171.	19	270.	26	370.	33	469.	40	569.
6	85.	13	185.	20	284.	27	384.	34	483.	41	583.
7	100.	14	199.	21	299.	28	398.	35	498.	42	597.

13. (Cont.)

$(KG)/(CM^2)$ to PSI
$1 \ kg/cm^2 = 14.2233 \ psi$

$\dfrac{kg}{cm^2}$	psi	$\dfrac{kg}{cm^2}$	psi	$\dfrac{kg}{cm^2}$	psi	$\dfrac{kg}{cm^2}$	psi	$\dfrac{kg}{cm^2}$	psi	$\dfrac{kg}{cm^2}$	psi
43	612.	57	810.	71	1009.	85	1208.	99	1408.	750	10667.
44	626.	58	824.	72	1024.	86	1223.	100	1422.	800	11378.
45	640.	59	839.	73	1038.	87	1237.	150	2133.	8850	12089.
46	654.	60	853.	74	1052.	88	1251.	200	2844.	900	12800.
47	668.	61	867.	75	1066.	89	1265.	250	3555.	950	13512.
48	682.	62	881.	76	1080.	90	1280.	300	4266.	1000	14223.
49	697.	63	896.	77	1095.	91	1294.	350	4978.		
50	711.	64	910.	78	1109.	92	1308.	400	5689.		
51	725.	65	924.	79	1123.	93	1322.	450	6400.		
52	739.	66	938.	80	1137.	94	1386.	500	7111.	1100	15645.
53	753.	67	952.	81	1152.	95	1351.	550	7822.	1200	17067.
54	768.	68	967.	82	1166.	96	1365.	600	8533.	1300	18489.
55	782.	69	981.	83	1180.	97	1379.	650	9245.	1400	19911.
56	796.	70	995.	84	1194.	98	1393.	700	9956.	1500	21333.

Table 14. Conversion Factors

To convert	Into	Multiply by
atm	cm of mercury	76.0
atm	ft of water (at $4°C$)	33.90
atm	in. of mercury (at $0°C$)	29.92
atm	kg/cm^2	1.0333
atm	$lb/in.^2$	14.70
atm	$tons/ft^2$	1.058
	B	
Btu	ft-lb	778.3
Btu	g-cal	252.0
Btu	J	1,054.8
Btu	k-cal	0.2520
Btu	kW-hr	2.928×10^{-4}
Btu/hr	g-cal/sec	0.0700
Btu/hr	W	0.2931
	C	
cm	ft	3.281×10^{-2}
cm	in.	0.3937
cm	miles	6.214×10^{-6}

14. (Cont.)

To convert	Into	Multiply by
cm	mm	10.0
cm/sec	ft/min	1.1969
cm/sec	ft/sec	0.03281
cm/sec	k/hr	0.036
cm/sec	m/min	0.6
cc	ft^3	3.531×10^{-5}
cc	$in.^3$	0.06102
cc	m^3	10^{-6}
cc	gal (U.S. liq.)	2.642×10^{-4}
cc	l	0.001
ft^3	$in.^3$	1,728.0
ft^3	$m.^3$	0.02832
ft^3	gal (U.S. liq.)	7.48052
ft^3	l	28.32
ft^3/min	cm^3/sec	472.0
ft^3/min	gal/sec	0.1247
ft^3/min	l/sec	0.4720
$in.^3$	cc	16.39
$in.^3$	ft^3	5.787×10^{-4}
ft^3/lb	cm^3/g	62.43
$in.^3$/oz.	cc/g	0.577
$in.^3$	gal	4.329×10^{-3}
$in.^3$	l	0.01639
m^3	ft^3	35.31
m^3	$in.^3$	61,023.0
m^3	gal (U.S. liq)	264.2

D

Days	min	1,440.0
Days	sec	86,400.0
Degrees (angle)	rad	0.01745
deg/sec	rad/sec	0.01745
Dynes	g	1.020×10^{-3}
Dynes	J/cm	10^{-7}
Dynes	lb	2.248×10^6

F

Ft	km	3.048×10^4
Ft	m	0.3048
Ft of Water	atm	0.02950
Ft of Water	in. of mercury	0.8826
Ft of Water	kg/cm^2	0.03048

To convert	Into	Multiply by
Ft of Water	lb/in.^{2}	0.4335
Ft/min	cm/sec	0.5080
Ft/min	m/min	0.3048
Ft/sec	cm/sec	30.48
Ft/sec	m/min	18.29
Ft-lb	Btu	1.286×10^{-3}
Ft-lb	g-cal	0.3238
Ft-lb	hp-hr	5.050×10^{-7}
Ft-lb	J	1.356
Ft-lb	k-hr	3.766×10^{-7}
Ft-lb/min	Btu/min	1.286×10^{-3}
Ft-lb/min	hp	3.030×10^{-5}
Ft-lb/min	kg-cal/min	3.24×10^{-4}
Ft-lb/min	kW	2.260×10^{-5}

G

To convert	Into	Multiply by
Gal	cc	3,785.0
Gal	ft^3	0.1337
Gal	in.^{3}	231.0
Gal (liq. Br. Imp.)	gal (U.S. liq.)	1.20095
Gal (U.S.)	gal (Imp.)	0.83267
Gal of Water	lb of water	8.3453
Gal/min	ft^3/sec	2.228×10^{-3}
Gal/min	l/sec	0.06308
Gal/min	l/min	3.785
G	oz (avdp)	0.03527
G	lb	2.205×10^{-3}
G/cm	lb/in.	5.600×10^{-3}
G/cc	lb/ft^3	62.43
G/cc	lb/ft^3	0.03613
G/cc	oz/in.^{3}	0.5781
G/cc	lb/in.^{3}	0.03613
gr/cm^2	lb/ft^2	2.0481

H

To convert	Into	Multiply by
hp	Btu/min	42.44
hp	ft-lb/sec	550.0
hp	W	745.7
hp-hr	Btu	2,547.
hp-hr	kg-cal	641.1
hp-hr	k-hr	0.7457

14. (Cont.)

To convert	Into	Multiply by
	I	
in.	cm	2.540
in. of mercury	atm	0.03342
in. of mercury	kg/cm^2	0.03453
in. of mercury	lb/in.^{2}	0.4912
in. of Water (at 4°C)	atm	2.458×10^{-3}
in. of Water (at 4°C)	in. of mercury	0.07355
in. of Water (at 4°C)	kg/cm^2	2.540×10^{-3}
in. of Water (at 4°C)	lb/in.^{2}	0.03613
	J	
J	Btu	9.480×10^{-4}
J/cm	g	1.020×10^{4}
J/cm	lb	22.48
	K	
Kg/cm^2	atm	0.9678
Kg/cm^2	ft of water	32.81
Kg/cm^2	in. of mercury	28.96
Kg/cm^2	lb/in.^{2}	14.22
Kg-cal	Btu	3.968
Kg-cal	J	4,186.
Kg-cal	kW-hr	1.163×10^{-3}
Kg-m	Btu	9.294×10^{-3}
Kg-m	k-hr	2.723×10^{-6}
kW	Btu/min	56.92
kW	hp	1.341
kW	kg-cal/min	14.34
kW-hr	Btu	3,413.
kW-hr	hp	1.341
kW-hr	kg-cal	860.5
	L	
l	in.^{3}	61.02
l	gal (U.S. liq.)	0.2642
l/min	ft^2/sec	5.886×10^{-4}
l/min	gal/sec	4.403×10^{-3}
	M	
m/min	cm/sec	1.667
m/min	miles/hr	0.03728

To convert	Into	Multiply by
m/sec	ft/sec	3.281
m/sec	k/min	0.06
m/sec	miles/hr	2.237
m-kg	cm-dynes	9.807×10^7
m-kg	lb-ft	7.233
miles (statute)	ft	5,280.
miles (statute)	k	1.609
miles (statute)	yards	1,760.
miles/hr	cm/sec	44.70
miles/hr	ft/min	88.
miles/hr	m/min	26.82
miles/min	cm/sec	2,682.
miles/min	nautical knots/min	0.8684
mm	ft	3.281×10^{-3}
mm	in.	0.03937
mils	cm	2.540×10^{-3}
mils	in.	0.001

O

To convert	Into	Multiply by
oz	g	28.3495
oz	lb	0.0625
oz (fluid)	in.^{3}	1.805
oz (fluid)	l	0.02957
oz/in.^{2}	lb/in.^{2}	0.0625
oz/in.^{3}	g/cc	1.733

P

To convert	Into	Multiply by
lb	dynes	44.4823×10^4
lb	g	453.59
lb of water	ft^3	0.01602
lb of water	in.^{3}	27.68
lb of water	gal	0.1198
lb of water/min	ft^3/sec	2.670×10^{-4}
lb-ft	cm-dynes	1.356×10^7
lb-ft	m-kg	0.1383
lb/ft^3	g/cc	0.01602
lb/ft^3	kg/m^3	16.02
lb/in.^{3}	g/cc	27.68
lb/in.	g/cm	178.6
lb/in.^{2}	atm	0.06804
lb/in.^{2}	ft of water	2.307
lb/in.^{2}	in. of mercury	2.036

14. (Cont.)

To convert	Into	Multiply by
$lb/in.^2$	kg/m^2	703.1
$lb/in.^2$	kg/cm^2	0.07031

Q

quarts (liq.)	cc	946.4
quarts (liq.)	$in.^3$	57.75
quarts (liq.)	l	0.9463

S

cm^2	circular mils	1.973×10^5
cm^2	ft^2	1.076×10^{-3}
cm^2	$in.^2$	0.1550
ft^2	cm^2	929.0
$in.^2$	cm^2	6.452
m^2	ft^2	10.76
m^2	$in.^2$	1,550.
m^2	$yards^2$	1.196
mm^2	circular mils	1,973.
mm^2	$in.^2$	1.550×10^{-3}
$yards^2$	cm^2	8,361.

T

tons (long)	lb	2,240.
tons (metric)	k	1,000.
tons (metric)	lb	2,205.
tons (short)	kg	907.18
tons (short)	lb	2,000.
tons (short)	tons (long)	0.89287
tons (short)	tons (metric)	0.9078

W

W	Btu/hr	3.413
W	erg/sec	10^7
W	ft-lb/min	44.27
W	hp	1.341×10^{-3}
W	kg-cal/min	0.01433
W-hr	Btu	3.413
W-hr	g-cal	859.85
W-hr	hp-hr	1.341×10^{-3}
W-hr	kg-m	367.2

Y

yards	cm	91.44

INDEX